EMS Textbooks in Mathematics

EMS Textbooks in Mathematics is a series of books aimed at students or professional mathematicians seeking an introduction into a particular field. The individual volumes are intended not only to provide relevant techniques, results, and applications, but also to afford insight into the motivations and ideas behind the theory. Suitably designed exercises help to master the subject and prepare the reader for the study of more advanced and specialized literature.

Jørn Justesen and Tom Høholdt, *A Course In Error-Correcting Codes*
Markus Stroppel, *Locally Compact Groups*
Peter Kunkel and Volker Mehrmann, *Differential-Algebraic Equations*
Dorothee D. Haroske and Hans Triebel, *Distributions, Sobolev Spaces, Elliptic Equations*
Thomas Timmermann, *An Invitation to Quantum Groups and Duality*
Oleg Bogopolski, *Introduction to Group Theory*
Marek Jarnicki und Peter Pflug, *First Steps in Several Complex Variables: Reinhardt Domains*
Tammo tom Dieck, *Algebraic Topology*

Mauro C. Beltrametti
Ettore Carletti
Dionisio Gallarati
Giacomo Monti Bragadin

Lectures on Curves, Surfaces and Projective Varieties

A Classical View of Algebraic Geometry

Translated from the Italian
by Francis Sullivan

Authors:

Mauro C. Beltrametti
Ettore Carletti
Dionisio Gallarati
Giacomo Monti Bragadin

Dipartimento di Matematica
Università degli Studi di Genova
Via Dodecaneso, 35
16146 Genova
Italy

Translator:

Francis Sullivan
Dipartimento di Matematica Pura ed Applicata
Università di Padova
Via Trieste, 63
35121 Padova
Italy

Originally published in Italian by Bollati Boringhieri, Torino, under the title
"Letture su curve, superficie e varietà proiettive speciali – Un'introduzione alla geometria algebrica".

© 2003 Bollati Boringhieri editore s.r.l., Torino, corso Vittorio Emanuele II, 86

2000 Mathematical Subject Classification (primary; secondary): 14-01; 14E05, 14E07, 14H50, 14J26, 14J70, 14M99, 14N05

Key words: Algebraic geometry, projective varieties, curves, surfaces, special varieties

ISBN 978-3-03719-064-7

The Swiss National Library lists this publication in The Swiss Book, the Swiss national bibliography, and the detailed bibliographic data are available on the Internet at http://www.helveticat.ch.

This work is subject to copyright. All rights are reserved, whether the whole or part of the material is concerned, specifically the rights of translation, reprinting, re-use of illustrations, recitation, broadcasting, reproduction on microfilms or in other ways, and storage in data banks. For any kind of use permission of the copyright owner must be obtained.

© 2009 European Mathematical Society

    Contact address:

    European Mathematical Society Publishing House
    Seminar for Applied Mathematics
    ETH-Zentrum FLI C4
    CH-8092 Zürich
    Switzerland

    Phone: +41 (0)44 632 34 36
    Email: info@ems-ph.org
    Homepage: www.ems-ph.org

Typeset using the author's TEX files: I. Zimmermann, Freiburg
Printed in Germany

9 8 7 6 5 4 3 2 1

To my daughter Barbara
Mauro C. Beltrametti

In memory of my father
Ettore Carletti

To my son Mario
Dionisio Gallarati

To my children Alessandro and Margherita
Giacomo Monti Bragadin

Preface

*Ché se la voce tua sarà molesta
nel primo gusto, vital nutrimento
lascerà poi, quando sarà digesta.*

(Dante, Paradiso, XVII, 130–132)

This book consists of lectures on classical algebraic geometry, that is, the methods and results created by the great geometers of the late nineteenth and early twentieth centuries.

This book is aimed at students of the last two years of an undergraduate program in mathematics: it contains some rather advanced topics that could form material for specialized courses and which are suitable for the final two years of undergraduate study, as well as interesting topics for a senior thesis. The book will be welcomed by teachers and students of algebraic geometry who are seeking a clear and panoramic path leading from the basic facts about linear subspaces, conics and quadrics, learned in courses on linear algebra and advanced calculus to a systematic discussion of classical algebraic varieties and the tools needed to study them.

The topics chosen throw light on the intuitive concepts that were the starting point for much contemporary research, and should therefore, in our opinion, make up part of the cultural baggage of any young student intending to work in algebraic geometry. Our hope is that this text, which can be a first step in recovering an important and fascinating patrimony of mathematical ideas, will stimulate in some readers the desire to look into the original works of the great geometers of the past, and perhaps even to find therein motivation for significant new research.

Another reason which induced us to write this book is the observation that many young researchers, though able to obtain significant results by using the sophisticated techniques presently available, can also encounter notable difficulty when faced with questions for which classical methods are particularly indicated. This book combines the more classical and intuitive approach with the more formally rigorous and modern approach, and so contributes to filling a gap in the literature.

This book, which we consider new and certainly different from texts published in the last fifty years, is the text we would have liked on our desk when we began our studies; it is our hope that it will serve as a useful introduction to Algebraic Geometry along classical lines.

The ideal use for this text could well be to provide a solid preliminary course to be mastered before approaching more advanced and abstract books. Thus we lay a firm classical foundation for understanding modern expositions such as Hartshorne [50], Mumford [68], Liu [65], or also Dolgachev's forthcoming treatise [34]. Our text can also be considered as a more modern version of Walker's classic book [113], but greatly enriched with respect to the latter by the discussion of important classes of higher dimensional varieties as mentioned above.

Prerequisites. We suppose that the reader knows the foundational elements of Projective Geometry, and the geometry of projective space and its subspaces. These are topics ordinarily encountered in the first two years of undergraduate programs in mathematics. The basic references for these topics are the classic treatment of Cremona [31] and the texts of Berger [13] and Hodge and Pedoe [52, Vols. 1, 2]. The introductory text [10] by the authors of the present volume is also useful. For the convenience of the reader in the purely introductory Chapter 1 we have given a concise review of those facts that will be most frequently used in the sequel.

Moreover to understand the book, in addition to a few elementary facts from Analysis, the reader should also be familiar with the basic structures of Algebra (groups, rings, polynomial rings, ideals, prime and maximal ideals, integral domains and fields, the characteristic of a ring), as well as extensions of fields (algebraic and transcendental elements, minimal polynomials, algebraically closed fields) as found in the texts of [35] or [75].

Possible "Itineraries". The book contains several "itineraries" that could suggest or constitute topics for different advanced undergraduate courses in mathematics, and also for graduate level courses. Here are some more precise indications, which also offer a view of the topics treated here.

- Chapters 2 and 3 can be the introduction to any course in algebraic geometry. They contain the essential notions regarding algebraic and projective sets: the Hilbert Nullstellensatz, morphisms and rational maps, dimension, simple points and singular points of an algebraic set, tangent spaces and tangent cones, the order of a projective variety. If one then adds the brief comments on elimination theory in Chapter 4, one has enough material for a semester course.

- Chapter 5 is dedicated to hypersurfaces in $\mathbb{P}^n$ with particular attention to algebraic plane curves and surfaces in $\mathbb{P}^3$. It assumes only the rudiments of the geometry of projective space and a thorough familiarity with projective coordinates. The topics covered in this chapter, suitably amplified and accompanied by the exercises given in Sections 5.7, 5.8, can in themselves form the program for a course that probably requires more than a semester, especially if one adds the first two paragraphs of Chapter 9 which are dedicated to quadratic transformations between planes and their most important applications, for example the proof of the existence of a plane model with only ordinary singularities for any algebraic curve.

- Chapter 6, which deals with linear systems of algebraic hypersurfaces in $\mathbb{P}^n$, contains topics necessary for the subsequent chapters. Veronese varieties and mappings are introduced, as well as the notion of the blowing up of $\mathbb{P}^n$ with center a subvariety of codimension ≥ 2.

- The program for a specialized one semester course for advanced undergraduates could be furnished by Chapter 7 and the first two sections of Chapter 9, which are dedicated respectively to algebraic curves in $\mathbb{P}^n$ (with particular attention to rational curves and the curves on a quadric in $\mathbb{P}^3$) and to quadratic transformations between planes. The genus of a curve is introduced, and its birational nature is placed in evidence. Adding the remaining results discussed in Chapter 9, which has some originality with respect to the existing literature on Cremona transformations, would give rise to a full year course.

- Chapter 8 is the natural continuation and completion of Chapter 7, and also makes use of some results from the first two sections of Chapter 9. It deals with the theory of linear series on an algebraic curve, including an extensive discussion on the Riemann–Roch theorem, and an approach to the classification of algebraic curves in $\mathbb{P}^n$ in terms of properties of the canonical series and the canonical curve. This chapter was largely inspired by Severi's classic text [100], where the so called "quick method" for studying the geometry of algebraic curves is expounded. The content of this chapter would give rise to a one semester course.

- Chapter 10 can furnish material for a one semester course for students who already have a good mastery of the geometry of hyperspaces ([52, Vol. 1, Chapter V]), of plane projective curves (Chapter 5, Section 5.7) and of Cremona transformations between planes (Chapter 9, Sections 9.1, 9.3). Thus this chapter is well adapted for an upper-undergraduate level course in mathematics or also for graduate level courses. Nevertheless, the methods used are rather elementary. Among the topics to which the most space is dedicated, we mention the rational normal ruled surfaces, the Veronese surface, and the Steiner surface. Some of the surfaces already described in the last section of Chapter 5 are here rediscovered and seen in a new light. They are studied together with other surfaces that occupy an important place in algebraic-projective geometry.

- Veronese varieties, Segre varieties, and Grassmann varieties are discussed in Chapter 6, Section 6.7, Chapter 11 and Chapter 12 respectively. They constitute examples of special varieties that every student of geometry should know. These topics too could be part of an advanced course or graduate course. Among other things, they might well suggest topics for research projects or a senior thesis.

- The numerous exercises of the text are in part distributed throughout the various chapters, and in part collected in Chapter 13. They can be quite useful to young graduates who are preparing for admission to a doctoral program or a position as research assistant, or also to high school teachers preparing to qualify for promotion. The easier exercises are merely stated; others, almost always new to this text, offer various levels of difficulty. Most of them are accompanied by a complete solution, but in some cases the method of solution is merely suggested.

Sources. In addition to the already cited text [10] to which the present volume may be seen as the natural successor, classical references and sources of inspiration for part of the material contained here are the books by Bertini [14], [15], Castelnuovo [25], Comessatti [27], Enriques and Chisini [36], Fano and Terracini [38], Hodge and Pedoe [52], B. Segre [81], Semple and Roth [92], and C. Segre's memoir [83]. We have also been influenced by more modern texts like Shafarevich [103] and Harris [48], and, with particular reference to the topics regarding algebraic sets, rational regular functions, and rational maps developed in the second chapter, by Reid's text [74].

Besides the texts mentioned above, in our opinion the very nice introductory texts of Musili [69] and Kunz [60] as well as Kempf's more advanced book [59] merit special mention. We also call the reader's attention to the charming "bibliographie commentée" in Dieudonné's text [33] which offers a panoramic view of the basic and advanced texts and the fundamental articles which have constituted the history and development of Algebraic Geometry, from the origins of Greek mathematics up to the late 1960s. The bibliography is rendered even more valuable, topic by topic, through interesting comments and historical notes illustrating an "excursus" that starts from Heath's interpretation of certain algebraic methods in Diophantus and arrives at Mumford's construction of the space of moduli for curves of a given genus.

Changes and improvements with respect to the Italian version. The present text offers some substantial changes and improvements with respect to the original Italian version [9]. Among the major changes are an entirely new chapter, Chapter 8, devoted to linear series on algebraic curves, a major revision to Chapter 2, the new Section 4.3, giving greater detail regarding intersection multiplicities in Chapter 4. Moreover a number of new exercises have been added throughout the book, including, in particular, a new final section of Chapter 13.

Among the minor changes there is a new final paragraph in Chapter 10 dealing with birational Cremona transformations between projective spaces of dimension 3. There are also numerous corrections of minor typographical and mathematical errors.

We thank the many colleagues and students who have had occasion to read parts of the Italian version of the book, thus contributing to the correction of errors and improving the exposition of the material. In particular, we wish to thank our friends and colleagues L. Bădescu, E. Catalisano, A. Del Padrone, A. Geramita, P. Ionescu, R. Pardini for their comments. We would like to thank I. Dolgachev who first encouraged us to consider the possibility of a translation of the original version of the book. We would also like to thank our friend and colleague A. Languasco for his invaluable assistance in resolving various problems involving the use of LaTeX.

We also wish to thank D. B. Leep for his careful reading of portions of the text and the useful suggestions for improvements in the presentation that he gave. Special thanks are also due to F. Sullivan for the translation and for helping to make that task a truly friendly interaction.

We are very grateful to Manfred Karbe and the European Mathematical Society Publishing House not only for their professional and courteous manner, but also for their unfailing warmth and encouragement that has gone well beyond mere professional courtesy. The authors and the translator would also like to thank Irene Zimmermann not only for her careful reading of the proofs and the many very helpful suggestions she gave for improving the clarity and fluidity of the text, but also for her great patience in waiting for the delayed arrival of its final version.

Finally, let us mention the web page

http://www.dima.unige.it/~beltrame/book.pdf (dvi)

where data and updates regarding the book will be collected and an "errata corrige" placed online.

Mauro C. Beltrametti
Ettore Carletti
Dionisio Gallarati
Giacomo Monti Bragadin

The authors and their translator
(from left to right: D. G., M. C. B., E. C., G. M. B., and F. S.)

Contents

Preface   vii

1 **Prerequisites**   1

2 **Algebraic Sets, Morphisms, and Rational Maps**   12
   2.1   Review of topology   12
   2.2   The correspondences V and I   15
   2.3   Morphisms   21
   2.4   Rational maps   24
   2.5   Projective algebraic sets   30
   2.6   Rational maps and birational equivalence   35
   2.7   Complements and exercises   43

3 **Geometric Properties of Algebraic Varieties**   51
   3.1   Tangent space, singularities and dimension   51
   3.2   Independence of polynomials. Essential parameters . . .   62
   3.3   Dimension of a projective variety   68
   3.4   Order of a projective variety, tangent cone and multiplicity   72

4 **Rudiments of Elimination Theory**   83
   4.1   Resultant of two polynomials   83
   4.2   Bézout's theorem for plane curves   89
   4.3   More on intersection multiplicity   90
   4.4   Elimination of several variables   99
   4.5   Bézout's theorem   102

5 **Hypersurfaces in Projective Space**   106
   5.1   Generalities on hypersurfaces   106
   5.2   Multiple points of a hypersurface   108
   5.3   Algebraic envelopes   115
   5.4   Polarity with respect to a hypersurface   119
   5.5   Quadrics in projective space   126
   5.6   Complements on polars   134
   5.7   Plane curves   140
   5.8   Surfaces in $\mathbb{P}^3$ 150

6 Linear Systems — 166
- 6.1 Linear systems of hypersurfaces — 167
- 6.2 Hypersurfaces of a linear system that satisfy given conditions — 168
- 6.3 Base points of a linear system — 170
- 6.4 Jacobian loci — 177
- 6.5 Simple, composite, and reducible linear systems — 182
- 6.6 Rational mappings — 185
- 6.7 Projections and Veronese varieties — 189
- 6.8 Blow-ups — 192

7 Algebraic Curves — 197
- 7.1 Generalities — 197
- 7.2 The genus of an algebraic curve — 201
- 7.3 Curves on a quadric — 212
- 7.4 Rational curves — 218
- 7.5 Exercises on rational curves — 225

8 Linear Series on Algebraic Curves — 238
- 8.1 Divisors on an algebraic curve with ordinary singularities — 239
- 8.2 Linear series — 246
- 8.3 Linear equivalence — 248
- 8.4 Projective image of linear series — 251
- 8.5 Special linear series — 256
- 8.6 Adjoints and the Riemann–Roch theorem — 261
- 8.7 Properties of the canonical series and canonical curves — 270
- 8.8 Some results on algebraic correspondences between two curves — 274
- 8.9 Some remarks regarding moduli — 277
- 8.10 Complements and exercises — 284

9 Cremona Transformations — 292
- 9.1 Quadratic transformations between planes — 292
- 9.2 Resolution of the singularities of a plane algebraic curve — 297
- 9.3 Cremona transformations between planes — 307
- 9.4 Cremona transformations between projective spaces of dimension 3 — 317
- 9.5 Exercises — 322

10 Rational Surfaces — 340
- 10.1 Planar representation of rational surfaces — 340
- 10.2 Linearly normal surfaces and their projections — 349
- 10.3 Surfaces of minimal order — 354
- 10.4 The conics of a plane as points of $\mathbb{P}^5$ and the Veronese surface — 360
- 10.5 Complements and exercises — 364

11 Segre Varieties 386
11.1 The product of two projective lines 386
11.2 Segre morphism and Segre varieties 389
11.3 Segre product of varieties 392
11.4 Examples and exercises 395

12 Grassmann Varieties 399
12.1 The lines of $\mathbb{P}^3$ as points of a quadric in $\mathbb{P}^5$ 399
12.2 Complexes of lines in $\mathbb{P}^3$ 403
12.3 Congruences of lines in $\mathbb{P}^3$ 407
12.4 Ruled surfaces in $\mathbb{P}^3$ 408
12.5 Grassmann coordinates and Grassmann varieties 414
12.6 Further properties of $\mathbb{G}(1,n)$ and applications 422

13 Supplementary Exercises 433
13.1 Miscellaneous exercises 433
13.2 Further problems 454
13.3 Exercises on linear series on curves 457

Bibliography 467

Index 475

Chapter 1
Prerequisites

We assume that the reader is familiar with the fundamental notions of the projective geometry of hyperspaces, for which one may consult any classical treatise. Bertini's book [14] is the preferred reference, but the texts of Hodge and Pedoe [52, Vols. 1, 2] and of Cremona [31] also merit attention. We will refer to all of these for proofs and further developments.

Nevertheless, we believe that a rapid review of the facts that will be most frequently used here may be useful for the reader, and that is the goal of this first chapter. We assume that the base field K is algebraically closed and of characteristic zero. The reader may, should he so desire, assume that $K = \mathbb{C}$, the field of complex numbers, without substantial loss of generality in regard of the methods and results expounded throughout the book.

1.1.1. Let λ be an abscissa coordinate on a complex line r extended to include the point at infinity P_∞. If

$$A = \begin{pmatrix} a_{00} & a_{01} \\ a_{10} & a_{11} \end{pmatrix} \quad (1.1)$$

is a non-degenerate 2×2 matrix with complex entries (so that $\det(A) = a_{00}a_{11} - a_{01}a_{10} \neq 0$), the formula

$$\lambda' = \frac{a_{11}\lambda + a_{10}}{a_{01}\lambda + a_{00}} \quad (1.2)$$

furnishes a one-to-one correspondence between the numbers λ and λ' (including $\lambda = \infty$ and $\lambda' = \infty$), so that we can determine a point of r by assigning the value of λ'. Thus we could take λ' rather than λ as coordinate on r. We will say that λ' is a *projective coordinate*. In particular, if $a_{00} = a_{11} = 1$ and $a_{10} = a_{01} = 0$, then equation (1.2) becomes $\lambda' = \lambda$, and so the abscissa is a particular projective coordinate.

Equation (1.2) can be considered as a formula for passing from one projective coordinate λ on r to another projective coordinate λ' on r. Thus (1.2) is the transformation formula for projective coordinates.

If λ and λ' are projective coordinates on two lines r and r' (possibly coinciding), then (1.2) establishes a one-to-one correspondence $\omega: r \to r'$ which includes all points of the extended lines r and r' without exception. One says that ω is a *projectivity* or a *projective correspondence*.

If $\lambda_1, \lambda_2, \lambda_3, \lambda_4$ are (the projective coordinates of) four points of r the element $(\in \mathbb{C} \cup \{\infty\})$

$$\frac{(\lambda_3 - \lambda_1)(\lambda_4 - \lambda_2)}{(\lambda_3 - \lambda_2)(\lambda_4 - \lambda_1)}$$

is called the *cross ratio* of the four points (or of the four numbers). It will be denoted by $(\lambda_1, \lambda_2, \lambda_3, \lambda_4)$. The cross ratio depends on the order in which the points are taken, but it is easy to check that one obtains the same number if one interchanges any two of the four numbers, and simultaneously interchanges the remaining two. This implies that the 24 possible permutations of the λ_i yield only 6 distinct values of the cross ratio. If one of these values is k, then the six cross ratios that one obtains from the four points are

$$k, \quad \frac{1}{k}, \quad 1-k, \quad \frac{1}{1-k}, \quad \frac{k}{k-1}, \quad \frac{k-1}{k}.$$

If $k = -1$ or $k = \frac{1}{2}$ or $k = 2$, then the six cross ratios reduce to only three $(-1, \frac{1}{2}, 2)$ and one says that $\lambda_1, \lambda_2, \lambda_3, \lambda_4$ constitute a *harmonic set*. If $k = -1$ we will say that $\lambda_1, \lambda_2, \lambda_3, \lambda_4$ form a *harmonic quadruple*, or that the elements $\lambda_1, \lambda_2, \lambda_3, \lambda_4$ (in the given order) form a *harmonic group* (or *harmonic range*).

If one has $k^2 + k + 1 = 0$ the six numbers reduce to only two, and $\lambda_1, \lambda_2, \lambda_3, \lambda_4$ is an *equianharmonic quadruple*.

The number

$$J = J(k) := \frac{(k+1)^2(2k-1)^2(k-2)^2}{(k^2-k+1)^3}$$

does not depend on the order in which the four points are taken, and is called the *absolute invariant* of the quadruple $\lambda_1, \lambda_2, \lambda_3, \lambda_4$.

An easy calculation shows that the cross ratio of four points is invariant under coordinate transformations and projectivities. Indeed, if $a_{00}a_{11} - a_{01}a_{10} \neq 0$ one has

$$(\lambda_1, \lambda_2, \lambda_3, \lambda_4) = \left(\frac{a_{11}\lambda_1 + a_{10}}{a_{01}\lambda_1 + a_{00}}, \frac{a_{11}\lambda_2 + a_{10}}{a_{01}\lambda_2 + a_{00}}, \frac{a_{11}\lambda_3 + a_{10}}{a_{01}\lambda_3 + a_{00}}, \frac{a_{11}\lambda_4 + a_{10}}{a_{01}\lambda_4 + a_{00}} \right).$$

If λ_1, λ_2, and λ_3 are three distinct points of the line r and λ is a variable point of r we set

$$\mu = (\lambda_1, \lambda_2, \lambda_3, \lambda). \tag{1.3}$$

Equation (1.3) has the form (1.2) and so μ is a projective coordinate on r. It follows that

$$(\lambda_1, \lambda_2, \lambda_3, \lambda_1) = \infty, \quad (\lambda_1, \lambda_2, \lambda_3, \lambda_2) = 0, \quad (\lambda_1, \lambda_2, \lambda_3, \lambda_3) = 1$$

and therefore one can choose projective coordinates on r in such a way that any three arbitrarily assigned distinct points have coordinates $\infty, 0, 1$. Such a choice is unique since $(\infty, 0, 1, \lambda) = \lim_{\mu \to \infty}(\mu, 0, 1, \lambda) = \lambda$.

If λ is a projective coordinate on the line r, then to each point of r we can associate two numbers x_0, x_1 (not both zero) such that $\lambda = \frac{x_1}{x_0}$. They are defined by the point only up to a non-zero proportionality factor, since if a is an arbitrary non-zero complex number the two ordered pairs (x_0, x_1) and (ax_0, ax_1) give the same point. Moreover, the line r may be identified with the collection of such ordered pairs (x_0, x_1) (defined, that is, up to a non-zero factor).

We will say that x_0, x_1 are *homogeneous projective coordinates* on the line. To indicate that the coordinates of P are x_0, x_1 we write $P = [x_0, x_1]$.

The points $A_0 = [1, 0]$, $A_1 = [0, 1]$ (called the *fundamental points* of the coordinates) and $U = [1, 1]$ (called the *unit point*), that is, the points $\lambda = \infty$, $\lambda = 0$ and $\lambda = 1$, constitute a *reference system* $\mathcal{S} = \{A_0, A_1, U\}$ (cf. 1.1.5).

To determine a projective system of coordinates on r we can therefore take three arbitrary distinct points A_0, A_1, U of r and impose the condition that they have coordinates $(1, 0)$, $(0, 1)$, and $(1, 1)$.

If we introduce homogeneous projective coordinates, and set $\lambda = \frac{x_1}{x_0}$, $\lambda' = \frac{x_1'}{x_0'}$, equation (1.2) becomes

$$\frac{x_1'}{x_0'} = \frac{a_{11}x_1 + a_{10}x_0}{a_{01}x_1 + a_{00}x_0}$$

and therefore, if ρ is a non-zero factor,

$$\begin{cases} \rho x_0' = a_{00}x_0 + a_{01}x_1, \\ \rho x_1' = a_{10}x_0 + a_{11}x_1. \end{cases} \tag{1.4}$$

If we set $\xi = \begin{pmatrix} x_0 \\ x_1 \end{pmatrix}$, $\xi' = \begin{pmatrix} x_0' \\ x_1' \end{pmatrix}$, then instead of equation (1.4) we have the corresponding matrix equation

$$\rho \xi' = A\xi.$$

1.1.2 (Projectivities of a line into itself). Let r be a line and ω_1 and ω_2 two projectivities of r. The map $\omega_1 \circ \omega_2 \colon r \to r$ defined by setting $\omega_1 \circ \omega_2(\lambda) = \omega_1(\omega_2(\lambda))$ is a projectivity called the *product* of ω_1 and ω_2. Under this product the projectivities of r form a (non-commutative) group which has the *identity map* on r as its neutral element. To indicate that ω is the identity we will write $\omega = 1$.

A non-identity projectivity ($\omega \neq 1$) has two *fixed points* u, v, that is, points such that $\omega(u) = u$, $\omega(v) = v$. If u and v are distinct, then the cross ratio $(u, v, \lambda, \omega(\lambda))$ is constant, that is, it does not depend on λ. The resulting constant is called the *characteristic* or *multiplier* of ω. Projectivities of characteristic -1 are of special interest. They are called *involutions* and may be characterized as non-identity projectivities for which $\omega^2 = 1$.

Given a pair of points $\varphi_0(x_0, x_1) = 0$, $\varphi_1(x_0, x_1) = 0$ on the line, one sees easily that the ∞^1 pairs of points given by the equation

$$\lambda_0 \varphi_0(x_0, x_1) + \lambda_1 \varphi_1(x_0, x_1) = 0, \tag{1.5}$$

with φ_0, φ_1 forms of degree two, are corresponding pairs in an involution. More generally, an *involution of order r* consists of the set of all r-tuples of points given by an equation of the type (1.5) with φ_0, φ_1 homogeneous polynomials of degree r.

The fixed points of an involution are usually referred to as *double points*.

1.1.3 (Algebraic correspondences between lines). If r and r' are two lines and λ and λ' are projective coordinates respectively on r and r', an equation

$$f(\lambda, \lambda') = 0 \qquad (1.6)$$

with f a polynomial, defines an algebraic correspondence $\omega \colon r \to r'$.

Projectivities are particular examples of algebraic correspondences. If we set $\lambda = \frac{x_1}{x_0}, \lambda' = \frac{x_1'}{x_0'}$ in equation (1.6), and eliminate the denominators, we obtain a *bihomogeneous* equation for an algebraic correspondence ω, that is, an equation of the form

$$\varphi(x_0, x_1; x_0', x_1') = 0, \qquad (1.7)$$

where φ is a homogeneous polynomial with respect to each pair of variables (x_0, x_1) and (x_0', x_1').

If m is the degree of φ with respect to x_0, x_1 and n is its degree with respect to x_0', x_1', then one says that φ is an (m, n) *algebraic correspondence* or of *indices* m and n. To each $\lambda \in r$ such a correspondence associates n points $\lambda' \in r'$, and to each $\lambda \in r'$ are associated m corresponding points in r.

If $r = r'$ one says that λ is a *fixed point* for an algebraic correspondence if it coincides with (at least) one of its corresponding points. *Chasles' correspondence principle* asserts that the number of fixed points for a non-identity (m, n) algebraic correspondence on a line r is equal to $m + n$.

1.1.4. Everything said up to now can be repeated without change when one considers a fundamental form of the first type (a pencil of lines, a pencil of planes, etc.) rather than a line r, or, more generally, a simply infinite algebraic entity whose elements can be put into a one-to-one correspondence with $\mathbb{P}^1$. For example, the set of points of a conic, those of the conics belonging to a pencil, or those of the lines belonging to one of the two systems of lines of a quadric.

1.1.5. The discussion carried out for a line can be extended to the plane or ordinary space, or, more generally, to the *projective space* $\mathbb{P}^n$ of *dimension n*, which we will also denote by S_n. By definition S_n is the set of ordered homogeneous $(n + 1)$-tuples of complex numbers (not all zero) $(x_0, x_1, \ldots, x_n)$. These $(n + 1)$-tuples are the points of S_n, and $x_0, x_1, \ldots, x_n$ are projective and homogeneous coordinates for the points of S_n.

The points having a single non-zero coordinate, namely the points

$$A_0 = [1, 0, \ldots, 0], \ A_1 = [0, 1, \ldots, 0], \ \ldots, \ A_n = [0, 0, \ldots, 1],$$

are called the *fundamental points* of the coordinates. They are the vertices of the *fundamental* $(n + 1)$-*hedron* (or *pyramid*). The point $U = [1, 1, \ldots, 1]$ is the *unit point*. Together these points form the *reference system* $\mathcal{S} = \{A_0, A_1, \ldots, A_n, U\}$ for the homogeneous projective coordinates x.

To change projective homogeneous coordinates in S_n means replacing the x's with x''s related to the x's by a non-degenerate linear relation

$$\rho x'_i = a_{i0}x_0 + a_{i1}x_1 + \cdots + a_{in}x_n, \quad i = 0, 1, \ldots, n; \quad \det(A) = \det(a_{ij}) \neq 0. \tag{1.8}$$

The change of coordinate equations (1.8) can also be regarded as the formulas relating the coordinates of two corresponding points under a mapping ω of the space S_n (with x-coordinates) to the space S'_n (with x'-coordinates). Such a correspondence is bijective (since $\det(A) \neq 0$) and is called a *non-degenerate projectivity* or *non-degenerate projective correspondence*.

1.1.6 (Quadrangles and quadrilaterals). We define a (plane) *quadrangle* or *quadrangular set* to be any set of four coplanar points, no three of which are collinear.

The dual figure of a quadrangle is the plane *quadrilateral* or *quadrilateral set*. It consists of four lines in a plane, no three of which belong to a pencil (of concurrent lines).

The *complete quadrangle* is the plane figure composed of four points (called *vertices*) no three of which are collinear, and six lines (called *sides*), each of which passes through two vertices. Two sides not containing a common vertex are said to be *opposite*; the three points of intersection of the pairs of opposite sides are called *diagonal points*, and constitute the vertices of the *diagonal triangle*.

1.1.7. For each index α, let U_α be the set of points P of S_n with $x_\alpha \neq 0$, $\alpha = 0, 1, \ldots, n$. Thus one has $n + 1$ subsets $U_0, U_1, \ldots, U_n$ which cover S_n and which we will call *standard affine charts*. If $x_0, x_1, \ldots, x_n$ are the homogeneous projective coordinates in S_n we can take the quotients $\frac{x_0}{x_\alpha}, \ldots, \frac{x_{\alpha-1}}{x_\alpha}, \frac{x_{\alpha+1}}{x_\alpha}, \ldots, \frac{x_n}{x_\alpha}$ as (projective, non-homogeneous) coordinates in U_α.

1.1.8. Let us consider $k + 1$ points

$$P_0 = [x_0^{(0)}, x_1^{(0)}, \ldots, x_n^{(0)}],$$
$$P_1 = [x_0^{(1)}, x_1^{(1)}, \ldots, x_n^{(1)}], \ldots, P_k = [x_0^{(k)}, x_1^{(k)}, \ldots, x_n^{(k)}]$$

in S_n. We will write $P = \lambda_0 P_0 + \cdots + \lambda_k P_k$ to indicate the point whose coordinates are obtained by forming the linear combination with parameters $\lambda_0, \ldots, \lambda_k$ (not all zero) of the coordinates of the points P_j, $j = 0, 1, \ldots, k$. Thus, if $(x_0, x_1, \ldots, x_n)$ are the coordinates of P, one has

$$\rho x_i = \lambda_0 x_i^{(0)} + \cdots + \lambda_k x_i^{(k)} \quad (\rho \neq 0; \ i = 0, 1, \ldots, n).$$

6   Chapter 1. Prerequisites

We will denote the set of points P of the form $\lambda_0 P_0 + \lambda_1 P_1 + \cdots + \lambda_k P_k$ by $J(P_0, P_1, \ldots, P_k)$. It is clear that $J(P_0, P_1, \ldots, P_k)$ does not depend on the choice of coordinates used for the individual points P_j.

Suppose that we have chosen coordinates for each of the points P_j; for example, in such a way that each has first non-zero coordinate equal to 1. There are then two possibilities:

(1) There is a point P in $J(P_0, \ldots, P_k)$ which does *not* uniquely determine the homogeneous $(k+1)$-tuple $\lambda_0, \lambda_1, \ldots, \lambda_k$; that is, there are two non-proportional $(k+1)$-tuples $(\lambda_0, \ldots, \lambda_k)$ and $(\mu_0, \ldots, \mu_k)$ such that

$$\sum_{j=0}^{k} \lambda_j P_j = \sum_{j=0}^{k} \mu_j P_j.$$

This implies that one of the points P_j is a linear combination of the others, and so could be deleted without changing $J(P_0, \ldots, P_k)$.

(2) Each point $P \in J(P_0, \ldots, P_k)$ uniquely determines the homogeneous $(k+1)$-tuple $\lambda_0, \lambda_1, \ldots, \lambda_k$ of its coefficients λ. In this case $J(P_0, \ldots, P_k)$ can be identified with the set of homogeneous $(k+1)$-tuples $(\lambda_0, \ldots, \lambda_k)$, and is therefore a k-dimensional projective space. We will call it the *linear subspace* S_k of S_n determined (or spanned) by the points $P_0, P_1, \ldots, P_k$. The reference system in S_k for the projective and homogeneous coordinates λ is $\{P_0, P_1, \ldots, P_k, \sum_{j=0}^{k} P_j\}$.

In case (1) we will say that the points P_j are *linearly dependent*. In case (2) the points P_j are *linearly independent*. One sees immediately that the necessary and sufficient condition for the points P_j to be linearly independent is that the matrix $(x_i^{(j)})$ formed from their coordinates have rank $k+1$.

If $P = [x_0, x_1, \ldots, x_n]$ is an arbitrary point of S_n one has $P = x_0 A_0 + x_1 A_1 + \cdots + x_n A_n$, whence S_n coincides with the subspace spanned by the $n+1$ linearly independent points $A_0, \ldots, A_n$.

The spaces S_0, S_1, S_2 in S_n are respectively the *points*, *lines*, and *planes* of S_n. The subspaces S_{n-1} are called *hyperplanes* of S_n. By convention, the empty set is also a subspace of S_n and its dimension is -1.

We remark explicitly that $k+1$ is the maximal number of linearly independent points that can lie in S_k. Furthermore, if $P_0, P_1, \ldots, P_k$ are not linearly independent, then one can choose a linearly independent subset of size $h+1 < k+1$, such that $J(P_0, P_1, \ldots, P_h)$ is the subspace S_h defined by the original $k+1$ dependent points.

The *intersection* of subspaces is a subspace. The minimal (that is, of minimal dimension) subspace that contains two or more subspaces $S^{(1)}, S^{(2)}, \ldots, S^{(t)}$, or the intersection of all the subspaces that contains all the $S^{(j)}$, is called the *conjunction* (or, more simply and more frequently, the *join*) of the spaces $S^{(1)}, S^{(2)}, \ldots, S^{(t)}$. It is denoted by $J(S^{(1)}, S^{(2)}, \ldots, S^{(t)})$.

If S_h and S_k are two subspaces of dimension h and k, and if S_i and S_c are their intersection and join respectively, then one has

$$h + k = c + i \quad \text{(Grassmann's formula)}.$$

Two subspaces S_h and S_k are said to be *skew* if they have no point in common, and hence have an S_{h+k+1} as join. Two skew subspaces S_h and S_k which have S_n as join (so that $h + k = n - 1$) are said to be *complementary*. If S_h and S_k are complementary, then on taking $k + 1$ independent points in S_k and $h + 1$ independent points in S_h one obtains $n + 1 = h + k + 2$ independent points which span the entire space S_n.

1.1.9. We consider $k + 1$ linearly independent points $P_0, P_1, \ldots, P_k$ and we let $P = [x_0, x_1, \ldots, x_n]$ be a point of the space S_k which they span. The coordinates of P are linear combinations of those of $P_0, P_1, \ldots, P_k$ and so the $(k + 2) \times (n + 1)$ matrix formed from the coordinates of the $k + 2$ points $P, P_0, P_1, \ldots, P_k$ has rank $k + 1$ (and indeed that is the necessary and sufficient condition to have $P \in J(P_0, P_1, \ldots, P_k)$). This implies that $(x_0, x_1, \ldots, x_n)$ is a solution to a system of $n - k$ linearly independent homogeneous linear equations. Conversely, any solution $P = [x_0, x_1, \ldots, x_n]$ of a system of $n - k$ linearly independent homogeneous linear equations in $k + 1$ unknowns can be written as a linear combination of $n + 1 - (n - k) = k + 1$ independent solutions of the system, and so such solutions constitute the points of an S_k.

In particular, a hyperplane S_{n-1} of S_n is the locus of points of S_n that satisfy a linear homogeneous equation (the equation of the S_{n-1} in the reference system $\mathcal{S} = \{A_0, A_1, \ldots, A_n; U\}$)

$$u_0 x_0 + u_1 x_1 + \cdots + u_n x_n = 0. \tag{1.9}$$

This hyperplane can be identified with the homogeneous $(n + 1)$-tuple of complex numbers (not all zero) $u_0, u_1, \ldots, u_n$. Thus, the hyperplanes of S_n are the points of an n-dimensional projective space which we denote S_n^*, and which is called the *dual* of S_n. One has $(S_n^*)^* = S_n$.

In S_1, S_2, and S_3 the hyperplanes are respectively the points, lines, and planes.

The notion of linear independence extends naturally to the dual space S_n^*. If $L_0(x) = 0, \ldots, L_k(x) = 0$ are the equations of $k + 1$ linearly independent hyperplanes, then the subspace S_k^* of S_n^* determined by them consists of those hyperplanes having equation of the form $\lambda_0 L_0(x) + \cdots + \lambda_k L_k(x) = 0$. Thus the space S_k^* consists precisely of the hyperplanes that pass through (that is, contain) the space S_{n-k-1} common to the hyperplanes $L_0(x) = 0, \ldots, L_k(x) = 0$, namely S_{n-k-1} with equations $L_0(x) = L_1(x) = \cdots = L_k(x) = 0$. The space S_k^* is called the *k-dimensional star* with *center* or *axis* the common S_{n-k-1}.

In S_2 the stars S_1^* are pencils of lines, while in S_3 the stars S_1^* and the stars S_2^* are respectively pencils and stars of planes.

The reference system S^* for the coordinates $u_0, u_1, \ldots, u_n$ in S_n^* has as its fundamental points the $n+1$ hyperplanes α_i which, in the reference system S, have equations $x_i = 0$ ($i = 0, 1, \ldots, n$). Moreover, the unit point is the hyperplane which has equation $\sum_{i=0}^n x_i = 0$ in the reference system S. Two reference systems S and S^* (one for the points and the other for the hyperplanes *of the same S_n*) related in this way are said to be *associated* with each other. The hyperplanes α_i are the face hyperplanes of S, that is, $\alpha_i = J(A_0, A_1, \ldots, A_{i-1}, A_{i+1}, \ldots, A_n)$.

If one chooses associated reference systems for the points and hyperplanes of S_n, the equation (1.9) is the condition of incidence (or belonging) of a point and a hyperplane. If one holds the x's fixed in (1.9) and allows the u's to vary, one has the equation satisfied by the hyperplanes which are incident with x (that is, which pass through x). Thus one has the equation of the point x. More generally, any geometric procedure regarding points and hyperplanes which leads to a "positional" property will always have a double interpretation according to whether one considers the variables to be the point coordinates or the hyperplane coordinates. Thus in addition to every positional property that one proves to hold for the spaces $S_0, S_1, \ldots, S_0^*, S_1^*, \ldots$ one will also have the *dual* property for the spaces $S_0^*, S_1^*, \ldots, S_0, S_1, \ldots$. This is the *duality principle*.

1.1.10. To project a point, or, more generally, a space S_a from a space S_k means to consider the join space $J(S_a, S_k)$ of S_a and S_k, a space that we will call the *projecting space* of S_a from S_k. To project S_a from S_k onto a space S_h means to take the intersection of the projecting space $J(S_a, S_k)$ with S_h. The space S_k is also called the *center* of the projection (cf. §3.4.5).

1.1.11. Let ω be a non-degenerate projectivity between two projective spaces S_n and S_n'. One sees immediately that when a point varies in a hyperplane of one of the two space the corresponding point also runs over a hyperplane in the other space in such manner that ω induces a projectivity between the dual spaces S_n^* and $(S_n')^*$.

If the two spaces are *superposed* (that is, if S_n and S_n' are two distinct copies of a common projective space, cf. [52, Vol. 1, Chapter VIII, § 1]) we may consider not only the projectivities (called *homographies* or *collineations*) that send points into other points, but also the projectivities (called *reciprocities* or *correlations*) that send points into hyperplanes, that is, projectivities $\omega: S_n \to S_n^*$ (which induce projectivities $S_n^* \to S_n$).

In the case of a homography between two superposed spaces the search for *fixed points*, that is for points P such that $\omega(P) = P$, is of interest, for example, for the classification of homographies. If $x_i' = \sum_{j=0}^n a_{ij} x_j$ are the equations of the homography, and x is a fixed point, there must be a complex number $\rho \neq 0$ such that $x_i' = \rho x_i$, that is, $\rho x_i = \sum_{j=0}^n a_{ij} x_j$, $i = 0, \ldots, n$. To have fixed points one must find a ρ such that this system of linear homogeneous equations have non-trivial solutions: ρ must be a root $\bar{\rho}$ of the characteristic equation

$$\det(A - \rho I) = \begin{vmatrix} a_{00} - \rho & a_{01} & \cdots & a_{0n} \\ a_{10} & a_{11} - \rho & \cdots & a_{1n} \\ \vdots & \vdots & \vdots & \vdots \\ a_{n0} & a_{n1} & \cdots & a_{nn} - \rho \end{vmatrix},$$

that is, one of the eigenvalues $\bar{\rho}$ of the matrix $A = (a_{ij})$ which are certainly all non-zero since A is non-degenerate. All the points of the linear subspace with equations

$$\bar{\rho} x_i = \sum_{j=0}^{n} a_{ij} x_j, \quad i = 0, 1, \ldots, n,$$

will be fixed points. The dimension of that subspace depends on the rank of the matrix $A - \bar{\rho} I$. Thus every eigenvalue of A leads to a subspace of fixed points.

If the matrix A has $n + 1$ distinct eigenvalues $\rho_1, \ldots, \rho_{n+1}$ so that the matrices $A - \rho_t I, t = 1, \ldots, n+1$, all have rank n (which is the most general situation) the projectivity ω will then have $n + 1$ fixed points. It is easy to prove the following fundamental result.

Theorem 1.1.12 (Fundamental Theorem for Projectivities). *Given two $(n + 2)$-tuples of independent points $\{P_1, \ldots, P_{n+2}\}$ in S_n and $\{Q_1, \ldots, Q_{n+2}\}$ in S'_n, there exists one and only one projectivity $\omega: S_n \to S'_n$ such that $\omega(P_t) = Q_t$, $t = 1, \ldots, n+2$.*

This statement is equivalent to stating that a projectivity between two superposed S_n's having $n + 2$ fixed points is the identity.

If we effect a change of coordinates in S_n (associated to a matrix T), and if y, y' are the new coordinates of x and x', then the homography ω will be represented by an equation $y'_i = \sum_{j=0}^{n} b_{ij} y_j$ with $B = (b_{ij})$ a matrix similar to A ($B = TAT^{-1}$). Hence we may assume that the matrix A be in Jordan canonical form (see, for example, the appendix of [9]). Thus we find exactly as many types of homographies as there are types of Jordan canonical forms.

For example, when $n = 1$ the Jordan canonical forms are, for $\alpha, \beta \in \mathbb{C}$,

$$\begin{pmatrix} \alpha & 0 \\ 0 & \beta \end{pmatrix}, \quad \begin{pmatrix} \alpha & 0 \\ 1 & \alpha \end{pmatrix}, \quad \begin{pmatrix} \alpha & 0 \\ 0 & \alpha \end{pmatrix}.$$

For $n = 2$ the Jordan canonical forms are, for $\alpha, \beta, \gamma \in \mathbb{C}$,

$$\begin{pmatrix} \alpha & 0 & 0 \\ 0 & \beta & 0 \\ 0 & 0 & \gamma \end{pmatrix}, \quad \begin{pmatrix} \alpha & 0 & 0 \\ 0 & \alpha & 0 \\ 0 & 0 & \beta \end{pmatrix}, \quad \begin{pmatrix} \alpha & 0 & 0 \\ 1 & \alpha & 0 \\ 0 & 0 & \beta \end{pmatrix},$$

$$\begin{pmatrix} \alpha & 0 & 0 \\ 1 & \alpha & 0 \\ 0 & 0 & \alpha \end{pmatrix}, \quad \begin{pmatrix} \alpha & 0 & 0 \\ 1 & \alpha & 0 \\ 0 & 1 & \alpha \end{pmatrix}, \quad \begin{pmatrix} \alpha & 0 & 0 \\ 0 & \alpha & 0 \\ 0 & 0 & \alpha \end{pmatrix}.$$

1.1.13. Let $\omega: S_n \to S_n^*$ be a reciprocity with equation $\rho u_i = \sum_{j=0}^n a_{ij} x_j$, $i = 0, \ldots, n$. We will say that two points $P = P(x)$ and $P' = P'(x')$ of S_n are *reciprocals* if P' belongs to the hyperplane corresponding to P (and then P will belong to the hyperplane corresponding to P'). The algebraic form for this condition is $\sum_{i=0}^n u_i x_i' = 0$ or $\sum_{i=0}^n \left(\sum_{j=0}^n a_{ij} x_j \right) x_i' = 0$, or also

$$\sum_{i,j=0}^n a_{ij} x_i' x_j = 0. \tag{1.10}$$

This bilinear equation expresses the fact that the two points P and P' are reciprocal, which means that each belongs to the hyperplane corresponding to the other. If we fix the x's, the equation is that of the hyperplane corresponding to the point $P(x)$, while fixing the x''s it is the equation of the hyperplane corresponding to $P'(x')$. Thus the reciprocity ω is represented by the bilinear equation (1.10).

The *involutory reciprocities* are particularly noteworthy, namely those reciprocities for which to each point $P(x)$ thought of as lying either in S_n or S_n^* there corresponds the same hyperplane of S_n^* or S_n. For this to happen it is necessary that there exist a $\rho \neq 0$ such that one has the identity

$$\sum_{i,j=0}^n a_{ij} x_i x_j' = \rho \sum_{i,j=0}^n a_{ij} x_i' x_j \ (= \rho \sum_{i,j=0}^n a_{ji} x_i x_j').$$

One must then have, for each pair of indices i, j, that $a_{ij} = \rho a_{ji}$ and this implies that the matrix $A = (a_{ij})$ be either symmetric ($\rho = 1$) or anti-symmetric (skew-symmetric, $\rho = -1$). Since an anti-symmetric matrix of odd degree is necessarily degenerate, the anti-symmetric case is possible only if n is odd (so $n+1$ is even). In the case of an odd dimensional space the involutory reciprocities with $\rho = -1$ are called *null systems* or *null polarities*. Under a null polarity each point belongs to its corresponding hyperplane. The symmetric case is possible for all n and one then finds a reciprocity in which the auto-reciprocal points are precisely those which are zeros of the quadratic form $\sum_{i,j=0}^n a_{ij} x_i x_j$. Such a reciprocity is called a *polarity with respect to the quadric* whose equation is $\sum_{i,j=0}^n a_{ij} x_i x_j = 0$.

1.1.14. Particularly simple examples of projectivities between two superposed S_n's are the so called *projectivities of general type* for which it is possible to choose a representation with A being a diagonal matrix. For such a projectivity the spaces which are loci of the fixed points are linearly independent, and they have as join the entire space S_n.

If, in particular, there are two subspaces S_h and $S_{h'}$ which are loci of fixed points and of complementary dimension (that is, $h + h' = n - 1$) then the projectivity is said to be a *biaxial homography*, and the two subspaces are its *axes*. The line r that joins two corresponding points under a biaxial homography ω is supported by the axes (that is, it intersects both axes), and is therefore fixed (that is, $\omega(A) \in r$ for every $A \in r$), and on r the homography ω induces a projectivity that has as fixed points the intersections of r with the axes. The characteristic of this projectivity on r does not depend on the pair P, P' of corresponding points, and is therefore an invariant $c(\omega)$ of ω called the *characteristic* of ω (cf. §1.1.2). If $c(\omega) = -1$, we say that ω is a *harmonic biaxial homography*.

A biaxial homography having as axes a point O and a hyperplane H is called a *homology* of center O and axis H. If, moreover, $c(\omega) = -1$ the homography ω is said to be a *harmonic homology*.

Chapter 2
Algebraic Sets, Morphisms, and Rational Maps

This is an introductory chapter containing basic notions regarding affine and projective algebraic sets, the Zariski topology, as well as morphisms and rational maps. These topics are discussed in Sections 2.2, 2.6.

In Section 2.1 we recall some preliminary topological definitions which are useful for handling the topics subsequently developed. In Section 2.7 we give a number of exercises.

The texts of Reid [74] (whose framework we follow), [72] and the first chapter of Hartshorne's book [50] constitute excellent references for the contents of this chapter. Musili's book [69] is another good reference to the material discussed in this chapter. For any background result from algebra that we use, we also refer to [35] and [120].

We assume through out the chapter, except for explicit mention to the contrary, that the base ring K is a commutative algebraically closed field of characteristic zero. Usually K will be the complex field $\mathbb{C}$. We use the usual set theoretic notation and terminology.

2.1 Review of topology

For the convenience of the reader we briefly recall some elementary notions of topology which are necessary in the sequel. For further information and proofs of properties that are only stated here we refer the reader to [58] or [18].

2.1.1 Topological spaces. A *topology* on a set X is a family τ of subsets of X satisfying the following properties.

(1) $\emptyset, X$ belong to τ;

(2) τ is stable under arbitrary unions: if $U_i \in \tau$ for all $i \in I$ then $\bigcup_{i \in I} U_i \in \tau$;

(3) τ is stable under finite intersections: if $U_i \in \tau$ for all i in the finite set I, then $\bigcap_{i \in I} \in \tau$.

The elements of τ are called *open subsets* of X and (1)–(3) are called the *axioms for open subsets*.

We say that the pair (X, τ) is a *topological space*; often we consider τ as implicitly understood and speak of X alone as a topological space.

A subfamily $\mathcal{B}$ of τ is a *base* of τ if every open subset is a union of elements of $\mathcal{B}$.

The *closed subsets* of (X, τ) are the complements of the open subsets: that is, $A \subset X$ is closed if and only if $X \setminus A$ is open. The family $\mathcal{F}$ of closed subsets satisfies the following properties.

(1) $\emptyset, X$ belong to $\mathcal{F}$;

(2) $\mathcal{F}$ is stable under finite unions: if $F_i \in \mathcal{F}$ for all i in the finite set I then $\bigcup_{i \in I} F_i \in \mathcal{F}$;

(3) $\mathcal{F}$ is stable under arbitrary intersections: if $U_i \in \mathcal{F}$ for all $i \in I$, then $\bigcap_{i \in I} \in \mathcal{F}$.

The verification of these properties is an easy application of De Morgan's rules [58, p. 4].

These properties characterize the topology in the sense that if one gives a family $\mathcal{F}$ satisfying them, then there exists a unique topology τ on X such that $\mathcal{F}$ is the collection of closed subsets of τ. Obviously it suffices to define the open subsets of τ as the complements of the subsets comprising $\mathcal{F}$. The three properties listed just above are called the *axioms for closed subsets*, and one can define a topology on X by specifying the closed subsets.

The collection of all topologies on X is partially ordered by the relation $\prec$ of *fineness*: $\sigma \prec \tau$ (σ *less fine (or coarser)* than τ) if every open subset in σ is also an open subset of τ, or equivalently if every closed subset of σ is a closed subset of τ.

A *neighborhood* of a point $x \in X$ is a set V such that there is an open subset U of X with $x \in U \subset V$. For every subset $A \subset X$ one defines the two subsets

$$\mathring{A} = \{x \in X \mid A \text{ is a neighborhood of } x\}$$

and

$$\bar{A} = \{x \in X \mid U \cap A \neq \emptyset \text{ for each neighborhood } U \text{ of } x\}$$

which are called respectively the *interior* and the *closure* of A. The points of $\mathring{A}$ are called the *interior points* of A; those of $\bar{A}$ are called *adherent points* of A.

The interior of A is the union of all the open sets of X which are contained in A, or, equivalently, the largest open set of X which is contained in A. Dually, the closure $\bar{A}$ of A is the intersection of all the closed sets of X that contain A, or, equivalently, the smallest closed set of X that contains A.

A subset $A \subset X$ is *dense* if $\bar{A} = X$; this happens if and only if A intersects every non-empty open set of X.

If $A \subset X$, one defines a topology τ_A on A by decreeing that the open sets of τ_A are precisely the intersections with A of open sets of τ, that is, $M \subset A$ is open in the topology τ_A if and only if there is an open set U of τ such that $M = A \cap U$. The topology τ_A is called the *topology induced* on A by τ or the *relative topology* on A induced by τ, and (A, τ_A) is said to be a *subspace* of (X, τ).

2.1.2 Continuity, compactness, and connectedness. Consider two topological spaces (X, τ) and (Y, σ) and let $f: X \to Y$ be a mapping. We say that f is *continuous at the point* $x \in X$ if for every neighborhood V of $f(x)$ there is a neighborhood U of x such that $f(U) \subset V$; we say that f is *continuous* if it is continuous at every point of X. The following conditions are equivalent:

(1) f is continuous;

(2) for every open $A \subset Y$ the subset $f^{-1}(A)$ is open in X;

(3) for every closed $A \subset Y$ the subset $f^{-1}(A)$ is closed in X;

(4) for all $x \in X$ and for every neighborhood V of $f(x)$ the set $f^{-1}(V)$ is a neighborhood of x.

An application $f: X \to Y$ is said to be a *homeomorphism* if it is continuous, invertible (bijective), and the inverse f^{-1} is also continuous. If f is continuous and invertible, an equivalent condition for f to be a homeomorphism is that f be an *open mapping* (respectively, *closed mapping*), that is, for each open (respectively, closed) subset U of X the subset $f(U)$ is open (respectively, closed) in Y.

Topological properties preserved by continuous maps are particularly important: that is, the properties such that if they hold for a space X they also hold for any space Y which is the image of X under a continuous mapping.

A topological space (X, τ) is *compact* if from every open covering of X one can extract a finite subcover; that is, if whenever one has $X = \bigcup_{i \in I} U_i$ with all U_i open in τ, there is a finite subset $J \subset I$ such that $X = \bigcup_{j \in J} U_j$. A subset $A \subset X$ is compact if it is compact in the topology induced by τ on A.

A topological space is *connected* if there does not exist any proper non-empty subset of X which is both open and closed, or equivalently, if it is not possible to obtain X as a union of two disjoint non-empty open subsets. A subset $A \subset X$ is said to be connected if it is connected in the topology induced by τ on A. A maximal connected subset of X is called a *connected component* of X, that is, if it is not properly contained in any larger connected subset of X. The connected components of X are closed subsets and form a partition of X into disjoint subsets.

2.1.3 Product topology and quotient topology. Let X and Y be two topological spaces, and let $X \times Y$ be their cartesian product as sets with $p: X \times Y \to X$ and $q: X \times Y \to Y$ the canonical projection maps onto the first and second factors respectively. The *product topology* on $X \times Y$ is the coarsest topology with respect to which the projection maps p and q are continuous. The set $X \times Y$ with the product topology is called the *topological product* of X and Y. The product topology on $X \times Y$ has as a basis the family of products $U \times V$ where U is open in X and V is open in Y. Indeed, if $\mathcal{B}_X$ and $\mathcal{B}_Y$ are bases for the opens of X and Y respectively,

then
$$\mathcal{B}_{X\times Y} = \{U \times V \mid U \in \mathcal{B}_X,\ V \in \mathcal{B}_Y\}$$
is a basis for the open sets of the topological product $X \times Y$.

Now let $\sim$ be an equivalence relation on X and let $\pi : X \to X/\!\sim$ be the natural projection on the quotient. If τ is a topology on X, the subsets M of $X/\!\sim$ such that $\pi^{-1}(M)$ is an open set of τ are the open sets of a topology on $X/\!\sim$. That topology is called the *quotient topology*, and is the finest topology on $X/\!\sim$ such that the projection π is continuous. The open subsets of the quotient topology are the images under π of saturated open subsets of X, namely the open subsets that are unions of equivalence classes.

2.2 The correspondences V and I

Let $\mathbb{A}^n := \mathbb{A}^n(K)$ be an n-dimensional affine space over the field K and let $y_1,\ldots,y_n$ be affine coordinates in $\mathbb{A}^n$. If $\mathfrak{a}$ is an ideal of the polynomial ring $K[Y_1,\ldots,Y_n]$ we consider the "correspondence V" which associates the subset $V(\mathfrak{a})$ to the ideal $\mathfrak{a}$ where
$$V(\mathfrak{a}) = \{y \in \mathbb{A}^n \mid f(y) = 0 \text{ for all } f \in \mathfrak{a}\}.$$

The set $V(\mathfrak{a})$ is the locus of the zeros of the polynomials in $\mathfrak{a}$. Since $\mathfrak{a}$ is finitely generated, $V(\mathfrak{a})$ is the locus of zeros of a finite number of polynomials $f_j \in K[Y_1,\ldots,Y_n]$, $j = 1,\ldots,m$.

The subsets X of $\mathbb{A}^n$ of the type $X = V(\mathfrak{a})$ are called *(affine) algebraic sets*. An algebraic set is said to be *irreducible* if there is no decomposition $X = X_1 \cup X_2$ with X_1, X_2 algebraic sets strictly contained in X.

The correspondence V satisfies the following formal properties (where $\mathfrak{b}$ and $\mathfrak{a}_i$ indicate ideals of $A = K[Y_1,\ldots,Y_n]$).

(1) $V((0)) = \mathbb{A}^n$; $V(A) = \emptyset$ (the empty set is not considered irreducible);

(2) $\mathfrak{a} \subset \mathfrak{b} \Longrightarrow V(\mathfrak{b}) \subset V(\mathfrak{a})$;

(3) $V(\mathfrak{a} \cap \mathfrak{b}) = V(\mathfrak{a}) \cup V(\mathfrak{b}) = V(\mathfrak{ab})$, where $\mathfrak{ab}$ means the product of ideals;

(4) $V\left(\sum_{i \in I} \mathfrak{a}_i\right) = \bigcap_{i \in I} V(\mathfrak{a}_i)$ (recall that the ideal sum, even if not finite, of the ideals $\mathfrak{a}_i$ consists of all finite sums of elements of the $\mathfrak{a}_i$).

All the preceding properties are almost obvious except for (3), the inclusion $V(\mathfrak{a} \cap \mathfrak{b}) \subset V(\mathfrak{a}) \cup V(\mathfrak{b})$, which can be proved as follows. Let $x \in V(\mathfrak{a} \cap \mathfrak{b})$ and suppose that $x \notin V(\mathfrak{a}) \cup V(\mathfrak{b})$. Then there exist $f \in \mathfrak{a}$ and $g \in \mathfrak{b}$ such that $f(x) \neq 0$ and $g(x) \neq 0$. Thus $fg \in \mathfrak{a} \cap \mathfrak{b}$ but $fg(x) = f(x)g(x) \neq 0$, which contradicts the assumption that $x \in V(\mathfrak{a} \cap \mathfrak{b})$.

Let $P = (a_1, \ldots, a_n) \in \mathbb{A}^n$. We call $\mathfrak{m}_P := (Y_1 - a_1, \ldots, Y_n - a_n)$ the *ideal of* P. It is easy to see that $\mathfrak{m}_P$ is a maximal ideal of $K[Y_1, \ldots, Y_n]$, and that the homomorphism $\sigma \colon K[Y_1, \ldots, Y_n] \to K$, defined by $\sigma(f) = f(a_1, \ldots, a_n)$, induces an isomorphism $K[Y_1, \ldots, Y_n]/\mathfrak{m}_P \cong K$. One has $P = V(\mathfrak{m}_P)$.

If $\mathfrak{a}$ is an ideal of $K[Y_1, \ldots, Y_n]$ the *radical* of $\mathfrak{a}$ is the ideal $\sqrt{\mathfrak{a}}$ defined by

$$\sqrt{\mathfrak{a}} := \{g \in K[Y_1, \ldots, Y_n] \mid g^t \in \mathfrak{a} \text{ for some integer } t \geq 1\}.$$

One sees immediately that

$$V(\mathfrak{a}) = V(\sqrt{\mathfrak{a}}).$$

Indeed, we have $\mathfrak{a} \subset \sqrt{\mathfrak{a}}$ whence $V(\sqrt{\mathfrak{a}}) \subset V(\mathfrak{a})$. Conversely, let $y \in V(\mathfrak{a})$ and $g \in \sqrt{\mathfrak{a}}$. Then $g^t(y) = g(y)^t = 0$ so that $g(y) = 0$ and hence $y \in V(\sqrt{\mathfrak{a}})$.

Example 2.2.1. Recall that a polynomial $f \in K[Y_1, \ldots, Y_n]$ is *irreducible* if it is not a constant and if whenever $f = f_1 f_2$ with $f_1, f_2 \in K[Y_1, \ldots, Y_n]$, then one of f_1 and f_2 is a constant.

An algebraic set X given by a single equation $f = 0$ (that is, associated to the principal ideal (f)) is called a *hypersurface* in $\mathbb{A}^n$. If $n = 2$ it is a plane affine curve, if $n = 3$ it is an affine surface, etc. If K is algebraically closed such a hypersurface is irreducible if and only if f is a power of an irreducible polynomial, as follows from the Hilbert Nullstellensatz 2.2.2. If f is a polynomial of degree 1 (respectively of degree 2) we will say that X is a *hyperplane* (respectively a *quadric* or *hyperquadric* of $\mathbb{A}^n$).

If X is a subset of $\mathbb{A}^n$, we consider the "correspondence I" which to X associates the ideal $I(X) \subset K[Y_1, \ldots, Y_n]$ defined by

$$I(X) := \{f \in K[Y_1, \ldots, Y_n] \mid f(x) = 0 \text{ for all } x \in X\}.$$

The ideal $I(X)$ is a *radical ideal*, namely

$$I(X) = \sqrt{I(X)} := \{f \in K[Y_1, \ldots, Y_n] \mid f^t \in I(X) \text{ for some integer } t \geq 1\}.$$

It is surely obvious that $I(X) \subset \sqrt{I(X)}$. Moreover, $f \in \sqrt{I(X)}$ if and only if $f^t(x) = 0$ for all $x \in X$ and for some positive integer t. But this is equivalent to $(f(x))^t = 0$ and thus to $f(x) = 0$ for all $x \in X$, and so $f \in I(X)$.

The correspondence I enjoys the following additional properties, where X, Y and X_i denote subsets of $\mathbb{A}^n$ (cf. [120, Theorem 14, p. 38]).

(1) $I(\emptyset) = K[Y_1, \ldots, Y_n]$; $I(\mathbb{A}^n(K)) = (0)$;

(2) $X \subset Y \implies I(Y) \subset I(X)$;

(3) $I\left(\bigcup_{i \in I} X_i\right) = \bigcap_{i \in I} I(X_i)$.

2.2. The correspondences V and I

As far as the composition $V \circ I$ is concerned, one has the following.

- If $\mathfrak{a}$ is an ideal of $K[Y_1, \ldots, Y_n]$ and $X \subset \mathbb{A}^n$ then

$$X \subset V(I(X)) \quad \text{and} \quad \mathfrak{a} \subset I(V(\mathfrak{a})).$$

In particular, if $X = V(\mathfrak{a})$ is an algebraic set one has

$$X = V(I(X)).$$

The composition $V \circ I$ of the correspondences V and I is thus the identity on algebraic sets.

The inclusion $\mathfrak{a} \subset I(V(\mathfrak{a}))$ can be a strict inclusion and so, in particular, the composition $I \circ V$ is not the identity. In this regard see Theorem 2.2.2 below which shows how the composition $I \circ V$ is the identity when restricted to radical ideals. If K is not algebraically closed it suffices to consider a non-constant polynomial f with no roots in K. For example, if $K = \mathbb{R}$ one can take $f = Y_1^2 + 1$. Then $\mathfrak{a} = (f) \subset K[Y_1, \ldots, Y_n]$ and $\mathfrak{a} \neq K[Y_1, \ldots, Y_n]$ because $1 \notin \mathfrak{a}$. However, $V(\mathfrak{a}) = \emptyset$ and so $I(V(\mathfrak{a})) = K[Y_1, \ldots, Y_n]$. If K is algebraically closed it suffices to observe that $(f^t) \neq I(V(f^t))$ whenever $t \geq 2$: indeed, $V(f^t) = V(f)$ and so $f \in I(V(f)) = I(V(f^t))$. But $f \notin (f^t)$.

The following fundamental theorem holds. We propose here a quick proof (of statement (1)) which is due to Kaplansky and which we heard from P. Ionescu. For further details and complete proofs see, for instance, the two texts of Reid [74], [75] or Shafarevich's book [103].

Theorem 2.2.2 (Hilbert Nullstellensatz). *Let K be an uncountable algebraically closed field (in particular $K = \mathbb{C}$). Then*

(1) *every maximal ideal $\mathfrak{m}$ of $A := K[Y_1, \ldots, Y_n]$ is of the form $\mathfrak{m} = (Y_1 - a_1, \ldots, Y_n - a_n)$ for some point $P = (a_1, \ldots, a_n) \in \mathbb{A}^n(K)$;*

(2) *if $\mathfrak{a}$ is an ideal of A, $\mathfrak{a} \neq A$, one has $V(\mathfrak{a}) \neq \emptyset$;*

(3) *for every ideal $\mathfrak{a} \subset A$ one has $I(V(\mathfrak{a})) = \sqrt{\mathfrak{a}}$ (hence in particular $\mathfrak{m}$ is the ideal $I(P)$ of all polynomials that vanish at P).*

Proof. Write $B = K[Y_1, \ldots, Y_n]/\mathfrak{m}$, with $\mathfrak{m}$ a maximal ideal of $K[Y_1, \ldots, Y_n]$. Since B is a field generated over K by the classes of all monomials in $Y_1, \ldots, Y_n$ it follows that the dimension of B, as a K-vector space, is at most countable.

Let now $b \in B \setminus K$ be an arbitrary element. We have to prove that b is algebraic over K. To this end consider the family $\{\frac{1}{b-t}\}_{t \in K}$ of elements of the field B. Since K is uncountable, this is in fact an uncountable family of elements of B. Since $\dim_K(B)$ is at most countable, the elements of this family are linearly dependent over K, i.e., there exist finitely many non-zero elements $\lambda_1, \ldots, \lambda_s \in K$ and elements $t_1, \ldots, t_s \in K$, $s \geq 1$, such that

$$\frac{\lambda_1}{b - t_1} + \cdots + \frac{\lambda_s}{b - t_s} = 0.$$

Clearing denominators we get a non-zero polynomial $f(T) \in K[T]$ of degree ≥ 1 such that $f(b) = 0$, i.e., b is algebraic over K. To see this, observe that $f(t_i) \neq 0$, $i = 1, \ldots, s$. Since K is algebraically closed, we conclude that $B = K$.

Now, set $a_i := Y_i \mod \mathfrak{m}$, $i = 1, \ldots, n$, and let $P := (a_1, \ldots, a_n) \in \mathbb{A}^n(K)$. It is clear that the polynomials $Y_i - a_i$ belong to the kernel of the quotient map $K[Y_1, \ldots, Y_n] \to K[Y_1, \ldots, Y_n]/\mathfrak{m} = B$, $i = 1, \ldots, n$. This implies that $\mathfrak{m}_P \subseteq \mathfrak{m}$ and, since $\mathfrak{m}_P$ is maximal, we get $\mathfrak{m} = \mathfrak{m}_P$. □

If $X = V(\mathfrak{a})$ is an algebraic set of $\mathbb{A}^n$ we will say that the quotient ring $K[X] = K[Y_1, \ldots, Y_n]/I(X)$ is the *coordinate ring* of X. We also say that the pair $(X, K[X])$ is an *affine algebraic variety*. It is immediately clear that

- X is irreducible if and only if the ideal $I(X)$ is a prime ideal, that is, if and only if its coordinate ring is an integral domain.

Indeed, let $f_1, f_2 \notin I(X)$ and let X_i be the subset of X consisting of the points at which f_i vanishes, $i = 1, 2$. If X is irreducible one then has that the union $X_1 \cup X_2$ is strictly contained in X. Thus if $x \in X - (X_1 \cup X_2)$ we must have $f_1 f_2(x) \neq 0$ and so $f_1 f_2 \notin I(X)$, which is to say that $I(X)$ is a prime ideal. Conversely, suppose that X is reducible, $X = X_1 \cup X_2$, and let $f_1, f_2 \notin I(X)$ be such that $f_1(X_1) = 0 = f_2(X_2)$ (i.e., f_i vanishes at each point of X_i for $i = 1, 2$). It follows that $f_1 f_2 \in I(X)$ and so $I(X)$ is not prime.

In particular, by the preceding arguments, a hypersurface $X = V(f)$ is irreducible if and only if the polynomial f is a power of an irreducible polynomial.

2.2.3 The Zariski topology on an affine variety. In view of the properties of the correspondence V, it is clear that the algebraic sets $X \subset \mathbb{A}^n$ form the set of closed subsets of a topology on $\mathbb{A}^n$, called the *Zariski topology* on $\mathbb{A}^n$.

A basis for the open subsets of $\mathbb{A}^n$ is given by the open sets

$$\mathbb{A}^n_f := \mathbb{A}^n - V(f) = \{y \in \mathbb{A}^n \mid f(y) \neq 0\},$$

with $f(y) \in K[Y_1, \ldots, Y_n]$. Open subsets of this type, complements of hypersurfaces in $\mathbb{A}^n$, are called *principal open* or *basic open* subsets of $\mathbb{A}^n$. We will also say that a *principal closed* subset is the complement $V(f)$ of $\mathbb{A}^n_f$ in $\mathbb{A}^n$. Note also that $\mathbb{A}^n_{f_1 f_2} = \mathbb{A}^n_{f_1} \cap \mathbb{A}^n_{f_2}$.

Let us now examine some properties of the Zariski topology.

2.2. The correspondences V and I

- $\mathbb{A}^n$ is not a Hausdorff space, because for every pair of non-empty open sets U_1, U_2 one has $U_1 \cap U_2 \neq \emptyset$. Indeed, if $U_1 \cap U_2 = \emptyset$, one would have $\mathbb{A}^n = \complement_{\mathbb{A}^n} U_1 \cup \complement_{\mathbb{A}^n} U_2$ (where $\complement_{\mathbb{A}^n} U$ denotes the complement of U in $\mathbb{A}^n$). Since $\mathbb{A}^n$ is irreducible one would have either $\complement_{\mathbb{A}^n} U_1 = \emptyset$ or $\complement_{\mathbb{A}^n} U_2 = \emptyset$. It follows that every non-empty open set is dense in $\mathbb{A}^n$.

- $\mathbb{A}^n$ is a Fréchet space (i.e., T_1) in the sense that if P and Q are any two distinct points, each of the two is contained in an open set which does not contain the other. In fact, if f is a polynomial satisfying $f(P) \neq 0$ but $f(Q) = 0$, then the open set which is the complement of the algebraic set $V(f)$ contains P but not Q.

- $\mathbb{A}^n$ is compact, that is, every open cover $\mathbb{A}^n = \bigcup_\alpha U_\alpha$ admits a finite subcover $\mathbb{A}^n = \bigcup_{i=1}^h U_i$.

  Since every ideal $\mathfrak{a}$ of $K[Y_1, \ldots, Y_n]$ is generated by a finite number of polynomials, one has that every closed subset of $\mathbb{A}^n$ is the intersection of a finite number of principal closed subsets and every open subset is a finite union of principal open subsets. It is then easy to see that from every open cover of $\mathbb{A}^n$ one can extract a finite subcover. By the above remarks it suffices to show this for a covering by principal open subsets. Let then $\mathbb{A}^n = \bigcup_\alpha \mathbb{A}^n_{f_\alpha}$. It follows that $\bigcap_\alpha (V(f_\alpha)) = \emptyset$ and therefore if $\mathfrak{a}$ is the ideal generated by all the polynomials f_α one has $\mathfrak{a} = K[Y_1, \ldots, Y_n]$. Hence $1 \in K$ is a polynomial linear combination of a finite number of elements $f_1, \ldots, f_h$ of $\mathfrak{a}$. This implies that $V(f_1, \ldots, f_h) = \emptyset$ and thus that $\mathbb{A}^n = \bigcup_{i=1}^h \mathbb{A}^n_{f_i}$.

  A more general argument, which does not make use of the Hilbert Nullstellensatz, goes as follows. From $\mathbb{A}^n = \bigcup_\alpha \mathbb{A}^n_{f_\alpha}$ we get

  $$V\left(\sum_\alpha (f_\alpha)\right) = \bigcap_\alpha (V(f_\alpha)) = \emptyset.$$

  Since $K[Y_1, \ldots, Y_n]$ is noetherian, there exists a finite set of indices $\{\alpha_1, \ldots, \alpha_h\}$ such that $V(f_{\alpha_1}, \ldots, f_{\alpha_h}) = V\left(\sum_{i=1}^h (f_{\alpha_i})\right) = \emptyset$. Thus we have $\mathbb{A}^n = \bigcup_{i=1}^h \mathbb{A}^n_{f_{\alpha_i}}$.

- If $K = \mathbb{R}$ or $\mathbb{C}$, the Zariski topology in $\mathbb{A}^n$ is coarser than the usual euclidean topology (where, as usual, $\mathbb{C}$ is identified with $\mathbb{R}^2$ via $a + ib = (a, b)$). In fact the Zariski closed sets are also closed for the euclidean topology, since polynomial functions are continuous.

If $X \subset \mathbb{A}^n$ is an algebraic set, we can consider the *Zariski topology on X*, namely the topology on X induced by the Zariski topology on $\mathbb{A}^n$. For each $f \in K[X]$ we set

$$X_f := X - V(f) = \{x \in X \mid f(x) \neq 0\}.$$

20   Chapter 2. Algebraic Sets, Morphisms, and Rational Maps

The open sets X_f are called *principal open* subsets of X, and form a base for the open subsets of the Zariski topology on X.

Remark–Example 2.2.4. We discuss a few more properties and examples.

(1) The closed sets of $\mathbb{A}^1$ are the finite subsets. The closed subsets of $\mathbb{A}^2$ are finite unions of isolated points and algebraic curves.

(2) If $\mathfrak{a} \subset K[T_1, \ldots, T_m]$ is an ideal, we consider the extended ideal

$$\mathfrak{a}^e := \mathfrak{a} K[Y_1, \ldots, Y_n, T_1, \ldots, T_m].$$

The algebraic set $V(\mathfrak{a}^e)$ in $\mathbb{A}^{n+m}$ is called a *cylinder*.

(3) If $X = V(\mathfrak{a})$ is an algebraic subset of $\mathbb{A}^n$, $\mathfrak{a} \subset K[Y_1, \ldots, Y_n]$, then the *projection* of X on the space $\mathbb{A}^m$, where the coordinates of $\mathbb{A}^m$ are $Y_1, \ldots, Y_m$ for $m < n$, is the algebraic set associated to the contracted ideal $\mathfrak{a}^c = \mathfrak{a} \cap K[Y_1, \ldots, Y_m]$ (cf. Section 4.4).

(4) Let $X_1 \subset \mathbb{A}^r$ and $X_2 \subset \mathbb{A}^s$ be algebraic sets, and let $T_1, \ldots, T_r$ and $Y_1, \ldots, Y_s$ be coordinates in $\mathbb{A}^r$ and $\mathbb{A}^s$ respectively. Let $\mathfrak{a}_1 = (f_1, \ldots, f_\lambda) = I(X_1) \subset K[T_1, \ldots, T_r]$ and $\mathfrak{a}_2 = (g_1, \ldots, g_\mu) = I(X_2) \subset K[Y_1, \ldots, Y_s]$. In the product space $\mathbb{A}^r \times \mathbb{A}^s = \mathbb{A}^{r+s}$ with coordinates $T_1, \ldots, T_r, Y_1, \ldots, Y_s$ the *product* $X := X_1 \times X_2$ is the algebraic set of $\mathbb{A}^{r+s}$ associated to the ideal

$$\mathfrak{a} := \mathfrak{a}_1^e + \mathfrak{a}_2^e = (f_1, \ldots, f_\lambda, g_1, \ldots, g_\mu) \subset K[T_1, \ldots, T_r, Y_1, \ldots, Y_s],$$

that is, $I(X) = \sqrt{\mathfrak{a}}$.

(5) (Zariski topology on a product) For each pair $r, s \in \mathbb{N}$, the product topology on $\mathbb{A}^r \times \mathbb{A}^s$ with respect to the Zariski topologies on $\mathbb{A}^r$ and $\mathbb{A}^s$ is strictly coarser than the Zariski topology on $\mathbb{A}^{r+s}$.

Indeed, if $\mathfrak{a}$ and $\mathfrak{b}$ are ideals of $K[T_1, \ldots, T_r]$ and $K[Y_1, \ldots, Y_s]$ respectively, then one has

$$(\mathbb{A}^r \setminus V(\mathfrak{a})) \times (\mathbb{A}^s \setminus V(\mathfrak{b}))$$
$$= [(\mathbb{A}^r \times \mathbb{A}^s) \setminus (V(\mathfrak{a}) \times \mathbb{A}^s)] \cap [(\mathbb{A}^r \times \mathbb{A}^s) \setminus (\mathbb{A}^r \times V(\mathfrak{b}))]$$
$$= [(\mathbb{A}^r \times \mathbb{A}^s) \setminus V(\mathfrak{a}^e)] \cap [(\mathbb{A}^r \times \mathbb{A}^s) \setminus V(\mathfrak{b}^e)].$$

Thus the open sets of the basis of standard open subsets of the product topology are open subsets in the Zariski topology on $\mathbb{A}^{r+s}$.

However, not every open subset in the Zariski topology on $\mathbb{A}^{r+s}$ is an open subset in the product topology. For example, the complement of $V(x - y)$ in $\mathbb{A}^2_{(x,y)}$ is not open in $\mathbb{A}^1 \times \mathbb{A}^1$ since $V(x - y)$ is not closed in $\mathbb{A}^1 \times \mathbb{A}^1$.

Let $X_1 \subseteq \mathbb{A}^r$ and $X_2 \subseteq \mathbb{A}^s$ be algebraic sets. The *Zariski topology on* $X_1 \times X_2$ is the topology induced on $X_1 \times X_2$ by the Zariski topology on $\mathbb{A}^{r+s}$. From the preceding remarks we deduce that this topology is, in general, finer than the product topology on $X_1 \times X_2$ with respect to the Zariski topologies on X_1 and X_2.

See Section 11.2 for the case of the products of projective spaces.

(6) If X is a subset of $\mathbb{A}^n$, $V(I(X))$ is the *Zariski closure* of X. One has $X = V(I(X))$ if and only if X is an algebraic set.

2.3 Morphisms

The contents of this section and the next one have been essentially taken from [74].

Let $X \subset \mathbb{A}^n$ be an irreducible algebraic subset and $K[X] = K[Y_1, \ldots, Y_n]/I(X)$ its coordinate ring. We use $\bar{F}$ to denote the class of a polynomial $F \in K[Y_1, \ldots, Y_n]$ modulo the ideal $I(X)$.

Given a polynomial $F \in K[Y_1, \ldots, Y_n]$ and putting $f = \bar{F}$ we have $F'(x) = F(x)$ for every polynomial $F' \in \bar{F}$ and for every $x \in X$. Thus the *polynomial function*

$$f : X \to K, \quad f(x) := F(x), \ x \in X,$$

is defined. Note that a polynomial function $f : X \to K$ is a continuous map if we consider the Zariski topology on both X and K, where K is identified with $\mathbb{A}^1(K)$. (If $K = \mathbb{R}$ or $\mathbb{C}$, the same is true for the euclidean topology.)

The set R of all such polynomial functions on X has a natural ring structure and the map $F \mapsto f$ defines a surjective ring homomorphism

$$K[Y_1, \ldots, Y_n] \to R \to 0$$

with kernel $I(X)$. Thus one has an isomorphism

$$K[X] \cong R,$$

which expresses the coordinate ring of X as a ring of polynomial functions (defined on all of X) with values in K.

Now let $\mathbb{A}^n$ and $\mathbb{A}^m$ be two affine spaces with coordinate rings $K[Y_1, \ldots, Y_n]$ and $K[T_1, \ldots, T_m]$ respectively. We say that a map $\phi : \mathbb{A}^n \to \mathbb{A}^m$ is a *morphism* (or *regular map*) if there are m polynomials $F_1, \ldots, F_m \in K[Y_1, \ldots, Y_n]$ such that for each $y = (y_1, \ldots, y_n) \in \mathbb{A}^n$ one has

$$\phi(y) = (t_1, \ldots, t_m) \quad \text{with } t_j = F_j(y_1, \ldots, y_n) \text{ for } j = 1, \ldots, m.$$

We also say that ϕ is the "polynomial function given by $F_1, \ldots, F_m$" and that

$$T_j = F_j(Y_1, \ldots, Y_n), \quad j = 1, \ldots, m,$$

are its equations.

If $X \subset \mathbb{A}^n$, $W \subset \mathbb{A}^m$ are algebraic sets, we say that a map $\phi: X \to W$ is a *morphism* (or *regular map*) if it is the restriction to X of a morphism $\Phi: \mathbb{A}^n \to \mathbb{A}^m$ such that $\Phi(X) \subset W$. Thus, if $\phi: X \to W$ is a morphism there exist m polynomial functions $f_1, \ldots, f_m \in K[X]$ such that for every $x \in X$ one has $\phi(x) = (f_1(x), \ldots, f_m(x)) \in W$ (and we say that "ϕ is the polynomial function given by $f_1, \ldots, f_m$").

We say that a morphism $\phi: X \to W$ is an *isomorphism* if ϕ is bijective and the inverse ϕ^{-1} is a morphism. If there exists an isomorphism $\phi: X \to W$ we say that X and W are *isomorphic*.

Note that a morphism $\phi: X \to W$ is a continuous map if we consider the Zariski topology on X and W. Furthermore, if ϕ is an isomorphism, then it is a homeomorphism.

Example 2.3.1. A bijective morphism need not be an isomorphism. For example, let C_1 and C_2 be the two plane curves with equations $y - 1 = 0$ and $y^3 - x^2 = 0$ respectively. Let $\phi: C_1 \to C_2$ be the map that sends the point $P \in C_1$ to its projection P' from the origin O onto C_2. One sees that ϕ is bijective, in particular $\phi(A) = O$ only for $A = (0, 1)$. If $(x, 1)$ are the coordinates of P, the coordinates of P' are $x' = x^3$, $y' = x^2$, and so ϕ is a morphism. But ϕ^{-1} is not a morphism because, given $P' = (x', y') \in C_2$, the coordinates of $P = \phi^{-1}(P')$ are $(\frac{x'}{y'}, 1)$ and $\frac{x'}{y'} \notin K[x', y']$.

Exercise 2.3.2. Let $\phi: X \to W$ be a morphism of algebraic sets with $X \subset \mathbb{A}^n$ and $W \subset \mathbb{A}^m$. With the preceding notation, and considering the coordinates $T_1, \ldots, T_m$ in $\mathbb{A}^m$ as polynomial functions, ϕ is a polynomial function given by $f_1, \ldots, f_m \in K[X]$ if and only if $f_j = T_j \circ \phi \in K[X]$ for $j = 1, \ldots, m$, that is, if and only if the diagram

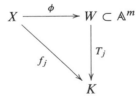

is commutative.

We observe that for $x \in X$ the jth component of $\phi(x)$ is $T_j \circ \phi(x)$, and therefore for each j if we put $f_j = T_j \circ \phi$ we have

$$\phi(x) = (T_1 \circ \phi(x), \ldots, T_m \circ \phi(x)) = (f_1(x), \ldots, f_m(x)).$$

Thus ϕ is the morphism given by $f_1, \ldots, f_m$.

Conversely, if $\phi(x) = (f_1(x), \ldots, f_m(x))$, for $x \in X$, we then have $T_j \circ \phi(x) = f_j(x)$ for all $x \in X$ and $j = 1, \ldots, m$, which is to say that $T_j \circ \phi = f_j \in K[X]$.

2.3. Morphisms

Theorem 2.3.3. *Let $X \subset \mathbb{A}^n$, $W \subset \mathbb{A}^m$ be algebraic sets, and $Y_1, \ldots, Y_n$ and $T_1, \ldots, T_m$ coordinates in $\mathbb{A}^n$ and $\mathbb{A}^m$ respectively. Then the following holds:*

(1) *A morphism $\phi \colon X \to W$ induces a K-algebra homomorphism $\phi^* \colon K[W] \to K[X]$.*

(2) *Conversely, every homomorphism of K-algebras $\theta \colon K[W] \to K[X]$ is of the type $\theta = \phi^*$ with $\phi \colon X \to W$ a uniquely determined morphism.*

(3) *If $\phi \colon X \to W$ and $\psi \colon W \to Z$ are morphisms of algebraic sets, then the morphisms $(\psi \circ \phi)^*$ and $\phi^* \circ \psi^*$ coincide as morphisms of K-algebras; that is, $(\psi \circ \phi)^* = \phi^* \circ \psi^* \colon K[Z] \to K[X]$.*

Proof. Let $g \in K[W]$, that is, let $g \colon W \to K$ be a polynomial function defined on all of W. Set $\phi^*(g) = g \circ \phi$. We show that $g \circ \phi \in K[X]$. To see this it suffices to note the following facts.

a) There exist $F_1, \ldots, F_m \in K[Y_1, \ldots, Y_n]$ such that for all $x \in X$ one has $\phi(x) = (f_1(x), \ldots, f_m(x))$ with $F_j(x) = f_j(x)$ and f_j the class of F_j in $K[X]$, $j = 1, \ldots, m$.

b) Let $G \in K[T_1, \ldots, T_m]$ be such that $G(w) = g(w)$ for all $w \in W$ (so that g is the class of G in $K[W]$). We then have that $P := G(F_1, \ldots, F_m) \in K[Y_1, \ldots, Y_n]$ and its class in $K[X]$ is $G(f_1, \ldots, f_n)$. Moreover, for each $x \in X$ we have

$$(g \circ \phi)(x) = G(f_1(x), \ldots, f_m(x)) = G(f_1, \ldots, f_m)(x) = P(x).$$

Thus $g \circ \phi \in K[X]$. It is then easy to see that ϕ^* is a K-algebra homomorphism, and this proves (1).

To prove (2), we consider the class t_j of the coordinate T_j in $\mathbb{A}^m$ as a function on W. We set $\theta(t_j) = \theta_j$, $j = 1, \ldots, m$. Since θ is a K-algebra homomorphism (and $K[W] = K[t_1, \ldots, t_m]$), we have, for each $g \in K[W]$, $\theta(g) = g(\theta_1, \ldots, \theta_m)$ and so $\theta(g)(x) = g(\theta_1(x), \ldots, \theta_m(x))$ for all $x \in X$. Let $\phi \colon X \to \mathbb{A}^m$ be the morphism defined by $\phi(x) := (\theta_1(x), \ldots, \theta_m(x))$ for all $x \in X$. By what has just been said it follows that for all $g \in K[W]$ we have $g \circ \phi = \theta(g)$, which means that $\theta = \phi^*$ is induced by ϕ.

To conclude it suffices to prove that $\operatorname{Im}(\phi) \subset W$. Let $y \in \operatorname{Im}(\phi)$, that is, $y = (\theta_1(x), \ldots, \theta_m(x))$ for some $x \in X$, and let $F \in I(W)$. Then the class f of F in $K[W]$ is zero whence $F(t_1, \ldots, t_m) = f(t_1, \ldots, t_m) = 0$ in $K[W]$. Hence $\theta(F(t_1, \ldots, t_m)) = 0$ in $K[X]$. But

$$0 = \theta(F(t_1, \ldots, t_m)) = F(\theta(t_1), \ldots, \theta(t_m)) = F(\theta_1, \ldots, \theta_m).$$

The θ_j's belong to $K[X]$ and $F(\theta_1, \ldots, \theta_m) \in K[X]$ is by definition the function $x \mapsto F(\theta_1(x), \ldots, \theta_m(x))$. Finally, for each $x \in X$ and for each $F \in I(W)$, the coordinates $(\theta_1(x), \ldots, \theta_m(x))$ of y satisfy the condition $F(\theta_1(x), \ldots, \theta_m(x)) = 0$, from which it follows that $y \in W$.

Statement (3) is merely the property of associativity for composition of mappings. For each $h \in K[X]$ one has

$$(\psi \circ \phi)^*(h) = h \circ (\psi \circ \phi) = (h \circ \psi) \circ \phi$$
$$= \psi^*(h) \circ \phi = \phi^*(\psi^*(h)) = (\phi^* \circ \psi^*)(h). \qquad \square$$

Corollary 2.3.4. *A morphism $\phi \colon X \to W$ of algebraic sets is an isomorphism if and only if $\phi^* \colon K[W] \to K[X]$ is an isomorphism of K-algebras.*

Proof. In fact we have $(\phi^{-1} \circ \phi)^* = \phi^* \circ (\phi^{-1})^* = \mathrm{id}_{K[X]}$ and $(\phi \circ \phi^{-1})^* = (\phi^{-1})^* \circ \phi^* = \mathrm{id}_{K[W]}$. $\qquad \square$

2.4 Rational maps

We recall the definition of the field of fractions of an integral domain.

Definition 2.4.1. Let A be an integral domain. The *field of fractions* $\mathrm{Frac}(A)$ is the localization $S^{-1}(A)$ of A with respect to the multiplicatively closed set $S := A \setminus \{0\}$, that is

$$\mathrm{Frac}(A) = (A \times S)/\sim,$$

where "$\sim$" is the equivalence relation defined by $(a, s) \sim (a', s') \iff as' = a's$. Thus we have

$$\mathrm{Frac}(A) = \{\tfrac{a}{b} \mid a, b \in A, b \neq 0 \text{ and } \tfrac{a}{b} = \tfrac{a'}{b'} \iff ab' = a'b\}.$$

Let $X \subset \mathbb{A}^n$ be an irreducible algebraic set on which we consider the Zariski topology. Let $K[X]$ be the ring of coordinates and $K(X)$ the field of fractions of $K[X]$, that is, indicating by $\bar{F}$ the class modulo $I(X)$ of a polynomial $F \in K[Y_1, \ldots, Y_n]$,

$$K(X) = \{\tfrac{g}{h} \mid g, h \in K[X], h \neq 0, \tfrac{g}{h} = \tfrac{g'}{h'} \iff gh' = g'h\}.$$

An element $f \in K(X)$ is called a *rational function* on X. Let $U \subset X$ be an open and let $P \in U$. We say that a rational function $f \in K(X)$ is *regular at P* if there exists a neighborhood U_P of P such that

$$f = \frac{g}{h}, \quad g, h \in K[X], \ h(x) \neq 0 \text{ for all } x \in U_P. \qquad (2.1)$$

We say that f is *regular on U* if it is regular at each point of U.

2.4. Rational maps

We note explicitly that if f is regular in a point $P' \in U$ with $P \neq P'$ then we will also have that

$$f = \frac{g'}{h'}, \quad g', h' \in K[X], \ h'(x) \neq 0 \text{ for all } x \in U_{P'}, \tag{2.2}$$

with $U_{P'}$ a suitable neighborhood of P'. Obviously it follows that $gh' = hg'$ on $U_P \cap U_{P'}$, and (2.1) and (2.2) are said to be *local representations* of f in P and P' respectively.

The set dom(f) of points $x \in X$ where f is regular is called the *domain* or *domain of definition* of f. If $x \in \text{dom}(f)$ there then exist an open subset U_x of X containing x and a local representation $f = \frac{g}{h}$ with $h \neq 0$ in U_x. Obviously $U_x \subset \text{dom}(f)$. Note that $f : \text{dom}(f) \to K$ is a continuous map in the Zariski topology.

We now discuss some properties of rational functions.

- A rational function $f \in K(X)$ which is regular on an open set U does not, in general, have a global representation valid on all of U.

Example 2.4.2. Let $X \subset \mathbb{A}^4(\mathbb{C})$ be the quadric with equation $Y_1 Y_2 - Y_3 Y_4 = 0$. If we let y_i denote the class of Y_i in $\mathbb{C}[X]$, $i = 1, \ldots, 4$, we have $y_1 y_2 - y_3 y_4 = 0$ in $\mathbb{C}[X]$. We consider the rational function

$$f = \frac{y_1}{y_3} = \frac{y_4}{y_2} \in \mathbb{C}(X).$$

More precisely, $\frac{y_1}{y_3}$ is a representation for f on the open set $U_3 := \{y_3 \neq 0\}$ while $\frac{y_4}{y_2}$ represents f on the open set $U_2 := \{y_2 \neq 0\}$. Hence f is defined on $U = U_2 \cup U_3$, but does not have a global representation on U.

Exercise 2.4.3. Let $f, f' \in K(X)$ be distinct rational functions. Then there exists a non-empty open subset $U \subset X$ such that $f(x) \neq f'(x)$ for all $x \in U$. Let

$$f = \frac{g}{h}, \quad f' = \frac{g'}{h'}$$

be defined on the principal open subsets X_h and $X_{h'}$ respectively. We have $gh' \neq g'h$ since $f \neq f'$. Consider the open subset

$$U := \{x \in X \mid (gh' - g'h)hh'(x) \neq 0\}.$$

Since $U \subset X_h$ and $U \subset X_{h'}$ both f and f' are defined on U, and for each $x \in U$ we have (since $(gh' - g'h)(x) \neq 0$)

$$f(x) = \frac{g(x)}{h(x)} \neq \frac{g'(x)}{h'(x)} = f'(x).$$

The following lemma shows that a rational function regular on all of X is a polynomial function.

Lemma 2.4.4. *Let K be an algebraically closed field and let $f \in K(X)$ be a rational function on an algebraic set X. Then:*

(1) *The domain $\mathrm{dom}(f)$ is a dense open subset of X.*

(2) $\mathrm{dom}(f) = X$ *if and only if* $f \in K[X]$.

Proof. We consider the "ideal $\mathfrak{a}$ of the denominators" of f defined by

$$\mathfrak{a} := \{h \in K[X] \mid fh \in K[X]\} = \{h \in K[X] \mid f = \tfrac{g}{h}, g \in K[X]\} \cup \{0\}.$$

One then has

$$X - \mathrm{dom}(f) = \{x \in X \mid h(x) = 0 \text{ for all } h \in \mathfrak{a}\} = V(\mathfrak{a}).$$

Thus $X - \mathrm{dom}(f)$ is an algebraic set and so $\mathrm{dom}(f) = X - V(\mathfrak{a})$ is a Zariski open subset of X; in particular it is a dense open subset and we have $\overline{\mathrm{dom}(f)} = X$. Furthermore, cf. Theorem 2.2.2,

$$\mathrm{dom}(f) = X \iff V(\mathfrak{a}) = \emptyset \iff 1 \in \mathfrak{a} \iff f \in K[X]. \qquad \square$$

If $X \subset \mathbb{A}^n$, $W \subset \mathbb{A}^m$ are algebraic sets, we say that a map $\phi \colon X \to \mathbb{A}^m$ is a *rational map* or *rational transformation* if there exist rational functions $f_1, \ldots, f_m \in K(X)$ such that

$$\phi(x) = (f_1(x), \ldots, f_m(x)) \quad \text{for every } x \in \bigcap_{j=1}^{m} \mathrm{dom}(f_j). \qquad (2.3)$$

By definition ϕ is defined on the open subset

$$\mathrm{dom}(\phi) := \bigcap_{j=1}^{m} \mathrm{dom}(f_j),$$

which we call the *domain* of ϕ. We will also say that ϕ is *regular at the points* $x \in \mathrm{dom}(\phi)$.

If $\phi(\mathrm{dom}(\phi)) \subset W$ then $\phi \colon X \to W$ is a rational map between the two algebraic sets X and W. The map $\phi \colon X \to W$ is *dominant* if $\phi(\mathrm{dom}(\phi))$ is dense in W, that is, if $\overline{\phi(\mathrm{dom}(\phi))} = W$.

We note that, given two rational maps $\phi \colon X \to W$, $\psi \colon W \to Z$ between algebraic sets, one can then consider the rational map $\psi \circ \phi \colon X \to Z$, the *composition* of ϕ and ψ, whenever $\phi(\mathrm{dom}(\phi)) \cap \mathrm{dom}(\psi) \neq \emptyset$. In particular, this is always the case if ϕ is dominant.

Remark 2.4.5. Let $\phi\colon X \to W$, with $\phi(X)$ dense in W, and $\psi\colon W \to Z$ be rational transformations between algebraic sets. If $\mathrm{Im}(\psi)$ is dense in Z then also $\mathrm{Im}(\psi \circ \phi)$ is dense in Z. It follows that $(\psi \circ \phi)^* = \phi^* \circ \psi^*$.

Remark 2.4.6. In the preceding notation, let $\phi\colon X \to W$ be a rational map defined as in (2.3). Each $g \in K[W]$ is of the form $g = G$ modulo $I(W)$ for some $G \in K[T_1, \ldots, T_m]$ and $g \circ \phi = F(f_1, \ldots, f_m)$ is a well-defined element of $K(X)$. Thus, exactly as in the case of morphisms, one has a morphism of K-algebras $\phi^*\colon K[W] \to K(X)$. However, if $h \in \ker(\phi^*) \neq (0)$ then $\phi^*(g/h)$ is not defined and so ϕ^* *does not admit an extension* to a homomorphism of K-algebras $K(W) \to K(X)$, except precisely in the case in which $\ker(\phi^*) = (0)$. In this regard we have the following fact.

- If $\phi\colon X \to W$ is a dominant rational map, the homomorphism $\phi^*\colon K[W] \to K(X)$ is injective (and so admits an extension to $\phi^*\colon K(W) \to K(X)$).

Indeed, if $g = G$ modulo $I(W)$ in $K[W]$ then

$$\phi^*(g) = G(f_1, \ldots, f_m).$$

Hence $\phi^*(g) = 0$ means that $G = 0$ on $\mathrm{Im}(\phi)$, that is,

$$\phi^*(g)(x) = G(f_1(x), \ldots, f_m(x)) = 0$$

for every $x \in X$. Hence $G = 0$ on W because $\overline{\mathrm{Im}(\phi)} = W$; that is to say, $G \in I(W)$ and so $g = 0$.

In the case of a morphism one has the following equivalence.

- Let $\phi\colon X \to W$ be a morphism of algebraic sets. Then $\phi^*\colon K[W] \to K[X]$ is injective if and only if ϕ is dominant.

Indeed, let $g \in K[W]$ be such that $g \circ \phi = \phi^*(g) = 0$, that is such that $(g \circ \phi)(x) = 0$ for all $x \in X$, or again, such that $G(f_1(x), \ldots, f_m(x)) = 0$ where the f_j are the classes in $K[X]$ of the polynomials $F_j \in K[Y_1, \ldots, Y_n]$ and g is the class of $G \in K[T_1, \ldots, T_m]$. Thus, G vanishes on a dense subset of W (since ϕ is dominant) and so vanishes on all of W. Therefore $G \in I(W)$, from which it follows that $g = 0$ and ϕ^* is injective. To prove the converse one notes that the kernel of ϕ^* consists of those polynomial functions $g \in K[W]$ such that $g \circ \phi = 0$, namely those $g \in K[W]$ that vanish on $\mathrm{Im}(\phi)$ and hence also on $\overline{\mathrm{Im}(\phi)}$. Since $\ker(\phi^*) = (0)$, we have that $g \in K[W]$ vanishes on $\overline{\mathrm{Im}(\phi)}$ if and only if it vanishes on W. From this it follows that $W = \overline{\mathrm{Im}(\phi)}$, since otherwise it would be possible to choose a point w in the open complement of $\overline{\mathrm{Im}(\phi)}$ in W, and $g \in K[W]$ such that $g(w) \neq 0$ (it suffices to take g to be a non-zero constant function).

Theorem 2.4.7. *Let $\phi\colon X \to W$ be a rational map between algebraic sets. Then the following holds:*

(1) *If ϕ is dominant, ϕ defines a homomorphism of K-algebras $\phi^*\colon K(W) \to K(X)$.*

(2) *Conversely, every homomorphism of K-algebras $\theta\colon K(W) \to K(X)$ is of the form $\theta = \phi^*$ with ϕ a dominant rational map.*

(3) *If $\phi\colon X \to W$ and $\psi\colon W \to Z$ are dominant rational maps, then the composition $(\psi \circ \phi)^* = \phi^* \circ \psi^* \colon K(Z) \to K(X)$ is a homomorphism of K-algebras.*

Proof. The first point follows immediately from Remark 2.4.6, and the proofs of (2) and (3) are slight modifications of the proofs for the corresponding statements in Theorem 2.3.3. $\square$

Let $\phi\colon X \to W$ be a dominant rational map between algebraic sets. We say that ϕ is a *birational isomorphism* (or *birational transformation*, or also that X and W are *birationally equivalent via ϕ*) if there exists a dominant rational map $\psi\colon W \to X$ which is inverse to ϕ, that is such that $\phi \circ \psi = \mathrm{id}_W$ and $\psi \circ \phi = \mathrm{id}_X$ (where defined).

From Theorem 2.4.7 and the definition just given one has:

Proposition 2.4.8. *Two algebraic sets X and W are birationally equivalent if and only if $K(X) \cong K(W)$.*

2.4.9 Morphism from an open set of an affine variety. Let X, W be affine varieties, and $U \subset X$ an open subset.

A *morphism* $\varphi\colon U \to W$ is a rational map $\varphi\colon X \to W$ such that $U \subset \mathrm{dom}(\varphi)$, so that φ is regular at every point $P \in U$.

If $U_1 \subset X$ and $U_2 \subset W$ are opens, then a morphism $\varphi\colon U_1 \to U_2$ is just a morphism $\varphi\colon U_1 \to W$ such that $\varphi(U_1) \subset U_2$. An isomorphism is a morphism which has a two-sided inverse morphism.

Note that if X, W are affine varieties, then by Lemma 2.4.4 (2),

$$\{\text{morphisms } \varphi\colon X \to W\} = \{\text{polynomial maps } \varphi\colon X \to W\};$$

the left-hand side of the equality consists of rational objects satisfying regularity conditions, whereas the right-hand side is defined more directly in terms of polynomials.

With regard to principal open sets, it is opportune to make the following observation.

- If $X \subset \mathbb{A}^n$ is an algebraic set and $f \in K[X]$, then X_f is isomorphic to an affine algebraic set $W \subset \mathbb{A}^{n+1}$.

Indeed, let $J = I(X) \subset K[Y_1, \ldots, Y_n]$ and choose $F \in K[Y_1, \ldots, Y_n]$ for which $f = F$ modulo $I(X)$. Consider the ideal $\mathfrak{a} = (J, Y_{n+1}F - 1) \subset K[Y_1, \ldots, Y_n, Y_{n+1}]$ and let $W := V(\mathfrak{a}) \subset \mathbb{A}^{n+1}$.

We observe that at every point x of X_f one has $f(x) \neq 0$. The two maps

$$\phi: W \to X_f, \quad (y_1, \ldots, y_n, y_{n+1}) \mapsto (y_1, \ldots, y_n),$$
$$\psi: X_f \to W, \quad (y_1, \ldots, y_n) \mapsto (y_1, \ldots, y_n, 1/f(y_1, \ldots, y_n)),$$

are mutually inverse morphisms; therefore one has an isomorphism $W \cong X_f$.

For example, if $X = \mathbb{A}^1$ and $f = y_1$, so that $X_f = X - \{0\}$, then $W \subset \mathbb{A}^2$ is the hyperbola of equation $y_1 y_2 = 1$ and the isomorphism $W \cong X_f$ is obtained by projection (Figure 2.1).

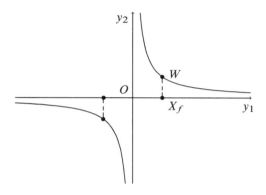

Figure 2.1

2.4.10 Birational equivalence of an algebraic set with a hypersurface.
Let us here anticipate an important fact to which we shall return in the sequel (cf. §2.6.11 and Remark 3.4.10).

- An algebraic set $X \subset \mathbb{A}^n$ is birationally equivalent to a hypersurface $V := V(F)$ in a suitable affine space.

In this regard we must recall a few results from the theory of fields, the proofs of which may be found, for example, in [74, II, §3].

Lemma 2.4.11 (Noether Normalization Lemma). *Let $X \subset \mathbb{A}^n$ be an algebraic set, and let $K[X] = K[y_1, \ldots, y_n]$ be its coordinate ring. Then there exist $m \leq n$ linear forms $L_1, \ldots, L_m$ in $y_1, \ldots, y_n$ such that*

(1) *$L_1, \ldots, L_m$ are algebraically independent over K;*

30   Chapter 2. Algebraic Sets, Morphisms, and Rational Maps

(2) *there exists a linear combination L_{m+1} of the y_i with coefficients in $K[L_1, \ldots, L_m]$ such that $K(X) \cong K(L_1, \ldots, L_m, L_{m+1})$.*

The statement of the preceding lemma may be paraphrased by saying that the extension $K \subset K(X)$ can be obtained as composition of a purely transcendental extension $K \subset K(L_1, \ldots, L_m)$ followed by an extension

$$K(L_1, \ldots, L_m) \subset K(L_1, \ldots, L_m)(L_{m+1}),$$

with L_{m+1} algebraic over $K(L_1, \ldots, L_m)$. Thus $K(X) \cong K(L_1, \ldots, L_m, L_{m+1})$ with *only one* relation of algebraic dependence among the generators. The geometric meaning of this fact is the basis for what we are in the process of proving. Indeed, by Lemma 2.4.11 there is a polynomial $f \in K(L_1, \ldots, L_m)[Y_{m+1}]$ such that $f(L_{m+1}) = 0$. Hence, eliminating the denominators, we have a polynomial $F \in K[Y_1, \ldots, Y_m, Y_{m+1}]$ such that $F(L_1, \ldots, L_m, L_{m+1}) = 0$. We consider the hypersurface $V := V(F) \subset \mathbb{A}^{m+1}$. One then has a morphism $\phi \colon X \to V \subset \mathbb{A}^{n+1}$ defined by

$$\phi(x) := (L_1(x), \ldots, L_m(x), L_{m+1}(x)), \quad x \in X.$$

By the above remarks, the field of fractions of X is $K(X) = K(L_1, \ldots, L_m, L_{m+1})$ whence X is birationally equivalent to V by Proposition 2.4.8.

2.5 Projective algebraic sets

It is useful to recall a few definitions (see also [75]). A *graded ring* is a ring R which is a direct sum

$$R = \bigoplus_{d \geq 0} R_d$$

where the R_d are subgroups of the abelian group of R such that for each pair of indices $d, d' \geq 0$ one has

$$R_d R_{d'} \subset R_{d+d'}.$$

Thus R_0 is a subring of R and for each $d \geq 0$ the subgroup R_d is an R_0-module. The elements of R_d are called the *homogeneous elements of degree d*. An ideal $\mathfrak{a}$ of R is said to be *homogeneous* if

$$\mathfrak{a} = \bigoplus_{d \geq 0} (\mathfrak{a} \cap R_d),$$

that is, for each $f \in \mathfrak{a}$ the decomposition $f = f_0 + f_1 + \cdots + f_r$ with each $f_j \in R_j$ satisfies the condition that $f_j \in \mathfrak{a}$ for $j = 0, 1, \ldots, r$.

2.5. Projective algebraic sets

Example 2.5.1. The ring of polynomials $K[T_1, \ldots, T_n]$ is graded in the obvious way: if one puts

$$R_d = \{\text{homogeneous polynomials of degree } d\},$$

then $R_d R_{d'} \subset R_{d+d'}$ and one has the direct sum decomposition

$$K[T_1, \ldots, T_n] = \bigoplus_{d \geq 0} R_d.$$

If $\mathfrak{a}$ is a homogeneous ideal of $K[T_1, \ldots, T_n]$ the quotient ring $K[T_1, \ldots, T_n]/\mathfrak{a}$ is also graded in a natural way by the grading induced by that of $K[T_1, \ldots, T_n]$. More precisely, if $\sigma : K[T_1, \ldots, T_n] \to K[T_1, \ldots, T_n]/\mathfrak{a}$ is the canonical projection, one has $\sigma(R_d)\sigma(R_{d'}) \subset \sigma(R_{d+d'})$ and the direct sum decomposition

$$K[T_1, \ldots, T_n]/\mathfrak{a} = \bigoplus_{d \geq 0} \sigma(R_d).$$

Hereafter we will use $\mathbb{P}^n := \mathbb{P}^n(K)$ to denote a projective space of dimension n over an algebraically closed field K, and we use $x_1, \ldots, x_{n+1}$ as homogeneous coordinates for $\mathbb{P}^n$.

Remark 2.5.2. A polynomial $f \in K[X_1, \ldots, X_{n+1}]$ vanishes at a point $x = [x_1, \ldots, x_{n+1}]$ of $\mathbb{P}^n$ if it is zero for all choices of the coordinates of x. Let $r := \deg(f)$. If $f = \sum_{i=0}^{r} f_i$, where the f_i are homogeneous polynomials of degree i for $i = 0, \ldots, r$, one then has

$$f(x) = 0 \quad \text{if and only if} \quad f_0(x) = f_1(x) = \cdots = f_r(x) = 0.$$

Indeed, if $f(x) = 0$ one has $f(\lambda x_1, \ldots, \lambda x_{n+1}) = \sum_{i=0}^{r} \lambda^i f_i(x_1, \ldots, x_{n+1}) = 0$ for every integer $\lambda > 0$. Hence the polynomial $p(\lambda) = \sum_{i=0}^{r} \lambda^i f_i(x) \in K[\lambda]$ has infinitely many zeros, which implies that $f_i(x) = 0$ for every index i, as stated.

Bearing this observation in mind, we define (as in the affine case, see Section 2.2) the correspondences V and I as follows. If J is a homogeneous ideal of $K[X_1, \ldots, X_{n+1}]$ and X is a subset of $\mathbb{P}^n$ we put

$$V(J) := \{x \in \mathbb{P}^n \mid f(x) = 0 \text{ for every homogeneous polynomial } f \in J\}$$

and

$$I(X) := \{f \in K[X_1, \ldots, X_{n+1}] \mid f(x) = 0 \text{ for all } x \in X\}.$$

It is easy to verify that $I(X)$ is a homogeneous ideal and that $I(X) = \sqrt{I(X)}$.

Let $x_1, \ldots, x_{n+1}$ be projective coordinates on $\mathbb{P}^n$. A *(projective) algebraic set* of $\mathbb{P}^n$ is a subset $X \subset \mathbb{P}^n$ of the form $X = V(J)$ with J a homogeneous ideal of $R = K[X_1, \ldots, X_{n+1}]$. The set X is therefore the locus of zeros of a finite

32   Chapter 2. Algebraic Sets, Morphisms, and Rational Maps

number of homogeneous polynomials $f_j \in R_j$, $j = 1, \ldots, m$, that is, $X = V(\mathfrak{a})$, where $\mathfrak{a} = (f_1, \ldots, f_m)$. An algebraic subset is said to be *irreducible* if there is no decomposition $X = X_1 \cup X_2$ with X_1 and X_2 algebraic sets strictly contained in X. As in the affine case, X is irreducible if and only if $I(X)$ is a prime ideal. We define a *projective variety* to be a pair $(X, K[X])$ where $X \subset \mathbb{P}^n$ is an algebraic set and $K[X] := K[X_1, \ldots, X_{n+1}]/I(X)$ is its *coordinate ring*.

We note explicitly that, as in the affine case (cf. Section 2.2), the homogeneous ideal J and its radical $\sqrt{J}$ define the same algebraic set $X = V(J) = V(\sqrt{J})$, and the same projective variety $(X, K[X])$ (cf. also Theorem 2.5.4). Hereafter it will sometimes be necessary to distinguish varieties associated to different ideals having the same radical (cf. Section 3.4).

Example 2.5.3. An algebraic set X given by a single homogeneous polynomial (form) $f = 0$ (that is, associated to the principal homogeneous ideal (f)) is said to be a *hypersurface* of $\mathbb{P}^n$. If $n = 2$ it is a projective plane curve, if $n = 3$ it is a projective surface, etc. Such a hypersurface X is irreducible if and only if f is a power of an irreducible polynomial (this is a consequence of the Hilbert Nullstellensatz, Theorem 2.2.2). If f is a form of degree 1 (respectively of degree 2) we shall say that X is a *hyperplane* (respectively a *quadric* or *hyperquadric*) of $\mathbb{P}^n$.

It is simple to verify that the correspondences V, I satisfy the same formal properties as in the affine case (cf. Section 2.2). In particular, $J \subset I(V(J))$ for every homogeneous ideal J and $V(I(X)) = X$ if X is a projective algebraic set.

There is, however, one fact to note. The improper ideal $(1) = K[X_1, \ldots, X_{n+1}]$ defines the empty set in $\mathbb{A}^{n+1}$ and therefore the empty set in $\mathbb{P}^n$. On the other hand, the ideal $(X_1, \ldots, X_{n+1})$ defines the origin in $\mathbb{A}^{n+1}$ and once again the empty set in $\mathbb{P}^n$, that is, in $\mathbb{P}^n$ one has $\emptyset = V((X_1, \ldots, X_{n+1}))$. The ideal $(X_1, \ldots, X_{n+1})$ is called the *irrelevant ideal*, and constitutes a "standard exception" in many statements of the projective theory.

The homogeneous version of the Hilbert Nullstellensatz becomes:

Theorem 2.5.4. *Let K be an algebraically closed field. Then for each homogeneous ideal $J \subset K[X_1, \ldots, X_{n+1}]$ one has*

(1) $V(J) = \emptyset$ *if and only if* $(X_1, \ldots, X_{n+1}) \subset \sqrt{J}$;

(2) *if* $V(J) \neq \emptyset$, *then* $I(V(J)) = \sqrt{J}$.

Proof. Let $\pi : \mathbb{A}^{n+1} \setminus \{(0, \ldots, 0)\} \to \mathbb{P}^n$ be the canonical projection that defines $\mathbb{P}^n$. If $J \subset K[X_1, \ldots, X_{n+1}]$ is a homogeneous ideal we write $V^a(J) \subset \mathbb{A}^{n+1}$ to indicate the affine algebraic set defined by J. Then, since J is homogeneous, $V^a(J)$ has the property that

$$(\alpha_1, \ldots, \alpha_{n+1}) \in V^a(J) \iff (\lambda\alpha_1, \ldots, \lambda\alpha_{n+1}) \in V^a(J), \text{ for all } \lambda \in K^*,$$

and $\mathbb{P}^n \supset V(J) = (V^a(J) - \{(0,\ldots,0)\})/\sim$, where $x \sim y$ iff $x = \lambda y$, $\lambda \in K^*$, for $x, y \in V^a(J) - \{(0,\ldots,0)\}$. Hence

$$V(J) = \emptyset \iff V^a(J) \subset \{(0,\ldots,0)\} \iff (X_1,\ldots,X_{n+1}) \subset \sqrt{J},$$

where the last implication follows from the affine version of the Nullstellensatz (Theorem 2.2.2). Furthermore, if $V(J) \neq \emptyset$ one has

$$f \in I(V(J)) \iff f \in I(V^a(J)) \iff f \in \sqrt{J}.$$ □

The affine algebraic set $V^a(J) \subset \mathbb{A}^{n+1}$ is called the *affine cone* over the projective algebraic set $X = V(J)$, and it is denoted by $C(X)$. If

$$\pi: \mathbb{A}^{n+1} \setminus \{(0,\ldots,0)\} \to \mathbb{P}^n$$

is the map defined by $(a_1,\ldots,a_{n+1}) \mapsto [a_1,\ldots,a_{n+1}]$ it follows that $C(X) = \pi^{-1}(X) \cup \{(0,\ldots,0)\}$.

Corollary 2.5.5. *The correspondences V and I determine mutually inverse bijections*

(1) *between the set of homogeneous radical ideals $J \subset K[X_1,\ldots,X_{n+1}]$ with $J \neq K[X_1,\ldots,X_{n+1}]$, $J \neq (X_1,\ldots,X_{n+1})$, and the collection of projective algebraic subsets $X \subset \mathbb{P}^n$;*

(2) *between the set of homogeneous prime ideals $J \subset K[X_1,\ldots,X_{n+1}]$ such that $J \neq K[X_1,\ldots,X_{n+1}]$, $J \neq (X_1,\ldots,X_{n+1})$, and the set of irreducible projective algebraic subsets $X \subset \mathbb{P}^n$.*

2.5.6 The Zariski topology on a projective variety. In strict analogy with the affine case, the algebraic subsets $X \subset \mathbb{P}^n$ are the closed subsets of a topology on $\mathbb{P}^n$: called the *Zariski topology* on $\mathbb{P}^n$. A base of open subsets is constituted by the *principal open* subsets

$$\mathbb{P}^n_f := \mathbb{P}^n - V(f) = \{x \in \mathbb{P}^n \mid f(x) \neq 0, \ f \text{ a homogeneous polynomial}\}.$$

The space $\mathbb{P}^n$ can be covered by $n+1$ particular principal open subsets, called *standard affine charts*,

$$U_i := \mathbb{P}^n_{X_i} = \{[x_1,\ldots,x_{n+1}] \in \mathbb{P}^n \mid x_i \neq 0\}, \quad i = 1,\ldots,n+1.$$

For each $i = 1,\ldots,n+1$ one associates to the point $[x_1, x_2,\ldots,x_{n+1}] \in U_i$ the point

$$\left(\frac{x_1}{x_i}, \frac{x_2}{x_i}, \ldots, \frac{x_{i-1}}{x_i}, \frac{x_{i+1}}{x_i}, \ldots, \frac{x_{n+1}}{x_i}\right) \in \mathbb{A}^n$$

to obtain a bijection between U_i and the affine space $\mathbb{A}^n$. We say that the $\frac{x_j}{x_i}$, $j \neq i$, are the non-homogeneous (affine) coordinates in U_i. The Zariski topology on each chart U_i is that induced by the Zariski topology on $\mathbb{P}^n$.

Let $X \subset \mathbb{P}^n$ be an algebraic set, and let $I(X)$ be the homogeneous ideal associated to it. We suppose for simplicity that X is not contained in any of the hyperplanes with equation $X_i = 0$, $i = 1, \ldots, n+1$. We know that $\mathbb{P}^n$ is covered by $n+1$ affine charts $U_i = \mathbb{P}^n_{X_i} = \mathbb{P}^n - \{X_i = 0\}$ with affine coordinates

$$y_1^{(i)} = \frac{x_1}{x_i}, \ldots, y_{i-1}^{(i)} = \frac{x_{i-1}}{x_i}, y_{i+1}^{(i)} = \frac{x_{i+1}}{x_i}, \ldots, y_{n+1}^{(i)} = \frac{x_{n+1}}{x_i}. \tag{2.4}$$

We set

$$X_{(i)} = X \cap U_i.$$

Then $X_{(i)} \subset \mathbb{A}^n$ is an affine algebraic set since, for example, for $i = n+1$ the point $P = [y_1^{(n+1)}, \ldots, y_n^{(n+1)}, 1] \in X_{(n+1)}$ if and only if $f(y_1^{(n+1)}, \ldots, y_n^{(n+1)}, 1) = 0$ for every polynomial $f \in I(X)$, and thus $X_{(n+1)}$ is the locus of the zeros of polynomials in the affine coordinates $(y_1^{(n+1)}, \ldots, y_n^{(n+1)})$. More precisely, the ideal of $X_{(n+1)}$ in $U_{n+1} \cong \mathbb{A}^n$ is

$$I(X_{(n+1)}) = \{f(X_1, \ldots, X_n, 1) \mid f \in I(X)\}$$

and

$$I(X)_d = \{X_{n+1}^d f\left(\tfrac{X_1}{X_{n+1}}, \ldots, \tfrac{X_n}{X_{n+1}}\right) \mid f \in I(X_{(n+1)}) \text{ with deg } f \leq d\},$$

where $I(X)_d$ is the degree d part of the homogeneous ideal $I(X)$.

We say that the $X_{(i)}$ are the *standard affine charts* for X. From what we have just seen it follows that the correspondence

$$X \mapsto X_{(i)} = X \cap U_i$$

defines a bijection

$$\{\text{algebraic subsets } X \subset \mathbb{P}^n \mid X \not\subset \{X_i = 0\}\}$$

$$\updownarrow$$

$$\{\text{algebraic subsets } X_{(i)} \subset U_i \cong \mathbb{A}^n\}.$$

2.5.7 Generic objects. Now that we have introduced the notion of algebraic set and the Zariski topology we can give the notion of a *generic* object. We sometimes use the world *general* with the same meaning. When a family of objects $\{X_p\}_{p \in \mathcal{P}}$ is parameterized by the points of an irreducible algebraic set $\mathcal{P}$ (that is, the objects of the family are in one-to-one correspondence with the points of $\mathcal{P}$), the statement

2.6. Rational maps and birational equivalence

"the generic object X_p has the property P" means that "the subset of points $p \in \mathcal{P}$ for which the corresponding object X_p has the property P is a non-empty open subset in the Zariski topology".

For example, we will say that "x is a generic point" of an algebraic set X to mean that the set of points of X from which we may choose the point x is a given open subset that depends on the context, that is, one must exclude that x be in the zero sets of certain polynomials not belonging to $I(X)$.

We shall often consider "generic" linear spaces S_r in $\mathbb{P}^n$; by this we will mean that the S_r vary in an open set of the Grassmann variety $\mathbb{G}(r,n)$ which parameterizes the r-dimensional linear subspaces of $\mathbb{P}^n$ (and for this we refer the reader to Chapter 12).

Again, given, for example, a point $p_0 \in \mathbb{P}^2$, we say that "a generic line $\ell \subset \mathbb{P}^2$ does not contain the point p_0" to express the fact that the set of lines containing p_0 is contained in a proper subvariety of the dual plane $\mathbb{P}^{2*}$ (which consists of all the lines in $\mathbb{P}^2$; see, for instance, [52, Vol. 1, Chapter V, §5]). Here is another example: we say that "the generic conic is non-degenerate" (namely the associated matrix has rank 3) to express the fact that the subset of conics in $\mathbb{P}^2$ can be parameterized by the points of $\mathbb{P}^5$, and that the subset consisting of the degenerate conics is contained in a proper subvariety of $\mathbb{P}^5$.

Exercise 2.5.8. Prove that, given a linear subspace $S_r \subset \mathbb{P}^n$ of dimension $r \le n-2$, the generic line of $\mathbb{P}^n$ does not intersect S_r.

2.6 Rational maps and birational equivalence

The contents of this paragraph are essentially taken from [74, III, §5]. Let X be an irreducible algebraic set and let $I(X) \subset K[X_1, \ldots, X_{n+1}]$ be the prime ideal associated to X. Unlike what happens in the affine case, a polynomial $F \in K[X_1, \ldots, X_{n+1}]$ can fail to define a polynomial function $\mathbb{P}^n \to K$. In order for F to define a polynomial function on $\mathbb{P}^n$ one must have that, for every $\lambda \in K$ and for every $x = [x_1, \ldots, x_{n+1}] \in \mathbb{P}^n$,

$$F(x_1, \ldots, x_{n+1}) = F(\lambda x_1, \ldots, \lambda x_{n+1})$$

and this happens only if F is homogeneous of degree zero, that is, constant. Similarly, if we set $f = F$ modulo $I(X)$ we see that f defines a polynomial function $X \to K$ only if F is homogeneous of degree zero.

A *rational function* is a function $X \to K$ defined by

$$f(x) = \frac{g(x)}{h(x)}, \quad x \in X,$$

where $g, h \in K[X_1, \ldots, X_{n+1}]$ are homogeneous polynomials of the *same* degree d. If $h(x) \ne 0$, the quotient $g(x)/h(x)$ is well defined, since for $0 \ne \lambda \in K$

one has
$$\frac{g(\lambda x_1,\ldots,\lambda x_{n+1})}{h(\lambda x_1,\ldots,\lambda x_{n+1})} = \frac{\lambda^d g(x_1,\ldots,x_{n+1})}{\lambda^d h(x_1,\ldots,x_{n+1})} = \frac{g(x_1,\ldots,x_{n+1})}{h(x_1,\ldots,x_{n+1})}.$$

Obviously g/h and g'/h' define the same rational function on X if and only if $h'g - g'h \in I(X)$. From this it follows that the set of all rational functions is a field, called the *field of fractions* of X,

$$K(X) := \{\tfrac{g}{h} \mid g, h \in K[X_1,\ldots,X_{n+1}],\ h \notin I(X)\} / \sim,$$

where g, h are homogeneous of the same degree and "$\sim$" is the equivalence relation defined by

$$\frac{g}{h} \sim \frac{g'}{h'} \iff h'g - g'h \in I(X).$$

The notion of regular rational function is given just as in the affine case. If $f \in K(X)$ is a rational function, we say that f is *regular* in a point $x \in X$ if there exists an expression $f = g/h$ with g, h homogeneous polynomials of the same degree such that $h(x) \neq 0$. The *domain* of f is

$$\mathrm{dom}(f) := \{x \in X \mid f \text{ is regular in } x\}.$$

We set
$$\mathcal{O}_{X,x} := \{f \in K(X) \text{ with } f \text{ regular at } x\}.$$

Then $\mathcal{O}_{X,x}$ is a subring of the field of fractions $K(X)$, called the *local ring* of X at x.

Proposition 2.6.1. *Let $X \subset \mathbb{P}^n$ be an algebraic set not contained in the hyperplane of equation $X_i = 0$ and let $X_{(i)} = X \cap \mathbb{P}^n_{X_i}$ be the corresponding affine chart. One then has an isomorphism of the fields of fractions $K(X) \cong K(X_{(i)})$, $i = 1,\ldots,n+1$.*

Proof. Suppose for example that $i = n+1$. If $g, h \in K[X_1,\ldots,X_{n+1}]$ are homogeneous polynomials of the same degree d and $h \notin I(X)$, then $g/h \in K(X)$ and the restriction to $X_{(n+1)}$ is the function

$$\frac{g(X_1/X_{n+1},\ldots,X_n/X_{n+1},1)}{h(X_1/X_{n+1},\ldots,X_n/X_{n+1},1)} \in K(X_{(n+1)}).$$

Thus one obtains a map $K(X) \to K(X_{(n+1)})$, and it is easy to see that it is an isomorphism of K-algebras. To construct the inverse map let $y_1 := y_1^{(n+1)},\ldots,y_n := y_n^{(n+1)}$ be the affine coordinates in $X_{(n+1)}$ and let

$$\frac{p(y_1,\ldots,y_n)}{q(y_1,\ldots,y_n)} \in K(X_{(n+1)})$$

be a rational function on $X_{(n+1)}$. Put $p = P$ modulo $I(X)$, $q = Q$ modulo $I(X)$ and introduce the homogeneous coordinates given by (2.4). Then the quotient P/Q may be rewritten as

$$\frac{P(X_1/X_{n+1}, \ldots, X_n/X_{n+1})}{Q(X_1/X_{n+1}, \ldots, X_n/X_{n+1})} = \frac{G(X_1, \ldots, X_{n+1})}{H(X_1, \ldots, X_{n+1})},$$

where G, H are homogeneous polynomials of the same degree in $K[X_1, \ldots, X_{n+1}]$. Passing to the quotient modulo $I(X)$, the fraction G/H gives rise to a rational function $g/h \in K(X)$. □

Rational maps between projective (or affine) varieties are defined by way of rational functions. If $X \subset \mathbb{P}^n$ is an irreducible algebraic set, then a *rational map* (or *rational transformation*) $\varphi \colon X \to \mathbb{A}^m$ is defined by setting

$$\varphi(x) := (f_1(x), \ldots, f_m(x)), \quad x \in X,$$

where $f_1, \ldots, f_m \in K(X)$. This map φ is well defined on the intersection $\bigcap_{j=1}^{m} \mathrm{dom}(f_j)$.

A *rational map* (or *rational transformation*) $\varphi \colon X \to \mathbb{P}^m$ is defined by setting

$$\varphi(x) := [f_1(x), \ldots, f_{m+1}(x)], \quad x \in X,$$

where $f_1, \ldots, f_{m+1} \in K(X)$ and it is well defined on the set, which is an open dense subset of X (cf. Lemma 2.4.4),

$$\bigcap_{i=1}^{m+1} \mathrm{dom}(f_i) - \{x \in X \mid f_1(x) = \cdots = f_{m+1}(x) = 0\}.$$

One notes that if $g \in K(X)$ is a non-zero element, then $gf_1, \ldots, gf_{m+1}$ define the same rational map. Then, assuming that the image of X is not contained in the hyperplane of $\mathbb{P}^m$ defined by $X_{m+1} = 0$, one can suppose that one has $f_{m+1} = 1$. From this it follows that there exists a bijection between the two sets (for the natural immersion $\mathbb{A}^m \subset \mathbb{P}^m$ see, for instance, the discussion in [13, Vol. I, Chapter 5])

$$\{\text{rational maps } \varphi \colon X \to \mathbb{A}^m \subset \mathbb{P}^m\}$$

and

$$\{\text{rational maps } \varphi \colon X \to \mathbb{P}^m \text{ such that } \varphi(X) \not\subset \{X_{m+1} = 0\}\},$$

inasmuch as each of these maps is given by m elements $f_1, \ldots, f_m \in K(X)$.

The preceding remarks are summarized in the following definition.

Definition 2.6.2. A rational map $\varphi \colon X \to \mathbb{P}^m$ is *regular* at a point $x \in X$ if there exists an expression $\varphi = (f_1, \ldots, f_{m+1})$, $f_i \in K(X)$, $i = 1, \ldots, m+1$, such that

38   Chapter 2. Algebraic Sets, Morphisms, and Rational Maps

a) the rational functions $f_1, \ldots, f_{m+1}$ are regular at x;

b) $f_i(x) \neq 0$ for at least one index i.

The set on which φ is regular is the *domain* of φ; it is an open subset of X and is denoted by $\operatorname{dom}(\varphi)$.

If $W \subset \mathbb{P}^m$ is an algebraic set and $\varphi(\operatorname{dom}(\varphi)) \subset W$, $\varphi \colon X \to W$ is a rational map between the two algebraic sets X and W. As in the affine case, we shall say that $\varphi \colon X \to W$ is *dominant* if $\varphi(\operatorname{dom}(\varphi))$ is dense in W, that is, if W is the closure of $\varphi(\operatorname{dom}(\varphi))$.

Note that if $\varphi = (f_1, \ldots, f_{m+1}) \colon X \to \mathbb{P}^m$ is a rational map then there is an open subset $U \subset X$ such that $\varphi_{|U} \colon U \to \mathbb{A}_{(i)}^m = \mathbb{P}_{X_i}^m \subset \mathbb{P}^m$ is a morphism: it suffices to take $U \subseteq \bigcap_j \operatorname{dom}(f_j/f_i)$, where $f_i \neq 0$. Then $\varphi_{|U}$ is the morphism given by $\{f_j/f_i\}$, $j = 1, \ldots, m+1$, $j \neq i$.

Definition 2.6.3. If $U \subseteq X$ is an open subset of a projective variety X, then a morphism $\varphi \colon U \to W$ is a rational map $\varphi \colon X \to W$ such that $U \subset \operatorname{dom}(\varphi)$. Thus, a morphism $\varphi \colon U \to W$ is a rational map which is regular on all of U.

Example 2.6.4 (Projection of a quadric from one of its points). The map $\pi \colon \mathbb{P}^3 \to \mathbb{P}^2$ defined by $[x_1, x_2, x_3, x_4] \mapsto [x_2, x_3, x_4]$ is a rational map, and indeed is a morphism away from the point $P_0 = [1, 0, 0, 0]$. Let $Q \subset \mathbb{P}^3$ be a quadric containing the point P_0. Each point P of $\mathbb{P}^2$ corresponds to the line ℓ of $\mathbb{P}^3$ that passes through P and P_0, and ℓ generally meets Q at P_0 and at a second point $\varphi(P)$. Putting $P \mapsto \varphi(P)$ we obtain a rational map $\varphi \colon \mathbb{P}^2 \to Q$.

For example, if Q has equation $X_1 X_4 = X_2 X_3$, then the restriction $\pi_{|Q} \colon Q \to \mathbb{P}^2$ has as its inverse the rational map $\varphi \colon \mathbb{P}^2 \to Q$ given by $[x_2, x_3, x_4] \mapsto \left[\frac{x_2 x_3}{x_4}, x_2, x_3, x_4\right]$.

As an exercise, determine $\operatorname{dom}(\pi)$ and $\operatorname{dom}(\varphi)$.

As in the affine case (cf. Section 2.4), we say that a dominant rational map $\varphi \colon X \to W$ between two projective varieties is a *birational isomorphism* or *birational transformation* (or also, that X and W are *birationally equivalent* or *birationally isomorphic via* φ) if there exists an inverse dominant rational map $\psi \colon W \to X$, that is, such that $\varphi \circ \psi = \operatorname{id}_W$, $\psi \circ \varphi = \operatorname{id}_X$ (where defined).

Proposition 2.6.5. *Let* $\varphi \colon X \to W$ *be a rational map between projective (or affine) varieties. The following three conditions are equivalent.*

(1) φ *is a birational equivalence.*

(2) φ *is dominant and* $\varphi^* \colon K(W) \to K(X)$ *is an isomorphism.*

(3) *There exist open sets* $X_0 \subset X$, $W_0 \subset W$ *such that* φ *restricted to* X_0 *is an isomorphism* $\varphi \colon X_0 \to W_0$.

Proof. The K-algebra homomorphism φ^* is defined exactly as in the affine case (cf. Theorem 2.4.7) and the equivalence (1) $\Leftrightarrow$ (2) is obtained as in Theorem 2.4.7 and Proposition 2.4.8.

The implication (3) $\Rightarrow$ (1) follows from the fact that an isomorphism $\varphi: X_0 \to W_0$ and its inverse $\varphi^{-1}: W_0 \to X_0$ give rise to a birational map $X \to W$.

The essential implication is (1) $\Rightarrow$ (3). We give a proof as in [74, p. 87]. By hypothesis, there exist mutually inverse rational maps $\varphi: X \to W$ and $\psi: W \to X$. We set
$$X' := \mathrm{dom}(\varphi) \subset X \quad \text{and} \quad \alpha := \varphi_{|X'}: X' \to W,$$
and similarly
$$W' := \mathrm{dom}(\psi) \subset W \quad \text{and} \quad \beta := \psi_{|W'}: W' \to X.$$

In the diagram
$$\beta^{-1}(X') \xrightarrow{\beta} X' \xrightarrow{\alpha} W$$
$$\cap$$
$$W$$

all the arrows are morphisms and the equality of the morphisms $\mathrm{id}_{W|\beta^{-1}(X')} = \alpha \circ \beta$ follows from the equality of the rational maps $\mathrm{id}_W = \varphi \circ \psi$. Thus
$$\alpha(\beta(x)) = x \quad \text{for all } x \in \beta^{-1}(X').$$

We set $X_0 := \alpha^{-1}\beta^{-1}(X')$ and $W_0 := \beta^{-1}\alpha^{-1}(W')$. Then by construction $\varphi: X_0 \to \beta^{-1}(X')$ is a morphism. On the other hand, $\beta^{-1}(X') \subset W_0$ since $x \in \beta^{-1}(X')$ implies that $\alpha(\beta(x)) = x$ and so $x \in \beta^{-1}\alpha^{-1}(W') = W_0$. It follows that $\varphi: X_0 \to W_0$ is a morphism. In the same way one proves that $\psi: W_0 \to X_0$ is a morphism. $\square$

The preceding proposition has an important consequence.

Corollary 2.6.6. *Given a projective (or affine) variety X, the following two conditions are equivalent.*

(1) *The field of fractions $K(X)$ is a purely transcendental extension of K, that is, $K(X) \cong K(t_1, \ldots, t_d)$ for some integer d.*

(2) *There is a dense open subset $X_0 \subset X$ which is isomorphic to a dense open subset $U_0 \subset \mathbb{A}^d$.*

A variety that satisfies the conditions of Corollary 2.6.6 is said to be *rational*. In particular condition (2) is the precise statement of the fact that a rational variety X can be parameterized by d independent variables (cf. Section 6.6).

We now give some further properties of rational transformations between projective varieties.

2.6.7 (Local representation of a rational transformation). Let X be a subvariety of $\mathbb{P}^n$. In the set of all rational transformations $\varphi \colon \mathbb{P}^n \to \mathbb{P}^m$ we introduce an equivalence relation with respect to X in the following way.

Let $\varphi_1, \varphi_2 \colon \mathbb{P}^n \to \mathbb{P}^m$ be two rational transformations. We say that $\varphi_1 \sim \varphi_2$ if for each point $x \in X$ at which φ_1 and φ_2 are both defined one has $\varphi_1(x) = \varphi_2(x)$. Note that the set $\mathrm{dom}(\varphi_1) \cap \mathrm{dom}(\varphi_2) \cap X$ of points of X in which φ_1 and φ_2 are both defined is an open subset of X (cf. Definition 2.6.2).

Let $[\varphi]$ be an equivalence class and $\phi \in [\varphi]$ one of its representatives (so that $\varphi = \phi$ on $U := \mathrm{dom}(\varphi) \cap \mathrm{dom}(\phi) \cap X$). The closure $X' = \overline{\phi(U)}$ does not depend on ϕ but only on the class $[\varphi]$; we shall say that $[\varphi]$ is a *rational transformation* $X \to X'$ and that X' is a *rational transform of the projective variety* X.

The rational transformation $[\varphi]$ is defined outside of its *exceptional set* $E := \bigcap_{\phi \in [\varphi]} (E_\phi \cap X)$, where $E_\phi := \mathbb{P}^n \setminus \mathrm{dom}(\phi)$. For every point $x \in X$ the image $[\varphi](x)$ is the image $\phi(x)$ under any rational transformation $\phi \in [\varphi]$ such that $x \in \mathrm{dom}(\phi)$.

From the fact that the ring of polynomials $K[X_1, X_2, \ldots, X_{n+1}]$ is a Noetherian ring one easily deduces that a finite number of representatives of $[\varphi]$ suffice to describe $[\varphi]$: that is, there exist a finite number h of representatives $\phi_1, \phi_2, \ldots, \phi_h$ of $[\varphi]$ such that for every $x \notin E$ one has $[\varphi](x) = \phi_j(x)$ for some $j = 1, \ldots, h$.

Example 2.6.8. Consider two surjective rational transformations $\phi_1, \phi_2 \colon \mathbb{P}^2 \to \mathbb{P}^1$, defined by

$$\phi_1(x) = [x_0, x_2], \quad \phi_2(x) = [x_2, x_1],$$

where $x = [x_0, x_1, x_2]$ is a point of $\mathbb{P}^2$. One sees immediately that they are equivalent with respect to the conic γ with equation $x_0 x_1 - x_2^2 = 0$. Hence we have

$$E_{\phi_1} = \mathbb{P}^2 \setminus \mathrm{dom}(\phi_1) = A_1 = [0, 1, 0], \quad E_{\phi_2} = \mathbb{P}^2 \setminus \mathrm{dom}(\phi_2) = A_0 = [1, 0, 0],$$

and thus $E = E_{\phi_1} \cap E_{\phi_2} \cap \gamma = \emptyset$. One then has

$$\phi_1(\gamma \setminus A_1) = \mathbb{P}^1 \setminus [0, 1], \quad \phi_2(\gamma \setminus A_2) = \mathbb{P}^1 \setminus [1, 0];$$

and therefore the rational transform γ' of γ is $\gamma' = \mathbb{P}^1$. Thus ϕ_1 and ϕ_2 are two representatives of an everywhere defined rational transformation $[\varphi] \colon \gamma \to \mathbb{P}^1$.

2.6.9 (The fibers of a rational transformation). If $\varphi \colon X \to W$ is a morphism between projective varieties, the *fiber* of φ at (or over) a point $w \in W$ is the inverse image $\varphi^{-1}(w)$ of w; it is a closed subset of X since φ is obviously a continuous map.

If $\varphi \colon X \to W$ is a rational transformation between projective varieties, $X \subset \mathbb{P}^n$, $W \subset \mathbb{P}^m$, let $U = \mathrm{dom}(\varphi)$ be the open set of X where φ is defined, and let $\varphi_U \colon U \to \varphi(U)$ be the restriction morphism of φ to U. If $w \in \varphi(U)$ we call the projective closure $\varphi^{-1}(w)$ of $\varphi_U^{-1}(w)$ in X the *fiber* of φ over w, that is, the closure of $\varphi_U^{-1}(w)$ in the Zariski topology on X. More precisely, $\varphi^{-1}(w)$ is the union of the inverse image $\varphi_U^{-1}(w)$ of w under φ_U and of the exceptional set $E_\varphi = X \setminus \mathrm{dom}(\varphi)$ of φ, cf. §2.1.1.

If the fiber $\varphi^{-1}(w)$ over a generic point w (which means variable in any open set) of $\varphi_U(U)$ contains only one point not belonging to E_φ, the transformation is birational. In this case if

$$Y_j = \varphi_j(X_1, \ldots, X_{n+1}), \quad j = 1, \ldots, m+1, \tag{2.5}$$

are the equations of a rational transformation $\mathbb{P}^n \to \mathbb{P}^m$ that determine φ (by restriction to X) and if $(f_1, \ldots, f_t)$ is a system of generators of the homogeneous ideal of X, the system of equations

$$\begin{cases} Y_j = \varphi_j(X_1, \ldots, X_{n+1}), \\ f_\alpha(X_1, \ldots, X_{n+1}) = 0, \end{cases} \tag{2.6}$$

where $j = 1, \ldots, m+1$ and $\alpha = 1, \ldots, t$, permits one to recover the $X_1, \ldots, X_{n+1}$ as algebraic and uniform functions (that is, "single valued") and hence as rational functions of the $Y_1, \ldots, Y_{m+1}$. Indeed, since the $X_1, \ldots, X_{n+1}$ are homogeneous coordinates, from (2.6) one can deduce (see, for instance, Exercise 13.1.1) formulas like the following:

$$\begin{cases} X_i = \theta_i(Y_1, \ldots, Y_{m+1}), \\ g_\beta(Y_1, \ldots, Y_{m+1}) = 0, \end{cases} \tag{2.7}$$

where θ_i, $i = 1, \ldots, n+1$, are homogeneous polynomials all of the same degree in the ring $K[Y_1, \ldots, Y_{m+1}]$ and the polynomials g_β, $\beta = 1, \ldots, s$, comprise a system of generators of the ideal of $W \subset \mathbb{P}^m$. Thus one has, together with $\varphi \colon X \to W$, also a birational transformation $\theta \colon W \to X$; moreover φ and θ are mutually inverse.

2.6.10 (Finite morphisms). Let $\varphi \colon X \to W$ be a dominant morphism between affine varieties. By what we have seen in Section 2.4, it defines an immersion $\varphi^* \colon K[W] \to K[X]$ and thus $K[W]$ can be regarded as a subring of $K[X]$. One says that φ is a *finite morphism* if $K[X]$ is an integral extension of $K[W]$.

Now let $\varphi \colon X \to W$ be a finite morphism (and so, by definition, also dominant) between affine varieties.

Since $K[W]$ can be viewed as a subring of $K[X]$ (via φ^*), we will use the same symbol to denote a function of $K[W]$ and its transform under φ^*, that is the "same" function regarded as an element of $K[X]$. An ideal $\mathfrak{a}$ of $K[W]$, generated by $g_1, \ldots, g_t$, gives rise to an ideal $\varphi^*(\mathfrak{a})$ of $K[X]$, generated by $\varphi^*(g_1), \ldots, \varphi^*(g_t)$, that is, by $g_1, \ldots, g_t$ in $K[X]$. Thus $\varphi^*(\mathfrak{a}) = \mathfrak{a} K[X]$ is the extended ideal of $\mathfrak{a}$. In particular if $\mathfrak{m}_w$ is the maximal ideal associated to the point $w \in W$, $\mathfrak{m}_w K[X]$ is the ideal of $K[X]$ whose zeros are the points of the fiber $\varphi^{-1}(w)$.

Let $X \subset \mathbb{A}^n$, $W \subset \mathbb{A}^m$. Then there exist m polynomial functions $f_1, \ldots, f_m$ such that $\varphi(x_1, \ldots, x_n) = (w_1, \ldots, w_m)$ with $w_j = f_j(x_1, \ldots, x_n)$. The coordinates $x_1, \ldots, x_n$ regarded as elements of $K[X] = K[x_1, \ldots, x_n]$ are integral over $K[W]$, and so for each $i = 1, \ldots, n$ one has an equation of the form

$$x_i^{r_i} + b_1^{(i)}(w) x_i^{r_i - 1} + \cdots + b_{r_i}^{(i)}(w) = 0,$$

where $w = (w_1, \ldots, w_m)$. This equation (which is a consequence of the equations $w_j = f_j(x_1, \ldots, x_n)$ that define the morphism) is satisfied by the ith coordinate of the points of the fiber $\varphi^{-1}(w)$. Hence these points are finite in number for each $w \in W$.

- A finite morphism $\varphi \colon X \to Y$ is surjective.

In fact, if there were $w \in W$ with $f^{-1}(w) = \emptyset$, by what we have just seen one would have the contradiction $\mathfrak{m}_w = K[X]$.

- (Finiteness is a local property) A morphism $\varphi \colon X \to W$ of projective (or affine) varieties is *finite* if there exists an affine open cover $\{U_\alpha\}$ of W such that $\varphi^{-1}(U_\alpha)$ is affine and the restriction $\varphi_\alpha \colon \varphi^{-1}(U_\alpha) \to U_\alpha$ is a finite morphism for each index α.

2.6.11 (Birational equivalence of a projective variety with a hypersurface). Among the transformations between two projective spaces one has in particular the projections.

Projecting the points of $\mathbb{P}^n$ from a subspace S_k of $\mathbb{P}^n$ onto a subspace $S_{k'}$ skew to it and of dual dimension (that is, $k + k' = n - 1$) one obtains a rational mapping $\varphi \colon \mathbb{P}^n \to S_{k'}$ defined by setting $\varphi(x) = J(x, S_k) \cap S_{k'}$ for all $x \notin S_k$. The exceptional set E_φ of the mapping φ coincides with S_k (cf. Exercise 2.7.37).

Let V_d be a variety of pure dimension d in $\mathbb{P}^n$, with $d < n$ (cf. Section 3.3). Let x be a point not belonging to V_d and let $\varphi \colon \mathbb{P}^n \to S_{n-1}$ be the projection of $\mathbb{P}^n$ from x onto a hyperplane S_{n-1}. The restriction φ_{V_d} of φ to V_d is the projection of V_d from x onto S_{n-1}. Since φ is a rational transformation having x as its only exceptional point (this means that $E_\varphi = \{x\}$) and $x \notin V_d$, the restriction $\varphi_{V_d} \colon V_d \to S_{n-1}$ is a morphism and has all its fibers finite (since otherwise V_d would contain the line joining x to the image $\varphi(P)$ of a point $P \in V_d$). Moreover, the image V' of V_d is a variety of pure dimension d.

If $d < n - 1$, we repeat the procedure: that is, we project $V'_d = \varphi(V_d)$ from a point of S_{n-1} not belonging to V'_d onto an $S_{n-2} \subset S_{n-1}$, and so on. After a finite number of projections we arrive at a surjective rational mapping $\pi \colon V_d \to \mathbb{P}^d$ with all fibers finite (cf. §2.6.10).

If the process of successive projections described above is terminated at the next to last step, one obtains a rational map $\sigma \colon V_d \to X_d$, which is surjective and has all fibers finite, where X_d is a hypersurface in $\mathbb{P}^{d+1}$. With a suitable choice for the successive centers of projection, one proves that it is always possible to arrange matters so that σ is a birational isomorphism (cf. §3.4.5). Therefore, every pure d-dimensional projective variety is the birational transform of a hypersurface in $\mathbb{P}^{d+1}$ (a fact discussed in §2.4.10 for the affine case and to which we will return at greater length in Remark 3.4.10) and can therefore be represented in the following form:

$$\begin{cases} x_i = \varphi_i(u_0, \ldots, u_{d+1}), & i = 0, \ldots, n, \\ f(u_0, \ldots, u_{d+1}) = 0, \end{cases} \tag{2.8}$$

where $\varphi_0, \varphi_1, \ldots, \varphi_n \in K[u_0, \ldots, u_{d+1}]$ are homogeneous polynomials all of the same degree, and $f = 0$ is the equation of the hypersurface $X_d \subset \mathbb{P}^{d+1}$.

If X_d is a hyperplane of $\mathbb{P}^{d+1}$ (and then one can suppose that it has equation $u_{d+1} = 0$) equation (2.8) is replaced by a representation of the type

$$x_i = \varphi_i(u_0, \ldots, u_d), \quad i = 0, 1, \ldots, n, \tag{2.9}$$

and V_d is a rational variety (i.e., V_d is the birational transform of a linear space).

One notes however that, given formulas like those of (2.9), they do not in general give a rational variety but only a *unirational variety*, that is, the rational transform of a linear space.

2.7 Complements and exercises

As usual, unless otherwise specified, K denotes an algebraically closed field.

2.7.1. A descending chain $X_1 \supseteq X_2 \supseteq \cdots$ of algebraic sets becomes stationary.

It suffices to note that the associated chain of ideals $I(X_1) \subseteq I(X_2) \subseteq \cdots$ is stationary and that $I(X) = I(X')$ implies $X = X'$.

2.7.2. Every algebraic set X is a finite union of irreducible algebraic sets.

Indeed, if for some X this were not true, that X would not be irreducible, and if $X = X_1 \cup X_2$ the reducibility property would also hold for at least one of the X_i, $i = 1, 2$. Suppose $X_2 = X_2' \cup X_3$; then in the same way $X_3 = X_3' \cup X_4$, and so on. Thus one arrives at a non-stationary sequence of strict inclusions $X \supset X_2 \supset X_3 \supset X_4 \supset \cdots$, which contradicts Exercise 2.7.1.

2.7.3. The decomposition of an algebraic set $X = X_1 \cup X_2 \cup \cdots \cup X_t$, with each X_i an irreducible algebraic set, is *reduced* if none of the X_i is superfluous. Every algebraic set X uniquely determines its reduced decomposition.

Let $X = X_1 \cup X_2 \cup \cdots \cup X_t = X_1' \cup X_2' \cup \cdots \cup X_{t'}'$. Since $X_j := X \cap X_j = (\cup_i X_i') \cap X_j = \cup_i(X_i' \cap X_j)$ and since X_j is irreducible, there is an index i such that $X_j \subset X_i'$. Similarly for some index h we have $X_i' \subset X_h$. Hence $X_j \subset X_i' \subset X_h$. From this it follows that $X_j (= X_h) = X_i'$, and so on.

2.7.4 (Primary decomposition). The decomposition of an affine algebraic set $X = V(\mathfrak{a})$ as a union of irreducible algebraic sets corresponds to the *primary decomposition* of the ideal $\mathfrak{a}$, that is, to the representation of $\mathfrak{a}$ as an intersection $\mathfrak{a} = \mathfrak{q}_1 \cap \cdots \cap \mathfrak{q}_h$ of primary ideals $\mathfrak{q}_j$. More precisely, one has $I(X) = \sqrt{\mathfrak{a}} = \mathfrak{p}_1 \cap \mathfrak{p}_2 \cap \cdots \cap \mathfrak{p}_h$ where $\mathfrak{p}_j = \sqrt{\mathfrak{q}_j}$ is a prime ideal and so $X = X_1 \cup X_2 \cup \cdots \cup X_h$, with $X_j = V(\mathfrak{p}_j)$, $j = 1, \ldots, h$.

2.7.5. If $\varphi \colon X \to Y$ is a morphism of algebraic sets and X' is a closed subset of X the restriction $\varphi_{|X'} \colon X' \to Y$ is a morphism.

2.7.6. Let $\alpha(t), \beta(t) \in K[t]$. Then $\varphi \colon \mathbb{A}^1 \to \mathbb{A}^2$ defined by $\varphi(t) = (\alpha(t), \beta(t))$ is a morphism. Verify that $\mathrm{Im}(\varphi)$ is a closed subset of $\mathbb{A}^2$.

2.7.7 (Frobenius morphism). Let $K = \mathbb{Z}_p = \mathbb{Z}/(p)$, and let p be a prime number. The *Frobenius morphism* $\varphi \colon \mathbb{A}^n \to \mathbb{A}^n$ is the morphism defined by $\varphi(x_1, \ldots, x_n) = (x_1^p, \ldots, x_n^p)$. If $X \subset \mathbb{A}^n$ is a closed subset, the Frobenius morphism maps X into X.

For each $f \in K[T_1, \ldots, T_n]$, we have $(f(x_1, \ldots, x_n))^p = f(x_1^p, \ldots, x_n^p)$. The points of X that have coordinates in K are precisely those points of X which are fixed under the action of φ. In fact, every $\lambda \in \mathbb{Z}_p$ satisfies the equation $T^p - T = 0$.

2.7.8. Let $\varphi \colon C \to \mathbb{A}^1$, where C is the hyperbola with equation $xy = 1$ in the affine plane and where φ is defined by $\varphi(x, y) = x$. Is the morphism φ surjective? Is it dominant?

2.7.9. Let $Y \subset X$ be closed subsets of $\mathbb{A}^n$. Then every regular function on Y is the restriction of a regular function on X. Hence, the inclusion $i \colon Y \hookrightarrow X$ induces a surjective morphism $i^* \colon K[X] \to K[Y]$.

2.7.10. Let $\varphi \colon X \to Y$ be a morphism between affine algebraic sets. If the induced morphism $\varphi^* \colon K[Y] \to K[X]$ is surjective, then φ is injective and $\mathrm{Im}(\varphi)$ is a closed subset of Y.

Suppose that $\varphi(x) = \varphi(x')$, with x and $x' \in X$. Then, since φ^* is surjective, every function $f \in K[X]$ necessarily assumes the same value at x and x'. This implies that $x = x'$.

The kernel of φ^* is an ideal of $K[Y]$, and if φ^* is surjective it follows that $\mathrm{Im}(\varphi) = V(\ker(\varphi))$ by way of the correspondence V.

2.7.11. In $\mathbb{A}^2$ we consider the curve with equation $y = P(x)$ where $P(x) \in K[x]$ is a polynomial. The projection $(x, y) \mapsto (x, 0)$ on the x-axis gives an isomorphism between this curve and $\mathbb{A}^1$.

2.7.12. The *diagonal* of an affine algebraic set X, $\Delta_X := \{(x, x) \mid x \in X\}$, is closed in $X \times X$.

The diagonal $\Delta_{\mathbb{A}^n}$ of $\mathbb{A}^n$ is the closed linear subvariety $V(T_1 - T_{n+1}, \ldots, T_n - T_{2n}) \subset \mathbb{A}^n \times \mathbb{A}^n$ and hence the diagonal Δ_X, of a closed subset X of $\mathbb{A}^n$, is the closed subset $(X \times X) \cap \Delta_{\mathbb{A}^n}$ of $X \times X$.

2.7.13. In the affine plane $\mathbb{A}^2$ we consider two curves C_1 and C_2 with equations $f_1 = 0$, $f_2 = 0$. The points they have in common are the solutions of the system $f_1(Y_1, Y_2) = 0$, $f_2(T_1, T_2) = 0$, $Y_1 = T_1$, $Y_2 = T_2$. But $f_1(Y_1, Y_2) = f_2(T_1, T_2) = 0$ are the equations of $C_1 \times C_2$ in the product space $\mathbb{A}^4 = \mathbb{A}^2 \times \mathbb{A}^2$ and $Y_1 = T_1$, $Y_2 = T_2$ are the equations of the diagonal Δ in $\mathbb{A}^4$. Thus the problem of finding the intersection of the two curves C_1, C_2 can be translated into the problem of finding the intersection of the closed subset $C_1 \times C_2$ with the linear subspace Δ.

More generally, let V_1, V_2 be algebraic subsets of $\mathbb{A}^n$ and consider the space $\mathbb{A}^{2n}$ regarded as a product $\mathbb{A}^n \times \mathbb{A}^n$. If $Y_1, \ldots, Y_n; T_1, \ldots, T_n$ are the coordinates in $\mathbb{A}^{2n}$, then the generators of the ideal of the diagonal Δ are $Y_i - T_j$, $i, j = 1, \ldots, n$. One then sees that the map $\varphi: V_1 \cap V_2 \to (V_1 \times V_2) \cap \Delta$ defined by $\varphi(v) = (v, v)$ is an isomorphism.

2.7.14. Consider in the affine plane the curve C with equation $x^2 = y^2 + y^3$ and the line ℓ with equation $y = 1$. If P is a point of ℓ and O is the coordinate origin, the line r_{OP} intersects C in three points, two (at least) of which always coincide with O. Discarding the two intersections that fall at O there remains a third, say $\varphi(P)$ (which could possibly itself also coincide with the origin O). In this way one obtains a morphism $\varphi: \ell \to C$. Prove that the associated morphism φ^* is injective.

2.7.15. Let X_1 and X_2 be algebraic sets and consider the projections $p_i : X_1 \times X_2 \to X_i$, $i = 1, 2$, defined by $p_1(x, y) = x$, $p_2(x, y) = y$. They are surjective but not injective. Verify that p_1^*, p_2^* are injective but not surjective morphisms.

2.7.16. An isomorphism $\varphi: X \to X$, with X an affine algebraic set, is said to be an *automorphism* of X. The automorphisms of $\mathbb{A}^1$ are precisely the maps of the form $\varphi_{a,b}: x \mapsto ax + b$, $a \neq 0$. They form a group.

2.7.17. The map $\varphi(x, y, z) := (x, y + \alpha(x), z + \beta(x, y))$, with α, β polynomials, is an automorphism of $\mathbb{A}^3$ (note that its inverse is $(x', y', z') \mapsto (x', y' - \alpha(x'), z' - \beta(x', y'))$). Similarly, the map $\varphi(x, y) := (x, y + \alpha(x))$ is an automorphism of $\mathbb{A}^2$. Verify that these automorphisms form a group.

2.7.18. Let $\alpha_1, \ldots, \alpha_n$ be polynomials in $K[x_1, \ldots, x_n]$. If

$$\varphi(x_1, \ldots, x_n) = (\alpha_1(x_1, \ldots, x_n), \ldots, \alpha_n(x_1, \ldots, x_n))$$

is an automorphism of $\mathbb{A}^n$ the determinant of the Jacobian matrix $\left(\frac{\partial \alpha_i}{\partial x_j}\right)$ belongs to K^*. The map that associates φ to $\det\left(\frac{\partial \alpha_i}{\partial x_j}\right) \in K^*$ is a homomorphism of the group of automorphisms of $\mathbb{A}^n$ into the multiplicative group K^*. (Note however that the converse does not hold in general. More precisely, if $\alpha_1, \ldots, \alpha_n \in K[X_1, \ldots, X_n]$ have Jacobian matrix with determinant in K^*, then, on setting $x := (x_1, \ldots, x_n)$, the map $x \mapsto (\alpha_1(x), \ldots, \alpha_n(x))$ does not necessarily define an automorphism of $\mathbb{A}^n$ when the characteristic of the base field is positive; in characteristic zero, it defines an automorphism of $\mathbb{A}^n$ in the case $n = 1$, and for $n \geq 2$ the question remains an open problem (the "Jacobian Conjecture"): for this see [109, Introduction].)

2.7.19. Study the morphism $\varphi: \mathbb{A}^2_{(x,y)} \to \mathbb{A}^2_{(x',y')}$ defined as follows: $\varphi(x, y) = (x, xy)$. Is φ an isomorphism? Is $\text{Im}(\varphi)$ an open set? Is $\text{Im}(\varphi)$ either closed or dense? (Note that $\text{Im}(\varphi) = \mathbb{A}^2 \setminus \{x' = 0\} \cup (0, 0)$.)

Study the restriction of φ to the parabola with equation $y = x^2$ or to the parabola with equation $y^2 = x$. Are these restrictions isomorphisms (between the parabola and its image)?

2.7.20. Consider two planes π and π' and let x_1, x_2 be coordinates in π, and y_1, y_2 coordinates in π'. Let moreover $C_1 \subset \pi$ and $C_2 \subset \pi'$ be two curves with equations $x_2 - x_1 - 1 = 0$ and $(y_2 - y_1 - 1)^2 - 4y_1 = 0$ respectively. Verify that $(x_1, x_2) \mapsto (y_1 = x_1^2, y_2 = x_2^2)$ defines an isomorphism $\varphi: C_1 \to C_2$. (Note that the inverse transformation φ^{-1} is $(y_1, y_2) \mapsto ((y_2 - y_1 - 1)/2, (y_2 - y_1 + 1)/2)$.)

2.7.21. Let X be an irreducible affine algebraic set and let x be a point of X. We denote the local ring of x by $K[X]_x$, namely, the localization of $K[X]$ in its maximal ideal $\mathfrak{m}_x$ formed by the functions in $K[X]$ which do not assume the value 0 at x. The functions of $K(X)$ that are regular at the point x are of the form $\frac{P}{Q}$ with $P, Q \in K[X]$, $Q(x) \neq 0$; therefore they are the elements of the local ring $K[X]_x$. It follows that $K[X] = \bigcap_{x \in X} K[X]_x$ (cf. Section 2.6).

2.7.22. An affine algebraic set X is said to be *unirational* if there exists a dominant rational transformation $\mathbb{A}^n \to X$ of some affine space $\mathbb{A}^n$ into X. If X is unirational there exists an integer d such that $K(X) \subset K(t_1, \ldots, t_d)$ (cf. Corollary 2.6.6).

2.7.23. An irreducible quadric X of $\mathbb{A}^n$ is rational; that is, a hypersurface with equation $F(T_1, \ldots, T_n) = 0$ with $F(T_1, \ldots, T_n)$ a polynomial of degree two is a rational variety. A birational isomorphism between the quadric X in $\mathbb{A}^n$ and $\mathbb{A}^{n-1}$ is obtained by way of the projection of X from any one of its non-singular points P, which means that P must be a point in which at least one of the first order partial derivatives of F does not vanish, cf. Section 3.1. If X passes through the origin $(0, \ldots, 0)$ and $F = A_1(T_1, \ldots, T_n) - A_2(T_1, \ldots, T_n)$ with the A_i homogeneous polynomials of degree i, $i = 1, 2$ (and $A_1 \neq 0$), a birational isomorphism $\varphi: \mathbb{A}^{n-1}_{(Y_1, \ldots, Y_{n-1})} \to X$ is given, for example, by $(y_1, \ldots, y_{n-1}) \mapsto (y_1 \theta, \ldots, y_{n-1} \theta, \theta)$, where $\theta = \frac{A_1(y_1, \ldots, y_{n-1}, 1)}{A_2(y_1, \ldots, y_{n-1}, 1)}$. If U_1 is the open subset of $\mathbb{A}^{n-1}$ complementary to the quadric of equation $A_2(Y_1, \ldots, Y_{n-1}, 1) = 0$ and U_2 is the open set formed by the points of X for which $T_n \neq 0$, then the restriction of $\varphi: U_1 \to U_2$ is bijective.

2.7.24. In the affine space $\mathbb{A}^3$ over a field K (of characteristic $0 \leq p \neq 3$) consider the surface $\mathcal{F}$ defined by the equation $x^3 + y^3 + z^3 = 1$. It is a rational surface.

Indeed, if r, s are two skew lines lying in $\mathcal{F}$ (for example those with equations: $x + y = z - 1 = 0$ and $x + \varepsilon y = z + \varepsilon = 0$, $\varepsilon^3 = 1$, $\varepsilon \neq 1$), a line transversal to r and s (that is, intersecting both) meets $\mathcal{F}$ in a point P (besides the two points in which it intersects r and s) and meets a fixed plane π in a point P'. Then $P \mapsto P'$ defines a birational isomorphism $\varphi : \mathcal{F} \to \pi$.

2.7.25. In $\mathbb{A}^2$ consider the curve C with equation $x^2 + y^2 = 1$ and $f = \frac{x}{y-1} \in K(C)$. For what points of C is the rational function f defined?

2.7.26. In what points of the plane curve C of equation $x^3 + y^3 - x = 0$ is the rational function $\frac{x}{x+y} \in K(C)$ defined?

2.7.27. In what points of the plane curve C of equation $y^2 = x^2 + x^3$ is the function $\frac{x}{y}$ regular? Prove that $\frac{x}{y} \notin K[C]$.

2.7.28. Prove that the plane curve C with equation $(x^2 + y^2)^2 = xy$ is rational.

Consider the circle γ_t of equation $x^2 + (y-t)^2 = t^2$ passing through the origin O and tangent there to the x-axis. Away from O, the intersection $C \cap \gamma_t$ consists of a single point P_t. Then $t \mapsto P_t$ defines a birational isomorphism of $\mathbb{A}^1$ with C.

2.7.29. Let A be a K-algebra and let X be an irreducible closed subset of some affine space $\mathbb{A}^n$ such that $A \cong K[X]$. Then A does not have 0-divisors and is finitely generated over K. Conversely, a K-algebra A with no 0-divisors and finitely generated over K is the ring of coordinates of some irreducible closed affine algebraic set.

Let $A = K[t_1, \ldots, t_n]$. The K-homomorphism $\sigma : K[T_1, \ldots, T_n] \to K[t_1, \ldots, t_n]$ defined by $T_i \mapsto t_i$ is surjective and hence $K[t_1, \ldots, t_n] \cong K[T_1, \ldots, T_n]/\ker(\sigma)$. The kernel $\ker(\sigma)$ is a prime ideal of $K[T_1, \ldots, T_n]$ inasmuch as $K[T_1, \ldots, T_n]$ is a domain. Thus A is the coordinate ring of the algebraic set associated to $\ker(\sigma)$.

2.7.30. An extension L of K is isomorphic to the field of rational functions of some irreducible affine algebraic set if and only if L is finitely generated over K.

Indeed, if $L = K(t_1, \ldots, t_n)$, then L is the field of fractions of $K[t_1, \ldots, t_n]$, that is $L = K(X)$, where X is the algebraic set having $K[t_1, \ldots, t_n]$ as coordinate ring.

2.7.31 (Lüroth's theorem, cf. Theorem 7.4.1). Let X be an affine algebraic set of dimension 1. If there exists a dominant rational transformation $\mathbb{A}^1 \to X$, then X is birationally isomorphic to $\mathbb{A}^1$ (that is, there exists a birational isomorphism $\mathbb{A}^1 \to X$).

Here we give the outline of an algebraic proof, based on the theory of fields. See Theorem 7.4.1, p. 218 for an elementary proof that uses elimination theory.

By hypothesis $K \subset K(X) \subseteq K(t)$ with t indeterminate and where the first inclusion is strict.

We observe that t is algebraic over $K(X)$ because an arbitrary element $\lambda \in K(X)$ has the form $\lambda = \frac{A(t)}{B(t)}$, with $A(t), B(t) \in K[t]$, and so t is a root of $\lambda B(T) - A(T) \in k(\lambda)[T] \subset K(X)[T]$.

Let $F \in K(X)[T]$ be the monic minimal polynomial of t over $K(X)$ and suppose that $\lambda = \frac{A(t)}{B(t)}$ is one of the coefficients of F and $\lambda \notin K$ (such a λ certainly exists). One proves that $\lambda B(T) - A(T)$ has degree not greater than (and so equal to) the degree n of F with respect to T. Thus $\lambda B(T) - A(T)$ is the minimal polynomial of t over $K(\lambda)$. Then

$$[K(t) : K(\lambda)] = [K(t) : K(X)][K(X) : K(\lambda)],$$

and so $n = n[K(X) : K(\lambda)]$. Therefore $[K(X) : K(\lambda)] = 1$ which implies that $K(X) = K(\lambda)$.

2.7.32. Let $X \subset \mathbb{P}^n$ be an algebraic set. Then $X = \emptyset$ if and only if there exists an integer $s \geq 0$ such that $(X_1, \ldots, X_{n+1})^s \subset I(X)$.

Let $\{F_1, \ldots, F_r\}$ be a homogeneous basis for the ideal $I(X)$, and let A be the polynomial ring $K\left[\frac{X_1}{X_i}, \frac{X_2}{X_i}, \ldots, \frac{X_{n+1}}{X_i}\right]$. In A consider the polynomials $F_j\left(\frac{X_1}{X_i}, \frac{X_2}{X_i}, \ldots, \frac{X_{n+1}}{X_i}\right)$, $j = 1, \ldots, r$; if $X = \emptyset$ they generate the whole ring A. Then

$$1 = G_1\left(\frac{X_1}{X_i}, \ldots\right) F_1\left(\frac{X_1}{X_i}, \ldots\right) + \cdots$$

and so, for an integer $s_i \geq 0$, and for each $i = 1, \ldots, n+1$,

$$X_i^{s_i} = G_1 F_1 + \cdots \in I(X).$$

2.7.33 (Quasi-projective varieties). A *quasi-projective variety* is an open subset V of a projective algebraic set $X \subset \mathbb{P}^n$. Thus, $V = A \setminus B$, A, B closed in $\mathbb{P}^n$, $B \subset A$.

2.7.34. Let X be a quasi-projective variety in $\mathbb{P}^n$, x a point of X and $f = \frac{P}{Q}$ a homogeneous rational function of degree zero with $Q(x) \neq 0$: f is a regular function at x. A function f regular at each point $x \in X$ is a regular function on X. The regular functions on X form a ring $K[X]$. In contrast to what happens in the case of closed algebraic sets, it is not necessarily true that $K[X]$ is a finitely generated K-algebra (see, for instance [2, Chapter 14]). If X is a projective algebraic set one has $K[X] = K$.

2.7.35. Each point x of a quasi-projective variety X has an *affine neighborhood*, that is a neighborhood isomorphic to an affine algebraic set.

Let $X \subset \mathbb{P}^n$ and suppose that x is contained in the chart U_1 of $\mathbb{P}^n$, that is $x = [x_1, \ldots, x_{n+1}]$ with $x_1 \neq 0$. Since X, as a quasi-projective variety, is of the form $X = A \setminus B$, A, B closed subsets of $\mathbb{P}^n$, it follows that $X \cap U_1 = W \setminus W_1$, where $W = A \cap U_1$, $W_1 = U_1 \cap B$ are closed subsets of U_1. Since $x \in X \cap U_1$, then $x \notin W_1$. Therefore there exists $f \in K[W]$ such that $f(W_1) = 0$ and $f(x) \neq 0$. The principal open affine W_1 of W is a neighborhood of x, isomorphic to an affine closed subset (cf. Section 2.3).

2.7. Complements and exercises

2.7.36. The composition of birational isomorphisms is a birational isomorphism.

Let X, W, Z be projective algebraic sets and $\alpha: X \to W$, $\beta: W \to Z$ birational isomorphisms. Then there exist two open subsets $X_0 \subset X$, $W_0 \subset W$ such that the restriction $\alpha_{|X_0}: X_0 \to W_0$ is an isomorphism. Similarly there exist two open sets $W_0' \subset W$, $Z_0 \subset Z$ such that $\beta_{|W_0'}: W_0' \to Z_0$ is an isomorphism. Then

$$\alpha^{-1}(W_0 \cap W_0') \cong W_0 \cap W_0' \cong \beta(W_0 \cap W_0').$$

This means that the restriction $\beta \circ \alpha_{|\alpha^{-1}(W_0 \cap W_0')}: \alpha^{-1}(W_0 \cap W_0') \to \beta(W_0 \cap W_0')$ is an isomorphism.

2.7.37 (Projections). In $\mathbb{P}^n$ let E be a linear space of dimension d, with equations $L_1 = L_2 = \cdots = L_{n-d} = 0$ where the L_i are linearly independent linear forms. The rational map $p: \mathbb{P}^n \to \mathbb{P}^{n-d-1}$ given by $p(x) = [L_1(x), \ldots, L_{n-d}(x)]$, $x \in \mathbb{P}^n$, is said to be the *projection from* E. It is regular at $\mathbb{P}^n \setminus E$. If X is an algebraic set of $\mathbb{P}^n$, the restriction $p_{|X}: X \to \mathbb{P}^{n-d-1}$ is a finite morphism if $X \cap E = \emptyset$; it is a rational mapping if $X \cap E$ is non-empty (and distinct from X).

2.7.38. The rational map $\varphi: \mathbb{P}^2 \to \mathbb{P}^2$, given by $[x_1, x_2, x_3] \mapsto [x_2 x_3, x_3 x_1, x_1 x_2]$, is a birational automorphism of $\mathbb{P}^2$ (that is, a birational isomorphism of $\mathbb{P}^2$ with itself). Prove that the inverse image $\varphi^{-1}(r)$ of a line r is a conic which describes a homaloidal net as r varies, that is, a net with three base points, cf. Section 9.1. Find two open subsets between which the restriction of φ is an isomorphism.

2.7.39. Let $X \subset \mathbb{P}^n$ be a projective variety. There exist forms of every order m which do not vanish on any irreducible component of X.

If $X = \bigcup_i X_i$, with irreducible components X_i and $x_i \in X_i$, consider a hyperplane containing none of the points x_i (which are finite in number) and its arbitrary powers.

2.7.40. Let $\varphi: X \to Y$ be a morphism of quasi-projective varieties X, Y. Then, its *graph* $\Gamma_\varphi := \{(x, \varphi(x)), x \in X\}$ is a closed subset of $X \times Y$, and is isomorphic to X.

One has $\Gamma_\varphi = (\varphi \times \mathrm{id}_Y)^{-1}(\Delta_Y)$, which is the inverse image of a closed set under a morphism and hence is closed. The restriction of the projection $p: X \times Y \to Y$ to Γ_φ is inverse to the graph morphism $X \to \Gamma_\varphi$, defined by $x \mapsto (x, \varphi(x))$, as required.

2.7.41. Let X be a projective variety, and Y quasi-projective. Then the second projection $p_2: X \times Y \to Y$ is a closed map, that is, it sends closed sets into closed sets.

This is a general topological fact. If X, Y are topological spaces with X compact, then the projection $p: X \times Y \to Y$ is a closed map (see for instance [18, Corollary 5, p. 103]).

2.7.42. Let X be a projective variety, Y a quasi-projective variety and $\varphi: X \to Y$ a morphism. Then the image $\mathrm{Im}(\varphi)$ is closed in Y.

Indeed, the graph Γ_φ is a closed subset of $X \times Y$ by Exercise 2.7.40; on the other hand $\text{Im}(\varphi) = p_2(\Gamma_\varphi)$, where $p_2 \colon \Gamma_\varphi \to Y$ denotes the projection on the second factor. Then $\text{Im}(\varphi)$ is closed by Exercise 2.7.41.

2.7.43. Let $X_1 \subset \mathbb{A}^m$ and $X_2 \subset \mathbb{A}^n$ be algebraic sets. Then $X_1 \times X_2$ is irreducible if and only if X_1 and X_2 are irreducible.

Chapter 3
Geometric Properties of Algebraic Varieties

In this chapter we discuss some fundamental properties of algebraic varieties which we will use in the sequel. By an *algebraic variety* over the base field K we mean an ordered pair $(X, K[X])$ where X is an affine or projective algebraic set and $K[X]$ is its coordinate ring. Unless otherwise specified, the base field K is algebraically closed and of characteristic zero. Usually K will be the complex field $\mathbb{C}$.

In Section 3.1 we define the tangent space of a variety at one of its points and we introduce the notions of singularity and dimension. Since these are local properties one can in practice assume that X is an affine algebraic set. In this section we have substantially followed the exposition of [74, III, §6]. In Section 3.2 we introduce the notion of independent polynomials and study a useful characterization of them in terms of the rational map they define.

In Section 3.3 we return to the concept of dimension, discussing some equivalent formulations in the case of a projective algebraic set. Here we have followed the discussion given in [48, Lecture 11], to which we refer the reader for further interesting examples.

In Section 3.4, making use of classical methods of projective geometry, we introduce and study the order of a projective variety, as well as the notions of tangent cone and multiplicity of a singular point.

3.1 Tangent space, singularities and dimension

In this section we assume that X is an affine algebraic set and we begin by considering the case of hypersurfaces. Let $f \in K[Y_1, \ldots, Y_n]$ be an irreducible polynomial, $f \notin K$, and put $X := V(f) \subset \mathbb{A}^n$. Let $x = (a_1, \ldots, a_n)$ be a point of X and ℓ a line that passes through x. Since $x \in X$, the coordinates of x are roots of the restriction of f to ℓ (in the sense specified in the course of the proof of the following proposition).

Proposition–Definition 3.1.1. *Let $X = V(f) \subset \mathbb{A}^n$ be an irreducible hypersurface. The point $x \in X$ is a multiple root of $f_{|\ell}$ if and only if the line ℓ is contained in the affine linear subspace $T_x(X) \subset \mathbb{A}^n$ defined by the equation*

$$\sum_{i=1}^{n} \frac{\partial f}{\partial Y_i}(x)(Y_i - a_i) = 0.$$

*The space $T_x(X)$ is called the **tangent space** to X at x. We say that every line contained in $T_x(X)$ and passing through x is **tangent** to X at x.*

If $(\partial f/\partial Y_i)(x) = 0$ *for each* $i = 1,\ldots,n$ *we say that each line* ℓ *passing through* x *is* **tangent** *to* X *at* x.

Proof. We consider parametric equations for ℓ of the form
$$Y_i = a_i + b_i t, \quad i = 1,\ldots,n,$$
where $x = (a_1,\ldots,a_n)$ and $(b_1,\ldots,b_n)$ is the direction vector of ℓ. Then
$$f_{|\ell} := f(\ldots, a_i + b_i t, \ldots) = g(t)$$
is a polynomial in t and $t = 0$ is a root of $g(t)$ (corresponding to the point x). Thus $t = 0$ is a multiple root of $g(t)$ if and only if $\frac{\partial g}{\partial t}(0) = 0$, that is, if and only if
$$\sum_{i=1}^{n} b_i \frac{\partial f}{\partial Y_i}(x) = 0.$$
This condition is equivalent to the fact that $\ell \subset T_x(X)$. $\square$

Definition 3.1.2. The point x is *non-singular* (or *regular*, or *simple*) for $X = V(f)$ if $(\partial f/\partial Y_i)(x) \neq 0$ for some $i = 1,\ldots,n$; otherwise x is a *singular point* (or *multiple*, or a *singularity*) for X.

The preceding definitions lead to the following conclusion:

- The tangent space $T_x(X)$ to a hypersurface X at one of its points x is an $(n-1)$-dimensional affine subspace of $\mathbb{A}^n$ if x is non-singular, and $T_x(X) = \mathbb{A}^n$ if $x \in X$ is singular.

Remark 3.1.3. Suppose that $K = \mathbb{R}$ or $K = \mathbb{C}$, and that $(\partial f/\partial Y_i)(x) \neq 0$ for example for $i = 1$. Consider the map $p \colon \mathbb{A}^n \to \mathbb{A}^n$ defined by $(Y_1,\ldots,Y_n) \mapsto (f, Y_2,\ldots,Y_n)$; the determinant of the Jacobian matrix
$$\begin{pmatrix} \frac{\partial f}{\partial Y_1}(x) & \frac{\partial f}{\partial Y_2}(x) & \cdots & \frac{\partial f}{\partial Y_n}(x) \\ 0 & 1 & \cdots & 0 \\ 0 & 0 & \cdots & 1 \end{pmatrix}$$
is non-zero at x. Thus, by the Inverse Function Theorem, there exists a neighborhood $U \subset \mathbb{A}^n$, $x \in U$, such that the restriction $p_{|U} \colon U \to p(U) \subset \mathbb{A}^n$ is a diffeomorphism of the neighborhood U with the open set $p(U)$ of $\mathbb{A}^n$ in the usual Euclidean topology of $\mathbb{R}^n$ or $\mathbb{C}^n$, that is, $p_{|U}$ is bijective and both p and p^{-1} are differentiable functions of real or complex variables. In other words, $(f, Y_2,\ldots,Y_n)$ is a new system of coordinates on $\mathbb{A}^n$ near to x. This implies that an euclidean neighborhood of x in the hypersurface X of equation $f = 0$ is diffeomorphic to an open set in $\mathbb{A}^{n-1}$ with coordinates $(Y_2,\ldots,Y_n)$. We express this fact by saying that close to the non-singular point x the non-singular variety X has $(Y_2,\ldots,Y_n)$ as *local parameters* (cf. paragraph 3.1.15).

Let us consider the set
$$\mathrm{Reg}(X) := \{x \in X,\ x \text{ non-singular}\}$$
of non-singular points of X.

Proposition 3.1.4. *Let $X = V(f) \subset \mathbb{A}^n$ be an irreducible hypersurface. Suppose that $K = \mathbb{C}$. Then the set $\mathrm{Reg}(X)$ is a dense open subset of X in the Zariski topology.*

Proof. The complement of $\mathrm{Reg}(X)$ is the set $\mathrm{Sing}(X)$ of singular points, which is defined by the equations
$$\frac{\partial f}{\partial Y_i} = 0, \quad i = 1, \ldots, n.$$

Hence $\mathrm{Sing}(X) = V\!\left(f, \frac{\partial f}{\partial Y_1}, \ldots, \frac{\partial f}{\partial Y_n}\right) \subset \mathbb{A}^n$ is a closed subset of X. But, X being irreducible (since f is), in order to prove that $\mathrm{Reg}(X)$ is a dense open subset it suffices to prove that it is non-empty (cf. Section 2.2).

We proceed by contradiction. Suppose that $X = V(f) = \mathrm{Sing}(X)$. Then each of the polynomials $\partial f/\partial Y_i$ must vanish on X, that is, $\partial f/\partial Y_i \in I(X) = \sqrt{(f)} = (f)$ (cf. Theorem 2.2.2). It follows that $\partial f/\partial Y_i$ is divisible by f in $K[Y_1, \ldots, Y_n]$; but considered as a polynomial in Y_i, $\partial f/\partial Y_i$ has degree strictly less than the degree of f. Hence if $\partial f/\partial Y_i$ is divisible by f, $\partial f/\partial Y_i$ it must necessarily be the zero polynomial. This is possible only if Y_i does not appear in f; and if this happens for every index i, then f is a constant, which we have excluded. □

We can now define the tangent space to an affine algebraic set X at one of its points, and study some properties related to the concept of dimension (see also Section 3.3 for further characterizations of the notion of dimension).

Definition 3.1.5. Let $X \subset \mathbb{A}^n$ be an affine algebraic set and $x = (a_1, \ldots, a_n)$ a point of X. For each $f \in K[Y_1, \ldots, Y_n]$ we set
$$f_x^{(1)} := \sum_{i=1}^{n} \frac{\partial f}{\partial Y_i}(x)(Y_i - a_i).$$

This is an affine linear polynomial, that is, linear plus a constant (*the first order part of the Taylor series development of f at x*). We define the *tangent space* $T_x(X)$ of X at x by setting
$$T_x(X) := \bigcap_{f \in I(X)} \{f_x^{(1)} = 0\}.$$

If $X = V(\mathfrak{a})$ (where we can always suppose that $\mathfrak{a}$ is a radical ideal and so $\mathfrak{a} = I(X)$, cf. Theorem 2.2.2) one sees immediately that the linear parts of the polynomials of $\mathfrak{a}$ generate an ideal $\mathfrak{a}^{(1)} := \{f_x^{(1)},\ f \in \mathfrak{a}\}$ and so
$$T_x(X) = V(\mathfrak{a}^{(1)}).$$

Let $\{f_1, \ldots, f_m\}$ be a set of generators of $I(X)$. Since the linear part of the sum of two polynomials is the sum of the linear parts of the two summands, one has that for each $g \in I(X)$, the linear part $g_x^{(1)}$ of g in x is a linear combination of those of the f_j, $j = 1, \ldots, m$. Therefore, $\mathfrak{a}^{(1)} = (f_{1,x}^{(1)}, \ldots, f_{m,x}^{(1)})$. Hence the definition of $T_x(X)$ becomes simply

$$T_x(X) = V(f_{1,x}^{(1)}, \ldots, f_{m,x}^{(1)}) = \bigcap_{j=1}^{m} \{f_{j,x}^{(1)} = 0\} \subset \mathbb{A}^n.$$

Proposition 3.1.6. *Given an algebraic set $X \subset \mathbb{A}^n$, the function $X \to \mathbb{N}$ defined by $x \mapsto \dim T_x(X)$, $x \in X$, is an upper semi-continuous function in the Zariski topology on X, that is, for every integer r, the subset*

$$S(r) := \{x \in X \mid \dim T_x(X) \geq r\} \subset X$$

is closed in X.

Proof. Let $\{f_1, \ldots, f_m\}$ be a set of generators for $I(X)$ and

$$T_x(X) = \bigcap_{j=1}^{m} \{f_{j,x}^{(1)} = 0\} \subset \mathbb{A}^n$$

the tangent space to X at x. Then $x \in S(r)$ if and only if the Jacobian matrix

$$\left(\frac{\partial(f_1, f_2, \ldots, f_m)}{\partial(Y_1, Y_2, \ldots, Y_n)}(x)\right) := \left(\frac{\partial f_j}{\partial Y_i}(x)\right)_{i=1,\ldots,n,\ j=1,\ldots,m} \tag{3.1}$$

has rank $\leq n - r$, that is, if and only if every minor of order $(n - r + 1) \times (n - r + 1)$ of the matrix (3.1) vanishes. On the other hand, every element $(\partial f_j / \partial Y_i)(x)$ of the matrix is a polynomial function of x. Thus every minor is the determinant of a matrix of polynomials, and so is itself a polynomial. From this it follows that $S(r) \subset X \subset \mathbb{A}^n$ is an algebraic set. $\square$

Corollary–Definition 3.1.7. *There exist an integer r and an open dense subset $X_0 \subset X \subset \mathbb{A}^n$ such that*

$$\dim T_x(X) = r \quad \text{for } x \in X_0 \text{ and } \dim T_x(X) \geq r \text{ for all } x \in X.$$

*We say that $r = \dim(X)$ is the **dimension** of X, and that $n - r = \mathrm{codim}_{\mathbb{A}^n}(X)$ is the **codimension** of X. A point $x \in X$ is said to be **non-singular** if $\dim T_x(X) = r$, and **singular** if $\dim T_x(X) > r$; the variety X is **non-singular** if each of its points is non-singular. The closed subset $\mathrm{Sing}(X)$, the locus of the singular points of X, is the **singular locus** of X; it is empty if X is non-singular.*

3.1. Tangent space, singularities and dimension   55

Proof. Let $r := \min_{x \in X}\{\dim T_x(X)\}$; then we obviously have $S(r) = X$ and the set $S(r+1)$ is strictly contained in X. Hence

$$S(r) - S(r+1) = \{x \in X \mid \dim T_x(X) = r\}$$

is open and non-empty. □

3.1.8 (Jacobian criterion). We remark explicitly that a sufficient condition for the point x to be simple on the affine variety $X \subset \mathbb{A}^n(K)$, the locus of zeros of the ideal $(f_1, f_2, \ldots, f_m)$, is that the rank at x of the Jacobian matrix $\left(\frac{\partial(f_1, f_2, \ldots, f_m)}{\partial(Y_1, Y_2, \ldots, Y_n)}\right)$ should be $n - \dim(X)$. This follows easily on noting that

$$n - \dim(X) \geq \varrho\left(\frac{\partial(g_1, \ldots, g_t)}{\partial(Y_1, \ldots, Y_n)}(x)\right) \geq \varrho\left(\frac{\partial(f_1, \ldots, f_m)}{\partial(Y_1, \ldots, Y_n)}(x)\right) = n - \dim(X),$$

where $(g_1, g_2, \ldots, g_t)$ is the ideal $I(X)$ of X.

This proposition is known as the "Jacobian criterion for simple points". Under the hypothesis that the field K is of characteristic zero, and so, in particular, if $K = \mathbb{R}$ or $K = \mathbb{C}$, the Jacobian criterion is a necessary and sufficient condition for non-singularity, provided that $(f_1, f_2, \ldots, f_m)$ is the ideal $I(X)$. The reader wishing to study this question in detail can consult Zariski's fundamental memoir [119], as well as [103, Chapter II] and [81, Chapter III].

As an immediate consequence, one has that if x is a simple point for X and F is an irreducible hypersurface passing simply through x and not containing X, the necessary and sufficient condition for x to be a multiple point of the variety $X \cap F$, the intersection of X with F, is that the tangent hyperplane to F at x should contain the tangent space to X at x (cf. §5.2.4). Indeed, if $f = 0$ is the equation of F, this is the necessary and sufficient condition to have

$$\varrho\left(\frac{\partial(f_1, \ldots, f_m, f)}{\partial(Y_1, Y_2, \ldots, Y_n)}(x)\right) = \varrho\left(\frac{\partial(f_1, f_2, \ldots, f_m)}{\partial(Y_1, Y_2, \ldots, Y_n)}(x)\right)$$
$$= n - \dim(X) = \mathrm{cod}_{\mathbb{A}^n}(X \cap F) - 1.$$

We shall also need several elementary notions from the theory of fields (which may be found, for example, in [62]).

Definition 3.1.9. If $k \subset K$ is an extension of fields, the *transcendence degree* of K over k is the maximal number of elements of K that are algebraically independent over k. It is indicated by $\mathrm{tr.\,deg.}_k K$.

More precisely, given $\alpha_1, \ldots, \alpha_m \in K$, we say that $\alpha_1, \ldots, \alpha_m$ are *algebraically independent* over k if they are not solutions of a common polynomial in $k[T]$. We say that $\alpha_1, \ldots, \alpha_m$ *generate the transcendental part of the extension* $k \subset K$ if K is an algebraic extension of $k(\alpha_1, \ldots, \alpha_m)$, where $k(\alpha_1, \ldots, \alpha_m)$ is the field of fractions of $k[\alpha_1, \ldots, \alpha_m]$ (i.e., of the k-algebra $k[\alpha_1, \ldots, \alpha_m]$ generated as a ring by k and $\alpha_1, \ldots, \alpha_m$). We say that $\alpha_1, \ldots, \alpha_m$ form a *transcendence basis* if they are algebraically independent over k and they generate the transcendental part of the extension $k \subset K$. It is not difficult to prove that a transcendence basis is a maximal set of algebraically independent elements of K over k, and also a minimal set of generators (of the transcendental part of the extension $k \subset K$), and that any two transcendence bases of K over k have the same number of elements.

3.1.10 The case of hypersurfaces. If $X = V(f) \subset \mathbb{A}^n$ is an (irreducible) hypersurface defined by a non-constant polynomial f, then $\dim(X) = n - 1$. Indeed, for each non-singular point $x \in X$ (such points form a dense open subset in view of Proposition 3.1.4), the tangent space is defined by the linear equation $f_x^{(1)} = 0$ and so $r = \min_{x \in X} \{\dim T_x(X)\} = n - 1$.

We now prove that $\text{tr.deg.}_K K(X) = n - 1$; from this it follows, in particular, that for a hypersurface X,

$$\dim(X) = \text{tr.deg.}_K K(X) = n - 1.$$

Consider the quotient mapping

$$\sigma : K[Y_1, \ldots, Y_n] \to K[X] = K[Y_1, \ldots, Y_n]/(f)$$

and let $y_i := \sigma(Y_i)$, $i = 1, \ldots, n$. Suppose, to fix our ideas, that the indeterminate Y_1 actually appears in f and consider the elements $y_2, \ldots, y_n \in K(X)$. If one had $\text{tr.deg.}_K K(X) < n - 1$, they would be algebraically dependent and so there would exist a polynomial $g(Y_2, \ldots, Y_n) \in K[Y_2, \ldots, Y_n]$ such that

$$g(y_2, \ldots, y_n) = 0,$$

that is, $g \in \ker(\sigma) = (f)$. But that is absurd because Y_1 does not appear in g.

Hence $\text{tr.deg.}_K K(X) \geq n - 1$. Since one certainly has $\text{tr.deg.}_K K(X) < n$, it follows that $\text{tr.deg.}_K K(X) = n - 1$.

The remainder of this section deals with the proof, via reduction to the case of hypersurfaces, of the fact that the equality $\dim(X) = \text{tr.deg.}_K K(X)$ holds for every algebraic set $X \subset \mathbb{A}^n$. The first thing to prove is that for a point $x \in X$, the tangent space $T_x(X)$, which, by what has just been seen, is defined in terms of a particular system of coordinates in $\mathbb{A}^n$, is really independent of the choice of coordinates.

3.1.11 Intrinsic nature of the tangent space. Let $x = (x_1, \ldots, x_n) \in X \subset \mathbb{A}^n$ be a point of an affine variety X. By means of the coordinate change

$$Y'_i := Y_i - a_i, \quad i = 1, \ldots, n,$$

we may suppose that $x = (0, \ldots, 0)$ is the coordinate origin. Then $T_x(X) \subset \mathbb{A}^n$ is a linear subspace of K^n. Let $\mathfrak{m}_x$ be the ideal of x in $K[X]$ and let us denote by $M_x = (Y_1, \ldots, Y_n) \subset K[Y_1, \ldots, Y_n]$ the ideal of x in $\mathbb{A}^n$. Then obviously

$$\mathfrak{m}_x \cong M_x/I(X).$$

Theorem 3.1.12. *Let $X \subset \mathbb{A}^n$ be an affine algebraic set and $x \in X$ one of its points. With the preceding notation,*

(1) *there is a natural isomorphism of vector spaces*

$$(T_x(X))^* \cong \mathfrak{m}_x/\mathfrak{m}_x^2,$$

where $(\ )^*$ denotes the dual of a vector space;

(2) *if $f \in K[X]$ is such that $f(x) \neq 0$, and $X_f \subset X$ is a principal affine open subset, then the natural map $T_x(X_f) \to T_x(X)$ is an isomorphism.*

Proof. Let $(K^n)^*$ be the vector space of linear forms on K^n. A basis for $(K^n)^*$ is $\{Y_1, \ldots, Y_n\}$. Since $x = (0, \ldots, 0)$, for each $f \in K[Y_1, \ldots, Y_n]$ the linear part $f_x^{(1)}$ is in a natural way an element of the dual vector space $(K^n)^*$. Consider the map

$$d : M_x \to (K^n)^*$$

defined by putting $d(f) := f_x^{(1)}$ for each $f \in M_x$.

The map d is obviously surjective. Indeed, the linear forms $Y_1, \ldots, Y_n \in (K^n)^*$ are the images of the elements $Y_1, \ldots, Y_n \in M_x$. Moreover $\ker(d) = M_x^2$ since $f_x^{(1)} = 0$ if and only if f has quadratic terms in $Y_1, \ldots, Y_n$ in minimal degree; that is, if and only if $f \in M_x^2$. Thus

$$M_x/M_x^2 \cong (K^n)^*.$$

This proves (1) in the particular case $X = \mathbb{A}^n$.

In the general case one has the restriction map $(K^n)^* \to (T_x(X))^*$, dual to the inclusion $T_x(X) \subset K^n$, which sends a linear form λ on K^n into its restriction to $T_x(X)$. By composition one obtains a map

$$D : M_x \to (K^n)^* \to (T_x(X))^*,$$

which is surjective since both factors are such. It suffices to prove that

$$\ker(D) = M_x^2 + I(X), \tag{3.2}$$

because from this it follows that

$$\mathfrak{m}_x/\mathfrak{m}_x^2 = M_x/(M_x^2 + I(X)) \cong T_x(X)^*.$$

To prove (3.2) one notes that $f \in \ker(D)$ if and only if $f_x^{(1)}|_{T_x(X)} = 0$, that is to say

$$f_x^{(1)} = \sum_i a_i g_{i,x}^{(1)} \quad \text{for some } g_i \in I(X), a_i \in K[Y_1, \ldots, Y_n]$$

(recall that $T_x(X) \subset K^n$ is the subspace defined by $\{g_x^{(1)} = 0, g \in I(X)\}$). The last condition is equivalent to

$$f - \sum_i a_i g_i \in M_x^2 \quad \text{for some } g_i \in I(X),$$

which means that $f \in M_x^2 + I(X)$.

To prove (2) one observes that $I(X_f) = (I(X), Tf - 1) \subset K[Y_1, \ldots, Y_n, T]$. Hence if $y = (a_1, \ldots, a_n, b) \in X_f$, then $T_y(X_f) \subset \mathbb{A}^{n+1}$ is defined by the equations that define $T_x(X) \subset \mathbb{A}^n$ with the addition of a linear equation in which T appears, of type $cT - f(x)b = 0$, for some constant $c \neq 0$. □

Corollary 3.1.13. *Let $X \subset \mathbb{A}^n$ be an affine algebraic set and $x \in X$ one of its points. The tangent space $T_x(X)$ depends only on a neighborhood of x.*

Furthermore, if $x \in X_0$ and $y \in W_0$, where X_0 and W_0 are open subsets in the affine varieties X and W respectively, and $\varphi: X_0 \to W_0$ is an isomorphism such that $\varphi(x) = y$, then there exists a natural isomorphism $T_x(X_0) \to T_y(W_0)$. Hence $\dim T_x(X_0) = \dim T_y(W_0)$.

In particular, if X and W are birationally equivalent, $\dim(X) = \dim(W)$.

Proof. By considering, if necessary, a smaller neighborhood of x in X, we may suppose X_0 to be isomorphic to an affine algebraic set (cf. §2.4.9). Then W_0 too is affine, and φ induces an isomorphism $K[X_0] \cong K[W_0]$ sending the ideal $\mathfrak{m}_x$ of x into the ideal $\mathfrak{m}_y$ of y. Therefore, $\mathfrak{m}_x/\mathfrak{m}_x^2 \cong \mathfrak{m}_y/\mathfrak{m}_y^2$, that is, $T_x(X_0) \cong T_y(W_0)$. □

Theorem 3.1.14. *For each affine algebraic set $X \subset \mathbb{A}^n$,*

$$\dim(X) = \operatorname{tr.deg.}_K K(X).$$

Proof. Equality holds for hypersurfaces, as observed in §3.1.10. Moreover, every affine variety is birationally equivalent to a hypersurface (cf. §2.4.10) and both terms of the required equality are the same for birational equivalent varieties. □

3.1.15 Local parameters. Let X be an affine variety of dimension n, $x \in X$ a non-singular point of X, $\mathcal{O}_{X,x}$ the local ring of x and $\mathfrak{m}_x \subset \mathcal{O}_{X,x}$ the maximal ideal. One says that $u_1, \ldots, u_n$ are *local parameters* at x if they form a basis of $\mathfrak{m}_x/\mathfrak{m}_x^2$. Given the isomorphism $d_x: \mathfrak{m}_x/\mathfrak{m}_x^2 \to (T_x(X))^*$ of Theorem 3.1.12 one has that the following data are equivalent.

3.1. Tangent space, singularities and dimension 59

(1) $u_1, \ldots, u_n$ are local parameters.

(2) $d_x u_1, \ldots, d_x u_n$ are linearly independent over $T_x(X)$.

(3) The system of linear equations $d_x u_1 = \cdots = d_x u_n = 0$ has only the trivial solution in $T_x(X)$.

Let $u_1, \ldots, u_n$ be local parameters at x (and thus rational functions on X, regular and zero at x). One can find an affine neighborhood U of x (and containing $\bigcap_i \text{dom}(u_i)$) such that $u_1, \ldots, u_n \in K[U]$, cf. Exercise 2.7.35. If F_i is a polynomial that determines the function u_i (i.e., $u_i = F_i$ modulo $I(U)$) on U and if X_i is the hypersurface of U of equation $F_i = 0$, one has $I(X) + (F_i) \subset I(X_i)$ (since $I(X) \subset I(X_i)$ and $F_i \in I(X_i)$) and so, bearing in mind the equations for the tangent space given in Definition 3.1.5, one has $T_x(X_i) \subset L_i$, where L_i is the subspace of $T_x(X)$ defined by the equation $d_x F_i = 0$. Since $\dim T_x(X) = n$, one has $\dim L_i = n - 1$ and so $\dim T_x(X_i) \leq n - 1$. On the other hand (cf. Corollary–Definition 3.1.7) $\dim T_x(X_i) \geq \dim X_i \geq n - 1$, and so $\dim T_x(X_i) = \dim X_i = n - 1$. Then $X_1, \ldots, X_n$ intersect (or *cut*) *transversally* at x, that is, x is non-singular for each of them and in some neighborhood of x one has $\bigcap_i X_i = \{x\}$. [Indeed, a component of $\bigcap_i X_i$ having positive dimension and passing through x would have tangent space in x of positive dimension and contained in all the spaces $T_x(X_i)$ which, rather, have in common only the point x in view of the preceding equivalent characterization of local parameters.]

If $u_1, \ldots, u_n$ are local parameters at the non-singular point x one has

$$\mathfrak{m}_x = (u_1, \ldots, u_n).$$

Indeed, let $U \subset \mathbb{A}^N$ be an affine neighborhood of x in which one has $\bigcap_i X_i = \{x\}$. If $T_1, \ldots, T_N$ are the coordinates in $\mathbb{A}^N$ and $t_1, \ldots, t_N$ are the corresponding functions on U and if $x = (0, \ldots, 0)$ one has

$$\mathfrak{m}_x = (t_1, \ldots, t_N) = (u_1, \ldots, u_n, t_1, \ldots, t_N).$$

By definition of local parameters one has $t_N = \lambda_1 u_1 + \cdots + \lambda_n u_n + \xi$ with $\xi \in \mathfrak{m}_x^2$. That is,

$$t_N = \lambda_1 u_1 + \cdots + \lambda_n u_n + \mu_1 u_1 + \cdots + \mu_n u_n + \mu'_1 t_1 + \cdots + \mu'_N t_N,$$

$\lambda_i \in K$, $\mu_i, \mu'_j \in \mathfrak{m}_x$. Hence

$$t_N(1 - \mu'_N) = (\lambda_1 + \mu_1)u_1 + \cdots + (\lambda_n + \mu_n)u_n + \mu'_1 t_1 + \cdots + \mu'_{N-1} t_{N-1}.$$

But $\mu'_N \in \mathfrak{m}_x$; so $1 - \mu'_N$ is invertible in $\mathcal{O}_{X,x}$ and therefore t_N belongs to the ideal $(u_1, \ldots, u_n, t_1, \ldots, t_{N-1})\mathcal{O}_{X,x}$. Hence, on iterating, one obtains the desired conclusion.

Exercise 3.1.16. Let $X \subset \mathbb{A}^n$, $Y \subset \mathbb{A}^m$ be affine algebraic sets and let $X \times Y \subset \mathbb{A}^{n+m}$ be their product. Prove that

$$\dim(X \times Y) = \dim(X) + \dim(Y).$$

Let $x_1, \ldots, x_n$ be coordinates in $\mathbb{A}^n$ regarded as elements of $K[X]$; and similarly let $y_1, \ldots, y_m$ in $K[Y]$. We put $d_1 = \dim(X)$, $d_2 = \dim(Y)$. If $\{x_1, \ldots, x_{d_1}\}$, $\{y_1, \ldots, y_{d_2}\}$ are transcendence bases respectively for $K(X)$ and $K(Y)$, the elements $x_1, \ldots, x_{d_1}, y_1, \ldots, y_{d_2}$ are algebraically independent over K. Indeed, were they not, there would exist a polynomial $F = F(X_1, \ldots, X_{d_1}, Y_1, \ldots, Y_{d_2})$ non-zero on $X \times Y$ such that, for each point $x = (x_1, \ldots, x_n)$ in X, the polynomial $F(x_1, \ldots, x_{d_1}, Y_1, \ldots, Y_{d_2}) \in K[Y_1, \ldots, Y_{d_2}]$ would vanish on Y and so would have all of its coefficients zero. Analogously, for each $y = (y_1, \ldots, y_m) \in Y$, the polynomial $F(X_1, \ldots, X_{d_1}, y_1, \ldots, y_{d_2})$ would have all of its coefficients equal to zero; but this contradicts the hypothesis on F. It then suffices to note that $K(X \times Y) = K(x_1, \ldots, x_n, y_1, \ldots, y_m)$ is an algebraic extension of $K(x_1, \ldots, x_{d_1}, y_1, \ldots, y_{d_2})$.

3.1.17 Singularities and tangent spaces for projective varieties. We begin by recalling the following formula for homogeneous functions due to Euler which we will use hereafter.

Exercise 3.1.18 (Euler's formula). Let ϕ be a homogeneous differentiable function of $x_1, x_2, x_3, \ldots$, and let m be the degree of ϕ. Then

$$\sum_i x_i \frac{\partial \phi}{\partial x_i} = m\phi. \qquad (3.3)$$

Conversely, every differentiable solution of this equation is homogeneous of degree m in $x_1, x_2, x_3, \ldots$.

Indeed, let $\phi = \phi(x_1, x_2, x_3, \ldots)$ be a homogeneous function of $x_1, x_2, x_3, \ldots$ of degree m and differentiable. From the equality

$$\phi(tx_1, tx_2, tx_3, \ldots) = t^m \phi(x_1, x_2, x_3, \ldots)$$

one has, on deriving both sides with respect to t, that

$$\sum_i x_i \phi_{x_i}(tx_1, tx_2, tx_3, \ldots) = mt^{m-1} \phi(x_1, x_2, x_3, \ldots)$$

from which (3.3) follows on putting $t = 1$.

Conversely, let $\phi = \phi(x_1, x_2, x_3, \ldots)$ be a solution of (3.3) so that one has the identity

$$\sum_i x_i \phi_{x_i}(x_1, x_2, x_3, \ldots) = m\phi(x_1, x_2, x_3, \ldots).$$

3.1. Tangent space, singularities and dimension

Therefore, one then also has, on replacing x_i by tx_i,

$$\sum_i tx_i \phi_{x_i}(tx_1, tx_2, tx_3, \ldots) = m\phi(tx_1, tx_2, tx_3, \ldots). \tag{3.4}$$

If we set $T(t) := \phi(tx_1, tx_2, tx_3, \ldots)$, then (3.4) may be rewritten in the form

$$t \frac{\partial T(t)}{\partial t} = mT(t)$$

or

$$\frac{\partial}{\partial t}\left(\frac{T(t)}{t^m}\right) = 0.$$

Therefore $T(t)/t^m = C$, where C is independent of t. Setting $t = 1$ one obtains $C = \phi(x_1, x_2, x_3, \ldots)$ and so $T(t) = t^m \phi(x_1, x_2, x_3, \ldots)$, that is,

$$\phi(tx_1, tx_2, tx_3, \ldots) = t^m \phi(x_1, x_2, x_3, \ldots).$$

This proves that ϕ is a homogeneous function of degree m.

3.1.19. It is hardly necessary to observe how the notion of singularity given in Corollary–Definition 3.1.7 extends to the projective case. If x is a point of a projective algebraic set $X \subset \mathbb{P}^n$ we know in fact that x is contained in a suitable affine neighborhood $X_0 \subset X$: it suffices to take, for example, X_0 to be a principal open subset containing x (cf. §2.5.6 and also Exercise 2.7.35). Then x is singular or non-singular according to its singularity or non-singularity for X_0. By Corollary 3.1.13 this fact does not depend on the choice of X_0.

We observe that if X is reducible, then each of its non-singular points belongs to only one of the irreducible components of X.

Let us consider the important case of hypersurfaces. Let then $X = V(f)$ with f a homogeneous form of degree r. Analogously to the affine case (cf. Definition 3.1.2) we have that a point $x = [x_1, \ldots, x_{n+1}] \in X$ is singular if and only if

$$\frac{\partial f}{\partial x_1}(x) = \cdots = \frac{\partial f}{\partial x_{n+1}}(x) = 0. \tag{3.5}$$

One notes, however, that in the projective case the conditions (3.5) are equivalent to the vanishing at the point x of n arbitrarily chosen of the $n+1$ partial derivatives $\partial f/\partial x_i$, $i = 1, \ldots, n+1$. Indeed, by Euler's formula (3.3) one has

$$rf = \sum_{i=1}^{n+1} x_i \frac{\partial f}{\partial x_i}$$

and thus the preceding assertion follows from the fact that $f(x) = 0$.

Thus we conclude that $x \in X = V(f)$ is singular if and only if $n+1$ arbitrarily chosen out of the following $n+2$ conditions hold:

$$f(x) = 0, \quad \frac{\partial f}{\partial x_1}(x) = \cdots = \frac{\partial f}{\partial x_{n+1}}(x) = 0.$$

If x is a non-singular point the tangent space to X at x is the hyperplane in $\mathbb{P}^n$ with equation

$$\sum_{i=1}^{n+1} x_i \frac{\partial f}{\partial x_i}(x) = 0. \tag{3.6}$$

Note also that if x belongs to the affine chart $X_{(i)} = X \cap \mathbb{P}^n_{X_i}$, then the hyperplane with equation (3.6) is the projective closure of the tangent hyperplane to $X_{(i)}$ at x (cf. §2.5.6).

In the general case, if $X \subset \mathbb{P}^n$ is an algebraic subset and $x \in X$ a point belonging to the affine chart $X_{(i)}$, the tangent space $T_x(X) \subset \mathbb{P}^n$ to X at x is the projective closure of the tangent space $T_x(X_{(i)})$ to $X_{(i)}$ at x. (Bear in mind the "local structure" of the tangent space expressed by Corollary 3.1.13.)

See also §5.2.4 for related questions.

3.2 Independence of polynomials. Essential parameters

Consider m polynomials $f_j(x_1, x_2, \ldots, x_n)$ in $K[x_1, x_2, \ldots, x_n]$. We will say that they are *independent* if there is no non-zero polynomial

$$\phi(y_1, y_2, \ldots, y_m) \in K[y_1, y_2, \ldots, y_m]$$

such that the polynomial $\phi(f_1, f_2, \ldots, f_m) \in K[x_1, x_2, \ldots, x_n]$ is identically zero. This is equivalent to saying that the morphism $\varphi: \mathbb{A}^n(K) \to \mathbb{A}^m(K)$, defined by the equations

$$\begin{cases} y_1 = f_1(x_1, x_2, \ldots, x_n), \\ y_2 = f_2(x_1, x_2, \ldots, x_n), \\ \vdots \\ y_m = f_m(x_1, x_2, \ldots, x_n), \end{cases}$$

is surjective. Indeed, if $\text{Im}(\varphi) = \mathbb{A}^m(K)$, then there are no *non-zero* polynomials that vanish in each point

$$(f_1(x_1, x_2, \ldots, x_n), f_2(x_1, x_2, \ldots, x_n), \ldots, f_m(x_1, x_2, \ldots, x_n)) \in \text{Im}(\varphi).$$

Conversely, if φ is not surjective, then the image V of φ is a closed algebraic subset of $\mathbb{A}^m(K)$ and if

$$g_1(y_1, y_2, \ldots, y_m), g_2(y_1, y_2, \ldots, y_m), \ldots, g_h(y_1, y_2, \ldots, y_m)$$

3.2. Independence of polynomials. Essential parameters

is a system of generators for the ideal $I(V)$ of polynomials vanishing on V one has that for each $x = (x_1, x_2, \ldots, x_n) \in \mathbb{A}^n(K)$,

$$g_t(f_1(x), f_2(x), \ldots, f_m(x)) = 0, \quad t = 1, \ldots, h,$$

and therefore $f_1, \ldots, f_m$ are dependent.

It is then obvious that the polynomials $f_1, f_2, \ldots, f_m$ can be independent only if $n \geq m$.

Consider the Jacobian matrix, with elements in $K[x_1, x_2, \ldots, x_n]$,

$$J = \left(\frac{\partial(f_1, f_2, \ldots, f_m)}{\partial(x_1, x_2, \ldots, x_n)}\right) := \left(\frac{\partial f_j}{\partial x_i}\right)_{j=1,\ldots,m,\ i=1,\ldots,n}$$

and denote by $\varrho(x)$ its rank in the point $x \in \mathbb{A}^n(K)$.

One sees immediately that if the polynomials f_j are dependent, then $\varrho(x) < m$ for each point $x \in \mathbb{A}^n(K)$. Indeed, in that case there exists a non-zero polynomial ϕ in $K[y_1, y_2, \ldots, y_m]$ and an identity of the form

$$\theta(x) = \phi(f_1(x), f_2(x), \ldots, f_m(x)) = 0,$$

for all $x \in \mathbb{A}^n(K)$. This identity also implies the following relations, for $i = 1, \ldots, n$,

$$\frac{\partial \theta}{\partial x_i}(x) = \left(\frac{\partial \phi}{\partial f_1}\frac{\partial f_1}{\partial x_i} + \frac{\partial \phi}{\partial f_2}\frac{\partial f_2}{\partial x_i} + \cdots + \frac{\partial \phi}{\partial f_m}\frac{\partial f_m}{\partial x_i}\right)(x) = 0.$$

However these can hold simultaneously only if the rank $\varrho(x)$ of J in x is less than m.

One notes that if $f_j(x_1, x_2, \ldots, x_n)$ are linear homogeneous polynomials, the Jacobian matrix is nothing but the matrix formed by their coefficients and one recovers a well-known theorem from theory of linear forms.

Conversely it is well known that if the Jacobian matrix J has rank $\varrho \leq m$, then $m - \varrho$ of the polynomials f_j are functions of the other remaining polynomials and the n "superabundant" parameters $x_1, x_2, \ldots, x_n$ (bound by relations between the f_j) can be replaced with *essential parameters* in number ϱ; and one has $\dim(V) = \varrho$. Compare, for example, [14, Chapter 9], and also [102, Parte I, pp. 255–262].

3.2.1 Tangent space as span of the derived points. We now describe a useful procedure for obtaining the tangent space.

Let $u_1, \ldots, u_{k+2}$ be affine coordinates in $\mathbb{A}^{k+2}(\mathbb{C})$; $T_1, \ldots, T_{n+1}$ affine coordinates in $\mathbb{A}^{n+1}(\mathbb{C})$; and $\theta : \mathbb{A}^{k+2} \to \mathbb{A}^{n+1}$ the map defined by the formulas

$$\begin{cases} T_1 = f_1(u_1, \ldots, u_{k+2}), \\ T_2 = f_2(u_1, \ldots, u_{k+2}), \\ \quad \vdots \\ T_{n+1} = f_{n+1}(u_1, \ldots, u_{k+2}), \end{cases} \quad (3.7)$$

with f_i functions having continuous first partial derivatives, $i = 1, \ldots, n+1$. Let $W \subset \mathbb{A}^{n+1}$ be the image (we consider W as parameterized subvariety of $\mathbb{A}^{n+1}$).

If the parameters $u_1, \ldots, u_{k+2}$ are essential (which means that the functions f_i can not be expressed in terms of $\varrho < k+2$ parameters) W is a variety of dimension $k+2$ (cf. Section 3.3). The analytic condition in order for this to occur is that one has

$$\operatorname{rank}\left(\frac{\partial(f_1, \ldots, f_{n+1})}{\partial(u_1, \ldots, u_{k+2})}\right) = k+2 \tag{3.8}$$

almost everywhere. If $P = (\bar{u}_1, \ldots, \bar{u}_{k+2})$ is a point at which (3.8) holds we will say that $(P, \theta(P))$ is a *regular pair* of the correspondence θ. In this case $\theta(P)$ is a non-singular point of W and the space S_{k+2} tangent at $\theta(P)$ to W is the space spanned by the $k+3$ points $\theta(P), \frac{\partial \theta(P)}{\partial u_1}, \ldots, \frac{\partial \theta(P)}{\partial u_{k+2}}$, where

$$\frac{\partial \theta(P)}{\partial u_j} := \left(\left(\frac{\partial f_1}{\partial u_j}\right)_P, \ldots, \left(\frac{\partial f_{n+1}}{\partial u_j}\right)_P\right), \quad j = 1, \ldots, k+2.$$

In $\mathbb{A}^{k+2}$ we take a line p containing P and we consider its image $\theta(p)$. If $(\zeta_1, \ldots, \zeta_{k+2})$ is a vector along p, the line p is the locus of the points $(\bar{u}_1 + \zeta_1 t, \ldots, \bar{u}_{k+2} + \zeta_{k+2} t)$ and the curve $\theta(p)$ is given parametrically by the equations

$$\begin{cases} T_1 = f_1(\bar{u}_1 + \zeta_1 t, \ldots, \bar{u}_{k+2} + \zeta_{k+2} t), \\ \quad \vdots \\ T_{n+1} = f_{n+1}(\bar{u}_1 + \zeta_1 t, \ldots, \bar{u}_{k+2} + \zeta_{k+2} t). \end{cases}$$

Moreover, a vector $\vec{v}$ tangent to $\theta(p)$ in the point

$$\theta(P) = (f_1(\bar{u}_1, \ldots, \bar{u}_{k+2}), \ldots, f_{n+1}(\bar{u}_1, \ldots, \bar{u}_{k+2}))$$

has components

$$\begin{cases} x_1 = \left(\frac{\partial f_1}{\partial u_1}\right)_P \zeta_1 + \cdots + \left(\frac{\partial f_1}{\partial u_{k+2}}\right)_P \zeta_{k+2}, \\ \quad \vdots \\ x_{n+1} = \left(\frac{\partial f_{n+1}}{\partial u_1}\right)_P \zeta_1 + \cdots + \left(\frac{\partial f_{n+1}}{\partial u_{k+2}}\right)_P \zeta_{k+2}. \end{cases}$$

Therefore,

$$\vec{v} = \zeta_1 \frac{\partial \theta(P)}{\partial u_1} + \cdots + \zeta_{k+2} \frac{\partial \theta(P)}{\partial u_{k+2}} \tag{3.9}$$

and $\zeta_1, \ldots, \zeta_{k+2}$ (which are coordinates of the line p in the star of lines of $\mathbb{A}^{k+2}$ passing through P) are the components of $\vec{v}$ with respect to the basis

$$\left(\frac{\partial \theta(P)}{\partial u_1}, \ldots, \frac{\partial \theta(P)}{\partial u_{k+2}}\right)$$

3.2. Independence of polynomials. Essential parameters

of the space of tangent vectors to W at $\theta(P)$. Thus we have established the following fact:

- If $(P, \theta(P))$ is a regular pair, θ induces a non-degenerate projectivity between the star of lines of $\mathbb{A}^{k+2}$ passing through P and the star of *tangent lines* at $\theta(P)$ to W (that is, corresponding to the tangent vectors to W at $\theta(P)$).

Now let Φ be a hypersurface of $\mathbb{A}^{k+2}$ with equation $\phi(u_1, \ldots, u_{k+2}) = 0$, passing through P and having a tangent hyperplane there, and let $\theta(\Phi) \subset W$ be its image.

Consider the lines p tangent at P to Φ. For these lines the vector $(\zeta_1, \ldots, \zeta_{k+2})$ satisfies the condition

$$\zeta_1 \left(\frac{\partial \phi}{\partial u_1}\right)_P + \cdots + \zeta_{k+2} \left(\frac{\partial \phi}{\partial u_{k+2}}\right)_P = 0. \tag{3.10}$$

To these vectors there correspond the vectors (3.9) whose components are connected by the relation (3.10). Thus in the space S_{k+2} one finds the tangent vectors at $\theta(P)$ to W to be a $(k+1)$-dimensional subspace, which is the space of vectors tangent at $\theta(P)$ to $\theta(\Phi)$. The equation of this space (in the space of tangent vectors at $\theta(P)$ to W) is (3.10).

If $\theta(P)$ is a simple point of W and if at P the hypersurface Φ is endowed with a tangent hyperplane, the tangent lines at $\theta(P)$ to $\theta(\Phi)$ thus form a linear space $T_{\theta(P)}$, of dimension $k + 1 = \dim(\Phi) = \dim(\theta(\Phi))$. One sees that $\theta(P)$ is a non-singular point for $\theta(\Phi)$, and $T_{\theta(P)}$ is the tangent space at $\theta(P)$ to $\theta(\Phi)$. We observe that if the functions $f_i, i = 1, \ldots, n+1$, are polynomials it is not necessary to limit ourselves to the complex field because one can define derivatives formally.

Suppose now that the functions $f_i, i = 1, \ldots, n+1$, are homogeneous polynomials all of the same degree m. The variety W is then a cone having as its vertex the coordinate origin. Indeed, if $Q = (\bar{t}_1, \ldots, \bar{t}_{n+1}) = \theta(P) = \theta(\bar{u}_1, \ldots, \bar{u}_{k+2})$ is a point of W (and so $\bar{t}_i = f_i(\bar{u}_1, \ldots, \bar{u}_{k+2})$, $i = 1, \ldots, n+1$) one has, for arbitrary λ,

$$f_i(\lambda \bar{u}_1, \ldots, \lambda \bar{u}_{k+2}) = \lambda^m f_i(\bar{u}_1, \ldots, \bar{u}_{k+2}) = \lambda^m \bar{t}_i, \quad i = 1, \ldots, n+1,$$

and therefore every point of the line joining the origin O to Q belongs to W.

The tangent space to W at Q is spanned by the points $Q, \frac{\partial Q}{\partial u_1}, \ldots, \frac{\partial Q}{\partial u_{k+2}}$, where

$$\frac{\partial Q}{\partial u_j} := \left(\left(\frac{\partial f_1}{\partial u_j}\right)_P, \ldots, \left(\frac{\partial f_{n+1}}{\partial u_j}\right)_P\right), \quad j = 1, \ldots, k+2.$$

On the other hand, by Euler's theorem on homogeneous functions one has (cf. Exercise 3.1.18)

$$mQ = \bar{u}_1 \frac{\partial Q}{\partial u_1} + \cdots + \bar{u}_{k+2} \frac{\partial Q}{\partial u_{k+2}},$$

66   Chapter 3. Geometric Properties of Algebraic Varieties

where mQ denotes the point with coordinates $(m\bar{t}_1, \ldots, m\bar{t}_{n+1})$, and hence the tangent space at Q to W passes through the origin and can be considered as the space spanned by the origin and the points $\frac{\partial Q}{\partial u_1}, \ldots, \frac{\partial Q}{\partial u_{k+2}}$.

Assuming this, let X be the projective algebraic variety (of dimension k, cf. Section 3.3) defined by the equations

$$T_i = f_i(u_1, \ldots, u_{k+2}), \quad i = 1, \ldots, n+1, \tag{3.11}$$

and

$$\phi(u_1, \ldots, u_{k+2}) = 0, \tag{3.12}$$

where $T_1, \ldots, T_{n+1}$ are projective homogeneous coordinates in $\mathbb{P}^n$ and $u_1, \ldots, u_{k+2}$ are projective homogeneous coordinates in $\mathbb{P}^{k+1}$.

If we interpret $T_1, \ldots, T_{n+1}$ as non-homogeneous affine coordinates in an affine space $\mathbb{A}^{n+1}$, and $u_1, \ldots, u_{k+2}$ as non-homogeneous affine coordinates in an affine space $\mathbb{A}^{k+2}$, then in $\mathbb{A}^{k+2}$ we have a cone Φ^* with vertex at the coordinate origin, and with equation (3.12), while in $\mathbb{A}^{n+1}$ we have a cone W^* with vertex at the coordinate origin and equation (3.11).

The lines of $\mathbb{A}^{n+1}$ passing through the origin are the points of a projective space $\mathbb{P}^n$ (which can be thought of as the "hyperplane at infinity", π_∞, of $\mathbb{A}^{n+1}$) and similarly the lines of $\mathbb{A}^{k+2}$ passing through the origin are the points of a projective space $\mathbb{P}^{k+1}$ ("hyperplane at infinity", σ_∞, of $\mathbb{A}^{k+2}$). The projective variety $W \subset \mathbb{P}^n$ with equations (3.11) is the section of the cone W^* by π_∞ and hence the tangent space to W at its generic point is defined by the $k+2$ *derived points*

$$\xi_j := \left[\frac{\partial f_1}{\partial u_j}, \ldots, \frac{\partial f_{n+1}}{\partial u_j} \right], \quad j = 1, \ldots, k+2;$$

and the projective hypersurface $\Phi \subset \mathbb{P}^{k+1}$ with equation (3.12) is the intersection of the cone Φ^* with σ_∞.

The *tangent vectors* to the variety X defined by the equations (3.11) and (3.12) in its generic point (that is, the directional vectors of the tangent lines in the generic point) are the vectors $\sum_{j=1}^{k+2} \zeta_j \xi_j$ with $\sum_{j=1}^{k+2} \left(\frac{\partial \phi}{\partial u_j} \right) \zeta_j = 0$ (cf. (3.10)).

Example 3.2.2. Notation as in §3.2.1. We wish to write the equations of the tangent line in the generic point Q of the projective curve $\mathcal{L}$ of $\mathbb{P}^5$ defined by the equations

$$\begin{cases} T_1 = u_1^2, \\ T_2 = 2u_1 u_2, \\ T_3 = u_2^2, \\ T_4 = 2u_1 u_3, \\ T_5 = 2u_2 u_3, \\ T_6 = u_3^2 \end{cases} \tag{3.13}$$

3.2. Independence of polynomials. Essential parameters

and
$$\phi(u_1, u_2, u_3) = 0, \tag{3.14}$$

where ϕ is an arbitrary homogeneous polynomial. We consider the *Veronese surface* $\mathcal{F}$ represented by (3.13) (cf. Example 10.2.1). The derived points are

$$\xi_1 = [u_1, u_2, 0, u_3, 0, 0], \quad \xi_2 = [0, u_1, u_2, 0, u_3, 0], \quad \xi_3 = [0, 0, 0, u_1, u_2, u_3].$$

Thus the generic point of the tangent plane to $\mathcal{F}$ at a point Q belonging to $\mathcal{L}$ (Q is the image, under (3.13), of a point $[u_1, u_2, u_3]$ such that $\phi(u_1, u_2, u_3) = 0$) is

$$[\zeta_1 u_1, \zeta_1 u_2 + \zeta_2 u_1, \zeta_2 u_2, \zeta_1 u_3 + \zeta_3 u_1, \zeta_2 u_3 + \zeta_3 u_2, \zeta_3 u_3].$$

In order that this point belongs to the tangent line to $\mathcal{L}$ at Q it is necessary that (cf. (3.10))

$$\zeta_1 \frac{\partial \phi}{\partial u_1} + \zeta_2 \frac{\partial \phi}{\partial u_2} + \zeta_3 \frac{\partial \phi}{\partial u_3} = 0.$$

For the line to be tangent to $\mathcal{L}$ at Q it is then sufficient to eliminate the parameters $\zeta_1, \zeta_2, \zeta_3$ from the equations

$$\begin{cases} T_1 = \zeta_1 u_1, \\ T_2 = \zeta_1 u_2 + \zeta_2 u_1, \\ T_3 = \zeta_2 u_2, \\ T_4 = \zeta_1 u_3 + \zeta_3 u_1, \\ T_5 = \zeta_2 u_3 + \zeta_3 u_2, \\ T_6 = \zeta_3 u_3, \\ 0 = \zeta_1 \frac{\partial \phi}{\partial u_1} + \zeta_2 \frac{\partial \phi}{\partial u_2} + \zeta_3 \frac{\partial \phi}{\partial u_3}. \end{cases}$$

The first, third, and sixth equation give $\zeta_1 = \frac{T_1}{u_1}$, $\zeta_2 = \frac{T_3}{u_2}$, $\zeta_3 = \frac{T_6}{u_3}$. Substituting into the remaining equations (and then eliminating the denominators) one finds the equations of four hyperplanes: the three hyperplanes

$$\begin{aligned} u_2^2 T_1 + u_1^2 T_3 - u_1 u_2 T_2 &= 0, \\ u_3^2 T_1 + u_1^2 T_6 - u_1 u_3 T_4 &= 0, \\ u_3^2 T_3 + u_2^2 T_6 - u_2 u_3 T_5 &= 0, \end{aligned} \tag{3.15}$$

whose intersection is the tangent plane π to $\mathcal{F}$ at Q, and the hyperplane

$$u_2 u_3 \frac{\partial \phi}{\partial u_1} T_1 + u_1 u_3 \frac{\partial \phi}{\partial u_2} T_3 + u_1 u_2 \frac{\partial \phi}{\partial u_3} T_6 = 0 \tag{3.16}$$

which intersects π along the tangent line to $\mathcal{L}$ at Q.

One notes that the hyperplane with equation (3.16) does not pass through the tangent plane to $\mathcal{F}$ at Q and is therefore independent of the hyperplanes (3.15). We verify that it does pass through Q. Indeed, one has, if $m = \deg(\phi)$,

$$u_2 u_3 \frac{\partial \phi}{\partial u_1} u_1^2 + u_1 u_3 \frac{\partial \phi}{\partial u_2} u_2^2 + u_1 u_2 \frac{\partial \phi}{\partial u_3} u_3^2 = u_1 u_2 u_3 \left(\sum_{i=1}^{3} u_i \frac{\partial \phi}{\partial u_i} \right)$$

$$= m u_1 u_2 u_3 \phi(u_1, u_2, u_3).$$

3.3 Dimension of a projective variety

Let $X \subset \mathbb{P}^n$ be a projective variety, which, except for explicit mention to the contrary, we shall suppose to be irreducible.

The discussion regarding the dimension of an affine variety carried out in Section 3.1 and Proposition 2.6.1 lead one naturally to define the *dimension* of X as the transcendence degree of its function field $K(X)$ over the base field K (which, as usual, we assume to be algebraically closed),

$$\dim(X) = \operatorname{tr.deg.}_K K(X). \tag{3.17}$$

In the sequel we will study the geometric meaning of (3.17), discussing some equivalent formulations expressed in terms of projective geometry.

We begin by proposing the following alternative definition of dimension.

Definition 3.3.1. The *dimension* of a variety $X \subset \mathbb{P}^n$ is the smallest integer k such that there exists a subspace $S_{n-k-1} \subset \mathbb{P}^n$ disjoint from X (or, equivalently, such that a generic $S_{n-k-1} \subset \mathbb{P}^n$ is disjoint from X).

One observes that given a generic S_{n-k}, a generic S_{n-k-1} contained in S_{n-k} is a generic S_{n-k-1} of $\mathbb{P}^n$. It follows that if $X \subset \mathbb{P}^n$ has dimension k, the generic S_{n-k} meets X in a finite number of points. Moreover, in the same way, the generic S_{n-k+1} meets X in a variety that consists of infinitely many points since otherwise the generic S_{n-k} would be disjoint from X. We can then express Definition 3.3.1 in the following equivalent form.

Definition 3.3.2. The *dimension* of a variety $X \subset \mathbb{P}^n$ is the integer k such that the generic $S_{n-k} \subset \mathbb{P}^n$ meets X in a finite number of points.

If $X \subset \mathbb{P}^n$ has dimension k, we write $\dim(X) = k$. One has $0 \le k \le n-1$. A 0-dimensional variety is a finite number of points. We will say *curve, surface, hypersurface*, to indicate varieties of dimension 1, 2, $n-1$ respectively. An S_r of $\mathbb{P}^n$ is a variety of dimension r.

3.3.3. Note that from either of the two definitions 3.3.1, 3.3.2, there follows the (apparently obvious) fact:

3.3. Dimension of a projective variety

- If X is an irreducible variety and Y is a variety properly contained in X, then $\dim(Y) < \dim(X)$.

This fact can also be seen directly as follows. Without loss of generality we may suppose that $Y \subset X$ are affine varieties in $\mathbb{A}^n$ and that Y too is irreducible.

Let $d = \dim(X)$ and $K(X) = K(x_1, \ldots, x_d, x_{d+1}, \ldots, x_n)$. No matter how one chooses $d + 1$ indices $i_1, i_2, \ldots, i_{d+1}$ the elements $x_{i_1}, x_{i_2}, \ldots, x_{i_{d+1}}$ satisfy a polynomial relation $f(x_{i_1}, x_{i_2}, \ldots, x_{i_{d+1}}) = 0$, which necessarily also holds on Y, and so $\dim(Y) \leq \dim(X)$.

Suppose that $\dim(Y) = \dim(X) = d$, and that there exists $0 \neq u \in K[X]$ with $u = 0$ on Y. Let $x_1, \ldots, x_d$ be algebraically independent coordinates on Y and thus also on X. The elements $u, x_1, \ldots, x_d \in K[X]$ are algebraically dependent and thus there is a polynomial relation $f(u, x_1, \ldots, x_d) = 0$. We can even choose the polynomial f so that $f(0, x_1, \ldots, x_d) \neq 0$. Over Y one has $f(0, x_1, \ldots, x_d) = 0$ (since $u = 0$ on Y) and thus, by the hypothesis on $x_1, \ldots, x_d$, the polynomial $f(0, x_1, \ldots, x_d)$ is identically zero. It follows that $f(0, x_1, \ldots, x_d) = 0$ on X as well; but this contradicts the independence of $x_1, \ldots, x_d$. Thus $u = 0$ on Y implies that $u = 0$ on X; therefore $Y = X$.

In the situation of 3.3.3, the difference $\dim(X) - \dim(Y)$ is called the *codimension* of Y in X. If $\dim(X) - \dim(Y) = 1$, the variety Y is called a *divisor* of X. In particular, a hypersurface $X \subset \mathbb{P}^n$ is a divisor of $\mathbb{P}^n$.

It is useful to make the following remark explicit:

Proposition 3.3.4. *Let $X \subset \mathbb{P}^n$ be a variety of dimension k. If S_r is a linear space of dimension $r \geq n - k$, then S_r meets X.*

Proof. Indeed, set $t = n - k - 1$ in Definition 3.3.1. Then t is the maximum of the dimensions of the subspaces of $\mathbb{P}^n$ which do not meet X. Thus, a linear space S_r meets X as soon as $r > n - k - 1$. $\square$

Remark 3.3.5. If S_r is a subspace of $\mathbb{P}^n$ and X is a variety contained in it, the dimension of X as a variety of $\mathbb{P}^n$ coincides with its dimension as a variety of S_r. Indeed, if k is the dimension of X as variety of S_r, an arbitrary S_{n-k} of $\mathbb{P}^n$ meets S_r in a space of dimension $\geq r - k$ and so meets X; and a generic S_{n-k-1} of $\mathbb{P}^n$ meets S_r in a generic S_{r-k-1} which does not meet X.

We now show that Definitions 3.3.1 and 3.3.2 are equivalent to (3.17).

Let $X \subset \mathbb{P}^n$ and let $k = \dim(X)$ be the dimension expressed by Definitions 3.3.1, 3.3.2. In view of Definition 3.3.1 there exists a space $S_{n-k-1} \subset \mathbb{P}^n$ disjoint from X. Consider the projection of $\mathbb{P}^n$ from that S_{n-k-1} onto a space $\mathbb{P}^k$ disjoint from S_{n-k-1} and let $\pi: X \to \mathbb{P}^k$ be the restriction. If x is a point of X, the join $\mathrm{J}(x, S_{n-k-1})$ is, by Grassmann's formula, 1.1.8, an S_{n-k} that meets $\mathbb{P}^k$ in a point. Then

$$\pi(x) := \mathrm{J}(x, \mathbb{P}^{n-k-1}) \cap \mathbb{P}^k$$

and, for a generic $x \in X$ (that is, a generic $y = \pi(x) \in \mathbb{P}^k$), the fiber $\pi^{-1}(\pi(x))$ consists of the finite number of points in which $S_{n-k} = \mathrm{J}(x, S_{n-k-1})$ meets X (cf. Definition 3.3.2 and the successive Proposition 3.4.8). This means that the map $\pi : X \to \mathbb{P}^k$ is generically finite.

The key point, which however is based on notions outside the scope of the present book, is an algebraic fact from field theory (for whose proof we refer the reader to [48, (7.16)]) which assures us that in the presence of a generically finite map $\pi : X \to \mathbb{P}^k$ one has

$$K(\mathbb{P}^k) = K(x_1, \ldots, x_k) \hookrightarrow K(X).$$

This inclusion of fields expresses the field of fractions $K(X)$ of X as a finite extension of $K(x_1, \ldots, x_k)$, that is, $K(X)$ is a $K(x_1, \ldots, x_k)$-vector space of finite dimension. From this it follows that $K(X)$ is an algebraic extension of $K(x_1, \ldots, x_k)$ (see, for example, [75] or [62]); that is, every element $a \in K(X)$ is the root of a polynomial $p(T) \in K(x_1, \ldots, x_k)[T]$ with coefficients in $K(x_1, \ldots, x_k)$. By definition (cf. Definition 3.1.9), this means that the transcendence degree of $K(X)$ over K is k. Thus we may conclude that Definitions 3.3.1 and 3.3.2 are geometric formulations of the notion of dimension equivalent to (3.17).

We extend the definition of dimension to include possibly reducible varieties by defining the dimension of an arbitrary variety as the maximum of the dimensions of its irreducible components.

We say that a variety X has *pure dimension* k if all the irreducible components of X have the same dimension k.

Exercise 3.3.6. Let $X \subset \mathbb{P}^n$ be a variety of dimension k and $S_{n-1} \subset \mathbb{P}^n$ a generic hyperplane; that is, not containing any irreducible component of X and such that the intersection $S_{n-1} \cap X$ is irreducible if X is irreducible (cf. Theorem 6.3.11). Then

$$\dim(S_{n-1} \cap X) = k - 1. \qquad (3.18)$$

Indeed, we set $W := S_{n-1} \cap X$. A generic S_{n-k-1} of S_{n-1} is also generic in $\mathbb{P}^n$ (because S_{n-1} is generic) and so does not intersect X; thus, it does not intersect W. Since $n - k - 1 = (n - 1) - (k - 1) - 1$, one concludes in virtue of Definition 3.3.1 that W as a subvariety of S_{n-1} (and so of S_n) has dimension $k - 1$.

Iterating the reasoning, one sees that the section of X by a space S_{n-k} is a 0-dimensional variety, that is, a finite number of points.

One notes that if $Z \subset \mathbb{P}^n$ is a hypersurface that does not contain any irreducible component of X, one has

$$\dim(Z \cap X) = k - 1,$$

which extends the relation (3.18). Indeed, let $f = f(x_0, \ldots, x_n)$ be the form that defines the hypersurface Z and put $X_1 := X \cap Z$. One has $\dim(X_1) <$

$\dim(X)$ by 3.3.3. Consider a form f_1, $\deg f_1 = \deg f$, that does not vanish on any irreducible component of X_1 (cf. Exercise 2.7.39) and let Z_1 be the hypersurface with equation $f_1 = 0$ and $X_2 = X_1 \cap Z_1$. Iterating the procedure one obtains a sequence

$$X = X_0 \supset X_1 \supset X_2 \supset \cdots,$$

with $X_{i+1} = X_i \cap Z_i$, $Z_i := \{f_i = 0\}$, $\deg f_i = \deg f$, $\dim X_{i+1} < \dim X_i$. Since $\dim X_0 = k$, the variety X_{k+1} is empty. This means that $f, f_1, \ldots, f_k$ do not have common zeros on X. We can obviously suppose that X is irreducible and consider the map $\varphi \colon X \to \mathbb{P}^k$ defined by $\varphi(x) = [f(x), f_1(x), \ldots, f_k(x)]$. On the other hand φ is a finite morphism (for this see Problem 13.1.16) and so $\dim(X) = \dim(\varphi(X)) = k$. But $\varphi(X)$ is a closed subset of $\mathbb{P}^k$; thus $\varphi(X) = \mathbb{P}^k$ again by 3.3.3. If one had $\dim(X_1) < \dim(X) - 1$, then the closed subset X_k would be empty and so the forms $f, f_1, \ldots, f_{k-1}$ would not have common zeros on X and so the point $[0, 0, \ldots, 0, 1] \in \mathbb{P}^k$ would not belong to $\varphi(X)$. Thus $\dim(X_1) = k - 1$. In fact, $\dim(X_i) = k - i$.

One should also note that if $X \subset \mathbb{A}^n$ is an affine variety of dimension k and $H \subset \mathbb{A}^n$ is a hyperplane not containing any irreducible component of X, by the preceding remarks and the definition of dimension it follows immediately that

$$\dim(H \cap X) = k - 1.$$

We now prove a fundamental result on the dimension of the intersection of varieties; we treat the affine and projective cases separately.

Theorem 3.3.7 (Affine case). *Let X, Y be irreducible subvarieties of dimensions s, t in $\mathbb{A}^n$. Then every irreducible component Z of $X \cap Y$ has dimension $\geq s + t - n$.*

Proof. If $Y \subseteq X$ the inequality is obvious, and so we suppose that $Y \not\subset X$.

Consider the product $X \times Y \subset \mathbb{A}^{2n}$, which is a variety of dimension $s + t$ (cf. Exercise 3.1.16). Let $\Delta := \{(x, x), x \in \mathbb{A}^n\} \subset \mathbb{A}^{2n}$ be the diagonal. Then $\mathbb{A}^n$ is isomorphic to Δ via the map $x \mapsto (x, x)$ and, under that isomorphism, $X \cap Y$ corresponds to $(X \times Y) \cap \Delta$. Since Δ has dimension n, and since $s + t - n = (s+t) + n - 2n$, we have reduced to proving the result for the two varieties $X \times Y$ and Δ in $\mathbb{A}^{2n}$.

Now, Δ is the intersection of exactly n affine hyperplanes in $\mathbb{A}^{2n}$, namely those with equations $x_1 - y_1 = 0, \ldots, x_n - y_n = 0$, where $x_1, \ldots, x_n, y_1, \ldots, y_n$ are the coordinates of $\mathbb{A}^{2n}$. Now n applications of Exercise 3.3.6 gives the desired conclusion. □

Theorem 3.3.8 (Projective case). *Let X, Y be irreducible subvarieties of dimensions s, t in $\mathbb{P}^n$. Then every irreducible component Z of $X \cap Y$ has dimension $\geq s + t - n$. Therefore, if $s + t - n \geq 0$, then $X \cap Y \neq \emptyset$.*

72 Chapter 3. Geometric Properties of Algebraic Varieties

Proof. The first part of the thesis follows from the definition of dimension and the preceding Theorem 3.3.7, since $\mathbb{P}^n$ is covered by affine spaces.

Let $C(X), C(Y)$ be the affine cones over X, Y in $\mathbb{A}^{n+1}$. Then $C(X), C(Y)$ have dimensions $s+1, t+1$ respectively. Moreover $C(X) \cap C(Y) \neq \emptyset$ since both contain the origin $O = (0, \ldots, 0)$. By Theorem 3.3.7,

$$\dim(C(X) \cap C(Y)) \geq (s+1) + (t+1) - (n+1) = s + t - n + 1 > 0.$$

Thus $C(X) \cap C(Y)$ contains some point $P \neq O$, and so $X \cap Y \neq \emptyset$. $\square$

3.4 Order of a projective variety, tangent cone and multiplicity

In Section 2.5 we defined a projective variety $X \subset \mathbb{P}^n$ as the locus of the zeros $X = V(\mathfrak{a})$ of a homogeneous ideal $\mathfrak{a}$ of the ring $K[T_1, \ldots, T_{n+1}]$. While the ideal $\mathfrak{a}$ determines the variety X, X does not determine the ideal; indeed two ideals that have the same radical have the same locus of zeros (cf. Section 2.2 and Corollary 2.5.5).

The attitude that we prefer to assume is that of considering as projective varieties the ordered pairs $(X, \mathfrak{a})$ formed by an algebraic set X and by an ideal $\mathfrak{a}$ for which $X = V(\mathfrak{a})$ is the set of its zeros. We will use the term (*projective*) scheme for the pair $(X, \mathfrak{a})$. In this way, if $\mathfrak{a}$ and $\mathfrak{b}$ are two different ideals that have the same radical, $(X, \mathfrak{a})$ and $(X, \mathfrak{b})$ are two different schemes having the same associated algebraic set $V(\mathfrak{a}) = V(\mathfrak{b})$ as *support*. In this setting, we will say that $(X, \mathfrak{a})$ is a *reduced* scheme (or also *reduced variety*) if $\mathfrak{a} = \sqrt{\mathfrak{a}}$. In essence this is the point of view of the classical geometers, for whom it was more than natural to distinguish, for example the hypersurface X with equation $f = 0$, from the hypersurface X' with equation $f^2 = 0$, and to say that X' is the double of X, i.e., X counted twice, and that every point of X is double for X'.

The classical geometers, when they thought of a variety X, in reality had in mind a system of algebraic equations that define it, and concepts like double or triple varieties, etc., were considered obvious. Yet they were indispensable in order to have fundamental instruments available, like, for example, the theorems of Bézout (cf. Sections 4.2, 4.5).

As far as the notion of order is concerned, we start by considering the simplest case, that of a hypersurface X, with equation

$$f(T_1, \ldots, T_{n+1}) = 0.$$

Define the *order* of X to be the degree r of the polynomial $f = f(T_1, \ldots, T_{n+1})$.

If f is irreducible the order r is nothing but the number of points common to X and a line not contained in X. In fact, these points are obtained by resolving an algebraic equation $g(t) = 0$, not identically zero, of degree r and each of the points corresponds to one of the solutions. Naturally, in agreement with what has been

3.4. Order of a projective variety, tangent cone and multiplicity

said above, each point must be counted with multiplicity equal to the multiplicity of the root of $g(t)$ to which it corresponds.

If the polynomial f is a power of an irreducible polynomial, $f = \varphi^\mu$, the equation $g(t) = 0$ has all of its μ-fold roots independent of the line and each such root furnishes a point which should be counted μ times. The order of the hypersurface X continues to be the number of its intersections with a line in virtue of the fact that $X = \mu F$ is the multiple of the hypersurface F with equation $\varphi = 0$.

In an affine chart, with $Y_1, Y_2, \ldots, Y_n$ as non-homogeneous coordinates, we consider a hypersurface and one of its points x, and we can certainly suppose that x coincides with the coordinate origin. We assume that there are no terms of degree $< m$ in the polynomial f, while there really are terms of degree m, and we write

$$f = f_m + f_{m+1} + \cdots,$$

where f_j is a homogeneous polynomial of degree j in $Y_1, \ldots, Y_n$. In this case we will say that x is a point of *multiplicity m* for the hypersurface X (which means that m is the order of vanishing of f in x) and that the hypersurface $TC_x(X)$ of equation $f_m = 0$ is the *tangent cone* to X at x.

The definition is justified by the fact that for every line not contained in X and passing through x the equation $g(t) = 0$ which gives its intersection with X has the root that corresponds to the point x as a root of multiplicity at least m (and in general exactly m), and so x should be counted at least m times in the group of intersections. The lines (generators) of the cone $TC_x(X)$ are exceptional in that for each of them the equation $g(t) = 0$ has the root furnished by the point x with multiplicity at least $m + 1$. Thus all the generators of $TC_x(X)$ can be considered to be tangent to X at x (cf. Proposition–Definition 3.1.1).

If $m = 1$ the point x is non-singular and the tangent cone coincides with the tangent space (hyperplane) at x.

The multiplicity of a point for a hypersurface thus coincides with the order of the tangent cone at the given point (for further details see Section 5.2).

Example 3.4.1. Let X be an algebraic variety and P one of its points. If P is singular on X, the tangent space $T_P(X)$ to X at P (cf. Section 3.1) does not furnish a good description of the local geometry of X at P. In particular, if $X \subset \mathbb{A}^2 := \mathbb{A}^2(K)$ is a plane curve and P a singular point of X, the tangent space coincides with the tangent space $T_P(\mathbb{A}^2) = \mathbb{A}^2$ of the ambient affine space $\mathbb{A}^2$ at P.

The "tangent cone" furnishes a better description of the local structure of a variety at its singular points. For example, if $X \subset \mathbb{A}^2_{(x,y)}$ is the curve with equation $y^2 - x^2(x+1) = 0$ the tangent cone is the union of the lines with equations $x \pm y = 0$, tangent to the two branches of X at $O = (0, 0)$ (and O is a *node* for X). Similarly, the tangent cone to the curve X with equation $y^2 - x^3 = 0$ is the line $y = 0$, counted twice (and O is a *cusp* for X) (see also §9.2.5). In each of the two cases O is a double point; that is, of multiplicity two, which is equal to the order of the tangent cone.

74   Chapter 3. Geometric Properties of Algebraic Varieties

Let $X = V(\mathfrak{a}) \subset \mathbb{A}^n$ be an affine algebraic set and let x be a point of X. After a suitable change of coordinates if necessary, we may suppose that x coincides with the coordinate origin.

Definition 3.4.2. Let $X = V(\mathfrak{a}) \subset \mathbb{A}^n$ be an affine algebraic set, containing the coordinate origin $x = (0, \ldots, 0)$. We define the *tangent cone* to X at x to be the subvariety $TC_x(X)$ of $\mathbb{A}^n$ defined by $TC_x(X) = (V(\mathfrak{a}^*), \mathfrak{a}^*)$, where $\mathfrak{a}^*$ is the homogeneous ideal generated by all the homogeneous polynomials f^* which are initial forms of the polynomials $f \in \mathfrak{a}$.

Abuse of language. For brevity, and when there is no chance of confusion, hereafter we will sometimes write "tangent cone" rather than "support of the tangent cone".

3.4.3 Intrinsic nature of the tangent cone. One might think that the preceding definition depends on the particular immersion of X in $\mathbb{A}^n$. To render that definition intrinsic it is useful to make use of the notion of the associated graded ring of an ideal in a commutative ring.

Let then A be a ring and $\mathfrak{a}$ an ideal of A. We consider the graded abelian group

$$G_A(\mathfrak{a}) = \bigoplus_{d \geq 0} \mathfrak{a}^d / \mathfrak{a}^{d+1}, \quad \mathfrak{a}^0 = A,$$

where the elements of $\mathfrak{a}^d / \mathfrak{a}^{d+1}$ are considered as homogeneous elements of degree d. It is possible to define a multiplication in $G_A(\mathfrak{a})$ in the following way. If $\bar{x} \in \mathfrak{a}^d / \mathfrak{a}^{d+1}$ and $\bar{y} \in \mathfrak{a}^{d'} / \mathfrak{a}^{d'+1}$ one has $x \in \mathfrak{a}^d$ and $y \in \mathfrak{a}^{d'}$ and so $xy \in \mathfrak{a}^{d+d'}$. We then set

$$\bar{x}\,\bar{y} = \overline{xy} \in \mathfrak{a}^{d+d'} / \mathfrak{a}^{d+d'+1}.$$

It is easy to see that this operation is well defined, associative, commutative, and distributive with respect to addition. In this way $G_A(\mathfrak{a})$ becomes a graded ring which we will call *graded ring associated to A with respect to the ideal $\mathfrak{a}$*.

Now suppose that $\mathfrak{a}$ is finitely generated and let $\mathfrak{a} = (a_1, \ldots, a_s)$. Then a basis for $\mathfrak{a}^d$ is given by the monomials of the type $a_1^{i_1} \ldots a_s^{i_s}$ with $i_1 + \cdots + i_s = d$. It then follows that

$$\overline{a_1}^{i_1} \ldots \overline{a_s}^{i_s} = \overline{a_1^{i_1} \ldots a_s^{i_s}} \in \mathfrak{a}^d / \mathfrak{a}^{d+1},$$

where $\overline{a_i} \in \mathfrak{a}/\mathfrak{a}^2$. Thus one sees that $G_A(\mathfrak{a})$ is generated over $A/\mathfrak{a}$ by the classes $\overline{a_i}$ of the a_i modulo $\mathfrak{a}^2$. If we then set $\xi_i = a_i \mod \mathfrak{a}^2$, we will have

$$G_A(\mathfrak{a}) = A/\mathfrak{a}[\xi_1, \ldots, \xi_s].$$

It is then immediate to consider the homomorphism

$$\varphi \colon A/\mathfrak{a}[Y_1, \ldots, Y_s] \to A/\mathfrak{a}[\xi_1, \ldots, \xi_s] = G_A(\mathfrak{a}),$$

3.4. Order of a projective variety, tangent cone and multiplicity

defined by $\varphi(Y_i) = \xi_i$, for $i = 1, \ldots, s$. The homomorphism φ is homogeneous of degree zero and surjective, and the kernel $\ker(\varphi)$ is the homogeneous ideal of $A/\mathfrak{a}[Y_1, \ldots, Y_s]$ that has as generators the forms $\bar{F}(Y_1, \ldots, Y_s)$ in $A/\mathfrak{a}[Y_1, \ldots, Y_s]$ such that $\bar{F}(\xi_1, \ldots, \xi_s) = 0$, namely, such that $F(a_1, \ldots, a_s) \in \mathfrak{a}^{r+1}$, where r is the degree of F. Thus, one has

$$G_A(\mathfrak{a}) = A/\mathfrak{a}[\xi_1, \ldots, \xi_s] \cong A/\mathfrak{a}[Y_1, \ldots, Y_s]/\ker(\varphi). \tag{3.19}$$

Let then $X = V(\mathfrak{a}) \subset \mathbb{A}^n$ be an affine variety containing the coordinate origin $x = (0, \ldots, 0)$. Let $K[Y_1, \ldots, Y_n]$ be the ring of polynomials and let $\mathfrak{m} = (Y_1, \ldots, Y_n)/\mathfrak{a}$ be the maximal ideal that defines the origin as a point of X in the ring $A = K[Y_1, \ldots, Y_n]/\mathfrak{a}$.

The intrinsic nature of the definition of the tangent cone given in Definition 3.4.2 is then expressed by the isomorphism

$$G_{A_\mathfrak{m}}(\mathfrak{m} A_\mathfrak{m}) \cong K[Y_1, \ldots, Y_n]/\ker(\varphi) \cong K[Y_1, \ldots, Y_n]/\mathfrak{a}^*, \tag{3.20}$$

where $\mathfrak{a}^*$ is the homogeneous ideal generated by all the homogeneous polynomials f^* that are initial forms of the polynomials $f \in \mathfrak{a}$. To prove (3.20), we observe that

$$G_{A_\mathfrak{m}}(\mathfrak{m} A_\mathfrak{m}) = \bigoplus_{d \geq 0} \mathfrak{m}^d A_\mathfrak{m}/\mathfrak{m}^{d+1} A_\mathfrak{m}.$$

Now, $A_\mathfrak{m}/\mathfrak{m} A_\mathfrak{m} \cong A/\mathfrak{m}$ and there is an isomorphism of $A/\mathfrak{m}$-vector spaces

$$(\mathfrak{m} A_\mathfrak{m})^d/(\mathfrak{m} A_\mathfrak{m})^{d+1} \cong \mathfrak{m}^d/\mathfrak{m}^{d+1}$$

defined by setting, for each $x \in \mathfrak{m}^d$,

$$\bar{x} = \text{class of } \frac{x}{1} \in \mathfrak{m}^d A_\mathfrak{m} \mod \mathfrak{m}^{d+1} A_\mathfrak{m}.$$

One has $\bar{x} = 0$ if and only if $\frac{x}{1} \in \mathfrak{m}^{d+1} A_\mathfrak{m}$, that is, if and only if there exists $t \notin \mathfrak{m}$ such that $xt \in \mathfrak{m}^{d+1}$; since $\mathfrak{m}^{d+1}$ is $\mathfrak{m}$-primary (i.e., $\sqrt{\mathfrak{m}^{d+1}} = \mathfrak{m}$) this is equivalent to $x \in \mathfrak{m}^{d+1}$. From this it then follows that $G_{A_\mathfrak{m}}(\mathfrak{m} A_\mathfrak{m}) \cong G_A(\mathfrak{m})$ and so, by (3.19),

$$G_{A_\mathfrak{m}}(\mathfrak{m} A_\mathfrak{m}) \cong A/\mathfrak{m}[Y_1, \ldots, Y_n]/\ker(\varphi) \cong K[Y_1, \ldots, Y_n]/\ker(\varphi).$$

There remains to prove that $\ker(\varphi) = \mathfrak{a}^*$, and, since these are homogeneous ideals, it suffices to prove that if $f(Y_1, \ldots, Y_n) \neq 0$ is a form of degree r in $K[Y_1, \ldots, Y_n]$ one has $f \in \mathfrak{a}^*$ if and only if $f \in \ker(\varphi)$. On the other hand, under the given hypotheses and setting $y_i = Y_i \mod \mathfrak{a}$, so that $\mathfrak{m} = (y_1, \ldots, y_n)$, one has the equivalences

$$f(Y_1, \ldots, Y_n) \in \ker(\varphi) \iff \bar{f}(y_1, \ldots, y_n) \in \mathfrak{m}^{r+1}$$
$$\iff f(Y_1, \ldots, Y_n) \in (Y_1, \ldots, Y_n)^{r+1} + \mathfrak{a}$$
$$\iff f(Y_1, \ldots, Y_n) \in \mathfrak{a}^*.$$

Remark 3.4.4. In contrast to what happens for the tangent space (cf. Definition 3.1.5), one notes that if $X = V(\mathfrak{a})$ with $\mathfrak{a} = (f_1, \ldots, f_m)$, the tangent cone at X is not always the intersection of the tangent cones to the individual hypersurfaces with equations $f_i = 0, i = 1, \ldots, m$. It suffices to observe that the initial form of a sum of polynomials does not necessarily belong to the ideal generated by the initial forms of the summands. For example, $C = V(\mathfrak{a}) \subset \mathbb{A}^3$, with $\mathfrak{a} = (x - y^2, z^3 - x)$, is an irreducible curve formed by the intersection of two cylinders. The tangent cone to each of the cylinders at the origin is the plane with equation $x = 0$ which can not be the tangent cone to C inasmuch as the initial form y^2 of $(x - y^2) + (z^3 - x)$ does not belong to the ideal (x) generated by the initial forms of the two summands. In order to have all the "information" possible regarding the origin as a point of the curve C we must keep in mind the tangent cones to all the surfaces that pass through C. For example, one must also consider the tangent cone (consisting of the plane $y = 0$ counted twice) of the surface with equation $y^2 = z^3$ which obviously contains C.

Another difficulty is related to the fact that even when $\mathfrak{a}$ is a prime ideal, the ideal $\mathfrak{a}^*$ of initial forms can very well not be prime. It is necessary to consider the primary decomposition of the ideal $\mathfrak{a}^*$, that is to write $\mathfrak{a}^* = \mathfrak{q}_1 \cap \cdots \cap \mathfrak{q}_h$ with $\mathfrak{q}_j$ primary ideals, $j = 1, \ldots, h$, see 2.7.4; to each of these is associated an irreducible component $V(\sqrt{\mathfrak{q}_j})$ of the tangent cone, which will be "counted" a suitable number of times (cf. Exercise 3.4.11 (2)).

If $X = V(\mathfrak{a})$ and $x = (0, \ldots, 0) \in X$, the ideal $\mathfrak{a}^{(1)} = \{f_x^{(1)}, f \in \mathfrak{a}\}$ generated by the linear forms $f_x^{(1)}$ of the polynomials $f \in \mathfrak{a}$ at x is obviously contained in the ideal $\mathfrak{a}^*$ generated by the initial forms of the polynomials $f \in \mathfrak{a}$. Thus one has
$$V(\mathfrak{a}^*) \subset T_x(X) = V(\mathfrak{a}^{(1)}) \subset \mathbb{A}^n,$$
that is, the (support $V(\mathfrak{a}^*)$ of the) tangent cone to X at x is a subvariety of the tangent space $T_x(X)$.

As in the case of tangent spaces (cf. Section 3.1), if $X \subset \mathbb{P}^n$ is a projective algebraic set and x is one of its points, we can choose an affine chart $\mathbb{A}^n \subset \mathbb{P}^n$ complementary to a hyperplane not containing x and consider the closure of the tangent cone $TC_x(X \cap \mathbb{A}^n) \subset \mathbb{A}^n$ in $\mathbb{P}^n$. In this way we obtain a projective variety that we call the *projective tangent cone* to X at x.

We now wish to extend the notion of order, tangent cone, and multiplicity of a point to varieties of arbitrary dimension in such a way as to retain the fact that the multiplicity of a singular point on a variety X is the order of the tangent cone in that point.

Unlike the case of hypersurfaces examined above, for projective varieties of arbitrary dimension the situation is not at all simple, and a rigorous algebraic treatment, for which we refer the reader for example to [48] and [49], requires tools which are outside the scope of this book.

3.4. Order of a projective variety, tangent cone and multiplicity   77

We will overcome the obstacle by using the methods of projective geometry. However, we will need some preliminary observations which are usually considered evident in classical texts and left to the reader's intuition, or possibly to verification in the case of examples.

3.4.5 Projection of a variety from linear spaces.
In $\mathbb{P}^n$ consider an irreducible variety X of dimension k and a generic subspace S_r of $\mathbb{P}^n$ of dimension $r \leq n-k-1$.

We define the *cone that projects X from S_r* (or the *projecting cone* of X from S_r) to be the locus V of the subspaces S_{r+1} that join the given S_r with the individual points of X. The S_r is called the *vertex* of V, and each space S_{r+1} is said to be a *generator* of V. We prove that

- the projecting cone V is an algebraic set of dimension $\dim(V) = r + k + 1$.

Suppose that the variety X is the locus of common zeros of the homogeneous polynomials $f_j(T_1, \ldots, T_{n+1})$ belonging to $K[T] := K[T_1, \ldots, T_{n+1}]$. That is, $X = V(\mathfrak{a})$ where $\mathfrak{a} = (\ldots, f_j(T), \ldots)$. Let S_r be the intersection of the $n-r$ independent hyperplanes with equations $H_q(T_1, \ldots, T_{n+1}) = 0, q = 1, \ldots, n-r$. The space S_{r+1} that joins S_r with a point $x = [a_1, \ldots, a_{n+1}] \in X$ has equations

$$\frac{H_1(T)}{H_1(x)} = \frac{H_2(T)}{H_2(x)} = \cdots = \frac{H_{n-r}(T)}{H_{n-r}(x)}; \qquad (3.21)$$

and the equations of V are obtained by eliminating the parameters $a_1, \ldots, a_{n+1}$ from the system consisting of the $n-r-1$ equations (3.21) and from the equations $f_j(a_1, \ldots, a_{n+1}) = 0$.

We will also say that the intersection $X' = V \cap S_{n-r-1}$ is the *projection* of X from S_r onto a subspace S_{n-r-1} skew to it.

We observe that if the center of projection S_r has equations $T_1 = T_2 = \cdots = T_{n-r} = 0$, the projection $X' = V(\mathfrak{b})$ of X from S_r onto the S_{n-r-1} with equations $T_{n-r+1} = \cdots = T_{n+1} = 0$ (where $T_1, T_2, \ldots, T_{n-r}$ are homogeneous coordinates) is the algebraic set associated to the contracted ideal $\mathfrak{b} = \mathfrak{a}^c := \mathfrak{a} \cap K[T_1, \ldots, T_{n-r}]$. The projecting cone V is the algebraic set defined by the same equations in $\mathbb{P}^n$, that is, $V = V(\mathfrak{b}K[T_1, \ldots, T_{n+1}])$, cf. §5.2.3.

As far as the dimension is concerned, we observe that a generic space $H = S_{n-r-k-2}$ does not meet S_r and is joined to S_r by a space $L = S_{n-k-1}$ which does not meet X (cf. Definition 3.3.1). Thus H does not meet V because if there were a point $A \in V \cap H$, the space L, that contains the join $J(A, S_r)$, would meet X (inasmuch as the join $J(A, S_r)$ is one of the spaces S_{r+1} of V and so contains a point of X). Thus, by Definition 3.3.1, we have that $\dim(V) \leq r + k + 1$.

To prove the equality, we consider a generic space $S_{n-r-k-1}$. The join space $\sigma := J(S_{n-r-k-1}, S_r)$ has dimension $n-k$ and so, again by Definition 3.3.1, meets X in at least one point P. The space $J(P, S_r)$, namely, the generator space of V that passes through P, and the space $S_{n-r-k-1}$, both contained in the space σ, have a point in common by Grassmann's formula, see 1.1.8. Thus, our $S_{n-r-k-1}$ meets V, and so $\dim(V) > r + k$.

Example 3.4.6. In $\mathbb{P}^5$ we consider the surface X which is the locus of zeros of the polynomials
$$T_1 T_2 - T_3 T_4 - T_5 T_6, \quad T_4 - T_5, \quad T_5 - T_6.$$

We project X from the line S_1 with equations $T_1 = T_2 = T_3 = T_4 = 0$ onto the subspace $S_3 = \mathbb{P}_{[T_1,T_2,T_3,T_4]}$ skew to it.

If $x = [a_1, \ldots, a_6]$ is a generic point of X, the join $S_2 = J(x, S_1)$ has equations
$$\frac{T_1}{a_1} = \frac{T_2}{a_2} = \frac{T_3}{a_3} = \frac{T_4}{a_4}.$$

Remembering that $a_1 a_2 - a_3 a_4 - a_5 a_6 = 0$ and $a_4 = a_5 = a_6$, we eliminate the parameters $a_1, \ldots, a_6$, to obtain the equation
$$T_1 T_2 - T_3 T_4 - T_4^2 = 0. \tag{3.22}$$

In this case the projecting cone is the hypersurface V of $\mathbb{P}^5$ with equation (3.22). The same equation, read in $S_3 = \mathbb{P}_{[T_1,T_2,T_3,T_4]}$, represents the projection X' of X from the line S_1 onto S_3.

We now consider the particular case $r = n - k - 2$ of the preceding discussion. Let σ be a generic S_{n-k} of $\mathbb{P}^n$. It meets X in a finite number of points, and so contains a finite number of chords of X consisting of the lines that join pairs of these points. A generic S_{n-k-1} of σ does not contain any of these lines and therefore a space S_{n-k-2} contained in it (and thus a generic S_{n-k-2} of $\mathbb{P}^n$) does not meet chords of X.

Thus we find spaces S_{n-k-1} passing through such S_{n-k-2} and having in common with X only a single point. This implies that the generic S_{n-k-1} generator of the cone V that projects X from S_{n-k-2} contains only one point of X (whereas particular generator spaces may very well meet X in more than a single point). In this case we will say that the projection π of X from the S_{n-k-2} onto a space S_{k+1} which is skew to it is *simple*. Then for a generic point x' belonging to the projection $X' = V \cap S_{k+1}$, the inverse image $\pi^{-1}(x')$ consists of the unique point in which $S_{n-k-1} = J(x', S_{n-k-2})$ meets X.

There can, however, be special spaces S_{n-k-2} from which X is projected multiply. If the generic generator space of V meets X in $\nu \geq 2$ points we say that the projection is *ν-fold*.

In the case $r = n - k - 2$ under consideration, the cone V that projects X from S_{n-k-2} onto a subspace S_{k+1} is a hypersurface of $\mathbb{P}^n$ and the projection $X' = V \cap S_{k+1}$ is a hypersurface in S_{k+1}. In this case, the elimination of the parameters $a_1, \ldots, a_{n+1}$ described above, where $x = [a_1, \ldots, a_{n+1}]$ is a generic

3.4. Order of a projective variety, tangent cone and multiplicity 79

point of X, leads to a single equation $\phi = 0$ which is the equation of the projecting cone V, and whose degree is the order of V.

We now define the order of a projective variety in terms of the orders of the projecting cones.

Definition 3.4.7. The *order* $\deg(X)$ of a k-dimensional projective variety $X \subset \mathbb{P}^n$ is the maximal order of the $(n-1)$-dimensional cones that one obtains by projecting X from spaces S_{n-k-2}.

Proposition 3.4.8. *The order* $\deg(X)$ *of a k-dimensional projective variety* $X \subset \mathbb{P}^n$ *is the number of points which X has in common with a generic S_{n-k}.*

Proof. Let S be a generic S_{n-k-2} and ℓ a line skew to it. The join space $J(S, \ell)$ is a generic S_{n-k} that, by Definition 3.3.2, meets X in a finite number ρ of points. The line ℓ meets the cone V that projects X from S in $m = \deg(V)$ points, each of which belongs to a generator space S_{n-k-1} of V. Since the projection π of X from S is simple (in view of the fact that S is generic) each of these spaces S_{n-k-1} contains one and only one point of X. These points are contained in our S_{n-k}; hence such a generic S_{n-k} meets X in (at least) m points, and so $\rho \geq m$. It follows that $\rho \geq \deg(X)$.

Conversely, let x be one of the points in which S_{n-k} meets X. The ρ spaces $J(x, S)$, which belong to the cone V, have dimension $n - k - 1$ and lie in a space S_{n-k} that contains ℓ, and so each has a point in common with ℓ. From this it follows that $\rho \leq m \leq \deg(X)$. $\square$

As a consequence of Proposition 3.4.8 we can reformulate the definition of order in the following equivalent fashion.

Definition 3.4.9. The *order* $\deg(X)$ of a k-dimensional algebraic set $X \subset \mathbb{P}^n$ is the order of the $(n-1)$-dimensional cone obtained by projecting X from a generic space S_{n-k-2}, that is, it coincides with the order of the hypersurface X' which is the projection of X from a generic S_{n-k-2} into a subspace S_{k+1}.

In the sequel we shall often use the notation X_k^d to indicate an algebraic variety of dimension k and order d.

Remark 3.4.10 (Explicit construction of a birational map between a projective variety and a hypersurface). In §2.4.10 we have proved the birational equivalence of any affine algebraic set with a hypersurface. In §2.6.11 we observed that as a consequence of essentially algebraic facts the analogous result also holds in the projective case.

If $X \subset \mathbb{P}^n$ is a projective algebraic set of dimension k, the construction of a simple projection $\pi : X \to X' \subset S_{k+1}$ of X from a generic S_{n-k-2} into a subspace S_{k+1} skew to it, discussed in this paragraph, constitutes a geometric proof of the birational equivalence of X with a hypersurface in $\mathbb{P}^{k+1}$.

We now propose two alternative and elementary definitions of multiplicity. Let $X = X_k^d \subset \mathbb{P}^n$ be a k-dimensional projective variety of order d. Let $\mathcal{L}$ be the set of S_{n-k-2} in $\mathbb{P}^n$ in general position with respect to X, namely such that

i) they do not contain points of X;

ii) from each of them X is projected simply.

With this as premise, let x be a point of X and $\pi_\sigma: X \to S_{k+1}$ the projection from a space $\sigma \in \mathcal{L}$. If $\mu_\sigma(x)$ is the multiplicity of the point $\pi_\sigma(x)$ for the hypersurface $\pi_\sigma(X) \subset S_{k+1}$, we take as the *multiplicity* of X at x (or of x for X) the integer

$$\mu_x(X) = \inf_{\sigma \in \mathcal{L}} \mu_\sigma(x).$$

Note that this definition agrees with that given above in the case of hypersurfaces ($k = n - 1$). Indeed, if $X = X_{n-1}$ is a hypersurface of $\mathbb{P}^n$ the projection $\pi: X \to S_{k+1} = S_n$ is the identity.

The following is another method (useful for explicit calculation) for defining multiplicity and tangent cone in a point. We say that $X = X_k^d$ has *multiplicity* $\mu_x(X) = s$ at x if the number of points different from x common to X and a generic linear space S_{n-k} passing through x is $d - s$. The union of those particular spaces S_{n-k} that have not more than $d - s - 1$ distinct points different from x in common with X constitute the *projective tangent cone* Γ_k, of order s, to X at x. One notes that this definition agrees with that given previously in the case of hypersurfaces. The fact that Γ_k has order s can be easily seen as follows. We consider a generic S_{n-k-2} and we project X and Γ_k onto a subspace S_{k+1}. In this way we find two hypersurfaces X' and Γ' that have their orders equal to that of X and to that of Γ_k respectively. Let $x' \in S_{k+1}$ be the projection of the s-fold point x of X. The cone Γ^* tangent to X' at x' thus has order s. But $\Gamma^* = \Gamma'$. Indeed, the generator lines of the cone Γ_k (with vertex x) are contained in spaces S_{n-k} that pass through x and that away from x have at most $d - s - 1$ intersections with X and are projected from the space S_{n-k-2} in S_{k+1} into the lines through x' and having at most $d - s - 1$ intersections with X' away from x'. This means that the generator lines of the cone project onto the generators of Γ'.

We note explicitly that the two definitions proposed are equivalent. Indeed, if $x \in X$ is a point of multiplicity s on the basis of the first definition, its image $\pi(x)$ under the projection $\pi: X \to X' \subset S_{k+1}$ from a generic subspace S_{n-k-2} onto a subspace S_{k+1} skew to it is an s-fold point for the hypersurface X' of S_{k+1}. Thus, a generic line ℓ of the space S_{k+1} has exactly $d - s$ points different from $\pi(x)$ in common with X'. Joining these points with S_{n-k-2} one obtains $d - s$ generic spaces S_{n-k-1} each of which meets the variety X in a point. Thus we have $d - s$ distinct points of X (and also distinct from x) which are the points different from x and common to X and to the (generic) space S_{n-k} which joins S_{n-k-2} with the

3.4. Order of a projective variety, tangent cone and multiplicity

line ℓ. Thus x is a point of multiplicity s on the basis of the second definition proposed. The verification of the converse is analogous.

Exercises 3.4.11. *Let $X = V(\mathfrak{a})$ be an algebraic set of the affine space $\mathbb{A}^3$ containing the origin $O = (0, 0, 0)$. Describe the tangent space $T_O(X)$ and the tangent cone $TC_O(X)$ at O in the two following cases:*

(1) $\mathfrak{a} = (x - y^2, x + yz + z^3)$; (2) $\mathfrak{a} = (xz - y^2, x^3 - yz, z^2 - x^2y)$.

(1) $\mathfrak{a} = (x - y^2, x + yz + z^3)$ is a prime ideal; $X = V(\mathfrak{a})$ is a curve of order 6 passing through the origin O. The tangent space $T_O(X)$ at O is the plane $x = 0$. Since the dimension of this space is bigger than $\dim(X)$, O is a multiple point for X. The ideal $\mathfrak{a}^*$ of the initial forms of the polynomials of $\mathfrak{a}$ is $(x, y^2 + yz) = (x, y) \cap (x, y + z)$ and thus the tangent cone $TC_O(X)$ is a pair of lines passing through O (and contained in $T_O(X)$).

Using the second definition proposed, it is easy to verify that O is a point of multiplicity 2 for X. In this regard, one notes that the points common to X and the plane $\pi: \rho z = \lambda x + \mu y$ passing through O are the solutions of the system of equations

$$x - y^2 = x + yz + z^3 = \rho z - \lambda x - \mu y = 0.$$

The result of the elimination of x and z from this system, that is, the result of elimination of z from the system

$$y^2 + yz + z^3 = \rho z - \lambda x - \mu y = 0,$$

is

$$y^2[\rho^2(\rho + \mu) + (\rho^2\lambda + \mu^3)y + 3\lambda\mu^2 y^2 + 3\lambda^2\mu y^3 + \lambda^3 y^4] = 0.$$

Therefore there are four intersections of X with π distinct from O if π is generic; there are instead only three if $\rho(\rho + \mu) = 0$, that is, if π passes through the z-axis or through the line $x = y + z = 0$. Thus one finds two pencils of planes that have as axes the two lines constituting the tangent cone (cf. Section 5.3).

(2) The ideal $\mathfrak{a}^{(1)}$ of the linear forms of the polynomials of $\mathfrak{a}$ is the ideal (0); hence $T_O(X) = \mathbb{A}^3$.

The ideal $\mathfrak{a}^*$ of the initial forms of the polynomials of $\mathfrak{a}$ is $\mathfrak{a}^* = (y^2 - xz, yz, z^2)$. Since $\sqrt{\mathfrak{a}^*} = (y, z)$ the tangent cone $TC_O(X)$ at O is the x-axis counted a suitable number of times.

It is useful to note that the smallest integer ν such that $(\sqrt{\mathfrak{a}^*})^\nu \subset \mathfrak{a}^*$ is $\nu = 3$. Indeed, $y^2 \notin \mathfrak{a}^*$ and so $\nu \geq 3$. Furthermore, $y^3 = y(y^2 - xz) + x(yz) \in \mathfrak{a}^*$. Then, if $A, B \in K[x, y, z]$ are two arbitrary polynomials we have

$$(Ay + Bz)^3 = A^3 y^3 + (3A^2 B y)yz + (3AB^2 y + B^3 z)z^2 \in \mathfrak{a}^*.$$

The variety $X = V(\mathfrak{a})$ is the monomial curve of 5th order locus of the point $P(t) = (t^3, t^4, t^5)$. To see this it suffices to observe that if $x \neq 0$, and thus

$y, z \neq 0$, then, setting $t := \frac{y}{x}$, one has $t = \frac{z}{y}$ (since $y^2 = xz$ in the ring of coordinates $K[X]$) and hence

$$t^2 = \frac{y}{x}\frac{z}{y} = \frac{z}{x}.$$

Thus

$$y = \frac{z^2}{x^2} = t^4; \quad x = y\frac{x}{y} = t^3; \quad z = ty = t^5.$$

Standard arguments from the algebra of polynomials, which we give here for completeness, show that the ideal $\mathfrak{a}$ is prime, inasmuch as it coincides with the kernel of the morphism $\psi : K[x, y, z] \to K[t]$ defined by $\psi(x) = t^3$, $\psi(y) = t^4$, $\psi(z) = t^5$; in particular X is an irreducible curve.

The inclusion $\mathfrak{a} \subset \ker(\psi)$ is obvious. To prove the opposite inclusion we set $\alpha = xz - y^2$, $\beta = yz - x^3$, $\gamma = z^2 - x^2y$. An arbitrary $f \in K[x, y, z]$ can be written in the form

$$f = A(x, y, z^2) + zB(x, y, z^2) = A + \lambda z + xzL + yzM,$$

with $\lambda \in K$, $A, L, M \in K[x, y, z^2]$. Since

$$z^2 = x^2y + \gamma, \quad xz = y^2 + \alpha, \quad yz = x^3 + \beta,$$

we have

$$f = \lambda z + g(x, y) + \theta, \quad \text{with } g \in K[x, y], \ \theta \in \mathfrak{a} = (\alpha, \beta, \gamma).$$

If $f \in \ker(\psi)$ one obtains

$$0 = \lambda t^5 + g(t^3, t^4)$$

and therefore $\lambda = 0$ and $g(x, y)$ is the zero polynomial (because no summand of $g(t^3, t^4)$ can be of degree 5). Thus $\ker(\psi) = \mathfrak{a}$.

The curve X has the origin O as a triple point. In fact, the planes passing through O have at most two distinct points in common with X; exactly two if the plane is generic. The planes of the pencil $\lambda y + \mu x = 0$ are exceptional in that the generic plane of this pencil meets X in only one point different from O; the plane $z = 0$ does not meet the curve away from O. Thus one has a triple point at O with three tangents lines that coincide there with the x-axis, and with the plane $z = 0$ as osculating plane.

One also notes that X constitutes an example of a curve which is not a complete intersection in $\mathbb{A}^3$ (cf. Section 7.1 and the note 5.8.5), which is however a set-theoretical complete intersection of the two surfaces $y^2 - xz = 0$, $x^5 + z^3 - 2x^2yz = 0$ which have the same tangent plane at every simple point of X (cf. [71] and also [19]).

Chapter 4
Rudiments of Elimination Theory

In Section 4.1 we introduce the Euler–Sylvester resultant of two polynomials and we recall some of its basic properties. As an application, in Section 4.2 we define the intersection multiplicity of two algebraic coplanar curves and we prove Bézout's theorem (Theorem 4.2.1), which gives the numbers of points common to two such curves.

In Section 4.3, by using another interpretation of the resultant, we show that the intersection multiplicity of two coplanar curves is independent of the system of projective coordinates chosen.

In Section 4.4 we discuss a procedure for the elimination of an indeterminate first proposed by Kronecker.

In Section 4.5 we introduce the intersection multiplicity in higher dimension and we state Bézout's theorem in its full generality.

4.1 Resultant of two polynomials

Let A be an integral domain with identity and of characteristic zero. Suppose furthermore that A is a *factorial ring*, that is, an integral domain such that every non-zero element admits a unique factorization (up to units) as a product of irreducible elements. An element $a \neq 0$ of A is said to be *irreducible* if it is not invertible and if for all $b, c \in A$ such that $a = bc$, either b or c is invertible. It is known that the ring $A[X]$ of polynomials in the indeterminate X is also factorial.

Consider two non-zero polynomials $f, g \in A[X]$ of degrees n and m that have a non-zero common divisor $h \in A[X]$. Setting $f = hf_1, g = -hg_1$ one then has

$$fg_1 + gf_1 = 0. \tag{4.1}$$

If h has degree $\geq s$ the degree of f_1 will then be $\leq n - s$ and that of g_1 will be $\leq m - s$.

Conversely, suppose that there exist two non-zero polynomials f_1 of degree $\leq n - s$ and g_1 of degree $\leq m - s$ (with $s > 0$) such that one has (4.1). Each of the irreducible divisors of f is a divisor of the product gf_1 and so, $A[X]$ being factorial, either it divides f_1 or it divides g. Since $\deg(f_1) \leq n - s < n = \deg(f)$, some divisor of f divides g, whence f and g have a common divisor of degree $\geq s$. Thus we have established the following fact:

- A necessary and sufficient condition for two (non-zero) polynomials f and g to have a non-zero common divisor h of degree $\geq s$ is that there exist two

84   Chapter 4. Rudiments of Elimination Theory

non-zero polynomials f_1 of degree $\leq n - s$ and g_1 of degree $\leq m - s$ such that $fg_1 + gf_1 = 0$.

Now, if $f = a_0 X^n + a_1 X^{n-1} + \cdots + a_n$, $g = b_0 X^m + b_1 X^{m-1} + \cdots + b_m$ with $a_0 b_0 \neq 0$, the necessary and sufficient condition for the existence of two non-zero polynomials

$$f_1 = p_0 X^{n-s} + p_1 X^{n-s-1} + \cdots + p_{n-s}, \quad g_1 = q_0 X^{m-s} + q_1 X^{m-s-1} + \cdots + q_{m-s}$$

such that (4.1) holds, or, equivalently, such that $q_0, \ldots, q_{m-s}, p_0, \ldots, p_{n-s}$ satisfy the homogeneous system

$$\begin{cases} a_0 q_0 + b_0 p_0 = 0, \\ a_1 q_0 + a_0 q_1 + b_1 p_0 + b_0 p_1 = 0, \\ \quad \vdots \\ a_n q_{m-s} + b_m p_{n-s} = 0 \end{cases}$$

is that the matrix

$$\begin{pmatrix} a_0 & 0 & 0 & \cdots & 0 & b_0 & 0 & 0 & \cdots & 0 \\ a_1 & a_0 & 0 & \cdots & 0 & b_1 & b_0 & 0 & \cdots & 0 \\ a_2 & a_1 & a_0 & \cdots & 0 & b_2 & b_1 & b_0 & \cdots & 0 \\ \vdots & \vdots & \vdots & & \vdots & \vdots & \vdots & & & \vdots \\ 0 & 0 & 0 & \cdots & a_n & 0 & 0 & 0 & \cdots & b_m \end{pmatrix}$$

with $n + m - s + 1$ rows and $(n - s + 1) + (m - s + 1) = m + n - 2s + 2$ columns has rank $< m + n - 2s + 2$. In particular, in the case $s = 1$, one has:

- The necessary and sufficient condition in order that f and g should have a common divisor of degree > 0 is the vanishing of the determinant of order $m + n$,

$$\mathcal{R}(f, g) = \begin{vmatrix} a_0 & a_1 & a_2 & \cdots & a_n & 0 & \cdots & \cdots & \cdots & 0 \\ 0 & a_0 & a_1 & a_2 & \cdots & & a_n & \cdots & \cdots & 0 \\ \vdots & \vdots & \vdots & & \vdots & \vdots & & \vdots & \vdots & \vdots \\ 0 & . & . & . & \cdots & \cdots & \cdots & \cdots & a_n & 0 \\ b_0 & b_1 & b_2 & \cdots & \cdots & b_m & \cdots & \cdots & \cdots & 0 \\ 0 & b_0 & b_1 & b_2 & \cdots & \cdots & b_m & 0 & \cdots & 0 \\ \vdots & \vdots & \vdots & & \vdots & \vdots & & \vdots & \vdots & \vdots \\ 0 & . & . & . & . & \cdots & \cdots & \cdots & \cdots & b_m \end{vmatrix}. \quad (4.2)$$

This determinant is called the Euler–Sylvester *resultant of the two polynomials f and g* obtained via *elimination of the indeterminate X*.

If $\mathcal{R}(f, g) = 0$ there exist two polynomials f_1 of degree $n - 1$ and g_1 of degree $m - 1$ such that
$$fg_1 + gf_1 = 0.$$

Lemma 4.1.1. *Let A be a factorial ring, and let $\mathcal{R}(f, g)$ be the Euler–Sylvester resultant of two polynomials $f, g \in A[X]$. Then*

(1) $\mathcal{R}(f, g) = 0$ *if and only if f and g have a common divisor of degree > 0;*

(2) $\mathcal{R}(f, g)$ *belongs to the ideal (f, g) generated by f and g.*

Proof. The above discussion shows statement (1).

To show (2), observe that if $p_i = a_{i0}X^t + a_{i1}X^{t-1} + \cdots + a_{it} \in A[X]$, $i = 0, \ldots, t$, are $t + 1$ polynomials of degree $\leq t$ one has

$$\begin{vmatrix} a_{00} & a_{01} & \cdots & a_{0t} \\ a_{10} & a_{11} & \cdots & a_{1t} \\ \vdots & \vdots & & \vdots \\ a_{t0} & a_{t1} & \cdots & a_{tt} \end{vmatrix} = \begin{vmatrix} a_{00} & a_{01} & \cdots & a_{0\,t-1} & p_0 \\ a_{10} & a_{11} & \cdots & a_{1\,t-1} & p_1 \\ \vdots & \vdots & & \vdots & \vdots \\ a_{t0} & a_{t1} & \cdots & a_{t\,t-1} & p_t \end{vmatrix} = \sum_{i=0}^{t} H_i\, p_i$$

with $H_i \in A$.

On the other hand $\mathcal{R}(f, g)$ is the determinant of the coefficients of the $m + n$ polynomials of degree $t \leq m + n - 1$,

$$p_i = X^i f \text{ if } i = 0, \ldots, m - 1; \quad p_i = X^{i-m} g \text{ if } i = m, \ldots, m + n - 1.$$

Thus one has

$$\mathcal{R}(f, g) = H_0(f) + \cdots + H_{m-1}(X^{m-1} f) + H_{m+1}(g) + \cdots + H_{m+n-1}(X^{n-1} g),$$

with $H_i \in A$, $i = 0, \ldots, m + n - 1$, and hence

$$\mathcal{R}(f, g) = Pf + Qg \quad \text{with } P, Q \in A[X]. \tag{4.3}$$

Therefore one has
$$\mathcal{R}(f, g) \in A \cap (f, g),$$
where (f, g) denotes the ideal of $A[X]$ generated by f, g. □

The *resultant ideal* of f, g is the ideal of A generated by $\mathcal{R}(f, g)$. By the preceding lemma one then has

$$(\mathcal{R}(f, g)) \subset A \cap (f, g).$$

Note that if A is a field the following equality holds (cf. Section 4.4):

$$(\mathcal{R}(f, g)) = A \cap (f, g).$$

Indeed, in that case the element $\mathcal{R}(f, g)$ is invertible and so by (4.3) one has $(f, g) = A[X]$.

The determinant $\mathcal{R}(f, g)$ is a sum of products of the type

$$\pm a_{i_1} \ldots a_{i_m} b_{j_1} \ldots b_{j_n};$$

the *weight* of a product of this type is the sum $i_1 + \cdots + i_m + j_1 + \cdots + j_n$. One sees easily that all the summands of $\mathcal{R}(f, g)$ have the same weight.

We will denote the element of $\mathcal{R}(f, g)$ that belongs to row α and column β by (α, β).

If $\alpha \leq m$ one has $(\alpha, \beta) = a_{\beta-\alpha}$ $((\alpha, \beta) = 0$ if $\alpha > \beta)$; if $\alpha > m$ one has instead $(\alpha, \beta) = b_{\beta-\alpha+m}$ $((\alpha, \beta) = 0$ if $\beta - \alpha + m < 0)$. An arbitrary non-zero summand appearing in the development $\sum \pm (r_1, s_1)(r_2, s_2) \ldots (r_{m+n}, s_{m+n})$ of the determinant $\mathcal{R}(f, g)$ is a product of $m + n$ elements (r, s) such that each row and each column contains one of the factors. Therefore, the weight of an arbitrary non-zero summand is

$$(s_1 - r_1 + \cdots + s_m - r_m) + (s_{m+1} - r_{m+1} + m + \cdots + s_{m+n} - r_{m+n} + m),$$

namely

$$\sum_{i=1}^{m+n} s_i - \sum_{j=1}^{m+n} r_j + mn = mn.$$

We conclude that the resultant $\mathcal{R}(f, g)$ is *isobaric of weight mn*.

For further properties of the determinant $\mathcal{R}(f, g)$ we refer the reader, for example, to [62, V, §10] and to [35, 14.1].

4.1.2 The homogeneous case. We add a few observations with regard to the homogeneous case.

(1) Let f, g be two binary forms (i.e., homogeneous polynomials of $K[x_0, x_1]$) of degrees n, m respectively:

$$f(x_0, x_1) = \sum_{i=0}^{n} a_i x_0^{n-i} x_1^i, \quad g(x_0, x_1) = \sum_{j=0}^{m} b_j x_0^{m-j} x_1^j.$$

Assume that f, g have positive degree with respect to x_0 and let us consider f, g as elements of $A[x_0]$, where $A := K[x_1]$. We denote by $\mathcal{R}(f, g, x_0)$ the Euler–Sylvester resultant of $f, g \in A[x_0]$. Then by Lemma 4.1.1 we know that

(a) the polynomials f, g have a common divisor in A of positive degree if and only if $\mathcal{R}(f, g, x_0)$ is the zero polynomial in A;

(b) $\mathcal{R}(f, g, x_0) \in (f, g) \cap A$.

Moreover, one has
$$\mathcal{R}(f, g, x_0) = x_1^{mn} \mathcal{R}(f, g),$$
where $\mathcal{R}(f, g)$ is defined by (4.2).

To see this, write

$$\mathcal{R}(f, g, x_0) := \begin{vmatrix} a_0 & a_1x_1 & a_2x_1^2 & \cdots & a_nx_1^n & 0 & \cdots & \cdots & \cdots & 0 \\ 0 & a_0 & a_1x_1 & a_2x_1^2 & \cdots & & a_nx_1^n & \cdots & \cdots & 0 \\ \vdots & \vdots & \vdots & & \vdots & & \vdots & \vdots & & \vdots \\ 0 & \cdot & \cdot & \cdot & \cdots & \cdots & \cdots & \cdots & & a_nx_1^n \\ b_0 & b_1x_1 & b_2x_1^2 & \cdots & \cdots & b_mx_1^m & \cdots & \cdots & & 0 \\ 0 & b_0 & b_1x_1 & b_2x_1^2 & \cdots & & b_mx_1^m & 0 & \cdots & 0 \\ \vdots & \vdots & \vdots & & \vdots & & \vdots & \vdots & & \vdots \\ 0 & \cdot & \cdot & \cdot & \cdots & \cdots & \cdots & \cdots & & b_mx_1^m \end{vmatrix}.$$

Multiply the second row by x_1, the third by x_1^2, and so on, the mth by x_1^{m-1}, the $(m+2)^{\text{nd}}$ by x_1, the $(m+3)^{\text{rd}}$ by x_1^2, and so on, the $(m+n)^{\text{th}}$ by x_1^{n-1}, to get

$$\Delta := \begin{vmatrix} a_0 & a_1x_1 & a_2x_1^2 & \cdots & a_nx_1^n & 0 & \cdots & \cdots & \cdots & 0 \\ 0 & a_0x_1 & a_1x_1^2 & a_2x_1^3 & \cdots & & a_nx_1^{n+1} & \cdots & \cdots & 0 \\ \vdots & \vdots & \vdots & & \vdots & & \vdots & \vdots & & \vdots \\ 0 & \cdot & \cdot & \cdot & \cdots & \cdots & \cdots & \cdots & & a_nx_1^{n+m-1} \\ b_0 & b_1x_1 & b_2x_1^2 & \cdots & \cdots & b_mx_1^m & \cdots & \cdots & & 0 \\ 0 & b_0x_1 & b_1x_1^2 & b_2x_1^3 & \cdots & & b_mx_1^{m+1} & 0 & \cdots & 0 \\ \vdots & \vdots & \vdots & & \vdots & & \vdots & \vdots & & \vdots \\ 0 & \cdot & \cdot & \cdot & \cdots & \cdots & \cdots & \cdots & & b_mx_1^{m+n-1} \end{vmatrix}.$$

Thus
$$\Delta = \mathcal{R}(f, g, x_0) x_1^{1+2+\cdots+m-1} x_1^{1+2+\cdots+n-1} = \mathcal{R}(f, g, x_0) x_1^{\frac{m(m-1)}{2} + \frac{n(n-1)}{2}}.$$

On the other hand,
$$\Delta = x_1^{1+2+\cdots+(m+n-1)} \mathcal{R}(f, g) = x_1^{\frac{(m+n)(m+n-1)}{2}} \mathcal{R}(f, g).$$

Comparing the two equalities we have the result.

Similar conclusions hold by interchanging x_0 and x_1.

(2) If $f_1, f_2, \ldots, f_h$ are homogeneous polynomials in $K[x_0, \ldots, x_r]$ of degrees $d_1, d_2, \ldots, d_h$ and if $d = \max\{d_1, \ldots, d_h\}$, consider, together with the system of equations
$$f_1 = f_2 = \cdots = f_h = 0, \qquad (4.4)$$

Chapter 4. Rudiments of Elimination Theory

also the system that one obtains by substituting for every equation $f_j = 0$, $j = 1, \ldots, h$, the $r + 1$ equations (all homogeneous of degree d):

$$x_i^{d-d_j} f_j = 0, \quad i = 0, \ldots, r. \tag{4.5}$$

It is clear that every solution of the system (4.4) is also a solution of (4.5) and one sees immediately that conversely every proper solution of the system (4.5) is also a solution of the system (4.4).

With these facts as premise, let $\varphi_1, \varphi_2, \ldots, \varphi_h \in K[x_0, \ldots, x_r]$ be forms of the same degree d, and suppose that one wishes to eliminate the variable x_0 from the system of equations

$$\varphi_1 = \varphi_2 = \cdots = \varphi_h = 0. \tag{4.6}$$

We set

$$x_0 = y_0 \quad \text{and} \quad x_i = y_1 x_i', \quad i = 1, \ldots, r.$$

If

$$\varphi_j = \alpha_{0j} x_0^d + \alpha_{1j}(x_1, \ldots, x_r) x_0^{d-1} + \cdots + \alpha_{dj}(x_1, \ldots, x_r),$$

$\alpha_{sj} \in K[x_1, \ldots, x_r]$ being homogeneous polynomials of degree s, $s = 0, \ldots, d$, one has that for each $j = 1, \ldots, h$:

$$\varphi_j' := \varphi_j(y_0, y_1 x_1', \ldots, y_1 x_r')$$
$$= \alpha_{0j} y_0^d + \alpha_{1j}(x_1', \ldots, x_r') y_1 y_0^{d-1} + \cdots + \alpha_{dj}(x_1', \ldots, x_r') y_1^d,$$

and so the polynomials $\varphi_j' \in K[y_0, y_1, x_1', \ldots, x_r']$ are binary forms belonging to the ring $A[y_0, y_1]$ where $A = K[x_1', \ldots, x_r']$.

We construct the resultant system of these h binary forms by writing the Euler–Sylvester determinant $\mathcal{R}(F, G)$ ($= \mathcal{R}(F, G, y_0)$ in our previous notation) of the two binary forms

$$F(y_0, y_1) = \sum_{j=1}^{h} \lambda_j \varphi_j' \quad \text{and} \quad G(y_0, y_1) = \sum_{j=1}^{h} \mu_j \varphi_j'.$$

It is a bihomogeneous polynomial in the two sets of indeterminates $\lambda = (\lambda_1, \ldots, \lambda_h)$, $\mu = (\mu_1, \ldots, \mu_h)$ with coefficients in $K[x_1', \ldots, x_r']$. On imposing the conditions that all the coefficients of $\mathcal{R}(F, G)$ be zero one obtains a system of homogeneous equations

$$\theta_t(x_1', \ldots, x_r') = 0, \quad t = 1, 2, \ldots. \tag{4.7}$$

If $\bar{x}_1', \ldots, \bar{x}_r'$ is a solution of the system (4.7), there exists a common zero for the binary forms φ_j'; that is, there exist $\bar{y}_0', \bar{y}_1'$ such that

$$\varphi_j(\bar{y}_0', \bar{y}_1' \bar{x}_1', \ldots, \bar{y}_1' \bar{x}_r') = 0, \quad j = 1, \ldots, h.$$

But this means that $(\bar{y}'_0, \bar{y}'_1\bar{x}'_1, \ldots, \bar{y}'_1\bar{x}'_r)$ is a solution of the system (4.6).

Since the polynomials θ_t in (4.7) are homogeneous and $x_i = y_1 x'_i, i = 1, \ldots, r$, the system of equations (4.6) is equivalent to the system

$$\theta_t(x_1, \ldots, x_r) = 0, \quad t = 1, 2, \ldots, \tag{4.8}$$

which therefore is a resultant system of the system (4.6). The equations (4.8) represent the cone V projecting the variety $X = V(\varphi_1, \ldots, \varphi_h)$, that is, $V = V(\ldots, \theta_t, \ldots)$.

4.2 Bézout's theorem for plane curves

As a simple application of the tools introduced in Section 4.1 we prove Bézout's theorem which gives the number of points common to two coplanar algebraic curves.

In the projective plane $\mathbb{P}^2 = \mathbb{P}^2(K)$ over an algebraically closed field K, let C^n, C^m be two algebraic curves of orders n and m and equations $f = 0$ and $g = 0$ respectively and not having common components.

We fix a system of homogeneous projective coordinates x_0, x_1, x_2 choosing as the point $A_0 = [1, 0, 0]$ a point not belonging to $C^n \cup C^m$ and we write the equations $f = 0, g = 0$ ordering them with respect to x_0:

$$f(x_0, x_1, x_2) = a_0 x_0^n + a_1(x_1, x_2) x_0^{n-1} + \cdots + a_n(x_1, x_2) = 0,$$
$$g(x_0, x_1, x_2) = b_0 x_0^m + b_1(x_1, x_2) x_0^{m-1} + \cdots + b_m(x_1, x_2) = 0,$$

where a_i, b_j are homogeneous polynomials of degree equal to the corresponding index and $a_0 b_0 \neq 0$.

We set $A = K[x_1, x_2]$ and consider the polynomials f, g as elements of the ring $A[x_0]$. Elimination of the indeterminate x_0 gives a resultant polynomial $R(x_1, x_2) := \mathcal{R}(f, g, x_0)$ isobaric of weight mn; and since the coefficients a_i, b_j are homogeneous polynomials of degree equal to their respective indices, $R(x_1, x_2) \in K[x_1, x_2]$ is a homogeneous polynomial of degree mn. It is not the zero of A because otherwise the two polynomials f, g would have a common divisor $h(x_0, x_1, x_2)$ of degree ≥ 1 with respect to x_0 in $A[x_0] = K[x_0, x_1, x_2]$ which contradicts the hypothesis that the two curves C^n, C^m have no common components.

If y_1, y_2 are two elements of K such that $R(y_1, y_2) = 0$, the two polynomials in $K[x_0]$

$$f(x_0, y_1, y_2) = a_0 x_0^n + a_1(y_1, y_2) x_0^{n-1} + \cdots + a_n(y_1, y_2),$$
$$g(x_0, y_1, y_2) = b_0 x_0^m + b_1(y_1, y_2) x_0^{m-1} + \cdots + b_m(y_1, y_2)$$

have a common divisor of positive degree (in x_0) and thus at least one common root y_0. Then there exists at least one point $[y_0, y_1, y_2]$ common to the two curves

90   Chapter 4. Rudiments of Elimination Theory

and having y_1 and y_2 as last two coordinates (that is, belonging to the line with equation $y_2 x_1 - y_1 x_2 = 0$), so that $y_2 x_1 - y_1 x_2$ is one of the linear factors of the binary form $R(x_1, x_2)$.

Thus we may conclude that the equation $R(x_1, x_2) = 0$ represents the union of the lines passing through A_0 and containing points common to the two curves. Each of these lines (which are finite in number, since there exists only a finite number of pairs (y_1, y_2) such that $R(y_1, y_2) = 0$) contains only a finite number of common points of the two curves (because otherwise it would be a component common to each of them) and therefore C^n and C^m have a finite number of points in common.

If the point A_0 has been chosen outside of the lines that contain pairs of these points, each linear factor of $R(x_1, x_2)$ furnishes a single common point of the two curves, and so the number of distinct common points of C^n and C^m equals the number of distinct linear factors of $R(x_1, x_2)$ and is therefore $\leq mn$, since $R(x_1, x_2)$ is a homogeneous polynomial of degree mn.

We denote the multiplicity with which a linear factor corresponding to a common point P of the two curves appears in the factorization of $R(x_1, x_2)$ by $m_P(C^n, C^m)$. This non-negative integer $m_P(C^n, C^m)$ is called the *intersection multiplicity* of the two curves in P. In Section 4.3 (see in particular Corollary 4.3.10) we will show that it does not depend on the choice of coordinate system.

If we agree to count each common point of the two curves with multiplicity equal to the intersection multiplicity of the two curves at the given common point, one thereby obtains Bézout's theorem for plane curves (cf. Theorem 4.5.2).

Theorem 4.2.1. *Let C^n and C^m be curves in $\mathbb{P}^2$ of orders n and m respectively. Then*
$$mn = \sum_{P \in C^n \cap C^m} m_P(C^n, C^m).$$

It would be easy to prove that if P is r-fold for one of the two curves and s-fold for the other, the intersection multiplicity at P satisfies $m_P(C^n, C^m) \geq rs$; with equality when the two curves present the *simple case* at P, namely, they do not have any common tangent at P. If instead t is the number of common tangents at P, then
$$m_P(C^n, C^m) \geq rs + t. \tag{4.9}$$

In this regard see [87].

4.3 More on intersection multiplicity

We would like to thank our colleague L. Bădescu for calling our attention to the results discussed in this section and for letting us freely use [3] from which the content of the section is taken. We also refer to [62, Chapter V].

4.3. More on intersection multiplicity

The aim of this section is to prove that the intersection multiplicity $m_P(C^n, C^m)$ defined above depends only on the curves C^n and C^m and $P \in C^n \cap C^m$, and not on the projective system of coordinates chosen.

The algebraic result that will be used is the following.

Theorem 4.3.1. *Let $A = K[Y]$ be the polynomial ring in one indeterminate Y over a field K and let $f, g \in A[X]$ be two monic polynomials in X with coefficients in A and without non-constant common factors. Then:*

$$\dim_K(A[X]/(f, g)) = \dim_K(A/(\mathcal{R}(f, g))),$$

where $\mathcal{R}(f, g)$ is the resultant of f and g.

To prove Theorem 4.3.1 we first need another interpretation of the resultant. We shall use the following simple lemma.

Lemma 4.3.2. *Let $h(X_1, \ldots, X_n) \in \mathbb{Z}[X_1, \ldots, X_n]$ be a polynomial with integral coefficients. If $h(X_1, X_1, X_3, \ldots, X_n) = 0$ (that is, if h becomes zero when we substitute X_1 for X_2 and leave the other X_i fixed, $i \neq 2$), then $X_1 - X_2$ divides h in $\mathbb{Z}[X_1, \ldots, X_n]$.*

Proof. It is an easy consequence of Ruffini's theorem and we left it to the reader. □

Let now be $v_0, t_1, \ldots, t_n, w_0, u_1, \ldots, u_m$ be independent variables over $\mathbb{Z}$ and consider the polynomials in $\mathbb{Z}[v_0, t_1, \ldots, t_n, w_0, u_1, \ldots, u_m][X]$:

$$f_v = v_0(X - t_1) \ldots (X - t_n) = v_0 X^n + v_1 X^{n-1} + \cdots + v_n,$$

$$g_w = w_0(X - u_1) \ldots (X - u_m) = w_0 X^m + w_1 X^{m-1} + \cdots + w_m.$$

Thus

$$v_i = (-1)^i v_0 s_i(t_1, \ldots, t_n) \quad \text{and} \quad w_j = (-1)^j w_0 s_j(u_1, \ldots, u_m),$$

where $s_i(t_1, \ldots, t_n)$ and $s_j(u_1, \ldots, u_m)$ are the i-th and the j-th elementary symmetric polynomials, $i = 1, \ldots, n, j = 1, \ldots, m$. Then it is easy to prove that

$$v_0, v_1, \ldots, v_n, w_0, w_1, \ldots, w_m$$

are still algebraically independent over $\mathbb{Z}$.

Proposition 4.3.3. *Under the above notation one has*

$$\mathcal{R}(f_v, g_w) = v_0^m w_0^n \prod_{i=1}^{n} \prod_{j=1}^{m} (t_i - u_j).$$

92   Chapter 4. Rudiments of Elimination Theory

Proof. Denote by Θ the right-hand side of the equality in the statement of the proposition, and set $\mathbb{Z}[v, w] := \mathbb{Z}[v_0, v_1, \ldots, v_n, w_0, w_1, \ldots, w_m]$. Since $\mathcal{R}(f_v, g_w) =: R(v_0, v_1, \ldots, v_n, w_0, w_1, \ldots, w_m) \in \mathbb{Z}[v, w]$ is homogeneous of degree m in the variables $v_0, \ldots, v_n$ and homogeneous of degree n in $w_0, w_1, \ldots, w_m$, we get

$$\mathcal{R}(f_v, g_w) = v_0^m w_0^n h(t_1, \ldots, t_n, u_1, \ldots, u_m) = R(v_0, t_1, \ldots, t_n, w_0, u_1, \ldots, u_m),$$

with
$$h(t_1, \ldots, t_n, u_1, \ldots, u_m) \in \mathbb{Z}[t_1, \ldots, t_n, u_1, \ldots, u_m].$$

By the discussion made in Section 4.1, we see that the resultant vanishes when we substitute t_i for u_j, $i = 1, \ldots, n$, $j = 1, \ldots, m$. Therefore by Lemma 4.3.2 the element $t_i - u_j$ (which is a prime element of $\mathbb{Z}[v_0, t_1, \ldots, t_n, w_0, u_1, \ldots, u_m]$) divides the polynomial $R(v_0, v_1, \ldots, v_n, w_0, w_1, \ldots, w_m)$. Since for different pairs (i, j) and (i', j'), $t_i - u_j$ and $t_{i'} - u_{j'}$ are coprime, it follows that Θ divides $R(v_0, t, w_0, u) := R(v_0, t_1, \ldots, t_n, w_0, u_1, \ldots, u_m)$.

From $\Theta := v_0^m w_0^n \prod_{i=1}^{n} \prod_{j=1}^{m} (t_i - u_j)$ and the equality

$$\prod_{i=1}^{n} g_w(t_i) = w_0^n \prod_{i=1}^{n} \prod_{j=1}^{m} (t_i - u_j),$$

we get

$$\Theta = v_0^m \prod_{i=1}^{n} g_w(t_i) = v_0^m \prod_{i=1}^{n} (w_0 t_i^m + w_1 t_i^{m-1} + \cdots + w_m). \quad (4.10)$$

Similarly,

$$\Theta = (-1)^{nm} w_0^n \prod_{j=1}^{m} f_v(u_j) = (-1)^{nm} w_0^n \prod_{j=1}^{m} (v_0 u_j^n + v_1 u_j^{n-1} + \cdots + v_n). \quad (4.11)$$

From (4.10) we see that Θ is homogeneous of degree n in $w_0, \ldots, w_m$, and from (4.11) we see that Θ is homogeneous of degree m in $v_0, \ldots, v_n$. Since the polynomial $R(v_0, t, w_0, u)$ has exactly the same homogeneity properties, and is divisible by Θ, it follows that $R(v_0, t, w_0, u) = k\Theta$, with $k \in \mathbb{Z}$. Since both $R(v_0, t, w_0, u)$ and Θ have a monomial $v_0^m w_m^n$ occurring in them with coefficient 1, it follows that $k = 1$. The proposition is proved. □

Corollary 4.3.4. *Let $f, g \in A[X]$ be two polynomials with coefficients in a factorial ring A. Assume that K is a field containing A such that these polynomials have all their roots in K, i.e.,*

$$f = a_0 X^n + a_1 X^{n-1} + \cdots + a_n = a_0(X - \lambda_1) \ldots (X - \lambda_n),$$

$$g = b_0 X^m + b_1 X^{m-1} + \cdots + b_m = b_0 (X - \mu_1) \ldots (X - \mu_m),$$

with $\lambda_i, \mu_j \in K$, $i = 1, \ldots, n$, $j = 1, \ldots, m$. Then

$$\mathcal{R}(f, g) = a_0^m b_0^n \prod_{i=1}^n \prod_{j=1}^m (\lambda_i - \mu_j).$$

Proof. We use the notation introduced above. From the universal property of polynomial rings it follows that there is a unique homomorphism of rings

$$\varphi : \mathbb{Z}[v_0, t_1, \ldots, t_n, w_0, u_1, \ldots, u_m] \to K$$

such that $\varphi(v_0) = a_0$, $\varphi(w_0) = b_0$, $\varphi(t_i) = \lambda_i$, $i = 1, \ldots, n$, and $\varphi(u_j) = \mu_j$, $j = 1, \ldots, m$. It also follows that $\varphi(v_i) = a_i$, $i = 1, \ldots, n$, and $\varphi(w_j) = b_j$, $j = 1, \ldots, m$. The homomorphism φ extends uniquely to a homomorphism of rings

$$\bar\varphi : \mathbb{Z}[v_0, t_1, \ldots, t_n, w_0, u_1, \ldots, u_m][X] \to K[X]$$

such that the restriction of $\bar\varphi$ to $\mathbb{Z}[v_0, t_1, \ldots, t_n, w_0, u_1, \ldots, u_m]$ coincides with φ and $\bar\varphi(X) = X$. Then $\bar\varphi(f_v) = f$ and $\bar\varphi(g_w) = g$, whence $\bar\varphi(\mathcal{R}(f_v, g_w)) = \mathcal{R}(f, g)$. Now the conclusion follows from Proposition 4.3.3. □

Proposition 4.3.5. *Let $A = K[Y]$ be the polynomial ring with coefficients in a field K in the indeterminate Y and let M be a free A-module of rank $n \geq 1$. Let $\varphi : M \to M$ be an injective homomorphism of A-modules. Then*

$$\dim_K(M/\varphi(M)) = \dim_K(A/\det(\varphi)).$$

Proof. Let $e_1, \ldots, e_n$ be a basis of M, and let $\Omega = (\alpha_{ij})_{i,j=1,\ldots,n}$ be the matrix associated to φ with respect to this basis. Since A is a principal ideal domain (a domain in which every ideal can be generated by one element), there are two invertible $n \times n$ matrices Ω_1 and Ω_2 with coefficients in A such that

$$\Omega_1 \Omega \, \Omega_2 = \begin{pmatrix} \delta_1 & 0 & \cdots & 0 \\ 0 & \delta_2 & \cdots & 0 \\ \vdots & \vdots & \ddots & \vdots \\ 0 & 0 & \cdots & \delta_n \end{pmatrix},$$

with $\delta_1, \ldots, \delta_n \in A \setminus \{0\}$ and δ_i divides δ_{i+1}, $i = 1, \ldots, n-1$ (see [62, XV, §2]). If $\psi : M \to M$ is the homomorphism associated to the matrix $\Omega_1 \Omega \, \Omega_2$ with respect to the basis $e_1, \ldots, e_n$, we have

$$\dim_K(M/\psi(M)) = \dim_K(M/\varphi(M)),$$

because the matrices Ω_1 and Ω_2 are invertible. Further, $\det(\psi) = \det(\Omega_1 \Omega \Omega_2) = \lambda \det(\Omega)$, with $\lambda = \det(\Omega_1)\det(\Omega_2) \in K \setminus \{0\}$. Therefore

$$\dim_K(A/\det(\varphi)) = \dim_K(A/\det(\psi)).$$

Thus we may replace φ for ψ, i.e., there is no loss of generality if we assume that the matrix of φ with respect to the basis $e_1, \ldots, e_n$ has the diagonal form

$$\begin{pmatrix} \delta_1 & 0 & \cdots & 0 \\ 0 & \delta_2 & \cdots & 0 \\ \vdots & \vdots & \ddots & \vdots \\ 0 & 0 & \cdots & \delta_n \end{pmatrix},$$

with $\delta_1, \ldots, \delta_n \in A \setminus \{0\}$ and δ_i divides δ_{i+1}, $i = 1, \ldots, n-1$. It follows that $M/\varphi(M)$ is generated by the classes $x_i := e_i \mod \varphi(M)$, $i = 1, \ldots, n$. Moreover, it follows easily that $M/\varphi(M) = Ax_1 \oplus \cdots \oplus Ax_n$ and $Ax_i \cong A/\delta_i A$, $i = 1, \ldots, n$, whence

$$\dim_K(M/\varphi(M)) = \sum_{i=1}^n \dim_K(A/\delta_i A) = \sum_{i=1}^n \deg(\delta_i). \tag{4.12}$$

On the other hand,

$$\det(\varphi) = \begin{vmatrix} \delta_1 & 0 & \cdots & 0 \\ 0 & \delta_2 & \cdots & 0 \\ \vdots & \vdots & \ddots & \vdots \\ 0 & 0 & \cdots & \delta_n \end{vmatrix} = \delta_1 \delta_2 \ldots \delta_n,$$

and therefore

$$\dim_K(A/(\det(\varphi)) = \dim_K(A/(\delta_1 \delta_2 \ldots \delta_n))$$
$$= \deg(\delta_1 \delta_2 \ldots \delta_n) = \sum_{i=1}^n \deg(\delta_i). \tag{4.13}$$

Comparing (4.12) and (4.13) we get the result. □

Proof of Theorem 4.3.1. Write

$$f = X^n + a_1 X^{n-1} + \cdots + a_n, \quad a_i = a_i(Y) \in A = K[Y], \quad i = 1, \ldots, n, \tag{4.14}$$

and

$$g = X^m + b_1 X^{m-1} + \cdots + b_m, \quad b_j = b_j(Y) \in A = K[Y], \quad j = 1, \ldots, m.$$

4.3. More on intersection multiplicity

Denote by $\bar{x} \in A[X]/(f)$ the class of X. Since f is monic,

$$\{1, \bar{x}, \bar{x}^2, \ldots, \bar{x}^{n-1}\}$$

is a basis of the A-module $M := A[X]/(f)$. Moreover, (4.14) yields

$$\bar{x}^n + a_1 \bar{x}^{n-1} + \cdots + a_n = 0. \tag{4.15}$$

Let K' be a field which contains $A = K[Y]$ as a subring such that the polynomials f and g have all the roots in K'. Let $t_1, \ldots, t_n \in K'$ be the roots of f and $s_1, \ldots, s_m \in K'$ the roots of g. Then

$$A[X]/(f, g) \cong M/g'M, \quad \text{where } g' = g \mod (f) \in M.$$

We shall first prove the result for $m = 1$. Thus $g = X - s$, with $s = s_1 \in A$, whence $g' = \bar{x} - s$. Then the multiplication (denoted "·") by $g' = \bar{x} - s$ yields in M the relations

$$(\bar{x} - s) \cdot 1 = (-s) \cdot 1 + 1 \cdot \bar{x} + 0 \cdot \bar{x}^2 + \cdots + 0 \cdot \bar{x}^{n-1},$$
$$(\bar{x} - s) \cdot \bar{x} = 0 \cdot 1 + (-s) \cdot \bar{x} + 1 \cdot \bar{x}^2 + \cdots + 0 \cdot \bar{x}^{n-1},$$
$$\vdots$$
$$(\bar{x} - s) \cdot \bar{x}^{n-2} = 0 \cdot 1 + 0 \cdot \bar{x} + \cdots + (-s) \cdot \bar{x}^{n-2} + 1 \cdot \bar{x}^{n-1},$$
$$(\bar{x} - s) \cdot \bar{x}^{n-1} = (-a_n) \cdot 1 + (-a_{n-1})\bar{x} + \cdots + (-a_2) \cdot \bar{x}^{n-2} + (-s - a_1) \cdot \bar{x}^{n-1}.$$

The last equality follows from (4.15). Thus the matrix associated to the injective map of A-modules $\varphi: M \to M$ defined by $\varphi(x) = g' \cdot x$, for each $x \in M$, is the following:

$$\Omega(s) = \begin{pmatrix} -s & 1 & 0 & \cdots & 0 & 0 \\ 0 & -s & 1 & \cdots & 0 & 0 \\ 0 & 0 & -s & \cdots & 0 & 0 \\ \vdots & \vdots & \vdots & \ddots & \vdots & \vdots \\ 0 & 0 & 0 & \cdots & -s & 1 \\ -a_n & -a_{n-1} & -a_{n-2} & \cdots & -a_2 & -(s + a_1) \end{pmatrix}$$

Then an easy calculation yields $\det(\Omega(s)) = (-1)^n f(s)$, whence by Corollary 4.3.4, one has $\det(\varphi) = \det(\Omega(s)) = \mathcal{R}(f, g)$. Then by Proposition 4.3.5 we get the conclusion in the case $m = \deg(g) = 1$.

If $m \geq 2$, set $\bar{M} := K'[X]/(f)K'[X]$. Then $\bar{M}$ is a K'-vector space with the basis the classes of $1, X, \ldots, X^{n-1}$ modulo the extended ideal $(f)K'[X]$, and $M \subseteq \bar{M}$. (In fact, $\bar{M} \cong M \otimes_A K'$, although we do not need this.) Then the morphism φ is the restriction to M of the composition map $\varphi_1 \circ \varphi_2 \circ \cdots \circ \varphi_m$, where the map $\varphi_j : \bar{M} \to \bar{M}$ is defined by $\varphi_j(y) = (\bar{x} - s_j)y$, for each $y \in \bar{M}$. By the case $m = 1$, we know that $\det(\varphi_j) = (-1)^n f(s_j)$, whence

$$\det(\varphi) = \prod_{j=1}^{m} \det(\varphi_j) = (-1)^{nm} \prod_{j=1}^{m} f(s_j)$$

$$= (-1)^{nm} \prod_{i=1}^{n}\prod_{j=1}^{m} (t_i - s_j) = (-1)^{nm} \mathcal{R}(f, g),$$

where the last equality follows from Corollary 4.3.4. Then we conclude Theorem 4.3.1 again by Proposition 4.3.5. □

Remark 4.3.6. Proposition 4.3.5 remains still valid if we replace $A = K[Y]$ by the localization $K[Y]_{(Y-c)}$ defined by

$$K[Y]_{(Y-c)} := \{\tfrac{P(Y)}{Q(Y)} \mid Q(c) \neq 0\},$$

where c is an arbitrary element of K. Indeed, the ring $K[Y]_{(Y-c)}$ is still Euclidean, in fact it is a discrete valuation ring, see Definition 4.3.7 below. Using this remark, from the proof of Theorem 4.3.1 it follows that Theorem 4.3.1 remains valid if we replace $A = K[Y]$ by the localization $K[Y]_{(Y-c)}$.

We now need the concept of discrete valuation ring.

Definition 4.3.7. Let A be a local Noetherian domain which is not a field, and let $\mathfrak{m}$ be the maximal ideal of A. We say that A is a *discrete valuation ring* (DVR for short) if $\mathfrak{m}$ is generated by one element t. Any such element t is called a *uniformising parameter*, or sometimes *regular parameter*, for A.

Example 4.3.8. Let K be a field and let $c \in K$ be an arbitrary element. Let

$$A := \{\tfrac{P}{Q} \text{ with } P, Q \in K[Y] \text{ and } Q(c) \neq 0\}.$$

Then A is the fraction ring of the polynomial ring $K[Y]$ in the indeterminate Y with respect to the multiplicative system $S_c := \{Q \in K[Y] \mid Q(c) \neq 0\}$. Indeed, since the ideal $(Y - c)$ is maximal in $K[Y]$, the ideal $\mathfrak{m} := (Y - c)A$ is maximal in A. Moreover, since $S_c = K[Y] \setminus (Y - c)$, $\mathfrak{m}$ is the unique maximal ideal of A, that is, $(A, \mathfrak{m})$ is a local ring, with maximal ideal $\mathfrak{m}$ generated by $t = Y - c$ (in other words, $Y - c$ is a uniformising parameter for A). It follows that $(A, \mathfrak{m})$ is a discrete valuation ring.

We are now ready to prove that the intersection multiplicity $m_P(C^n, C^m)$ introduced in Section 4.2 is independent of the system of projective coordinates chosen.

4.3. More on intersection multiplicity

Corollary 4.3.9. *Let C^n and C^m be two curves in the projective plane $\mathbb{P}^2$ of degrees n and m and equations $F = 0$, $G = 0$ respectively, and having no common irreducible components. Let $P \in C^n \cap C^m$. Then the intersection multiplicity $m_P(C^n, C^m)$ can be calculated by the formula*

$$m_P(C^n, C^m) = \dim_K(\mathcal{O}_{\mathbb{P}^2, P}/(f, g)),$$

where f and g are local equations of C^n and C^m around P (i.e., if $P \in \mathbb{P}^2_{x_i} = \mathbb{P}^2 \setminus \{x_i = 0\}$ then we can take $f = F/x_i^n$ and $g = G/x_i^m$).

Proof. According to the proof of Theorem 4.2.1 we can choose a system of projective coordinates of $\mathbb{P}^2$ such that the point $P_0 = [1, 0, 0]$ does not belong to $C^n \cup C^m$ and then the equations of C^n and C^m are of the following form:

$$F(x_0, x_1, x_2) = a_0 x_0^n + a_1(x_1, x_2) x_0^{n-1} + \cdots + a_n(x_1, x_2),$$
$$G(x_0, x_1, x_2) = b_0 x_0^m + b_1(x_1, x_2) x_0^{m-1} + \cdots + b_m(x_1, x_2),$$

with $a_0, b_0 \in K \setminus \{0\}$ and $a_i(x_1, x_2)$ and $b_j(x_1, x_2)$ homogeneous polynomials of degrees i and j respectively. Therefore there is no loss of generality in assuming $a_0 = b_0 = 1$. Moreover, we have $P \in \mathbb{P}^2_{x_1} \cup \mathbb{P}^2_{x_2}$ because the point $[1, 0, 0] \notin C^n \cap C^m$. If for instance $P \in \mathbb{P}^2_{x_2}$ then we can take $x_2 = 1$ in the above equations to get

$$f(x_0, x_1) := F(x_0, x_1, 1) = x_0^n + a_1(x_1, 1) x_0^{n-1} + \cdots + a_n(x_1, 1),$$
$$g(x_0, x_1) := G(x_0, x_1, 1) = x_0^m + b_1(x_1, 1) x_0^{m-1} + \cdots + b_m(x_1, 1).$$

Setting $x_0 = X$, $x_1 = Y$, $a_i(Y) = a_i(Y, 1)$ and $b_j(Y) = b_j(Y, 1)$, $i = 1, \ldots, n$ and $j = 1, \ldots, m$, we get

$$f = f(X, Y) = X^n + a_1(Y) X^{n-1} + \cdots + a_n(Y) \in A[X],$$
$$g = g(X, Y) = X^m + b_1(Y) X^{m-1} + \cdots + b_m(Y) \in A[X],$$

where $f, g \in A[X]$ are monic polynomials with coefficients in $A := K[Y]$. Since C^n and C^m have no irreducible components in common, f and g have no common non-constant factors. By Theorem 4.3.1,

$$\dim_K(A[X]/(f, g)) = \dim_K(A/(\mathcal{R}(f, g))),$$

where $\mathcal{R}(f, g)$ is the resultant of f and g. Now, according to Remark 4.3.6 we also have

$$\dim_K(A'[X]/(f, g)) = \dim_K(A'/(\mathcal{R}(f, g))), \qquad (4.16)$$

where c is a root of $\mathcal{R}(f, g)$ and $A' = K[Y]_{(Y-c)} = \{\frac{P(Y)}{Q(Y)} \mid Q(c) \neq 0\}$, which is a discrete valuation ring (see Example 4.3.8). Then clearly the right-hand side coincides with $m_P(C^n, C^m)$ defined in Section 4.2.

Now, to any point $P \in K^2$ it corresponds a maximal ideal $\mathfrak{m}_P$ in $A[X] = K[Y, X]$. Then the local ring $\mathcal{O}_{\mathbb{P}^2, P}$ is by definition $A[X]_{\mathfrak{m}_P} = K[Y, X]_{\mathfrak{m}_P}$. Since by the proof of Theorem 4.2.1, there is a bijective correspondence between the set $C^n \cap C^m$ and the roots of $\mathcal{R}(f, g)$, we infer that the ring $A'[X]/(f, g)$ is itself local. Let $c \in K$ be the unique root of $\mathcal{R}(f, g)$ which corresponds to the given point $P \in C^n \cap C^m$, say $P = (c_0, c)$, cf. the proof of Theorem 4.2.1. Then the rings $A'[X]/(f, g)$ and $\mathcal{O}_{\mathbb{P}^2, P}/(f, g)$ are fractions rings of $K[X, Y]/(f, g)$ with respect to the multiplicative systems $S_1 := \{h_1(Y) \in K[Y] \mid h_1(c) \neq 0\}$ and $S_2 := \{h_2(X, Y) \in K[X, Y] \mid h_2(c_0, c) \neq 0\}$ respectively. Obviously, $S_1 \subseteq S_2$. In particular, we get a canonical homomorphism of K-algebras

$$\varphi \colon A'[X]/(f, g) \to \mathcal{O}_{\mathbb{P}^2, P}/(f, g).$$

Then we claim that φ is in fact an isomorphism and, in particular,

$$\dim_K(A'[X]/(f, g)) = \dim_K(\mathcal{O}_{\mathbb{P}^2, P}/(f, g)). \tag{4.17}$$

To see this, first observe that $A'[X]/(f, g)$ and $\mathcal{O}_{\mathbb{P}^2, P}/(f, g)$ are rings having each just one prime ideal (the maximal ideal). Indeed, since $C^n \cap C^m$ is finite, the K-algebra $K[X, Y]/(f, g)$ has only finitely many prime ideals and all of them are maximal. Since $A'[X]/(f, g)$ is a local ring which is a fraction ring of $K[X, Y]/(f, g)$ it follows that $A'[X]/(f, g)$ has just one prime ideal (namely, the maximal ideal $\mathfrak{m}_{A'[X]/(f,g)}$). From this we deduce that $\varphi^{-1}(\mathfrak{m}_{\mathcal{O}_{\mathbb{P}^2, P}}) = \mathfrak{m}_{A'[X]/(f,g)}$, i.e., φ is a homomorphism of local rings. This latter fact implies that the class $[h_2]$ of every polynomial $h_2 \in S_2$ (that is, such that $h_2(P) \neq 0$) in $A'[X]/(f, g)$ is invertible in $A'[X]/(f, g)$. Finally, since $\mathcal{O}_{\mathbb{P}^2, P}/(f, g)$ is the fraction ring of $A'[X]/(f, g)$ with respect to the multiplicative system $\{[h_2] \mid h_2 \in S_2\}$, we get (4.17).

Thus, by combining (4.16) and (4.17) we get

$$m_P(C^n, C^m) = \dim_K(\mathcal{O}_{\mathbb{P}^2, P}/(f, g)),$$

which is exactly what we wanted. □

Corollary 4.3.10. *The intersection multiplicity $m_P(C^n, C^m)$ depends only on C^n, C^m and $P \in C^n \cap C^m$ and not on the system of projective coordinates chosen.*

Proof. We have to prove that if $\sigma \in \mathrm{PGL}_2(K)$ is a projective linear automorphism of $\mathbb{P}^2$ then $m_P(C^n, C^m) = m_{\sigma(P)}(\sigma(C^n), \sigma(C^m))$. Using Corollary 4.3.9 the conclusion is clear because the rings $\mathcal{O}_{\mathbb{P}^2, P}/(f, g)$ and $\mathcal{O}_{\mathbb{P}^2, \sigma(P)}/(f \circ \sigma, g \circ \sigma)$ are isomorphic as K-algebras (and hence also as K-vector spaces). □

Let us point out that Corollary 4.3.9 is also very efficient for computing explicitly the intersection multiplicities in many cases via the following result. To this end, we propose below some exercises.

Proposition 4.3.11. *Let C^n and C^m be two affine curves in $\mathbb{A}^2$ such that the intersection $C^n \cap C^m$ contains just one point P. Let f and g be the equations of C^n and C^m respectively. Then there is a canonical isomorphism of K-algebras*

$$K[X,Y]/(f,g) \cong \mathcal{O}_{\mathbb{A}^2,P}/(f,g).$$

Proof. We have a canonical homomorphism of K-algebras $\varphi \colon K[X,Y]/(f,g) \to \mathcal{O}_{\mathbb{P}^2,P}/(f,g)$. If $\mathfrak{m}_P$ is the maximal ideal of $K[X,Y]$ corresponding to P, then $\mathcal{O}_{\mathbb{P}^2,P} = S_P^{-1} K[X,Y]$, where $S_P = \{h \in K[X,Y] \mid h(P) \neq 0\} = K[X,Y] \setminus \mathfrak{m}_P$. Since by hypothesis $C^n \cap C^m = \{P\}$, then by the Hilbert Nullstellensatz 2.2.2, $\mathfrak{m}_P/(f,g)$ is the only prime ideal of $K[X,Y]/(f,g)$, and therefore $K[X,Y]/(f,g)$ is a local ring. It follows that $\varphi^{-1}(\mathfrak{m}_{\mathcal{O}_{\mathbb{P}^2,P}}/(f,g)) = \mathfrak{m}_P/(f,g)$ and, in particular, the class of every $h \in S_P$ in $K[X,Y]/(f,g)$ is invertible (cf. the proof of Corollary 4.3.9). This finishes the proof because $\mathcal{O}_{\mathbb{P}^2,P}/(f,g)$ is the fraction ring of $K[X,Y]/(f,g)$ with respect to the multiplicative system $\pi(S_P)$, where $\pi \colon K[X,Y] \to K[X,Y]/(f,g)$ is the canonical homomorphism of K-algebras. $\square$

Exercise 4.3.12. Let C, ℓ_1 and ℓ_2 be the curves in $\mathbb{A}^2$ of equations $x^2 + y^2 = 1$, $y = 1$ and $y = 0$ respectively. Find $C \cap \ell_1$, $C \cap \ell_2$ and the respective intersection multiplicities.

Exercise 4.3.13. Let C be the curve of $\mathbb{A}^2$ of equation $y^2 = x^2(x+1)$. Find the intersection multiplicities at the points of intersection of C with the axis $y = 0$.

Exercise 4.3.14. Find the intersection multiplicity of the curves of equations $y = x^n$ ($n \geq 2$) and $y = 0$ in $\mathbb{A}^2$ at the origin.

Exercise 4.3.15. Find the intersection multiplicity of the curves of equations $y^2 = x^3$ and $y^2 = x^2(x+1)$ in $\mathbb{A}^2$ at the origin.

4.4 Elimination of several variables

Let $\mathcal{J} = (f_1, \ldots, f_h)$ be a homogeneous ideal of $K[x_0, \ldots, x_r]$. We will assume that $\mathcal{J}$ is a radical ideal and that $X = V(\mathcal{J}) \subset \mathbb{P}^r$ is the projective variety associated to $\mathcal{J}$. Let A_i, $i = 0, \ldots, r$, be the $r+1$ fundamental points of a fixed system of reference in $\mathbb{P}^r$.

Let Γ be a conical hypersurface containing X and having the linear space $S_{r-m-1} = J(A_{m+1}, \ldots, A_r)$ as vertex, where $x_{m+1}, \ldots, x_r$ are projective homogeneous coordinates (cf. §5.2.3). If $f(x_0, \ldots, x_m) = 0$ is the equation of Γ we have $f \in \mathcal{J}$ and so $f \in \mathcal{J} \cap K[x_0, \ldots, x_m]$. Conversely, if $f \in \mathcal{J} \cap K[x_0, \ldots, x_m]$, the hypersurface with equation $f = 0$ contains $X = V(\mathcal{J})$ and is a cone with vertex the given S_{r-m-1}.

For each $m \leq r$, consider the contracted ideal

$$\mathfrak{J}_m := K[x_0, \ldots, x_m] \cap \mathfrak{J}. \tag{4.18}$$

One notes that $\mathfrak{J}_m$ is a homogeneous ideal of $K[x_0, \ldots, x_m]$. Indeed, if α_j are homogeneous polynomials of $K[x_0, \ldots, x_m]$ and $f = \sum_j \alpha_j$, then, in view of the homogeneity of $\mathfrak{J}$, it follows that for each index j,

$$f \in \mathfrak{J}_m \iff f \in \mathfrak{J} \iff \alpha_j \in \mathfrak{J} \iff \alpha_j \in \mathfrak{J}_m.$$

Moreover, obviously one has $\mathfrak{J}_r = \mathfrak{J}$ and

$$\mathfrak{J}_m = \mathfrak{J}_{m+1} \cap K[x_0, \ldots, x_m], \quad m \leq r - 1.$$

In §3.4.5 the cone projecting X from $S_{r-m-1} = \mathrm{J}(A_{m+1}, \ldots, A_r)$ was defined to be the algebraic set $V = V(\mathfrak{J}_m K[x_0, \ldots, x_r])$, and the projection X' of X from S_{r-m-1} onto the space $S_m = \mathrm{J}(A_0, \ldots, A_m)$ was defined as the intersection $X' = V \cap S_m$, that is, $X' = V(\mathfrak{J}_m) \subset S_m$.

Remark 4.4.1. Note that the space S_{r-m} that joins $S_{r-m-1} = \mathrm{J}(A_{m+1}, \ldots, A_r)$ with a point $P = [a_0, \ldots, a_m] \in X' \subset S_m$ contains the entire line joining P to a point $x \in X$ of which P is the projection, and so intersects X at x. This point x has coordinates $(a_0, \ldots, a_m, a_{m+1}, \ldots, a_r)$, whence the polynomials $f_i(a_0, \ldots, a_m, x_{m+1}, \ldots, x_r)$ in $K[x_{m+1}, \ldots, x_r]$ have at least $[a_{m+1}, \ldots, a_r] \in S_{r-m-1}$ as common zero.

The projecting cone V is a locus of zeros of polynomials in which the variables $x_{m+1}, \ldots, x_r$ do not appear. We shall say that $\mathfrak{J}_m$ is the *resultant ideal of the ideal* $\mathfrak{J} = (f_1, \ldots, f_h)$ with respect to the indeterminates $x_{m+1}, \ldots, x_r$, or also that $\mathfrak{J}_m$ is the *resultant of the elimination* of the variables $x_{m+1}, \ldots, x_r$ from the equations $f_i = 0, i = 1, \ldots, h$. The geometric interpretation of this procedure may be stated as follows:

- To eliminate the variables $x_{m+1}, \ldots, x_r$ from the equations $f_1 = \cdots = f_h = 0$ means to find the equations of the cone that projects $X = V(\mathfrak{J}), \mathfrak{J} = (f_1, \ldots, f_h)$, from $S_{r-m-1} = \mathrm{J}(A_{m+1}, \ldots, A_r)$ on $S_m = \mathrm{J}(A_0, \ldots, A_m)$.

Note that different orderings of the variables lead to different resultant ideals.

Since we have $\mathfrak{J}_m = \mathfrak{J}_{m+1} \cap K[x_0, \ldots, x_m]$, elimination of the indeterminates $x_{m+1}, \ldots, x_r$ can be realized by way of successive eliminations of a single coordinate; it is then sufficient to have available a procedure for the elimination of a single variable.

In general, the most modern and advantageous approach to the theory of elimination is that based on algorithms of computational type that make essential use of the notion of "Groebner basis" of a given ideal. For this we refer the reader to

[28, Chapter 3], to [35, Chapters 14, 15] and to [76] for a systematic exposition of the theory. Here we limit ourselves to explaining a procedure, suggested by Kronecker, for the elimination of a single indeterminate.

With the notation introduced in the course of this section, we eliminate, as an example, the variable x_r. If the original ideal $\mathcal{J} = (f_1, f_2)$ is generated by only two polynomials f_1, f_2, the resultant of the elimination of x_r is immediately obtained by writing the Euler–Sylvester determinant, after having ordered the two polynomials f_1, f_2 with respect to x_r and considering f_1, f_2 as elements of $A[x_r]$, where $A = K[x_0, \ldots, x_{r-1}]$ (cf. Section 4.1).

If the number of polynomials generating the ideal $\mathcal{J}$ is greater than two, one can make use of a trick, suggested by Kronecker, which reduces the question to the case of two equations.

We work over an affine chart, for example $U_0 = \{[x_0, \ldots, x_r] \mid x_0 \neq 0\}$. We set

$$\varphi_i(x_1, \ldots, x_r) = f_i(1, x_1, \ldots, x_r), \quad i = 1, \ldots, h,$$

and consider the system of equations

$$\varphi_1 = \varphi_2 = \cdots = \varphi_h = 0. \tag{4.19}$$

If d is the highest degree of the polynomials φ_i and if, for example, φ_1 has degree d, we can replace the polynomials φ_i with the polynomials $\varphi_i + \varphi_1$, all of degree d. Hence without loss of generality we may suppose that the polynomials φ_i are all of the same degree, $i = 1, \ldots, h$.

Every solution of the system (4.19) is a solution of the two equations

$$f := \lambda_1 \varphi_1 + \cdots + \lambda_h \varphi_h = 0 \quad \text{and} \quad g := \mu_1 \varphi_1 + \cdots + \mu_h \varphi_h = 0, \tag{4.20}$$

for any choice of coefficients $\lambda = (\lambda_1, \ldots, \lambda_h)$ and $\mu = (\mu_1, \ldots, \mu_h)$. Conversely, if $(y_1, \ldots, y_r)$ is a solution of the two equations (4.20) for every choice of the parameters λ and μ, then $(y_1, \ldots, y_r)$ is a zero of the ideal $(\varphi_1, \ldots, \varphi_h)$, which we will again denote by $\mathcal{J}$.

In order to find a system of equations for the cone $V = V(\mathcal{J} \cap K[x_1, \ldots, x_{r-1}])$ that projects $X = V(\mathcal{J})$ from the point $A_r = [0, \ldots, 0, 1] \notin U_0$, or equivalently, in order to find the points $[y_1, \ldots, y_{r-1}, 0]$ of the hyperplane of equation $x_r = 0$ for which there exists a solution of the system

$$\varphi_1(y_1, \ldots, y_{r-1}, x_r) = \cdots = \varphi_h(y_1, \ldots, y_{r-1}, x_r) = 0,$$

it then suffices to require that the resultant $\mathcal{R}(f, g)$ of the two polynomials given in equation (4.20) (considered as polynomials in the variable x_r with coefficients in $K[x_1, \ldots, x_{r-1}, \lambda, \mu]$) be zero for any choice of the parameters λ and μ.

The resultant $\mathcal{R}(f, g)$ is a bihomogeneous polynomial in the two series of variables λ, μ with coefficients in $K[x_1, \ldots, x_{r-1}]$. On requiring that the coefficients vanish we obtain a system of equations

$$\theta_j(x_1,\ldots,x_{r-1}) = 0, \quad j = 1, 2, \ldots,$$

for the cone V that we will also call the *resultant system* of the system (4.19). Bearing in mind that the ideal $\mathfrak{J}$ is a radical ideal, and since $\theta_j = 0$ on $V = V(\mathfrak{J} \cap K[x_1,\ldots,x_{r-1}])$, one has

$$\theta_j \in \sqrt{\mathfrak{J} \cap K[x_1,\ldots,x_{r-1}]} \subset \sqrt{\mathfrak{J}} = \mathfrak{J}$$

for every index j. In conclusion

$$\theta_j(x_1,\ldots,x_{r-1}) \in \mathfrak{J} \cap K[x_1,\ldots,x_{r-1}], \quad j = 1, 2, \ldots.$$

We observe explicitly that the polynomials θ_j need not be a minimal system of generators of the ideal $\mathfrak{J} \cap K[x_1,\ldots,x_{r-1}]$. On the other hand, the computational methods for elimination cited above do permit one to determine a minimal system of generators for the ideal $\mathfrak{J} \cap K[x_1,\ldots,x_{r-1}]$.

4.5 Bézout's theorem

One of the most important facts about the order of a given variety is Bézout's theorem, the general form of which we wish to discuss briefly in this section. We pattern our discussion on the exposition given in [48, Lecture 18]. In Section 4.2, the theorem was proved in the case of plane curves by using the concept of the resultant of two polynomials.

For our present purposes we must recall a few definitions. Suppose that X and Y are two algebraic sets in $\mathbb{P}^n$ and that $\dim(X) + \dim(Y) \geq n$. By Theorem 3.3.8 it follows that we then have $X \cap Y \neq \emptyset$; let p be a point of $X \cap Y$. We say that X and Y *intersect transversally at* p if they are non-singular at p with tangent spaces $T_p(X)$ and $T_p(Y)$ such that the join $\mathrm{J}(T_p(X), T_p(Y))$ is all of $\mathbb{P}^n$. If Z is an irreducible component of the intersection $X \cap Y$, we will say that X and Y *intersect transversally along* Z if X and Y intersect transversally at a generic point $p \in Z$. If this happens for every irreducible component of $X \cap Y$, we will say that X and Y *intersect transversally*.

We say that X and Y *intersect properly* in $\mathbb{P}^n$ if every irreducible component Z of $X \cap Y$ has dimension

$$\dim(Z) = \dim(X) + \dim(Y) - n.$$

Hence in particular $X \cap Y$ has "the expected dimension" (cf. Theorem 3.3.8).

The proof of the two forms of Bézout's theorem that follow will require some tools which are outside the scope of the present book, and so we refer the reader, for the general case, to the proof given in [48, Lecture 18].

Theorem 4.5.1 (Weak form). *Let X and Y be algebraic sets in $\mathbb{P}^n$ of pure dimensions k and k' respectively. Suppose that $k + k' \geq n$ and that X and Y intersect transversally. Then*
$$\deg(X \cap Y) = \deg(X) \deg(Y).$$
In particular, if X and Y intersect properly and $k + k' = n$, the equality assures that the intersection $X \cap Y$ consists of $\deg(X) \deg(Y)$ points.

To each pair of varieties $X, Y \subset \mathbb{P}^n$ that intersect properly, and to each irreducible variety $Z \subset \mathbb{P}^n$ of dimension $\dim(X) + \dim(Y) - n$, we can associate a non-negative integer $m_Z(X, Y)$, called the *intersection multiplicity* of X and Y along Z, which satisfies the following properties:

i) $m_Z(X, Y) \geq 1$ if $Z \subset X \cap Y$ (and $m_Z(X, Y) = 0$ if $Z \not\subset X \cap Y$);

ii) $m_Z(X, Y) = 1$ if and only if X and Y intersect transversally at a generic point $p \in Z$, which means that X and Y intersect transversally along Z;

iii) $m_Z(X, Y)$ is additive, that is,
$$m_Z(X \cup X', Y) = m_Z(X, Y) + m_Z(X', Y)$$
for each X and X' such that all three intersection multiplicities are defined and X and X' do not have common components.

Theorem 4.5.2 (Strong form). *Let X and Y be algebraic sets in $\mathbb{P}^n$ both of pure dimension which intersect properly. Then*
$$\deg(X) \deg(Y) = \sum_Z m_Z(X, Y) \deg(Z),$$
as Z varies over all the irreducible components of $X \cap Y$.

If X and Y intersect properly and $\dim(X) + \dim(Y) = n$ (and therefore $\dim(X \cap Y) = 0$) and p is a point of $X \cap Y$ then the intersection multiplicity $m_p(X, Y)$ can be described in a simple way as
$$m_p(X, Y) = \dim_K \left(\mathcal{O}_{\mathbb{P}^n, p} / (I(X) + I(Y)) \right),$$
where $\mathcal{O}_{\mathbb{P}^n, p}$ is the local ring of $\mathbb{P}^n$ at p, $I(X)$ and $I(Y)$ are the ideals that define X and Y in $\mathbb{P}^n$ and the dimension is that of the quotient ring considered as a vector space over K. (See Corollary 4.3.9 where the above equality is proved in the case of plane curves. See also Theorem 4.2.1 which is the special case for two plane curves of Theorem 4.5.2.)

In particular this is the case for two plane curves X, Y. The relation given above thus describes the intersection multiplicity of two plane curves X and Y in a point p which we have already met in Section 4.2.

See Section 5.7 for exercises regarding the calculation of the intersection multiplicity of plane curves and also Section 5.2 for a discussion of the useful case of the intersection multiplicity $m_p(X, Y)$ of a hypersurface X with a line Y at a point p.

Exercise 4.5.3. We use the same symbol to denote both a plane algebraic curve and its equation. Let f, g be two curves passing through a point P, φ a curve not passing through P, and ψ an arbitrary curve. One has

$$m_P(f, g) = m_P(\varphi f, g + \psi f).$$

A deeper study of intersection multiplicities is beyond the scope of the present book. However, making use of only those properties of intersection multiplicities already discussed, we obtain various consequences of Theorem 4.5.2.

Corollary 4.5.4. *Let X and Y be algebraic sets in $\mathbb{P}^n$ both of pure dimension and which intersect properly. Then*

$$\deg(X \cap Y) \leq \deg(X) \deg(Y).$$

Corollary 4.5.5. *Let X and Y be algebraic sets in $\mathbb{P}^n$ both of pure dimension and which intersect properly. Suppose that*

$$\deg(X \cap Y) = \deg(X) \deg(Y).$$

Then X and Y are both non-singular at the generic point of every irreducible component $X \cap Y$. In particular, if X and Y have complementary dimensions they are then necessarily non-singular at all the points of $X \cap Y$.

Proposition 4.5.6. *If $X \subset \mathbb{P}^n$ is an irreducible algebraic set of dimension k (not contained in any hyperplane of $\mathbb{P}^n$) then $\deg(X) \geq n - k + 1$.*

Proof. (Sketch) We consider a generic linear space $S_{n-k+1} \subset \mathbb{P}^n$. Then the intersection $X \cap S_{n-k+1}$ is an irreducible curve $C \subset \mathbb{P}^{n-k+1}$, not contained in any hyperplane of $\mathbb{P}^{n-k+1}$: for a complete proof of this fact, apparently almost obvious but actually non-trivial, we refer the reader to [48, (18.10)].

It follows that $\deg(X) = \deg(C)$. It is then sufficient to observe that for an irreducible algebraic curve C, belonging to an arbitrary projective space $\mathbb{P}^r$ but not contained in any hyperplane of $\mathbb{P}^r$, one has

$$\deg(C) \geq r.$$

Indeed, we consider r generic (and so independent) points $x_1, \ldots, x_r$ of C and the hyperplane H containing them. We have $\{x_1, \ldots, x_r\} \subset H \cap C$ and so $\deg(C) = \deg(H \cap C) \geq r$. □

We say that a variety $X_k^d \subset \mathbb{P}^n$ is of *minimal degree* if $d = n - k + 1$. Later we shall return to the detailed study of certain interesting classes of varieties of minimal degree: see Sections 7.4, 10.3. The study and classification of varieties of minimal degree constitutes an important and interesting classical problem; we shall return to this subject, in the case of surfaces, in Section 10.3 (see, for example, [82], [84], [8], [92], and also [48, (19.9)]).

Chapter 5
Hypersurfaces in Projective Space

This chapter is devoted to the study of a very important class of projective varieties, namely that of hypersurfaces, the subvarieties of a projective space $\mathbb{P}^n$, of maximal dimension $n-1$, which are defined as the locus of zeros of a homogeneous polynomial of a given degree d, the order of the hypersurface. By reconsidering the notion of multiplicity of a point of a hypersurface given in Section 3.4, we study in Section 5.2 the conditions for an arbitrary point of the space $\mathbb{P}^n$ to be a point of an assigned multiplicity s for a hypersurface X.

In Section 5.3 we consider algebraic envelopes, that is, the hypersurfaces of the dual projective space $\mathbb{P}^{n*}$; by duality one obtains also for algebraic envelopes the same properties previously studied for algebraic hypersurfaces. An important example of an algebraic envelope is constituted by the tangent hyperplanes to a given hypersurface (cf. Proposition 5.3.1).

In Section 5.4 we introduce the notion of polarity with respect to a hypersurface and we study its fundamental properties; polarity was considered and carefully discussed in [52, Vol. 2, Chapter XIII] (for the special cases of conics and quadrics see also [10, Chapters 6, 7]). Section 5.6 contains some useful complementary topics regarding polars; in particular, the notion of Hessian hypersurfaces and that of the class of a hypersurface.

Section 5.5 is dedicated to the simplest hypersurfaces in the space $\mathbb{P}^n$; namely the quadrics, defined by a homogeneous form of degree $d = 2$. They enjoy important geometric properties, the study of which is sketched here in the general case: we refer the reader to loc. cit. where the case of quadrics in $\mathbb{P}^3$ was extensively developed.

In Section 5.7 we consider the hypersurfaces of $\mathbb{P}^2$, that is, algebraic plane curves, and we study some of their properties via a series of exercises, giving particular prominence to the remarkable case of cubics.

Section 5.8 contains complements and exercises which illustrate some noteworthy properties of surfaces in $\mathbb{P}^3$.

5.1 Generalities on hypersurfaces

Let $K[T_0, \ldots, T_n] = \bigoplus_{d \geq 0} R_d$ be the graded ring of polynomials in $n+1$ indeterminates with coefficients in K. If $f \in R_r$ is a homogeneous polynomial of degree r, the locus $X = V(f)$ of the points $[x_0, \ldots, x_n] \in \mathbb{P}^n := \mathbb{P}^n(K)$ such that

$$f(x_0, \ldots, x_n) = 0, \tag{5.1}$$

that is, the locus of the zeros of f, is an *algebraic hypersurface* of *order* r.

The algebraicity and the order of a hypersurface are clearly projective properties, that is, they are invariant under projectivities.

If f is irreducible we say that X is *irreducible*; if however $f = f_1^{\mu_1} \ldots f_h^{\mu_h}$, with f_i distinct irreducible forms, $i = 1, \ldots, h$, we say that X is *reducible* (or *split*) and that the hypersurfaces $X_i := V(f_i)$ are its *irreducible components*. In that case X is composed of the hypersurfaces X_i counted with their respective multiplicities μ_i. We will write $X = \sum_{i=1}^{h} \mu_i X_i$. If the multiplicities μ_i are all equal to 1, we will say that X is *reduced*.

In particular, when $n = 1$, X is a finite set of points; if each of these points is counted according to its multiplicity then X consists of exactly $r = \deg(X)$ points.

Note that, since K is algebraically closed, the only irreducible hypersurfaces in $\mathbb{P}^1$ are those of first order, namely single points. Unless otherwise specified, all hypersurfaces we consider are supposed to be irreducible and reduced.

5.1.1 Sections of a hypersurface by linear spaces. Let $S_k = J(P_0, \ldots, P_k)$ be the linear subspace defined by the $k + 1$ independent points $P_0, \ldots, P_k$ and let $\lambda_0, \ldots, \lambda_k$ be the internal projective coordinates for this S_k with respect to the reference system $\{P_0, \ldots, P_k; \sum_{i=0}^{k} P_i\}$. Each point P of S_k has coordinates $T_0, \ldots, T_n$ that are linear forms in $\lambda_0, \ldots, \lambda_k$ and so the condition for P to belong to the hypersurface $X = V(f) \subset \mathbb{P}^n$ may be translated into the equation

$$g(\lambda_0, \lambda_1, \ldots, \lambda_k) = 0, \tag{5.2}$$

where $g \in K[\lambda_0, \ldots, \lambda_k]$ is a homogeneous polynomial of degree equal to that of f.

Equation (5.2) represents a hypersurface X' of S_k. The only exception is when S_k is contained in X, in which case (5.2) vanishes. Thus, the following fact holds:

- Every hypersurface X of order $r = \deg(X)$ in $\mathbb{P}^n$ is cut by a subspace S_k not contained in it in a hypersurface of S_k of the same order r. In particular, a line not situated on a hypersurface of order r has exactly r points in common with the hypersurface (provided that each point is counted with the correct multiplicity). If X is reduced, a generic line of $\mathbb{P}^n$ has exactly r distinct points in common with X (and thus one rediscovers, in the case of hypersurfaces, the statement of Proposition 3.4.8).

From this it follows that if $r + 1$ points of a line ℓ belong to a hypersurface X of order r, then the line ℓ lies on X.

We note that if X is reducible, so too are its sections by linear spaces S_k. But the converse does not hold in general. Indeed, it can happen that an irreducible hypersurface X intersects an S_k with the resulting hypersurface in S_k split. This always happens when $k = 1$ (and $r \geq 2$), inasmuch as a binary form (that is, a homogeneous polynomial in two indeterminates) of degree r is the product of r linear factors and hence a hypersurface of order r of a line S_1 consists of r points (not necessarily all distinct).

5.2 Multiple points of a hypersurface

In Section 3.4 we have already introduced the notion of the multiplicity of a point of a hypersurface X. In particular we have seen that if $X \subset \mathbb{A}^n$ is an algebraic hypersurface of an affine space in which $y_1, \ldots, y_n$ are non-homogeneous coordinates, the condition for the coordinate origin $x = (0, \ldots, 0)$ to be an s-fold point of X is that in the equation $f(y_1, \ldots, y_n) = 0$ of X there do not appear terms of degree $< s$.

Now we wish to indicate the condition for an arbitrary point $x = [a_0, \ldots, a_n]$ of $\mathbb{P}^n$ to be s-fold for the projective hypersurface that is the locus of the zeros of the homogeneous polynomial $f(x_0, \ldots, x_n)$. To this end it is necessary to consider the intersection multiplicity in x of X with a generic line passing through x (cf. Section 4.5). Indeed, if a hypersurface X is cut by *every* line passing through a point A in a hypersurface (that is, in a set of points) having A as at least an s-fold point, and thus if the intersection multiplicity at A of X with every line issuing from A is at least s, then A is at least an s-fold point for X; and it is exactly an s-fold point if for a generic line issuing from A the intersection multiplicity is precisely s. For this it suffices that there be a line for which that multiplicity is s.

Let us consider the hypersurface $X \subset \mathbb{P}^n$ of order r given by the equation

$$f(T_0, \ldots, T_n) = 0$$

and the line ℓ that contains the two points $A = [a_0, \ldots, a_n]$ and $P = [y_0, \ldots, y_n]$. The variable point on ℓ is

$$[\lambda a_0 + \mu y_0, \lambda a_1 + \mu y_1, \ldots, \lambda a_n + \mu y_n], \quad \lambda, \mu \in K, \ (\lambda, \mu) \neq (0, 0),$$

and it belongs to X if and only if

$$f(\lambda a_0 + \mu y_0, \lambda a_1 + \mu y_1, \ldots, \lambda a_n + \mu y_n) = 0. \tag{5.3}$$

We set $f(a) := f(a_0, \ldots, a_n)$, $f(y) := f(y_0, \ldots, y_n)$ and also

$$\Delta_y f(a) := y_0 \left(\frac{\partial f}{\partial T_0}\right)_A + \cdots + y_n \left(\frac{\partial f}{\partial T_n}\right)_A$$

$$:= \left(y_0 \frac{\partial}{\partial T_0} + \cdots + y_n \frac{\partial}{\partial T_n}\right) f(a),$$

$$\Delta_y^s f(a) := \left(y_0 \frac{\partial}{\partial T_0} + \cdots + y_n \frac{\partial}{\partial T_n}\right)^{(s)} f(a)$$

$$:= \sum_{i_1, \ldots, i_s} y_{i_1} y_{i_2} \cdots y_{i_s} \left(\frac{\partial^s f}{\partial T_{i_1} \partial T_{i_2} \cdots \partial T_{i_s}}\right)_A.$$

5.2. Multiple points of a hypersurface

In particular, $\Delta_y^1 f(a) = \Delta_y f(a)$. Then one can rewrite (5.3) in the form

$$\lambda^r f(a) + \lambda^{r-1}\mu \Delta_y f(a) + \frac{1}{2!}\lambda^{r-2}\mu^2 \Delta_y^2 f(a) + \cdots$$
$$\cdots + \frac{1}{s!}\lambda^{r-s}\mu^s \Delta_y^s f(a) + \cdots + \mu^r f(y) = 0. \quad (5.4)$$

Equation (5.4) is solved by the coordinates (λ, μ) of the points of the line r_{AP} that belong to the hypersurface X.

Suppose now that $A \in X$, that is, that $f(a) = 0$. Among the solutions (λ, μ) of (5.4) one then has $(1, 0)$ which gives the point A. The necessary and sufficient condition for this to be a simple solution, that is, of multiplicity 1, is that $\Delta_y f(a) \neq 0$ (so that μ appears to the first degree in (5.4)).

The intersection multiplicity of the line r_{AP} with X in the point A will be s if

$$\Delta_y f(a) = \Delta_y^2 f(a) = \cdots = \Delta_y^{s-1} f(a) = 0 \quad \text{and} \quad \Delta_y^s f(a) \neq 0. \quad (5.5)$$

Suppose that, on calculating all the various partial derivatives at the point A, the partial derivatives 0th (i.e., $f(a)$), 1st, 2nd, ..., $(s-1)^{\text{st}}$ of f all turn out to be 0, and for this, in view of Euler's formula for homogeneous functions (cf. Exercise 3.1.18), it is (necessary and) sufficient that all the partial derivatives of order $(s-1)$ should be zero. Then A is an s-fold point, or a point of *multiplicity s* for X; and the intersection multiplicity at A of X with the line r_{AP} is at least s. It is exactly s (and thus the number of points distinct from P which belong to both X and the line r_{AP} is $r - s$) if r_{AP} is a generic line issuing from A, that is, if P is a generic point of $\mathbb{P}^n$. More precisely, P must not belong to the hypersurface of order s with equation

$$\Delta_T^s f(a) = \sum_{i_1,\ldots,i_s} T_{i_1} T_{i_2} \cdots T_{i_s} \left(\frac{\partial^s f}{\partial T_{i_1} \partial T_{i_2} \cdots \partial T_{i_s}} \right)_A = 0. \quad (5.6)$$

Interpreting equation (5.6) as the condition for which a point P is such that the line r_{AP} has intersection multiplicity $> s$ at A with X, one sees that this hypersurface is the locus of lines issuing from A, namely, a cone with vertex A. This cone, of order s, is the (projective) *tangent cone* to X at A (cf. Section 3.4; see also the paragraphs 5.8.4 and 5.8.6 for the description of the tangent cone to a hypersurface in a point of an s-fold variety).

When the line r_{AP} is a generator of the cone, that is, when P satisfies condition (5.6), the intersection multiplicity of r_{AP} with X in the point A is greater than s (since one has $\Delta_y^s f(a) = 0$). It will be $s + h$, $h > 0$, when P, besides satisfying (5.6), also satisfies the equations

$$\Delta_y^{s+1} f(a) = \Delta_y^{s+2} f(a) = \cdots = \Delta_y^{s+h-1} f(a) = 0,$$

but not the equation

$$\Delta_y^{s+h} f(a) = 0.$$

If $s = 1$, that is, if $f(a_0, \ldots, a_n) = 0$, but not all the first partial derivatives of f at the point A vanish, then A is a non-singular point of X. In that case the tangent cone has equation

$$\sum_{i=0}^{n} T_i \left(\frac{\partial f}{\partial T_i}\right)_A = 0,$$

that is, it coincides with the tangent hyperplane to X at A (cf. §3.1.17).

We make two explicit remarks:

- If A is an s-fold point for X, then every space S_k passing through A intersects X in a hypersurface $X' := X \cap S_k \subset S_k$ that has at A multiplicity at least s. Indeed the intersection of X with a line ℓ passing through A and situated in S_k coincides with the intersection of the same line with X' (since $X \cap \ell = X \cap S_k \cap \ell = X' \cap \ell$). The multiplicity of A for X' is exactly s if the relevant S_k is not contained in the tangent cone to X at A.

- If $s = 1$, that is, when A is a simple point of X, every generic space S_k (that is, not contained in the tangent hyperplane to X at A) intersects X in a hypersurface $X' \subset S_k$ having A as a simple point. If, however, S_k is contained in the space S_{n-1} tangent to X at A, the point A will be at least double for the hypersurface section.

5.2.1 Fundamental point as multiple point.
What we have seen in the course of the first part of this section becomes particularly easy in the case in which the multiple point A of a hypersurface $X = V(f) \subset \mathbb{P}^n$ is one of the fundamental points for the coordinate reference system. It is always possible to reduce to this case via a change of coordinates.

To fix ideas, we suppose that $A = A_0 = [1, 0, \ldots, 0]$ and we redo the entire calculation after having ordered the polynomial f with respect to T_0:

$$f = f_0 T_0^r + f_1 T_0^{r-1} + f_2 T_0^{r-2} + \cdots + f_r = 0,$$

where $f_i = f_i(T_1, \ldots, T_n)$ is a homogeneous polynomial of degree i in only the indeterminates $T_1, \ldots, T_n$, $i = 0, \ldots, r$ (thus f_0 is a constant).

If $P = [y_0, \ldots, y_n]$, a variable point on the line $r_{A_0 P}$ is $[\lambda + \mu y_0, \mu y_1, \ldots, \mu y_n]$. Substituting in the expression for f (and having set $y = (y_1, \ldots, y_n)$), one finds the equation

$$f_0 \lambda^r + f_1(y)\lambda^{r-1}\mu + f_2(y)\lambda^{r-2}\mu^2 + \cdots$$
$$\cdots + f_{s-1}(y)\lambda^{r-s+1}\mu^{s-1} + f_s(y)\lambda^{r-s}\mu^s + \cdots + f_r(y)\mu^r = 0,$$

which gives the intersections of the line $r_{A_0 P}$ with X. In order for at least s of these intersections to coincide with A_0 no matter how the point P may be chosen, the

5.2. Multiple points of a hypersurface

latter equation must have $\mu = 0$ as an s-fold root, no matter how the $y_1, \ldots, y_n$ be chosen, and thus

$$f_0(T_1, \ldots, T_n) = f_1(T_1, \ldots, T_n) = \cdots = f_{s-1}(T_1, \ldots, T_n) = 0.$$

Thus a hypersurface of order r that has A_0 as an s-fold point must have equation of the form

$$f_s(T_1, \ldots, T_n)T_0^{r-s} + f_{s+1}(T_1, \ldots, T_n)T_0^{r-s-1} + \cdots + f_r(T_1, \ldots, T_n) = 0.$$

The points $P = [y_0, \ldots, y_n]$ which when joined to A_0 give the lines for which at least $s + 1$ of the intersections with the hypersurface X fall at A_0, are those that satisfy the condition $f_s(y) = 0$. Hence, they are the points of the cone with equation

$$f_s(T_1, \ldots, T_n) = 0.$$

Therefore, in order that A_0 be s-fold for the hypersurface X it is necessary and sufficient that in the equation for X the variable T_0 should appear with degree at most $r - s$; in that case the coefficient of T_0^{r-s}, set equal to zero, gives the equation of the tangent cone to X at A_0.

The generators common to the cones of equations

$$f_s(T_1, \ldots, T_n) = 0, \ f_{s+1}(T_1, \ldots, T_n) = 0, \ \ldots, \ f_{s+h}(T_1, \ldots, T_n) = 0 \quad (5.7)$$

are the lines issuing from A_0 for which (at least) $s + h + 1$ intersections with X are absorbed by A_0. In general these lines exist only if $h \leq n - 2$; and if $h = n - 2$ they are (in general) finite in number and are called *principal tangents* to X at A_0. If $h > n - 2$ the $h + 1$ cones (5.7) have in common (in general) only the vertex A_0.

Example 5.2.2. If A is a non-singular point of a surface X in $\mathbb{P}^3$, there are *two* principal tangents to X at A (in this case, with the preceding notation, $s = h = 1$ and the two principal tangents are the intersections of the tangent plane $f_1 = 0$ at A with the quadric cone of equation $f_2 = 0$). Each of these has intersection multiplicity at least 3 (and only 3 if A is generic) with the surface X at A (and not just 2 as happens for a generic line passing through A and lying in the tangent plane).

5.2.3 Conical hypersurfaces.
Let X be a hypersurface of order r endowed with an r-fold point A. Then X is a locus of lines issuing from A. Indeed, if P is an arbitrary point of X distinct from A, the line r_{AP} has at least $r + 1$ points in common with X (at least r at A plus the point P) and thus, since $\deg(X) = r$, the line r_{AP} is contained in X (cf. Section 3.4). In this case we say that X is a *conical hypersurface with vertex A* or a *cone with vertex A*.

112   Chapter 5. Hypersurfaces in Projective Space

If one intersects a conical hypersurface X with vertex A with a hyperplane H not passing through A one finds a hypersurface X' of H that has the same order as X; and X is the cone that projects X' from A (cf. Section 3.4).

Conversely, if X' is a hypersurface of order r of a hyperplane H, the locus X of the lines that join the single points of X' with a point A not belonging to H is a hypersurface of order r having A as an r-fold point, that is a cone of order r with vertex A. Indeed, let ℓ be a generic line of $\mathbb{P}^n$ and σ the plane $J(\ell, A)$ joining ℓ and A. The plane σ intersects H in a line ℓ' (generic in H) that meets X' in r points $Q_1, \ldots, Q_r$. The r lines r_{AQ_j}, $j = 1, \ldots, r$, that lie on X, are coplanar with ℓ and so the intersections of ℓ with X are the r points in which the lines r_{AQ_j} meet ℓ. Hence $\deg(X) = r$. If then Q is a point of H not belonging to X', the line r_{AQ}, that is, an arbitrary line of $\mathbb{P}^n$ passing through A and not belonging to X, does not meet X outside of A and therefore A is an r-fold point for X.

We note explicitly that the order of a conical hypersurface X can be defined as the number of generators of X that belong to a generic plane passing through the vertex.

The analytic representation of a conical hypersurface X having one of the fundamental coordinate points as vertex is particularly simple. Suppose for example that X has as vertex the point $A_0 = [1, 0, \ldots, 0]$. If r is the order of X the point A_0 is r-fold for X and so in the equation for X the variable T_0 must appear at most to degree $r - r = 0$, that is, the variable T_0 must be missing. The equation of the cone is thus of the form

$$f(T_1, \ldots, T_n) = 0. \tag{5.8}$$

This same equation can be interpreted as the equation of a hypersurface X' in the hyperplane $T_0 = 0$ (where $T_1, T_2, \ldots, T_n$ are the homogeneous projective coordinates) and X is the cone that projects X' from A_0. Naturally X' is represented in $\mathbb{P}^n$ by the equations $T_0 = f(T_1, \ldots, T_n) = 0$.

That an equation like (5.8) represents a cone of vertex A_0 can also be seen by observing that if $P = [0, x_1, x_2, \ldots, x_n]$ is a point satisfying the equation, this equation will also be satisfied by every point $[\lambda, \mu x_1, \mu x_2, \ldots, \mu x_n]$, $(\lambda, \mu) \neq (0, 0)$, of the line that joins P with A_0.

More generally, if $f(T_{k+1}, T_{k+2}, \ldots, T_n)$ is a polynomial of degree r in which the variables $T_0, T_1, \ldots, T_k$ are missing, the equation

$$f(T_{k+1}, T_{k+2}, \ldots, T_n) = 0 \tag{5.9}$$

represents in $\mathbb{P}^n$ the cone X projecting from the vertex $S_k = J(A_0, A_1, \ldots, A_k)$ the hypersurface X' represented in the space S_{n-k-1} ($T_0 = T_1 = \cdots = T_k = 0$) by the same equation (5.9), cf. Section 3.4. The hypersurface X' has as its equations in $\mathbb{P}^n$:

$$T_0 = T_1 = \cdots = T_k = f(T_{k+1}, T_{k+2}, \ldots, T_n) = 0.$$

5.2. Multiple points of a hypersurface

5.2.4 Sections of varieties by tangent hyperplanes. Let F, G be two hypersurfaces in $\mathbb{P}^r$ which have the same tangent hyperplane at the point A, a common simple point of both F and G. We will say that A is a *point of contact* of F and G, or also that at A the two hypersurfaces *touch* or are *tangent*.

It is easy to see that A is at least a double point for the intersection of F and G. We place ourselves in an affine chart containing A and in which $x_1, \ldots, x_n$ are coordinates (non-homogeneous) with origin A. If $\varphi_1 = 0$ is the hyperplane tangent to both the hypersurfaces in A let

$$f = \varphi_1 + \alpha_2 + \alpha_3 + \cdots = 0, \quad g = \varphi_1 + \beta_2 + \beta_3 + \cdots = 0$$

be the equations of F and G, where φ_1, α_j, β_k are homogeneous polynomials whose degree is expressed by the subscript. The intersection variety $\mathcal{L}$ of F and G can be obtained by intersecting one of the two hypersurfaces with the hypersurface of equation

$$f - g = (\alpha_2 - \beta_2) + (\alpha_3 - \beta_3) + \cdots = 0$$

which has A as at least a double point; thus A is at least a double point of $\mathcal{L}$.

In particular, the intersection X' of a hypersurface X having A as simple point with the hyperplane H tangent to it at A has the point A as a double point. This however is geometrically obvious because an arbitrary line of H issuing from A is tangent at A to X and so at least two of its intersections with X, so too with X', are absorbed by A. We now prove the following more general fact.

- If P is a simple point of a variety V_k of dimension k every hypersurface X having P as a simple point and *tangent* there to V_k (which means that the tangent hyperplane at P to X contains the space S_k tangent to V_k at P) intersects V_k in a variety V_{k-1} for which P is (at least) a double point. We will again say that P is a *point of contact* of X and V_k.

  In particular the section of V_k by a generic *tangent hyperplane* at P (that is, with a generic hyperplane of the star with center the space S_k tangent to V_k at P) is a variety V_{k-1} for which P is (at least) a double point.

Since we are dealing with a local question it suffices to consider the affine case and to suppose that V_k is a complete intersection of $n - k$ algebraic hypersurfaces F_j of $\mathbb{A}^n$, each having P as a simple point and tangent hyperplanes at P that are linearly independent. Assuming P to be the coordinate origin O and the tangent hyperplanes at O to the F_j to be the hyperplanes $x_j = 0$, one finds that V_k has equations of the form

$$x_j = f_j(x_1, \ldots, x_n), \quad j = 1, \ldots, n - k,$$

where the f_j are polynomials lacking terms of degree < 2, and $x_j - f_j = 0$ is the equation of the hypersurface F_j.

A hypersurface G having O as a simple point with tangent hyperplane passing through the space S_k tangent to V_k (that is, through the space S_k with equations $x_1 = x_2 = \cdots = x_{n-k} = 0$) has equation of the form

$$\lambda_1 x_1 + \lambda_2 x_2 + \cdots + \lambda_{n-k} x_{n-k} + \varphi = 0,$$

with φ a polynomial lacking terms of degree < 2. On the other hand, to intersect V_k with G is just to intersect V_k with the hypersurface of equation

$$\lambda_1 f_1(x_1, \ldots, x_n) + \lambda_2 f_2(x_1, \ldots, x_n) + \cdots + \lambda_{n-k} f_{n-k}(x_1, \ldots, x_n) + \varphi = 0$$

having a double point at O.

Remark 5.2.5. We know that the necessary and sufficient condition in order that a point P of a hypersurface F (with equation $f = 0$, f irreducible polynomial) in $\mathbb{P}^n$ should be simple for F is that

$$\left(\frac{\partial f}{\partial T_0}, \frac{\partial f}{\partial T_1}, \ldots, \frac{\partial f}{\partial T_n} \right)_P \neq (0, 0, \ldots, 0).$$

Analogously, if F_1, F_2 are two hypersurfaces passing simply through a point P (and with equations $f_1 = 0$, $f_2 = 0$), the necessary and sufficient condition in order that P be a simple point for their intersection variety V_{n-2}, that is, the necessary and sufficient condition in order that at P the two hypersurfaces do not have the same tangent hyperplane, is that the Jacobian matrix

$$\left(\frac{\partial(f_1, f_2)}{\partial(T_0, \ldots, T_n)} \right) = \begin{pmatrix} \frac{\partial f_1}{\partial T_0} & \cdots & \frac{\partial f_1}{\partial T_n} \\ \frac{\partial f_2}{\partial T_0} & \cdots & \frac{\partial f_2}{\partial T_n} \end{pmatrix}$$

should have rank 2 at P (cf. §3.1.8).

Exercise 5.2.6. Let X be a hypersurface of $\mathbb{P}^n$ and $H \subset \mathbb{P}^n$ a hyperplane that intersects X in a hypersurface (of H) having an s-fold point P. Then the multiplicity of P for X is $\leq s$.

As in §5.2.1, we may suppose that $P = [1, 0, \ldots, 0]$ and assume that H is the hyperplane $T_1 = 0$. If X has equation

$$f = f_0 T_0^r + f_1(T_1, \ldots, T_n) T_0^{r-1} + \cdots + f_r(T_1, \ldots, T_n) = 0,$$

the hypersurface $X' = X \cap H$ is represented in H by the equation

$$f_0 T_0^r + f_1(0, T_2, \ldots, T_n) T_0^{r-1} + \cdots + f_r(0, T_2, \ldots, T_n) = 0.$$

Since P is s-fold for X', we must have

$$f_0 = f_1(0, T_2, \ldots, T_n) = \cdots = f_{s-1}(0, T_2, \ldots, T_n) = 0 \qquad (5.10)$$

and
$$f_s(0, T_2, \ldots, T_n) \neq 0.$$

Hence $f_s(T_1, T_2, \ldots, T_n) \neq 0$ and so P is a point of multiplicity $\leq s$ for X (and P is exactly s-fold for X if $f_0(T_1, T_2, \ldots, T_n) = \cdots = f_{s-1}(T_1, T_2, \ldots, T_n) = 0$). Furthermore, we have:

- If the generic hyperplane passing through P intersects X in a hypersurface X' having P as an s-fold point, then P is also an s-fold point for X.

Indeed, by the above, the multiplicity of P for X is $\leq s$. On the other hand, if P is s-fold for the sections of X with s independent hyperplanes, then P is at least s-fold for X. Indeed, if for each $i = 1, 2, \ldots, s$ one has $f_j(T_1, \ldots, T_{i-1}, 0, T_{i+1}, \ldots, T_n) = 0$ ($j = 0, 1, \ldots, s-1$) the homogeneous polynomials $f_0, f_1, \ldots, f_{s-1}$ (which all have degree $< s$) are divisible by $T_1 T_2 \ldots T_s$ and are therefore identically zero.

5.3 Algebraic envelopes

In a projective space $\mathbb{P}^n$, where $x_0, x_1, \ldots, x_n$ are projective point coordinates, we choose the coordinates $u_0, u_1, \ldots, u_n$ for the hyperplanes so that the condition of membership point-hyperplane is $u_0 x_0 + u_1 x_1 + \cdots + u_n x_n = 0$ (cf. [52, Vol. 1, Chapter V, §5]).

Assuming the above choice, let Γ be an *algebraic envelope of class ν* of hyperplanes of $\mathbb{P}^n$, that is, the totality of the hyperplanes of $\mathbb{P}^n$ whose coordinates annihilate a homogeneous polynomial $\varphi(u_0, \ldots, u_n) \in K[u_0, \ldots, u_n]$ of degree ν.

As one sees via duality, the *class* of Γ is the number of hyperplanes of Γ that belong to a generic pencil, that is, passing through a generic S_{n-2}.

Everything that has been said regarding algebraic hypersurfaces in Sections 5.1 and 5.2 can be repeated via duality for algebraic envelopes (cf. [52, Vol. 1, Chapter IX, §7]); in particular, one can give the notions of a simple or multiple hyperplane for an algebraic envelope.

The following are examples of pairs of dual statements.

A point P is s-fold for an algebraic hypersurface X of order r if the number of points of X other than P which belong to a generic line through P is $r - s$.	A hyperplane Π is s-fold for an algebraic envelope Γ of class ν if the number of hyperplanes of Γ other than Π which pass through a generic S_{n-2} belonging to Π is $\nu - s$.

The points of the space which when joined with an s-fold point P of a hypersurface X of order r give lines ℓ such that the points of X distinct from P and belonging to ℓ are in number at most $r - s - 1$ form the points of an algebraic cone of order s. In particular, if $s = 1$, that is, if P is a simple point, this cone is a hyperplane, the tangent hyperplane to X at P.

The hyperplanes of the space that intersect with an s-fold hyperplane Π of an envelope Γ of class ν to give spaces S_{n-2} such that the hyperplanes of Γ distinct from Π and passing through the S_{n-2} are in number at most $\nu - s - 1$ are the hyperplanes of an algebraic envelope of class s. In particular, if $s = 1$, that is, if Π is a simple hyperplane, this envelope is a point which is said to be a *characteristic point* of Π.

If $f = 0$ is the equation of a hypersurface X, the equation of the tangent hyperplane to X at a simple point P is

$$\sum_{i=0}^{n} x_i \left(\frac{\partial f}{\partial x_i}\right)_P = 0.$$

If $\varphi = 0$ is the equation of an algebraic envelope Γ, the equation of the characteristic point of a simple hyperplane Π of Γ is

$$\sum_{i=0}^{n} u_i \left(\frac{\partial \varphi}{\partial u_i}\right)_\Pi = 0.$$

The study of an algebraic envelope Γ in the neighborhood of one of its hyperplanes Π turns out to be very easy if one assumes a projective reference system such that Π is one of the fundamental hyperplanes. For example, if $\Pi = [1, 0, \ldots, 0]$ is an s-fold hyperplane of Γ, the equation of Γ will have the following form:

$$u_0^{\nu-s} \theta_s(u_1, \ldots, u_n) + u_0^{\nu-s-1} \theta_{s+1}(u_1, \ldots, u_n) + \cdots + \theta_\nu(u_1, \ldots, u_n) = 0,$$

where the θ_j are forms of degree j, $j = s, \ldots, \nu$. The equation $\theta_s(u_1, \ldots, u_n) = 0$ represents an envelope Φ of class s, the dual figure to the tangent cone to a hypersurface at an s-fold point P (cf. §5.2.1). It consists of the hyperplanes that cut out on Π the spaces S_{n-2} of an envelope of class s (just as a cone of vertex P consists of the points that joined with P constitute the lines of a cone of order s).

Proposition 5.3.1. *The tangent hyperplanes to an algebraic hypersurface X form an algebraic envelope.*

Proof. If $f = 0$ is the equation of X and $u_0, \ldots, u_n$ are the coordinates of the tangent hyperplane at a simple point $P = [x_0, \ldots, x_n]$ of X, one has

$$\rho u_i = \frac{\partial f}{\partial x_i}, \quad i = 0, \ldots, n, \quad \text{and} \quad u_0 x_0 + \cdots + u_n x_n = 0. \tag{5.11}$$

The elimination of $x_0, \ldots, x_n$ from this system of equations leads to an algebraic equation $g^*(u_0, \ldots, u_n) = 0$ satisfied by the coordinates of the tangent hyperplane at every simple point of X.

5.3. Algebraic envelopes   117

The equation $g^*(u_0, \ldots, u_n) = 0$ is also satisfied by the coordinates of each hyperplane that passes through a multiple point of X. Indeed, $g^*(u_0, \ldots, u_n) = 0$ is the necessary and sufficient condition for the $n + 2$ equations (5.11) to admit solutions $\rho, x_0, \ldots, x_n$; and it is clear that they admit the solution $(\rho, x_0, \ldots, x_n)$ with $\rho = 0$ and $x_0, \ldots, x_n$ the coordinates of a multiple point P of X if $[u_0, \ldots, u_n]$ is an arbitrary hyperplane that passes through P.

If, for example, the singularities of X are isolated multiple points, the polynomial $g^*(u_0, \ldots, u_n)$ will be divisible by the linear factors (raised to suitable powers) that represent the multiple points of X. When these factors have been removed, there remains the equation $g(u_0, \ldots, u_n) = 0$ of the *envelope of the tangent hyperplanes* to X (cf. Example 5.3.3). □

Remark 5.3.2 (The dual hypersurface). Notation remains as in Proposition 5.3.1. On applying Euler's formula, see Exercise 3.1.18, equations (5.11) are seen to be equivalent to the system

$$\begin{cases} \rho u_i = \dfrac{\partial f}{\partial x_i}, & i = 0, \ldots, n, \\ f(x_0, \ldots, x_n) = 0. \end{cases} \quad (5.12)$$

Equations (5.12) show that the envelope of tangent hyperplanes to X is the image of the hypersurface X (with equation $f(x_0, \ldots, x_n) = 0$) under the rational map

$$\varphi : X \to \mathbb{P}^n_{[u_0, \ldots, u_n]}$$

defined by

$$x = [x_0, \ldots, x_n] \mapsto \left[\frac{\partial f}{\partial x_0}(x), \ldots, \frac{\partial f}{\partial x_n}(x) \right].$$

Moreover one sees immediately that

(1) $\varphi(X)$ is a point if and only if X is a hyperplane;

(2) if X is not a hyperplane: φ is regular at x if and only if x is a simple point for X.

We will say that $\varphi(X)$ is the *dual hypersurface* of X.

For example, the dual curve of a conic is a conic: see [52, Vol. 2, Chapter XIII, §2] for further details. See also Exercise 13.1.9 for the case of the dual curve of a plane cubic.

In similar fashion one obtains the dual version of Proposition 5.3.1: the characteristic points of the hyperplanes of an algebraic envelope are those of an algebraic hypersurface, called *the adhering hypersurface of the envelope*.

Consider in particular a plane algebraic curve C and the envelope of its tangents Γ. Every simple point P of C is the characteristic point of the tangent to C at P.

If C has singularities, the envelope Γ will be endowed with dual singularities. For example, to an ordinary s-fold point there corresponds via duality a line of Γ with s characteristic distinct points, that is a tangent line to C at s distinct points, and that will be called a *bitangent* if $s = 2$, *tritangent* if $s = 3, \ldots$. To a cusp of C there corresponds via duality an *inflectional line*, that is, a line p from every generic point of which there emerge $\nu - 2$ lines distinct from p, where ν is the class of Γ (and so p is a double line); the line p has a single characteristic point (at which the two characteristic points come to coincide).

Example 5.3.3 (Dual character of cusps and flexes). To further clarify the fact that cusps and flexes are mutually dual it can be useful to seek the equation of the envelope of tangents of plane cubic with equation $x_0^3 - x_1^2 x_2 = 0$, which has the cusp $[0, 0, 1]$ with tangent $x_2 = 0$ and the flex $[0, 1, 0]$ with tangent $x_1 = 0$ (see also Section 5.7).

One must eliminate x_0, x_1, x_2 and ρ from the four equations

$$\rho u_0 = 3x_0^2,$$
$$\rho u_1 = -2x_1 x_2,$$
$$\rho u_2 = -x_1^2,$$
$$u_0 x_0 + u_1 x_1 + u_2 x_2 = 0.$$

Setting the values of x_2 deduced from the second and fourth equations equal, one has

$$\frac{\rho u_1}{2x_1} = \frac{u_0 x_0 + u_1 x_1}{u_2}$$

or, in view of the third equation,

$$\rho u_1 u_2 = 2u_0 x_0 x_1 + 2u_1 x_1^2 = 2u_0 x_0 x_1 - 2\rho u_1 u_2,$$

that is,

$$3\rho u_1 u_2 = 2u_0 x_0 x_1.$$

Squaring both sides and bearing in mind the first and the third equations one has

$$9\rho^2 u_1^2 u_2^2 = 4u_0^2 x_0^2 x_1^2 = 4u_0^2 \left(\frac{\rho u_0}{3}\right)(-\rho u_2) = -\frac{4}{3}\rho^2 u_0^3 u_2,$$

and finally

$$g(u_0, u_1, u_2) = 4u_0^3 u_2 + 27 u_1^2 u_2^2 = 0.$$

On eliminating the linear factor u_2 (which corresponds to the singular point of C, that is the point $[0, 0, 1]$ whose equation is $u_2 = 0$) one has the equation of the envelope Γ of the tangents to C:

$$g(u_0, u_1, u_2) = 4u_0^3 + 27 u_1^2 u_2 = 0. \tag{5.13}$$

It has a double line $[0, 0, 1]$, that is the line with equation $x_2 = 0$, which is an inflexional tangent to C.

The simple line $[0, 1, 0]$ is the line, ℓ, of equation $x_1 = 0$. This line absorbs all three lines of the envelope emerging from its characteristic point $u_1 = 0$, because on setting $u_1 = 0$ in (5.13) one finds $u_0^3 = 0$. Thus ℓ is the cuspidal tangent of C.

Exercise 5.3.4. Determine the tangential equation, i.e., the equation of the envelope of the hyperplanes tangent to the quadric $\mathcal{Q}$ in $\mathbb{P}^n$ of equation $\sum_{i,j=0}^n a_{ij} x_i x_j = 0$.
The result of elimination of the parameters $x_0, x_1, \ldots, x_n, \rho$ from the equations

$$0 = u_0 x_0 + u_1 x_1 + \cdots + u_n x_n,$$
$$\rho u_0 = a_{00} x_0 + a_{01} x_1 + \cdots + a_{0n} x_n,$$
$$\vdots$$
$$\rho u_n = a_{n0} x_0 + a_{n1} x_1 + \cdots + a_{nn} x_n$$

is

$$\begin{vmatrix} 0 & u_0 & \cdots & u_n \\ u_0 & a_{00} & \cdots & a_{0n} \\ \vdots & \vdots & & \vdots \\ u_n & a_{n0} & \cdots & a_{nn} \end{vmatrix} = 0,$$

that is, $\sum_{i,j=0}^n A_{ij} u_i u_j = 0$, where A_{ij} are the algebraic complements (cofactors) of the elements a_{ij} of the matrix $A = (a_{ij})$ of the coefficients of $\mathcal{Q}$.

5.4 Polarity with respect to a hypersurface

In the projective space $\mathbb{P}^n = \mathbb{P}^n(K)$, with homogeneous coordinates $x_0, \ldots, x_n$, consider a hypersurface X of order $r > 1$ with equation

$$f(x_0, \ldots, x_n) = 0,$$

where f is a form of degree r with coefficients in K. Let $P = [y_0, \ldots, y_n]$ be a point of $\mathbb{P}^n$. The equation

$$\sum_{i=0}^n y_i \frac{\partial f}{\partial x_i} = 0 \qquad (5.14)$$

represents a hypersurface of order $r - 1$ that is said to be the *first polar* of P with respect to X (or also *first polar* of X with respect to P). We will denote it by $X_1(P)$.

Example 5.4.1. In the case of the plane $\mathbb{P}^2$, if $f(x_0, x_1, x_2) = 0$ is the equation of a non-singular conic, equation (5.14) is the equation of the polar line $\gamma(P)$ of the point $P = [y_0, y_1, y_2]$, where $\gamma \colon \mathbb{P}^2 \to \mathbb{P}^{2*}$ is the correlation defined by the matrix A of the coefficients of the equation $f(x_0, x_1, x_2) = 0$ (see also [52, Vol. 2, Chapter XIII, §2]).

Notation. If M and N are respectively matrices of type (p, q) and (r, q), henceforth we will use the symbol $M\,^t N$ to denote the usual matrix product of M with the transpose of N.

The following important result holds. It expresses the fact that the operation of "polarization with respect to the pole P", that is, the operation that maps f to $\Delta_y^1 f$, is covariant with respect to projectivities, so that the first polar of a point with respect to a hypersurface has a geometric (projective) meaning.

Theorem 5.4.2. *Let X be a hypersurface of $\mathbb{P}^n$ with equation $f(x_0, \ldots, x_n) = 0$. Let P be a point of $\mathbb{P}^n$ and $X_1(P)$ the first polar of P with respect to X. Let $\varphi: \mathbb{P}^n \to \mathbb{P}^n$ be a projectivity. Set $P^* := \varphi(P)$, $X^* = \varphi(X)$ and $(X_1(P))^* = \varphi(X_1(P))$. One then has*
$$(X_1(P))^* = X_1^*(P^*),$$
that is, the transform of $X_1(P)$ is the first polar of P^* with respect to X^*.

Proof. Suppose that φ is expressed in matricial form
$$\rho\,^t(x_0', \ldots, x_n') = A\,^t(x_0, \ldots, x_n),$$
where $\rho \in K$ is a non-zero constant and A is an invertible matrix in $M_{n+1}(K)$. Then, multiplying on the left by $\rho^{-1} A^{-1}$, one has
$$A^{-1\,t}(x_0', \ldots, x_n') = \rho^{-1\,t}(x_0, \ldots, x_n).$$
Thus, setting $\rho A^{-1} = (b_{ij})$, the equation of X^* becomes
$$f^*(x_0', \ldots, x_n') = f\left(\sum_{j=0}^n b_{0j} x_j', \sum_{j=0}^n b_{1j} x_j', \ldots, \sum_{j=0}^n b_{nj} x_j'\right) = 0$$
and so
$$\frac{\partial f^*}{\partial x_j'} = \sum_{i=0}^n \frac{\partial f}{\partial x_i} \frac{\partial x_i}{\partial x_j'} = \sum_{i=0}^n \frac{\partial f}{\partial x_i} b_{ij}.$$
From this it follows that
$$\sum_{j=0}^n x_j' \frac{\partial f^*}{\partial x_j'} = \sum_{j=0}^n x_j' \sum_{i=0}^n \frac{\partial f}{\partial x_i} b_{ij} = \sum_{i=0}^n \left(\sum_{j=0}^n b_{ij} x_j'\right) \frac{\partial f}{\partial x_i} = \sum_{i=0}^n x_i \frac{\partial f}{\partial x_i},$$
which gives the desired result. □

5.4. Polarity with respect to a hypersurface

As a consequence of the preceding theorem we may now conclude that in order to study properties of the polar hypersurfaces of points $P, Q, \ldots$ with respect to a given hypersurface X, of a given order r, there is no loss of generality in supposing that $P, Q, \ldots$ coincide with the vertices $A_0, A_1, \ldots$ of the fundamental $(n+1)$-hedron. For example, if $P = A_i$ is the point whose only non-zero coordinate is $y_i = 1$, one has $\Delta_y^1 = \frac{\partial}{\partial x_i}$ and so the first polar of A_i with respect to X has equation

$$\frac{\partial f}{\partial x_i} = 0.$$

The first polar of a point $Q = [z_0, \ldots, z_n]$ with respect to the first polar of $P = [y_0, \ldots, y_n]$ (with respect to X) is called the *second mixed polar of P, Q* (*with respect to X*) and has equation

$$\sum_{i=0}^{n} z_i \frac{\partial}{\partial x_i} \left(\sum_{j=0}^{n} y_j \frac{\partial f}{\partial x_j} \right) = 0,$$

that is

$$\sum_{i,j=0}^{n} z_i y_j \frac{\partial^2 f}{\partial x_i \partial x_j} = 0, \tag{5.15}$$

and clearly coincides with the second mixed polar of Q, P. In particular, assuming $Q = P$,

$$\sum_{i,j=0}^{n} y_i y_j \frac{\partial^2 f}{\partial x_i \partial x_j} = 0 \tag{5.16}$$

is the equation of the *second pure polar of P with respect to X*.

In analogous fashion one defines the successive polars $X_s(P), s = 2, \ldots, r-1$, which are hypersurfaces of order $r - s$; the equation of the sth polar of the point $P = [y_0, \ldots, y_n]$ is

$$\sum_{i_1, \ldots, i_s} y_{i_1} \cdots y_{i_s} \frac{\partial^s f}{\partial x_{i_1} \cdots \partial x_{i_s}} = 0. \tag{5.17}$$

The $(r-2)^{\text{nd}}$ polar $X_{r-2}(P)$ is also called the *polar quadric* of P. The $(r-1)^{\text{st}}$ polar $X_{r-1}(P)$ is also called the *polar hyperplane* of P. The hypersurface X is also called the 0th *polar*.

If $n = 1$, that is for a line, X and the successive polars are groups of points, and the polars are called *polar groups*.

The theory of polars may be efficiently formalized by using the operator

$$\Delta_a := a_0 \frac{\partial}{\partial x_0} + \cdots + a_n \frac{\partial}{\partial x_n},$$

where $a = (a_0, \ldots, a_n)$, and its symbolic powers

$$\Delta_a^0 = 1, \quad \Delta_a^1 = \Delta_a, \quad \Delta_a^2 = a_0^2 \frac{\partial^2}{\partial x_0^2} + 2a_0 a_1 \frac{\partial^2}{\partial x_0 \partial x_1} + \cdots.$$

If $f(x) = f(x_0, \ldots, x_n)$ is a homogeneous polynomial and $P = [y_0, \ldots, y_n]$ is a designated point in $\mathbb{P}^n$, the notations $\Delta_a f(x)$ (or $\Delta_a f$), $\Delta_a f(y)$ indicate respectively

$$\Delta_a f = \Delta_a f(x) = a_0 \frac{\partial f}{\partial x_0} + \cdots + a_n \frac{\partial f}{\partial x_n};$$

$$\Delta_a f(y) = \left(a_0 \frac{\partial f}{\partial x_0} + \cdots + a_n \frac{\partial f}{\partial x_n} \right)_P.$$

Equations (5.14) and (5.16) may be rewritten respectively in the form $\Delta_y^1 f(x) = 0$ and $\Delta_y^2 f(x) = 0$. In general the sth polar $X_s(P)$, $s = 2, \ldots, r-1$, has equation

$$\Delta_y^s f(x) = 0.$$

We say that a point P is *self-conjugate* (with respect to X) if P belongs to any one of its successive polars $X_s(P)$. In this regard one has the following result (cf. [52, Vol. 2, Chapter XIII, §2] for the case of conics).

Proposition 5.4.3. *Let $X = V(f)$ be a hypersurface of order $r > 1$ of $\mathbb{P}^n$. Then X is the locus of points of $\mathbb{P}^n$ self-conjugate with respect to X, that is, a point P belongs to any of it successive polars $X_s(P)$, $s = 1, \ldots, r-1$, if and only if P belongs to X.*

Proof. Let $P = [y_0, \ldots, y_n]$ and let $y = (y_0, \ldots, y_n)$. For $s = 1$ one has $\Delta_y^1 f(y) = r f(y)$ by Euler's formula and so $\Delta_y^1 f(y) = 0$ if and only if $f(y) = 0$. In general it suffices to observe that

$$\Delta_y^s f(y) = r(r-1)(r-2) \ldots (r-s+1) f(y). \qquad \square$$

Here are some properties of polars.

Proposition 5.4.4 (Permutability Theorem). *Let $X = V(f)$ be a hypersurface of order $r > 1$ in $\mathbb{P}^n$. Then the sth polar of a point P with respect to the tth polar (with respect to X) of a point Q coincides with the tth polar of Q with respect to the sth polar of P (with respect to X), that is*

$$(X_t(Q))_s(P) = (X_s(P))_t(Q).$$

Proof. Let $P = [y_0, \ldots, y_n]$, $y = (y_0, \ldots, y_n)$ and $Q = [z_0, \ldots, z_n]$, $z = (z_0, \ldots, z_n)$. It suffices to observe that

$$\Delta_y^s \Delta_z^t f = \Delta_z^t \Delta_y^s f,$$

where $f = 0$ is the equation of the hypersurface X.

5.4. Polarity with respect to a hypersurface

If $P = A_0$ and $Q = A_1$ this is nothing more than the property

$$\frac{\partial^s}{\partial x_0^s}\left(\frac{\partial^t f}{\partial x_1^t}\right) = \frac{\partial^t}{\partial x_1^t}\left(\frac{\partial^s f}{\partial x_0^s}\right) = \frac{\partial^{t+s} f}{\partial x_0^s \partial x_1^t}. \qquad \square$$

Proposition 5.4.5 (Section Theorem). *Let $X = V(f)$ be a hypersurface of order $r > 1$ in $\mathbb{P}^n$. Let S_h be a linear space of $\mathbb{P}^n$ not contained in X and let $X' := X \cap S_h$ be the hypersurface section of X by S_h. Let P be a point of S_h. Then the section by S_h of the sth polar of P with respect to X is the sth polar of P with respect to X', that is*

$$S_h \cap X_s(P) = X'_s(P).$$

Proof. One need only observe that if $P = A_0 = [1, 0, \ldots, 0]$ and the equations of S_h are $x_{h+1} = x_{h+2} = \cdots = x_n = 0$, one has

$$\left(\frac{\partial^s f}{\partial x_0^s}\right)_{x_{h+1}=x_{h+2}=\cdots=x_n=0} = \frac{\partial^s f(x_0, \ldots, x_h, 0, \ldots, 0)}{\partial^s x_0^s}. \qquad \square$$

The most important result in the theory of polarity is the following "reciprocity theorem".

Theorem 5.4.6 (Reciprocity Theorem). *Let X be a hypersurface of order $r > 1$ in $\mathbb{P}^n$. Given two points P, Q in $\mathbb{P}^n$, if the sth polar of P with respect to X passes through Q, $0 < s < r$, then the $(r-s)^{\text{th}}$ polar of Q with respect to X passes through P.*

Proof. We may assume that $P = A_0 = [1, 0, \ldots, 0]$ and $Q = A_1 = [0, 1, 0, \ldots, 0]$. Let $f = 0$ be the equation of X and let λ be the coefficient of $x_0^s x_1^{r-s}$ in f:

$$f = \cdots + \lambda x_0^s x_1^{r-s} + \cdots .$$

The equation of the sth polar $X_s(P)$ is

$$\frac{\partial^s f}{\partial x_0^s} = s! \lambda x_1^{r-s} + (\cdots) = 0,$$

where $(\cdots)$ stands for terms of degree $< r - s$ with respect to x_1. Similarly the equation of the $(r-s)^{\text{th}}$ polar $X_{r-s}(Q)$ is

$$\frac{\partial^{r-s} f}{\partial x_1^{r-s}} = (r-s)! \lambda x_0^s + (\cdots) = 0,$$

where $(\cdots)$ stands for terms of degree $< s$ with respect to x_0.

Therefore, $\lambda = 0$ is the necessary and sufficient condition in order that $X_s(P)$ pass through Q, and also the necessary and sufficient condition for $X_{r-s}(Q)$ to pass through P. $\qquad \square$

Remark 5.4.7 (Alternative proof of Theorem 5.4.6). Here we give an alternative proof of the preceding theorem which is independent of the fact that one can choose the coordinates in an opportune fashion. This argument also yields other consequences which will be considered in a subsequent observation (see Remark 5.4.8).

With the same notation as in Theorem 5.4.6, consider Taylor's formula for the polynomial $f(x_0, \ldots, x_n)$:

$$f(a_0 + b_0, a_1 + b_1, \ldots, a_n + b_n)$$
$$= f(a_0, \ldots, a_n) + b_0 \frac{\partial f}{\partial a_0} + \cdots + b_n \frac{\partial f}{\partial a_n}$$
$$+ \frac{1}{2!} \left(b_0^2 \frac{\partial^2 f}{\partial a_0^2} + 2 b_0 b_1 \frac{\partial^2 f}{\partial a_0 \partial a_1} + \cdots \right) + \cdots \quad (5.18)$$
$$+ \frac{1}{s!} \sum b_{i_1} \ldots b_{i_s} \frac{\partial^s f}{\partial a_{i_1} \ldots \partial a_{i_s}} + \cdots,$$

where, for simplicity of notation, $\frac{\partial f}{\partial a_i}$ indicates $\frac{\partial f}{\partial x_i}$ calculated at $(a_0, \ldots, a_n)$.

Putting $a_i = \lambda y_i$, $b_i = \mu z_i$, $i = 0, \ldots, n$, in (5.18) one obtains

$$f(\lambda y_0 + \mu z_0, \ldots, \lambda y_n + \mu z_n) = f(\lambda y_0, \ldots, \lambda y_n) + \cdots$$
$$+ \frac{1}{s!} \sum \mu z_{i_1} \ldots \mu z_{i_s} \left(\frac{\partial^s f}{\partial x_{i_1} \ldots \partial x_{i_s}} \right)_{x_i = \lambda y_i} + \cdots.$$

But f being a homogeneous polynomial of degree r we have

$$f(\lambda y_0 + \mu z_0, \ldots, \lambda y_n + \mu z_n) = \lambda^r f(y_0, \ldots, y_n) + \cdots$$
$$+ \frac{1}{s!} \lambda^{r-s} \mu^s \sum z_{i_1} \ldots z_{i_s} \frac{\partial^s f}{\partial y_{i_1} \ldots \partial y_{i_s}} + \cdots, \quad (5.19)$$

and, interchanging λ with μ and y with z,

$$f(\mu z_0 + \lambda y_0, \ldots, \mu z_n + \lambda y_n) = \mu^r f(z_0, \ldots, z_n) + \cdots$$
$$+ \frac{1}{s!} \mu^{r-s} \lambda^s \sum y_{i_1} \ldots y_{i_s} \frac{\partial^s f}{\partial z_{i_1} \ldots \partial z_{i_s}} + \cdots. \quad (5.20)$$

The left-hand sides of (5.19) and (5.20) coincide and so the right-hand sides of (5.19) and (5.20) are the same polynomials in λ and μ. In particular the condition that they have equal coefficients of $\lambda^s \mu^{r-s}$ may be written as:

$$\frac{1}{s!} \sum y_{i_1} \ldots y_{i_s} \frac{\partial^s f}{\partial z_{i_1} \ldots \partial z_{i_s}} = \frac{1}{(r-s)!} \sum z_{i_1} \ldots z_{i_{r-s}} \frac{\partial^{r-s} f}{\partial y_{i_1} \ldots \partial y_{i_{r-s}}}. \quad (5.21)$$

Now, the equation of the sth polar of the point $P = [y_0, \ldots, y_n]$ with respect to X is

$$\sum y_{i_1} \ldots y_{i_s} \frac{\partial^s f}{\partial x_{i_1} \ldots \partial x_{i_s}} = 0;$$

5.4. Polarity with respect to a hypersurface

and the equation of the $(r-s)^{\text{th}}$ polar of the point $Q = [z_0, \ldots, z_n]$ with respect to X is

$$\sum z_{i_1} \ldots z_{i_{r-s}} \frac{\partial^{r-s} f}{\partial x_{i_1} \ldots \partial x_{i_{r-s}}} = 0.$$

If the sth polar of P passes through Q then the left-hand side of (5.21) is zero and so too must be the right-hand side, that is, the $(r-s)^{\text{th}}$ polar of Q passes through P.

Remark 5.4.8 (Double method for reading the equation of a polar). Notation as in Theorem 5.4.6 and Remark 5.4.7. We set

$$\varphi_s(t, x) := \sum_{i_1, \ldots, i_s} t_{i_1} \ldots t_{i_s} \frac{\partial^s f}{\partial x_{i_1} \ldots \partial x_{i_s}}.$$

The polynomial $\varphi_s(t, x)$ is homogeneous of degree s with respect to $t = (t_0, \ldots, t_n)$ and of degree $r - s$ with respect to $x = (x_0, \ldots, x_n)$. The equation of the sth polar of $P = [y_0, \ldots, y_n]$ with respect to X is $\varphi_s(y, x) = 0$. It passes through $Q = [z_0, \ldots, z_n]$ if $\varphi_s(y, z) = 0$. Therefore $\varphi_s(y, z) = 0$ is the condition in order for Q to belong to the sth polar of P and so, by Theorem 5.4.6, it is also the condition for P to belong to the $(r-s)^{\text{th}}$ polar of Q. Hence $\varphi_s(x, z) = 0$ is the equation of the $(r-s)^{\text{th}}$ polar of Q. It follows that the same equation

$$\varphi_s(y, x) = 0,$$

of degree s with respect to $y = (y_0, \ldots, y_n)$ and of degree $r - s$ with respect to $x = (x_0, \ldots, x_n)$, represents both the sth polar of the point $[y_0, \ldots, y_n]$ when y is fixed (and so the variables are the x's) and the $(r-s)^{\text{th}}$ polar of the point $[x_0, \ldots, x_n]$ when x is fixed (and the y's vary).

In particular, consider the equation (of degree $r - 1$ in the x's)

$$\varphi_1(y, x) = y_0 \frac{\partial f}{\partial x_0} + \cdots + y_n \frac{\partial f}{\partial x_n} = 0$$

of the first polar of the point $[y_0, \ldots, y_n]$ with respect to X. On fixing $x := a = (a_0, \ldots, a_n)$ and allowing the y's to vary, one has the equation of the $(r-1)^{\text{th}}$ polar of the point $A = [a_0, \ldots, a_n]$:

$$y_0 \left(\frac{\partial f}{\partial x_0}\right)_a + \cdots + y_n \left(\frac{\partial f}{\partial x_n}\right)_a = 0. \tag{5.22}$$

If A is a point of the hypersurface X of equation $f(x_0, \ldots, x_n) = 0$, equation (5.22) is the equation of the tangent hyperplane at A to X. (Here we tacitly assume that A is a non-singular point for X: in this regard see the discussion in the following paragraph 5.4.10.) Thus we have proved the following fact.

Proposition 5.4.9. *Let X be a hypersurface of order r in $\mathbb{P}^n$ and let P be a non-singular point of X. Then the tangent hyperplane to X at P is the $(r-1)^{\text{st}}$ polar of P with respect to X; analytically, if $f = 0$ is the equation of X and $P = [y_0, \ldots, y_n]$,*

$$\sum_{i_1,\ldots,i_{r-1}} y_{i_1} \cdots y_{i_{r-1}} \frac{\partial^{r-1} f}{\partial x_{i_1} \cdots \partial x_{i_{r-1}}} = \sum_{i=0}^{n} \left(\frac{\partial f}{\partial x_i}\right)_P x_i.$$

5.4.10 The singular case. Let X be a hypersurface in $\mathbb{P}^n$ of order $r > 1$, with equation $f(x_0, \ldots, x_n) = 0$, and let P be a point of $\mathbb{P}^n$. Suppose that the sth polar of P with respect to X is indeterminate for $1 \leq s \leq r-1$. If we suppose that $P = A_0 = [1, 0, \ldots, 0]$, this means that $\frac{\partial^s f}{\partial x_0^s}$ is the null polynomial. It follows that the derivative $\frac{\partial^{s-1} f}{\partial x_0^{s-1}}$ does not depend on x_0, that the derivative $\frac{\partial^{s-2} f}{\partial x_0^{s-2}}$ is of degree one with respect to x_0 and so on, that f is of degree $s-1$ with respect to x_0. This means that P is a point of multiplicity $r - s + 1$ for X.

Analogously, one has that if the $(r - s + 1)^{\text{st}}$ polar of P with respect to X is indeterminate, then P is an s-fold point for X.

Conversely, if P is an s-fold point for X, then f may be written in the form

$$f = f_s(x_1, x_2, \ldots, x_n) x_0^{r-s} + f_{s+1}(x_1, x_2, \ldots, x_n) x_0^{r-s-1} + \cdots$$
$$\cdots + f_r(x_1, x_2, \ldots, x_n).$$

Thus one has

$$\frac{\partial^{r-s+1} f}{\partial x_0^{r-s+1}} = 0$$

and also that

$$\frac{\partial^{r-s} f}{\partial x_0^{r-s}} = (r-s)! f_s(x_1, x_2, \ldots, x_n).$$

Therefore we have:

Proposition 5.4.11. *Let X be a hypersurface of $\mathbb{P}^n$ of order $r > 1$. For each integer $s = 1, \ldots, r-1$, a necessary and sufficient condition in order for a point P to be s-fold for X is that the $(r - s + 1)^{\text{st}}$ polar of P with respect to X be indeterminate. In that case the $(r-s)^{\text{th}}$ polar of P with respect to X is the tangent cone to X at P.*

5.5 Quadrics in projective space

In this paragraph we consider the remarkable case of hypersurfaces of order $r = 2$ in $\mathbb{P}^n := \mathbb{P}^n(K)$. We define a *quadric* (or *hyperquadric*) in the projective space $\mathbb{P}^n$

as a hypersurface Q defined by a quadratic form

$$f(x_0, \ldots, x_n) = \sum_{i,j=0}^{n} a_{ij} x_i x_j = 0, \quad a_{ij} = a_{ji}, \tag{5.23}$$

with coefficients $a_{ij} \in K$. We will say that the symmetric matrix $A = (a_{ij}) \in M_{n+1}(K)$ is the *matrix associated* to the quadric Q. Thus, on setting $\xi := (x_0, \ldots, x_n)$, equation (5.23) may be rewritten in matrix form

$$\xi A \,{}^t\xi = 0. \tag{5.24}$$

If $\rho(A)$ is the rank of A we set

$$\rho = \rho(A) - 1.$$

If $\rho = n$, that is if $\det(A) \neq 0$, Q is *non-degenerate* (or *not specialized*). If $\rho = n - \lambda$ we will say that Q is λ *times specialized*.

5.5.1 Singular points of a quadric. The system of linear equations having A as its matrix of coefficients,

$$\eta A = (0, \ldots, 0) \tag{5.25}$$

has non-trivial solutions $\eta = (y_0, \ldots, y_n)$ only when Q is degenerate. More precisely, if Q is λ times specialized the solutions of (5.25) are the points of a space $S_{\lambda-1}$, called the *singular space* of Q.

The *singular points*, that is, the points of that space $S_{\lambda-1}$, are double points for Q because the system (5.25) may be rewritten in the form

$$\sum_{i=0}^{n} a_{ij} x_i = 0, \quad \text{that is,} \quad \frac{\partial f}{\partial x_j} = 0, \ j = 0, \ldots, n. \tag{5.26}$$

A quadric which is λ times specialized therefore has a double subspace $S_{\lambda-1}$ and is thus a cone having this $S_{\lambda-1}$ as vertex (cf. §5.2.3). We will say that it is an $S_{\lambda-1}$-*quadric cone*.

Exercise 5.5.2. Consider a quadric Q that is λ times specialized, and let r be a line of $\mathbb{P}^n$ that meets both the locus $S_{\lambda-1}$ of singular points of Q and its complement in Q. Then it is easy to prove directly that r is contained in Q.

Indeed, let $Z = [z_0, \ldots, z_n]$ be a point of Q, $Z \notin S_{\lambda-1}$, and let $r = r_{YZ}$ be a line that passes through Z and meets $S_{\lambda-1}$ in a point $Y = [y_0, \ldots, y_n]$. As u, v vary in K, the point $P = uY + vZ$ traces the line r. Consider the vectors $\eta = (y_0, \ldots, y_n), \zeta = (z_0, \ldots, z_n)$. By equation (5.24), $P \in Q$ if and only if

$$(u\eta + v\zeta) A \,{}^t(u\eta + v\zeta) = 0, \tag{5.27}$$

that is
$$u^2\eta A\,{}^t\eta + uv(\eta A\,{}^t\zeta + \zeta A\,{}^t\eta) + v^2\zeta A\,{}^t\zeta = 0;$$
or again, since $\eta A\,{}^t\zeta = \zeta A\,{}^t\eta$,
$$u^2\eta A\,{}^t\eta + 2uv\eta A\,{}^t\zeta + v^2\zeta A\,{}^t\zeta = 0.$$

Since Y is a singular point of $\mathcal{Q}$ we have $\eta A\,{}^t\eta = \eta A\,{}^t\zeta = 0$. On the other hand we also have $\zeta A\,{}^t\zeta = 0$ since $Z \in \mathcal{Q}$. Thus equation (5.27) hold for all $u, v \in K$, namely r is contained in $\mathcal{Q}$.

The following proposition shows that the study of specialized quadrics reduces to the study of non-degenerate quadrics.

Proposition 5.5.3. *Let $\mathcal{Q}$ be a quadric λ times specialized in $\mathbb{P}^n$ and let $S_{\lambda-1}$ be the linear space that is the locus of its singular points. Then there is a linear space $S_{n-\lambda}$ skew to $S_{\lambda-1}$ and a non-degenerate quadric $\mathcal{Q}_0$ in $S_{n-\lambda}$ such that $\mathcal{Q}$ is the cone with vertex $S_{\lambda-1}$ projecting $\mathcal{Q}_0$.*

Proof. Set $\rho = n - \lambda$. We may suppose that $S_{\lambda-1}$ is the space that joins the $n - \rho = \lambda$ vertices $A_{\rho+1}, A_{\rho+2}, \ldots, A_n$ of the reference $(n+1)$-hedron. Then the linear equations (5.26) that define $S_{\lambda-1}$ become $x_0 = x_1 = \cdots = x_\rho = 0$ and so the equation (5.23) of $\mathcal{Q}$ assumes the form

$$\sum_{i,h=0}^{\rho} a_{ih} x_i x_h = 0 \qquad (5.28)$$

with $\det((a_{ih})_{i,h=0,\ldots,\rho}) \neq 0$.

Since in equation (5.28) the variables $x_{\rho+1}, \ldots, x_n$ are missing, a λ times specialized quadric is the cone that projects from its vertex $S_{\lambda-1}$ a non-degenerate quadric $\mathcal{Q}_0$, with equation (5.28), in a space $S_\rho = S_{n-\lambda}$ skew to $S_{\lambda-1}$. $\square$

We note a final point:

- The necessary and sufficient condition in order that one of the fundamental points of the reference system be singular for $\mathcal{Q}$ is that one of the variables be missing from the equation of $\mathcal{Q}$. For example, the point $A_0 = [1, 0, \ldots, 0]$ is singular if and only if the variable x_0 is missing.

5.5.4 Polarity with respect to a quadric. The study of the polarity with respect to a hypersurface X becomes particularly simple in the case when X is a quadric $\mathcal{Q}$ so that one need only consider first polars, which are hyperplanes. One sees immediately that the *polar hyperplane* $\mathcal{Q}_1(P)$ of a point P with respect to $\mathcal{Q}$ is nothing more than the hyperplane corresponding to P in the involutory reciprocity γ of $\mathbb{P}^n$ having as matrix of coefficients the (symmetric) matrix A associated to $\mathcal{Q}$

(cf. [52, Vol. 2, Chapter XIII, §2]). In other words, $Q_1(P) = \gamma(P)$. The point P is called the *pole* of $\gamma(P)$.

The results given in Section 5.4 reduce simply, in the case $r = 2$, to the following.

(1) Given two points P_1, P_2, if the polar hyperplane of P_1 passes through P_2, the polar hyperplane of P_2 passes through P_1 (and the two points are said to be *reciprocal* or *conjugate* with respect to Q). Similarly, if the pole of a hyperplane π_1 belongs to a hyperplane π_2, that of π_2 belongs to π_1 (and the two hyperplanes are said to be *reciprocal* or *conjugate* with respect to Q).

(2) If P is a simple point of Q, the polar hyperplane of P is the tangent hyperplane in P to Q.

(3) If Q' is a quadric section of Q by a linear space L, and P is a point of L, the polar hyperplane of P with respect to Q' is the section of L by the polar hyperplane of P with respect to Q.

(4) If Q is λ times specialized and $S_{\lambda-1}$ is its vertex, the polar hyperplane of each point $P \in S_{\lambda-1}$ is indeterminate; and $S_{\lambda-1}$ belongs to the polar hyperplane of each point $P \notin S_{\lambda-1}$.

Furthermore,

a) The quadric Q is the locus of the self-conjugate points with respect to Q, that is, of the points P such that $P \in \gamma(P)$.

b) The polar hyperplane of a point $P \notin Q$ with respect to Q is the locus of the harmonic conjugates P' of P with respect to the pairs of points of Q collinear with P. Indeed, on each line ℓ starting from P (and not tangent to Q) the polarity γ induces a non-degenerate involution which sends P to the intersection of ℓ with the polar hyperplane of P. The fixed points of this involution are the intersections of ℓ with Q.

c) The line joining two mutually reciprocal points of Q is contained in Q. Indeed, if P is a point of Q, the polar hyperplane of P with respect to Q is the tangent hyperplane to Q there. A point R reciprocal to P thus belongs to the tangent hyperplane to Q at P and if R belongs to Q the line r_{PR} has three points in common with Q (two at P and one at R), and thus is contained in Q.

d) (Polar space of a given subspace) The polar hyperplanes of the points of a subspace S_h of $\mathbb{P}^n$ form a star Σ_h (of dimension h) whose center S_{n-h-1} is called the *polar space* of S_h. It is obtained by intersecting the polar hyperplanes of $h + 1$ linearly independent points P_j of S_h.

One sees immediately that S_h is the polar space of its polar S_{n-h-1}. Indeed, every point of S_{n-h-1} belongs to the polar hyperplanes of the points P_j and therefore, by the Reciprocity Theorem, its polar hyperplane passes through these points and so too through the space S_h that joins them.

Suppose that two mutually polar spaces S_h and S_{n-h-1} have a space S_i in common. Each point $R \in S_i$ is self-conjugate with respect to $\mathcal{Q}$. Indeed, if we regard R for example as a point of S_h one sees immediately that its polar hyperplane passes through the space S_{n-h-1} which is the polar space of S_h and so through the point R itself which belongs to that S_{n-h-1}. This means that

- if two mutually polar spaces are not skew, their intersection is contained in the quadric.

For the sequel it will be useful to consider the special case $n = 1$.

Remark 5.5.5. A non-specialized quadric $\mathcal{Q}$ of $\mathbb{P}^1$ is constituted by a pair of distinct points (since the square matrix of the quadratic form $f(x_0, x_1) = 0$ which defines $\mathcal{Q}$ is non-degenerate). Moreover, for each $P_0 \in \mathbb{P}^1$, if $P_1 = \mathcal{Q}_1(P_0)$ is the polar point of P_0 with respect to $\mathcal{Q}$, one has $P_0 = \mathcal{Q}_1(P_1)$ by the Reciprocity Theorem 5.4.6.

We will say that a set $\{P_0, \ldots, P_n\}$ of $n + 1$ linearly independent points constitutes a *self-polar* $(n + 1)$*-hedron* with respect to a quadric $\mathcal{Q}$ (or with respect to the polarity γ associated to $\mathcal{Q}$) if each of the points P_i coincides with the pole of the hyperplane H_i generated by the remaining points P_j, $j = 0, \ldots, n$, $j \neq i$, or equivalently (cf. Theorem 5.4.6)

$$\gamma(P_i) = H_i = \langle P_0, \ldots, \widehat{P_i}, \ldots, P_n \rangle.$$

We now prove the existence of a self-polar $(n + 1)$-hedron.

Lemma 5.5.6. *Let $\mathcal{Q}$ be a non-specialized quadric in $\mathbb{P}^n$. Then in $\mathbb{P}^n$ there exist infinitely many systems of $n + 1$ independent points $P_0, P_1, \ldots, P_n$ none of which belongs to $\mathcal{Q}$ and constituting a self-polar $(n + 1)$-hedron with respect to $\mathcal{Q}$.*

Proof. We proceed by induction on n. If $n = 1$ it suffices to take as P_0 an arbitrary point distinct from the two points comprising $\mathcal{Q}$ and as P_1 the polar point of P_0 with respect to $\mathcal{Q}$ (cf. Remark 5.5.5).

If $n > 1$ we take as P_0 an arbitrary point not belonging to $\mathcal{Q}$ and let $H_0 = \mathcal{Q}_1(P_0)$ be the polar hyperplane of P_0 with respect to $\mathcal{Q}$. Moreover let $\mathcal{Q}_0$ be the quadric section of H_0, $\mathcal{Q}_0 = \mathcal{Q} \cap H_0$. By the inductive hypothesis there exist n independent points $P_1, \ldots, P_n \in H_0$ that constitute (in H_0) a self-polar n-hedron with respect to the quadric $\mathcal{Q}_0$. In virtue of the Section Theorem 5.4.5, the polar hyperplane H_j of P_j with respect to $\mathcal{Q}$ intersects H_0 along the space S_{n-2} polar to P_j with respect to $\mathcal{Q}_0$, which is to say along the space joining the points $\{P_1, \ldots, \widehat{P_j}, \ldots, P_n\}$. Since H_j passes through P_0, H_j is the join of the n points $\{P_0, P_1, \ldots, \widehat{P_j}, \ldots, P_n\}$. $\square$

5.5. Quadrics in projective space

5.5.7 Reduction to canonical form of a quadric in $\mathbb{P}^n$. Let $\mathcal{Q}$ be a non-specialized quadric in $\mathbb{P}^n$, $A = (a_{ij}) \in M_{n+1}(K)$ the associated matrix and $\gamma\colon \mathbb{P}^n \to \mathbb{P}^{n*}$ the polarity associated to A. By Lemma 5.5.6 we can take as our fundamental points $A_0 = [1, 0, \ldots, 0], \ldots, A_n = [0, \ldots, 0, 1]$, the vertices of a self-polar $(n + 1)$-hedron with respect to $\mathcal{Q}$. Then the polar hyperplane $\gamma(A_i) = \langle A_0, \ldots, \widehat{A_i}, \ldots, A_n\rangle$ of A_i with respect to $\mathcal{Q}$ has equation $x_i = 0$, $i = 0, \ldots, n$. On the other hand $\gamma(A_i)$ has equation

$$x_{A_i} A\,{}^t(x_0, \ldots, x_n) = 0,$$

where x_{A_i} is the vector of the coordinates of A_i, $i = 0, \ldots, n$. Thus one must have

$$a_{ij} = 0, \quad i, j = 0, \ldots, n,\; i \neq j,$$

and so the equation of the quadric $\mathcal{Q}$ assumes the *diagonal form*

$$\sum_{i=0}^{n} \alpha_i x_i^2 = 0, \tag{5.29}$$

with all coefficients $\alpha_i = a_{ii}$ different from zero. By a suitable choice of the unit point one can then always suppose that (5.29) is rewritten in the *canonical form*

$$\sum_{i=0}^{n} x_i^2 = 0. \tag{5.30}$$

Exercise 5.5.8. As an exercise, we give a variant of the preceding argument for obtaining (5.29), (5.30).

We begin by considering the case $n = 1$. Let P_0, P_1 be two points of a line not belonging to $\mathcal{Q}$ and mutually reciprocal with respect to $\mathcal{Q}$, that is such that each is the polar point of the other (cf. Remark 5.5.5). Since P_0 and P_1 are distinct we may suppose that $P_0 = [1, 0]$ and $P_1 = [0, 1]$ and write the equation of $\mathcal{Q}$ in the form

$$ax_0^2 + bx_1^2 = 0$$

with $ab \neq 0$. It then suffices to change the unit point via a change of coordinates expressed by

$$x_0' = \alpha x_0, \quad x_1' = \beta x_1,$$

with α and β such that $\alpha^2 = a$, $\beta^2 = b$, in order to obtain equation (5.30).

Now let $n \geq 2$. We take a point P_0 not belonging to $\mathcal{Q}$ and its polar hyperplane $H_0 = \mathcal{Q}_1(P_0)$. Since P_0 does not belong to H_0 we may assume that the reference system is chosen so that $P_0 = A_0$ and H_0 has equation $x_0 = 0$. Hence we can suppose that the equation of $\mathcal{Q}$ is of the form

$$a_{00} x_0^2 + g(x_1, \ldots, x_n) = 0,$$

where $g = 0$ is the equation of a non-specialized quadric in $S_{n-1} = H_0$. One completes the proof easily by induction on n.

5.5.9 The linear subspaces of a quadric. Let $\mathcal{Q}$ be a non-specialized quadric in $\mathbb{P}^n$ and $S_h \subset \mathbb{P}^n$ a linear space. If S_h is contained in $\mathcal{Q}$, it belongs to its polar S_{n-h-1} (cf. §5.5.4, d)). Hence $h \leq n - h - 1$, that is,

$$h \leq \left[\frac{n-1}{2}\right],$$

where "[ ]" denotes the greatest integer function.

We now prove that $\mathcal{Q}$ contains linear subspaces S_h of maximal dimension $h = h(n) = \left[\frac{n-1}{2}\right]$.

The hyperplane T_P tangent to $\mathcal{Q}$ in one of its points P intersects $\mathcal{Q}$ in a cone Γ with vertex P which contains every linear space S_h lying on $\mathcal{Q}$ and passing through P.

Consider a hyperplane H not passing through P and let $\mathcal{Q}'$ be the quadric section of $\mathcal{Q}$ with $S_{n-2} = H \cap T_P$. Every maximal linear subspace of $\mathcal{Q}$ intersects this S_{n-2} in a space S_{h-1} of maximal dimension in $\mathcal{Q}'$: indeed, if $\mathcal{Q}'$ contained a linear space of dimension $> h - 1$, that space, joined with P, would give a space of dimension $> h$ lying in $\mathcal{Q}$. Conversely, each S_{h-1} lying on $\mathcal{Q}'$ gives, when joined with P, a space S_h belonging to $\mathcal{Q}$. Thus there is a bijection between the set of subspaces of maximal dimension of $\mathcal{Q}$ issuing from a fixed point and the totality of the subspaces of maximal dimension of the quadric sections $\mathcal{Q}' \subset S_{n-2}$. Therefore the maximal dimension $h(n)$ of the linear spaces lying on a quadric $\mathcal{Q} \subset \mathbb{P}^n$ verifies the relation

$$h(n) = 1 + h(n-2).$$

If $n = 2p + 1$ is odd, one has the $(p-1)$ relations

$$h(2p+1) = 1 + h(2p-1),$$
$$h(2p-1) = 1 + h(2p-3),$$
$$\vdots$$
$$h(5) = 1 + h(3).$$

Summing term by term, and bearing in mind that $h(3) = 1$, one obtains

$$h(n) = h(2p+1) = p - 1 + h(3) = p = \left[\frac{n-1}{2}\right].$$

Similarly, if $n = 2p$ is even, one has the $(p-1)$ relations

$$h(2p) = 1 + h(2p-2),$$
$$h(2p-2) = 1 + h(2p-4),$$
$$\vdots$$
$$h(4) = 1 + h(2).$$

5.5. Quadrics in projective space

Again summing term by term, and remembering that $h(2) = 0$, one obtains

$$h(n) = h(2p) = p - 1 + h(2) = p - 1 = \left[\frac{n-1}{2}\right].$$

The subspaces of maximal dimension $h(n)$ lying on $\mathcal{Q}$ are $\infty^{d(n)}$, where $d(n)$ is recursively defined by the relation

$$d(n) = d(n-2) + (n-1) - h(n), \tag{5.31}$$

on taking into account that $d(2) = 1$ (points on a conic are parameterized by one parameter). Indeed, if we consider a generic $S_{n-h(n)}$, it will cut $\mathcal{Q}$ in a quadric $\bar{\mathcal{Q}}$ and a generic linear subspace of maximal dimension of $\mathcal{Q}$ in a point. Thus it suffices to count the linear subspaces of maximal dimension of $\mathcal{Q}$ issuing from the points of $\bar{\mathcal{Q}}$ (which are $\infty^{n-h(n)-1}$). On the other hand, as has already been observed, the spaces of maximal dimension issuing from a fixed point on $\mathcal{Q}$ are $\infty^{d(n-2)}$. Thus $d(n) = d(n-2) + n - h(n) - 1$, that is (5.31).

Then, if $n = 2p + 1$ is odd, we have the relations

$$d(2p+1) = d(2p-1) + 2p - p = d(2p-1) + p,$$
$$d(2p-1) = d(2p-3) + p - 1,$$
$$\vdots$$
$$d(5) = d(3) + 2.$$

Summing term by term and recalling that $d(3) = 1$ (the lines on a quadric in $\mathbb{P}^3$ are ∞^1, which means that they depend on one parameter) one obtains

$$d(n) = d(2p+1) = 1 + 2 + \cdots + p = \frac{p(p+1)}{2} = \frac{n^2 - 1}{8}.$$

If $n = 2p$ is even, one has

$$d(2p) = d(2p-2) + 2p - p = d(2p-2) + p,$$
$$d(2p-2) = d(2p-4) + p - 1,$$
$$\vdots$$
$$d(4) = d(2) + 2.$$

Once again adding term by term and recalling that $d(2) = 1$, one finds that

$$d(n) = d(2p) = 1 + 2 + \cdots + p = \frac{p(p+1)}{2} = \frac{n(n+2)}{8}.$$

Note that, if n is odd, on $\mathcal{Q}$ there are two different systems of linear subspaces of maximal dimension (indeed, by way of successive intersections of the original quadric $\mathcal{Q}$ with linear subspaces of codimension 2, one obtains a non-specialized quadric $\mathcal{Q}'$ in $\mathbb{P}^3$, that contains two arrays of lines). The linear subspaces of maximal dimension of a quadric in a space of even dimension by contrast form a unique system (if fact, in the same way, in this case one obtains a conic $\mathcal{Q}'$, whose points constitute a unique system).

For example, a non-singular quadric in $\mathbb{P}^4$ contains a unique 3-dimensional system of lines. A quadric in $\mathbb{P}^5$ contains two different systems of planes, both of dimension 3.

Exercise 5.5.10. Let $\mathcal{Q}$ be a quadric in $\mathbb{P}^n$ tangent to a linear space S_h along a subspace S_{h-1}. Show that if $h \geq n/2$ then S_{h-1} contains a double (that is, singular for $\mathcal{Q}$) subspace of dimension $\geq 2h - n - 1$.

In fact, if $x_{h+1} = \cdots = x_n = 0$ are the equations of S_h and $x_h = x_{h+1} = \cdots = x_n = 0$ are those of S_{h-1}, then the equation of $\mathcal{Q}$ may be written in the form

$$x_h^2 + L_{h+1} x_{h+1} + \cdots + L_n x_n = 0,$$

with L_j linear forms. The linear space with equations $x_h = x_{h+1} = \cdots = x_n = L_{h+1} = \cdots = L_n = 0$ has dimension $2h - n - 1$, and all its points are double points for $\mathcal{Q}$.

5.6 Complements on polars

In the following discussion we will systematically employ the results of Sections 5.1, 5.2 and 5.4 without explicit reference.

Let X be a hypersurface in $\mathbb{P}^n$ of order r. We have seen in §5.4.10 that if P is an s-fold point of X, the $(r-s+1)^{\text{st}}, (r-s+2)^{\text{nd}}, \ldots$ polars of P with respect to X are indeterminate and the $(r-s)^{\text{th}}$ polar is the tangent cone to X at P. Now we consider the jth polar $X_j(P)$ of the s-fold point P, under the hypothesis that $j \leq r - s$.

If $P = A_0 = [1, 0, \ldots, 0]$, the hypersurface X has equation

$$f = x_0^{r-s} f_s(x_1, \ldots, x_n) + x_0^{r-s-1} f_{s+1}(x_1, \ldots, x_n) + \cdots + f_r(x_1, \ldots, x_n) = 0,$$

and $f_s(x_1, \ldots, x_n) = 0$ is the equation of the tangent cone $TC_P(X)$ to X at P. The jth polar $X_j(P)$ has equation

$$\frac{\partial^j f}{\partial x_0^j} = (r-s)(r-s-1) \ldots (r-s-j) f_s(x_1, \ldots, x_n) x_0^{r-s-j}$$

$$+ (r-s-1) \ldots (r-s-j-1) f_{s+1}(x_1, \ldots, x_n) x_0^{r-s-j-1} + \cdots = 0.$$

Recalling that $X_j(P)$ has order $r - j$, we have thus established the following fact:

5.6. Complements on polars

Lemma 5.6.1. *Let $X = V(f)$ be a hypersurface of order r in $\mathbb{P}^n$, P an s-fold point of X, $TC_P(X)$ the tangent cone to X at P. For each $j \leq r - s$ the jth polar $X_j(P)$ has P as an s-fold point with $TC_P(X)$ as tangent cone.*

Now let Q be an arbitrary point of $\mathbb{P}^n$ and $P \neq Q$ an s-fold point of X. Suppose that $Q = A_1 = [0, 1, \ldots, 0]$ and also that $P = A_0 = [1, 0, \ldots, 0]$. The first polar $X_1(P)$ of P has equation

$$\frac{\partial f}{\partial x_1} = x_0^{r-s} \frac{\partial f_s(x_1, \ldots, x_n)}{\partial x_1} + x_0^{r-s-1} \frac{\partial f_{s+1}(x_1, \ldots, x_n)}{\partial x_1} + \cdots$$
$$\cdots + \frac{\partial f_r(x_1, \ldots, x_n)}{\partial x_1} = 0$$

and so has multiplicity $r - 1 - (r - s) = s - 1$ at P, and its tangent cone at P, with equation $\frac{\partial f_s}{\partial x_1} = 0$, is the first polar of Q with respect to the tangent cone to X at P. But the case when the polynomial $f_s(x_1, \ldots, x_n)$ doesn't depend on x_1, so that $\frac{\partial f_s}{\partial x_1} = 0$, is an exception. If that happens, then the point P is at least s-fold for $X_1(Q)$ and the tangent cone to X at P has the line r_{PQ} as its vertex.

More generally, the jth polar $X_j(Q)$ of Q has equation

$$\frac{\partial^j f}{\partial x_1^j} = x_0^{r-s} \frac{\partial^j f_s}{\partial x_1^j} + x_0^{r-s-1} \frac{\partial^j f_{s+1}}{\partial x_1^j} + \cdots + \frac{\partial^j f_r}{\partial x_1^j} = 0$$

and one has the following result.

Lemma 5.6.2. *Let X be a hypersurface of order r in $\mathbb{P}^n$, P an s-fold point of X, and $TC_P(X)$ the tangent cone to X at P; let $Q \neq P$ be an arbitrary point of $\mathbb{P}^n$. The jth polar $X_j(Q)$ of Q generally has multiplicity $s - j$ ($= r - j - (r - s)$) in P, and has as its tangent cone in P the jth polar of Q with respect to $TC_P(X)$; in symbols*

$$TC_P(X_j(Q)) = (TC_P(X))_j(Q).$$

In the exceptional case in which the variable x_1 appears in $f_s(x_1, \ldots, x_n)$ only to a degree $< j$ (and $Q = A_1$) the multiplicity of $X_j(Q)$ at P is $> s - j$. In this case the tangent cone to X at P has the line r_{PQ} as at least an $(s-j)$-fold generator (that is, each point of the line is a point of multiplicity at least $s - j$ for X).

In particular Lemma 5.6.2 assures us that if P is a double point of X the first polar $X_1(Q)$ of Q with respect to X passes simply through P and has as its tangent hyperplane at P the polar hyperplane of Q with respect to the tangent cone $TC_P(X)$. This hyperplane is the locus of the harmonic conjugates of Q with respect to the pairs of points of $TC_P(X)$ that are collinear with Q, and so (if P is a double point with non-singular tangent cone) it is not tangent to $TC_P(X)$.

136   Chapter 5. Hypersurfaces in Projective Space

Now let X again be a hypersurface of order r in $\mathbb{P}^n$, $P = A_0$, and $Q = A_1$. If s_1, s_2 are positive integers such that $s := s_1 + s_2 < r$ (and so $r \geq 3$) we have

$$\frac{\partial^{s_1}}{\partial x_0^{s_1}} \left(\frac{\partial^{s_2} f}{\partial x_1^{s_2}} \right) = \frac{\partial^{s_2}}{\partial x_1^{s_2}} \left(\frac{\partial^{s_1} f}{\partial x_0^{s_1}} \right). \tag{5.32}$$

Now, the vanishing of the left-hand side of (5.32) means that in the equation $\frac{\partial^{s_2} f}{\partial x_1^{s_2}} = 0$ of the s_2^{th} polar $X_{s_2}(A_1)$ of A_1 the variable x_0 appears to degree $\leq s_1 - 1$, and therefore that A_0 is a point of multiplicity $\geq r - s_2 - (s_1 - 1) = r - s + 1$ for $X_{s_2}(A_1)$. The vanishing of the right-hand side of (5.32) means that in the equation $\frac{\partial^{s_1} f}{\partial x_0^{s_1}} = 0$ of the s_1^{th} polar $X_{s_1}(A_0)$ of A_0 the variable x_1 appears to degree $\leq s_2 - 1$, and thus that A_1 is a point of multiplicity $\geq r - s_1 - (s_2 - 1) = r - s + 1$ for $X_{s_1}(A_0)$. Thus we have:

Lemma 5.6.3. *Let X be a hypersurface of order r in $\mathbb{P}^n$, P, Q two points of $\mathbb{P}^n$ and s_1, s_2 positive integers such that $s := s_1 + s_2 < r$. If the s_2^{th} polar $X_{s_2}(Q)$ of Q has P as (at least) an $(r - s + 1)$-fold point, then the s_1^{th} polar $X_{s_1}(P)$ of P has Q as (at least) an $(r - s + 1)$-fold point.*

For example, if $s_1 = 1$ and $s_2 = r - 2$, we find that if the polar quadric $X_{r-2}(Q)$ of Q has P as a double point, the first polar $X_1(P)$ of P has a double point at Q.

5.6.4 The Hessian hypersurface. Let $X \subset \mathbb{P}^n$ be a hypersurface of order r and let $\mathcal{H}$ be the locus of points P of $\mathbb{P}^n$ whose quadric polar $X_{r-2}(P)$ with respect to X has a double point.

By Lemma 5.6.3, if P is a point of $\mathcal{H}$ and Q is the double point of its quadric polar $X_{r-2}(P)$, the first polar $X_1(Q)$ of Q has a double point at P; and conversely, if the first polar $X_1(Q)$ of Q has a double point at P, the quadric polar $X_{r-2}(P)$ of P has Q as a double point. Therefore, $\mathcal{H}$ can be defined as the locus of the double points for some first polar.

In order for the quadric polar of $P = [y_0, \ldots, y_n]$, which has equation (cf. (5.21) and Remark 5.4.8)

$$\sum_{i_1,\ldots,i_{r-2}} y_{i_1} \cdots y_{i_{r-2}} \frac{\partial^{r-2} f}{\partial x_{i_1} \cdots \partial x_{i_{r-2}}} = \sum_{i,j=0}^{n} x_i x_j \frac{\partial^2 f}{\partial y_i \partial y_j} = 0,$$

to have a double point it is necessary and sufficient that the determinant of the matrix of coefficients vanish (cf. Section 5.5). Thus, the locus $\mathcal{H}$ of points of $\mathbb{P}^n$ whose polar quadric has a double point, or also the locus of the double points of the first polars, is the hypersurface represented by the equation

$$\det \left(\frac{\partial^2 f}{\partial y_i \partial y_j} \right) = 0. \tag{5.33}$$

5.6. Complements on polars

This hypersurface is called the *Hessian hypersurface* of X and has order $\deg(\mathcal{H}) = (n+1)(r-2)$.

One sees immediately that $\mathcal{H}$ passes through every multiple point P of X. Indeed, if the multiplicity of P is $s \geq 3$, then at P all the second order partial derivatives of f vanish and so P satisfies (5.33). Moreover, we have seen that if P is only a double point for X, the polar quadric of P is the tangent cone to X at P and hence has a double point at P, which is to say $P \in \mathcal{H}$.

It is easy to prove that the Hessian hypersurface of X generally has multiplicity $m_P(\mathcal{H}) = (n+1)(s-2) + 2$ at an s-fold point P of X. In fact, suppose that $P = [1, 0, \ldots, 0]$ is an s-fold point of X, and so

$$f = x_0^{r-s} f_s(x_1, \ldots, x_n) + x_0^{r-s-1} f_{s+1}(x_1, \ldots, x_n) + \cdots + f_r(x_1, \ldots, x_n).$$

By observing the following table, in which position (i, j) shows the degree of $\frac{\partial^2 f}{\partial x_i \partial x_j}$ with respect to x_0,

$r-s-2$	$r-s-1$	$r-s-1$	$\ldots$	$r-s-1$
$r-s-1$	$r-s$	$r-s$	$\ldots$	$r-s$
$\vdots$	$\vdots$	$\vdots$		$\vdots$
$r-s-1$	$r-s$	$r-s$	$\ldots$	$r-s$

one sees immediately that the degree of $\mathcal{H}$ with respect to x_0 is in general

$$(r-s-2) + n(r-s) = 2(r-s-1) + (n-1)(r-s) = (r-s)(n+1) - 2.$$

It follows that

$$\deg(\mathcal{H}) - ((r-s)(n+1) - 2) = (r-2)(n+1) - (r-s)(n+1) + 2 = (n+1)(s-2) + 2,$$

that is, $m_P(\mathcal{H}) = (n+1)(s-2) + 2$.

We shall now see how the non-singular points of a hypersurface which belong to the Hessian may be characterized.

Let $P = A_0$ be a simple point of X and $x_1 = 0$ the tangent hyperplane to X at P, so that

$$f = x_0^{r-1} x_1 + x_0^{r-2} f_2(x_1, \ldots, x_n) + \cdots.$$

If $f_2 = \sum_{i,j=1}^n a_{ij} x_i x_j$, a direct calculation (long but elementary and which we omit for brevity) shows that the necessary and sufficient condition for the Hessian of X to pass through P is

$$\det(a_{ij}) = 0.$$

But this says that the quadric cone with equations $x_1 = f_2 = 0$ (the locus of the lines having intersection multiplicity > 2 with X at P) has a double generator, that is, the tangent hyperplane to X at P is tangent to the quadric cone $f_2 = 0$.

If P is a simple point of X belonging to the Hessian one says that P is a *parabolic point*. The parabolic points of a plane algebraic curve X are its flexes, which are in number (if X is non-singular) $3r(r-2)$. The parabolic points of a surface of order r in $\mathbb{P}^3$ are the points for which the two principal tangents coincide. In general they are the points of a curve, which is called the *parabolic curve* of the surface X, of order $4r(r-2)$.

The *locus of the parabolic points* is the intersection of the hypersurface X with its Hessian and is (in general) an $(n-2)$-dimensional variety of order $r(r-2)(n+1)$, cf. Corollary 4.5.5.

If X is a quadric (the case $r = 2$), the Hessian determinant is a constant $(= 2^{n+1} \det(A)$, where A is the matrix of the coefficients of the quadric) and so the points of a non-degenerate quadric are all non-parabolic, while the simple points of a degenerate (but irreducible) quadric are all parabolic (cf. Section 5.5).

5.6.5 The class of a hypersurface. As an application of the theory of polar hypersurfaces and of Bézout's theorem, we calculate the class of an algebraic hypersurface X of order r in $\mathbb{P}^n$. The *class* $v = v(X)$ of X is the number of tangent hyperplanes to X that belong to a generic pencil, that is, that pass through a generic S_{n-2}.

Let $S = S_{n-2}$ be a generic $(n-2)$-dimensional subspace, and let $P_1, \ldots, P_{n-1}$ be $n-1$ linearly independent points of S. Let Π be the hyperplane tangent to X at one of its non-singular points Q. If Π (which is the $(r-1)^{\text{th}}$ polar of Q with respect to X) passes through S, that is, if Π contains all the points P_j, the first polar of each of these points with respect to X passes through Q by the Reciprocity Theorem (Theorem 5.4.6). Conversely, if Q is a simple point of X belonging to the first polars of the points P_j, the tangent hyperplane to X at Q contains the points P_j and therefore passes through S.

The points Q in which X has tangent hyperplane passing through S are thus the simple points of X that belong to the first polars of the points P_j, $j = 1, \ldots, n-1$. Since these polars are hypersurfaces of order $r-1$, the number of such points Q, that is, the class of X, is, by Bézout's theorem (Theorem 4.5.2), $v(X) \leq r(r-1)^{n-1}$.

With respect to a reference system having among its fundamental points the points $Q = A_0, P_1 = A_1, \ldots, P_{n-1} = A_{n-1}$, the equation of the hyperplane Π is $x_n = 0$ and that of X is

$$f = x_0^{r-1} x_n + x_0^{r-2} f_2(x_1, \ldots, x_n) + \cdots = 0.$$

The first polars of the points A_j have equations

$$\frac{\partial f}{\partial x_j} = x_0^{r-2} \frac{\partial f_2(x_1, \ldots, x_n)}{\partial x_j} + \cdots = 0;$$

and $\frac{\partial f_2}{\partial x_j} = 0$ is the equation of the tangent hyperplane to $X_1(A_j)$ at Q, $j =$

$1, \ldots, n-1$. The hypothesis that S_{n-2} is generic implies that the polynomials $\frac{\partial f_2}{\partial x_j}$ are not null and therefore A_0 is a simple point of $X_1(A_j)$, for $j = 1, \ldots, n-1$.

The condition in order for the n hyperplanes $x_n = 0$, $\frac{\partial f_2}{\partial x_j} = 0$, $j = 1, \ldots, n-1$, not to be linearly independent is that the matrix of their coefficients have rank $\leq n-1$, namely that the matrix $\Lambda \in M_{n-1}(K)$ of the coefficients of the linear forms $f_2^{(j)}(x_1, \ldots, x_{n-1}, 0)$, have rank $\varrho(\Lambda) \leq n-2$, where $f_2^{(j)} := \frac{\partial f_2}{\partial x_j}$. Thus the cone with equations

$$f_2(x_1, \ldots, x_{n-1}, x_n) = x_n = 0,$$

consisting of the lines having multiplicity of intersection at least 3 with X at A_0, should have its locus of double points of dimension ≥ 1 (cf. (5.7) in §5.2.1).

Since the points for which this occurs are at most ∞^{n-2} (∞^{n-1} for the points of X and the relation $\varrho(\Lambda) \leq n-2$ imposes at least one condition) for a generic space S_{n-2} there do not pass tangent hyperplanes of this nature. (Indeed, in the dual space $\mathbb{P}^{n*}$ the pencil of hyperplanes is a line, and a generic line of $\mathbb{P}^{n*}$ does not meet a variety of dimension $\leq n-2$.)

Hence we may conclude that each non-singular point Q of X for which X has tangent hyperplane passing through a generic $S_{n-2} = J(P_1, \ldots, P_{n-1})$ is a non-singular point for the first polar $X_1(P_j)$ of each of the points P_j. Moreover, the n hypersurfaces $X, X_1(P_1), X_1(P_2), \ldots, X_1(P_{n-1})$ intersect transversally at Q (having there linearly independent tangent hyperplanes) and thus their intersection multiplicity at Q is $m_Q(X, X_1(P_1), X_1(P_2), \ldots, X_1(P_{n-1})) = 1$.

By Bézout's theorems 4.5.1 and 4.5.2, we may then conclude that if X is non-singular its class is $\nu(X) = r(r-1)^{n-1}$.

Suppose now that X has a *node* of the most general type, that is, a double point P whose tangent cone has no multiple generators. Let $P = A_0 = [1, 0, \ldots, 0]$ and let

$$f = x_0^{r-2} f_2(x_1, \ldots, x_n) + \cdots = 0$$

be the equation of X. The point $P = A_0$ is non-singular for $X_1(P_j)$ and $\frac{\partial f_2}{\partial x_j} = 0$ is the equation of the tangent hyperplane to $X_1(P_j)$ at P. The first polars $X_1(P_j)$ have intersection multiplicity

$$m_Q(X, X_1(P_1), \ldots, X_1(P_{n-1})) = 2$$

with X at Q. Indeed, by the hypotheses of generality assumed, the $n-1$ hyperplanes $\frac{\partial f_2}{\partial x_j} = 0$ tangent to the first polars $X_1(P_j)$, $j = 1, \ldots, n-1$, at $P_1 = A_0$ have in common a line ℓ (passing through A_0). This line is not contained in the cone with equation $f_2 = 0$ tangent to X at A_0, because otherwise from the identity (cf. Exercise 3.1.18)

$$2f_2 = x_1 \frac{\partial f_2}{\partial x_1} + x_2 \frac{\partial f_2}{\partial x_2} + \cdots + x_n \frac{\partial f_2}{\partial x_n}$$

and the fact that $x_n = 0$ is a generic hyperplane, one would have that also the hyperplane $\frac{\partial f_2}{\partial x_n} = 0$ would contain the line ℓ and so ℓ would be a singular generator of the cone $f_2 = 0$.

On the other hand, by what we have seen above, and in the case of a hypersurface of order r all of whose singularities are only isolated double points P of the most general type, it follows from Bézout's theorem (Theorem 4.5.2) that

$$\deg(X) \deg(X_1(P_1)) \ldots \deg(X_1(P_{n-1}))$$
$$= v(X) + \sum_P m_P(X, X_1(P_1), \ldots, X_1(P_{n-1})),$$

where the sum is extended over the isolated double points $P \in X$. Thus, if d is the number of these double points, the class of X is

$$v(X) = r(r-1)^{n-1} - 2d.$$

In analogous fashion one proves that if the singularities of X are only isolated multiple points of multiplicity s_i and of the most general type, then the class of X is

$$v(X) = r(r-1)^{n-1} - \sum_i s_i(s_i - 1)^{n-1}.$$

See [44], [43] for results regarding the class of an algebraic surface; and also the texts [108] and [115] for an exposition of other results on the notion of class for algebraic varieties.

5.7 Plane curves

The exercises of this section serve to illustrate some properties of plane curves related to the theory developed in this chapter. Exercises 5.7.13–5.7.20 are dedicated to the remarkable case of plane cubics, that is, plane algebraic curves of order 3; in this regard see also [113, Chapter III, §6]. We assume that $K = \mathbb{C}$.

We start by recalling a few definitions. Let P be a non-singular point of a plane algebraic curve X (of order $r > 1$) and let ℓ be the tangent to X at P. We say that P is a *flex* of X if the intersection multiplicity of ℓ and X at P is $m_P(X, \ell) \geq 3$. If $m_P(X, \ell) = 3$, P is an *ordinary flex* or *flex of the first kind*; if $m_P(X, \ell) = 2 + h$, P is a *flex of type h*. If $m_P(X, \ell) = 4$, that is, if P is a flex of the second type, one also says that P is a *point of undulation*.

An s-fold point P of a plane algebraic curve X is said to be *ordinary* if the tangent cone to X at P consists of s mutually distinct tangent lines such that the intersection multiplicity at P of each of these with X is $s + 1$ (that is, the minimum possible); in the case $s = 2$, an ordinary double point is also called a *node*.

A double point P having its two tangent lines coinciding with a single line ℓ is called an *(ordinary) cusp* if the intersection multiplicity of ℓ with X at P is $m_P(X, \ell) = 3$, the minimum possible.

In the rest of this section, $X \subset \mathbb{P}^2$ will denote an (irreducible) algebraic plane curve of order r with equation $f(x_0, x_1, x_2) = 0$.

5.7.1. Let X, X' be two plane cubics having a common point P as an ordinary cusp of both curves, and with the same cuspidal tangent. Prove that $m_P(X, X') = 6$.

In affine coordinates, if $P = (0, 0)$ is the common cusp and the line $y = 0$ is the common tangent, the two cubics have equations of the form

$$y^2 + f_3(x, y) = 0, \quad y^2 + g_3(x, y) = 0,$$

with f_3, g_3 homogeneous polynomials of degree 3. If C is the curve with equation $f_3 - g_3 = 0$, one has (cf. Exercise 4.5.3)

$$m_P(X, X') = m_P(X, C) = 6,$$

since P is a double point for X and a triple point for C. Note that the number of common tangents to the two cubics X, X' at P is $t = 2$, in agreement with relation (4.9).

5.7.2. In the plane π we consider an algebraic curve X, a point Q not belonging to X and a simple point P of X belonging to the first polar $X_1(Q)$ of Q. We know that if Q is a generic point one has $m_P(X, X_1(Q)) = 1$ (cf. §5.6.5). Prove that $m_P(X, X_1(Q)) > 1$ if and only if P is a flex of X.

If r is the order of X and if $Q = [0, 1, 0]$, $P = [1, 0, 0]$ then the equation of X has the form

$$f = x_0^{r-1} x_2 + x_0^{r-2}(ax_1^2 + bx_1 x_2 + cx_2^2) + x_0^{r-3} f_3(x_1, x_2) + \cdots,$$

with $a, b, c \in \mathbb{C}$ and f_3 a form of degree 3 in x_1, x_2. And in order for the first polar of Q, which has equation

$$\frac{\partial f}{\partial x_1} = x_0^{r-2}(2ax_1 + bx_2) + x_0^{r-3} \frac{\partial f_3}{\partial x_1} + \cdots = 0,$$

to be tangent at P to X, that is, that it have the line ℓ with equation $x_2 = 0$ as tangent at P, it is necessary and sufficient that $a = 0$, that is, that P be a flex (if $a = 0$ one has in fact $m_P(X, \ell) \geq 3$).

5.7.3. We know that if X is a plane algebraic curve and P is an s-fold point of X, the first polar $X_1(Q)$ of a generic point Q of the plane has P as an $(s-1)$-fold point (cf. Lemma 5.6.2). If P is an s-fold ordinary point for X, that is, with s distinct tangents, the two curves X and $X_1(Q)$ do not have any tangent in common at P and thus $m_P(X, X_1(Q)) = s(s-1)$.

Suppose that $P = [1, 0, 0]$, $Q = [0, 1, 0]$ and that $x_1 = 0$ is one of the s tangents to X at P, so that X has equation

$$f = x_0^{r-s} x_1 f_{s-1}(x_1, x_2) + x_0^{r-s-1} f_{s+1} + \cdots$$

with f_{s-1} a binary form not divisible by x_1. The first polar of Q has equation

$$\frac{\partial f}{\partial x_1} = x_0^{r-s} \left(f_{s-1} + x_1 \frac{\partial f_{s-1}}{\partial x_1} \right) + \cdots = 0$$

and the polynomial $f_{s-1} + x_1 \frac{\partial f_{s-1}}{\partial x_1}$ is not divisible by x_1. Thus a simple tangent of X is not tangent to $X_1(Q)$.

5.7.4. Let P be an s-fold point for X and Q a generic point of the plane. The first polar $X_1(Q)$ of Q with respect to X has at P multiplicity $m_P(X_1(Q)) \geq s - 1$; and we have $m_P(X_1(Q)) > s - 1$ if and only if the s tangents to X at P all coincide with the line r_{PQ}.

If $P = A_0 = [1, 0, 0]$ and $Q = A_1 = [0, 1, 0]$, the equation of X is

$$f = x_0^{r-s} f_s(x_1, x_2) + x_0^{r-s-1} f_{s+1}(x_1, x_2) + \cdots = 0,$$

and the equation of $X_1(Q)$ is

$$\frac{\partial f}{\partial x_1} = x_0^{r-s} \frac{\partial f_s}{\partial x_1} + x_0^{r-s-1} \frac{\partial f_{s+1}}{\partial x_1} + \cdots = 0.$$

That $X_1(Q)$ has multiplicity $\geq s - 1$ at P is a consequence of Lemma 5.6.2. The condition for P to be at least s-fold for $X_1(Q)$ is that $\frac{\partial f_s}{\partial x_1} = 0$ and so $f_s = x_2^s$; therefore, the s tangents to X at P coincide with the line $r_{PQ} : x_2 = 0$.

5.7.5. If P is an ordinary cusp for X, the first polar $X_1(Q)$ of a generic point $Q \in \mathbb{P}^2$ has a non-singular point at P with tangent that coincides with the cuspidal tangent, and at P the intersection multiplicity of X with $X_1(Q)$ is 3.

If $P = A_0 = [1, 0, 0]$ with cuspidal tangent having equation $x_1 = 0$, the equation of X is

$$f = x_0^{r-2} x_1^2 + x_0^{r-3} f_3(x_1, x_2) + \cdots = 0,$$

with f_3 not divisible by x_1, while that of $X_1(Q)$ is

$$\frac{\partial f}{\partial x_1} = 2x_0^{r-2} x_1 + x_0^{r-3} \frac{\partial f_3}{\partial x_1} + \cdots = 0.$$

From this it follows that P is non-singular for $X_1(Q)$ with tangent $x_1 = 0$.

In particular, the system of equations $f = \frac{\partial f}{\partial x_1} = 0$ is equivalent to the system $f = g = 0$, where

$$g := 2f - x_1 \frac{\partial f}{\partial x_1} = x_0^{r-3} \left(2f_3 - x_1 \frac{\partial f_3}{\partial x_1} \right) + \cdots.$$

The equation $g = 0$ represents a curve C having P as a triple point with all tangents distinct from the cuspidal tangent $x_1 = 0$ of X at P. It follows that $m_P(X, X_1(Q)) = m_P(X, C) = 3$ (cf. Section 4.2).

5.7.6. Let H be the Hessian curve of X. Prove that the following:

(1) An ordinary double point P of X is also a double point for H and the intersection multiplicity of X and H at P is 6.

(2) An ordinary cusp P of X is a triple point of H and two (and, in general only two) of the three tangents to H at P coincide with the cuspidal tangent of X. The intersection multiplicity of X and H at P is 8.

(1) If $P = A_0 = [1, 0, 0]$ and the equation of X is

$$f = x_0^{r-2} x_1 x_2 + x_0^{r-3} f_3(x_1, x_2) + \cdots = 0,$$

one finds that the Hessian equation is

$$H = (r-1)(r-2) x_1 x_2 x_0^{3r-8} + (r-1)\left(2(r-2) x_1 x_2 \frac{\partial^2 f_3}{\partial x_1 \partial x_2} - 2(r-1) f_3\right) x_0^{3r-9} + \cdots = 0.$$

Since $3r - 8 = 3(r-2) - 2 = \deg(H) - 2$, P is a double point of H. If we set

$$g := H - (r-1)(r-2) x_0^{2r-6} f = \left(2(r-1)(r-2) x_1 x_2 \frac{\partial^2 f_3}{\partial x_1 \partial x_2} - 2(r-1)^2 f_3\right) x_0^{3r-9} + \cdots,$$

we see that the system $f = H = 0$ is equivalent to $f = g = 0$, and the equation $g = 0$ represents a curve C that has at P a triple point ($3r - 9 = 3(r-2) - 3 = \deg(C) - 3$) without any tangent in common with X. It follows that $m_P(X, H) = m_P(X, C) = 6$.

(2) The equation of X is

$$f = x_0^{r-2} x_1^2 + x_0^{r-3} f_3(x_1, x_2) + \cdots = 0,$$

where $x_1 = 0$ is the equation of the cuspidal tangent at $P = [1, 0, 0]$. Calculation shows that the equation of the Hessian is

$$H = -2(r-1)(r-2) x_1^2 \frac{\partial^2 f_3}{\partial x_1^2} x_0^{3r-9} + \cdots = 0,$$

and so two of the three tangents to H at P coincide with the cuspidal tangent of X. On setting

$$g := H + 2(r-1)(r-2) \frac{\partial^2 f_3}{\partial x_1^2} x_0^{2r-7} f,$$

one finds that the system $f = H = 0$ is equivalent to $f = g = 0$, and the equation $g = 0$ represents a curve C that has (in general) a quadruple point without any tangent in common with X at P. Thus $m_P(X, H) = m_P(X, C) = 8$.

5.7.7. Show that a non-singular point P of X is a flex if and only if the polar conic P with respect to X is degenerate: one of its components is the tangent to X at P, and the other does not pass through P. The flexes of X are therefore the simple points of X through which the Hessian curve passes.

If $P = A_0 = [1, 0, 0]$, the equation of X is of the form

$$f = x_0^{r-1}(ax_1 + bx_2) + x_0^{r-2}(a_{11}x_1^2 + 2a_{12}x_1x_2 + a_{22}x_2^2) + \cdots = 0,$$

$a, b, a_{11}, a_{12}, a_{22} \in \mathbb{C}$. The equation of the polar conic of P is

$$\frac{\partial^{r-2} f}{\partial x_0^{r-2}} = (r-1)!(ax_1 + bx_2)x_0 + (r-2)!(a_{11}x_1^2 + 2a_{12}x_1x_2 + a_{22}x_2^2) = 0.$$

The necessary and sufficient condition in order that the polar conic of P be degenerate is

$$\begin{vmatrix} 0 & a & b \\ a & a_{11} & a_{12} \\ b & a_{12} & a_{22} \end{vmatrix} = -a_{11}b^2 + 2a_{12}ab - a_{22}a^2 = 0,$$

namely, that the coordinates of the point $[1, b, -a]$ should annull the quadratic form $f_2 := a_{11}x_1^2 + 2a_{12}x_1x_2 + a_{22}x_2^2$, and thus that the latter should be divisible by $ax_1 + bx_2$. If we set

$$a_{11}x_1^2 + 2a_{12}x_1x_2 + a_{22}x_2^2 = (ax_1 + bx_2)(cx_1 + dx_2), \quad c, d \in \mathbb{C},$$

we have

$$\frac{\partial^{r-2} f}{\partial x_0^{r-2}} = (r-2)!(ax_1 + bx_2)((r-1)x_0 + cx_1 + dx_2).$$

In particular, one component of the polar conic is the line ℓ tangent to X at P and the other component does not pass through P. On the other hand, P is a flex for X if and only if $ax_1 + bx_2$ divides the quadratic form f_2; indeed, this is equivalent to $m_P(X, \ell) \geq 3$.

Since the Hessian curve of X is the locus of the points $P \in \mathbb{P}^2$ whose polar quadric is degenerate (cf. §5.6.4), one has that the flexes of X are the simple points of X through which the Hessian curve passes.

5.7.8. Let P be an ordinary flex of X. Then X and its Hessian H meet transversally at P and so $m_P(X, H) = 1$.

Take P to be the point $[1, 0, 0]$ and the inflectional tangent to be the line with equation $x_2 = 0$, so that the equations of X and H may be written in the form

$$f = x_0^{r-1} x_2 + x_0^{r-2} x_2 f_1(x_1, x_2) + x_0^{r-3} f_3(x_1, x_2) + \cdots,$$

$$H = (r-1)\left((r-2)\left(\frac{\partial f_1}{\partial x_1}\right)^2 x_2 - (r-1)\frac{\partial^2 f_3}{\partial x_1^2}\right)x_0^{3r-7} + \cdots.$$

It then suffices to observe that P is a non-singular point for H ($3r - 7 = 3(r-2) - 1 = \deg(H) - 1$) and the tangent to H at P is distinct from the inflectional tangent of X inasmuch as $f_3(x_1, x_2)$ is not divisible by x_2 since P is an ordinary flex.

5.7.9. Suppose that X has only *Plückerian singularities*, that is, only nodes and ordinary cusps. If r, d, k are the order, the number of nodes and the number of cusps and v, δ, ρ are the dual characters that is, the class, the number of bitangents and the number of flexes, one has the following two relations:

$$v = r(r-1) - 2d - 3k, \tag{5.34}$$
$$\rho = 3r(r-2) - 6d - 8k, \tag{5.35}$$

known as *Plücker's formulas*. In addition to these relations, one also has the dual relations

$$r = v(v-1) - 2\delta - 3\rho, \tag{5.36}$$
$$k = 3v(v-2) - 6\delta - 8\rho. \tag{5.37}$$

These formulas are not independent: given any three of them one can deduce the fourth. Indeed, from (5.34), (5.35) or (5.36), (5.37) one obtains the relation

$$3r - k = 3v - \rho.$$

If $k = 0$, that is, if X has only nodes, the class is $v = r(r-1) - 2d$ (cf. §5.6.5). Relation (5.34) (for $k \geq 0$) then follows from Exercise 5.7.5.

We have seen in Exercise 5.7.7 that the flexes of X are the non-singular points of X which belong to the Hessian curve, whose order is $3(r-2)$. In the case $k > 0$ relation (5.35) then follows from Exercise 5.7.6 and Theorem 4.2.1. Relation (5.36) follows by duality from (5.34) (cf. Example 5.3.3).

5.7.10. A non-singular C^4 has 24 flexes and 28 bitangents.

This follows immediately from Plücker's formulas.

5.7.11. Show that any non-singular plane curve of order 4 can be represented by an equation of the type $\varphi_2^2 - \varphi_1 \varphi_3 = 0$, where $\varphi_j(x_0, x_1, x_2)$ denotes a homogeneous polynomial of degree j, $j = 1, 2, 3$.

Let C be a non-singular plane quartic with equation $f = 0$, $\varphi_1 = 0$ the equation of a bitangent ℓ to C and $\varphi_2 = 0$ that of a conic passing through the two points of tangency. In the pencil of quartics $f + \lambda \varphi_2^2 = 0$ there is a quartic split into the line ℓ and an additional cubic $\varphi_3 = 0$. To obtain that curve, it suffices to choose λ in such a way to ensure that the quartic $f + \lambda \varphi_2^2 = 0$ of the pencil passes through a point of ℓ distinct from the two points of contact of ℓ with C.

5.7.12. Prove that it is possible to choose three homogeneous polynomials $\varphi_2(x_0, x_1, x_2)$, $\varphi_3(x_0, x_1, x_2)$, $\varphi_4(x_0, x_1, x_2)$ of degrees 2, 3, 4 respectively so that the polynomial $\varphi_2 \varphi_4 - \varphi_3^2$ is the product of six linear forms.

Let γ be a conic with equation $\varphi_2(x_0, x_1, x_2) = 0$, and let C be a cubic with equation $\varphi_3(x_0, x_1, x_2) = 0$ that intersects γ in six distinct points P_i. Then let ℓ_i be the tangents to γ in the points P_i. In the pencil with equation $\varphi_3^2 - \lambda \ell_1 \ell_2 \ell_3 \ell_4 \ell_5 \ell_6 = 0$ we take the curve that contains a point of γ distinct from the points P_i. This curve splits into γ and a further quartic of equation $\varphi_4(x_0, x_1, x_2) = 0$. Thus, $\varphi_2 \varphi_4 - \varphi_3^2 = \ell_1 \ell_2 \ell_3 \ell_4 \ell_5 \ell_6$.

146   Chapter 5. Hypersurfaces in Projective Space

5.7.13. A plane cubic C can not have two multiple points in view of Proposition 7.2.7; it can have only a node or an ordinary cusp. Using Plücker's formulas (cf. 5.7.9) one sees that if C is non-singular its class is 6 and it has nine flexes. The class of a cubic with a node is 4 and there are three flexes. Finally, a cuspidal cubic has class 3 and only one flex. A cubic with (one) singular point is rational and one can easily find its rational parametrization by intersecting it with the pencil of lines having center at the singular point.

5.7.14. Show that the plane cubics that pass through eight generic points of the plane, also pass through a ninth point determined by the others.

The cubics in the plane form a linear system Σ of dimension 9, and passing through a point imposes one linear condition on the system (cf. Sections 6.1, 6.2). Hence the cubics of Σ that pass through eight generic points of the plane form a pencil Φ and so they also pass through the ninth base point of the pencil Φ.

5.7.15. Prove the following facts.

(1) If six of the nine base points of a pencil of cubics belong to a conic γ then the remaining three points are collinear.

(2) Let A, B, C and A', B', C' be two triples of collinear points on a plane cubic $\mathcal{C}$. The lines $r_{AA'}, r_{BB'}, r_{CC'}$ intersect $\mathcal{C}$ in three points A'', B'', C'' which are also collinear.

(3) The tangentials of three collinear points of a cubic $\mathcal{C}$ are collinear (the *tangential* of a point $P \in \mathcal{C}$ is the point in which the tangent at P again meets the cubic).

(4) The line joining two flexes of a plane cubic meets the cubic in a third flex.

Statement (1) is a simple consequence of 5.7.14. Indeed, the cubic composed of the conic γ and the line r that joins two of the three base points that do not belong to γ passes through eight base points of the pencil, and so r must also pass through the ninth point.

One could also reason as follows: the curve of the pencil that passes through a point P belonging to γ but distinct from the six base points that lie on γ contains γ as component (by Bézout's theorem); the residual component is a line that must contain the three remaining base points.

Statements (2), (3), (4) are special cases of (1). To prove (2) it suffices to consider the conic γ composed of the two lines that join the two triples of collinear points A, B, C and A', B', C'. Statement (3) is merely (2) in the particular case $A = A', B = B'$ and so $C = C'$. If then $A = A' = A''$ and $B = B' = B''$ one must also have $C = C' = C''$ and thus one obtains (4). For this last statement one can also reason as follows. Let L and M be two flexes of the cubic $\mathcal{C}$, r the line that joins them and N the remaining intersection of $\mathcal{C}$ with r. The pencil of cubics determined by the cubic $\mathcal{C}$ and the cubic that is split into the line r counted three times has as base points the points L, M, N each counted three times.

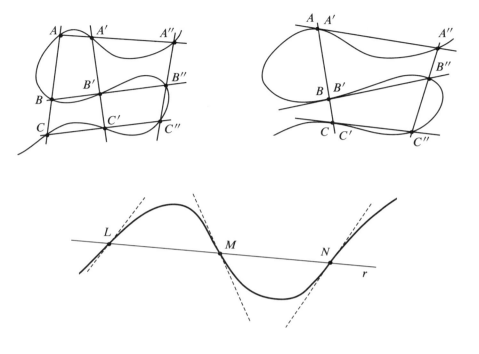

Figure 5.1

The cubic split into the three tangents of $\mathcal{C}$ at L, M, N contains eight of the nine base points of the pencil, and so must also contain the ninth, so that N too is a flex (Figure 5.1).

5.7.16. Another demonstration of Pascal's theorem on conics (cf. [13, Vol. II, Chapter 16]).

Let $A_1, A_2, A_3, A_4, A_5, A_6$ be six points of a conic γ. The two triples of lines $f = (r_{A_1 A_2}, r_{A_3 A_4}, r_{A_5 A_6})$, $g = (r_{A_4 A_5}, r_{A_6 A_1}, r_{A_2 A_3})$ define a pencil of cubics whose base points are the six points A_i, together with the three points $L = r_{A_1 A_2} \cap r_{A_4 A_5}$, $M = r_{A_2 A_3} \cap r_{A_5 A_6}$ and $N = r_{A_3 A_4} \cap r_{A_1 A_6}$ (Figure 5.2). Since the points A_i belong to a conic, the three points L, M, N are collinear by Exercise 5.7.15 (1).

5.7.17 (G. Salmon's theorem [36, Vol. I, pp. 271–272]). Let $\mathcal{C}$ be a non-singular cubic. Two of the six tangents that issue from a point M of $\mathcal{C}$ coincide with the tangent at M. Prove that the absolute invariant of the other four tangents does not depend on M (cf. 1.1.1). It is called the *modulus* of the cubic.

On $\mathcal{C}$ we take two points M, M' and let A be the remaining intersection of $\mathcal{C}$ with the line $r_{MM'}$, t a line issuing from A and tangent to $\mathcal{C}$ elsewhere and T the relative point of tangency. Consider a line r issuing from M that intersects the cubic in two points P, Q and

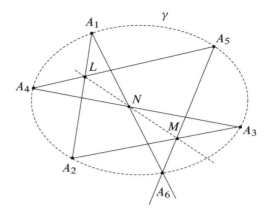

Figure 5.2

let P', Q' be the remaining intersections of the cubic with the lines r_{TP}, r_{TQ}. The two lines r, t intersect $\mathcal{C}$ in the two triples of points M, P, Q and A, T, T. Therefore, the three points M', P', Q' belong to a line r' by Exercise 5.7.15 (2).

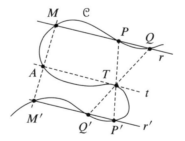

Figure 5.3

Thus one has a one-to-one algebraic correspondence between the two pencils of lines centered at M and M', which is therefore a projectivity: two lines like r and r' correspond to each other. To the four tangents issuing from M there correspond the four tangents issuing from M' (note that the class of $\mathcal{C}$ is 6, and if P is a point of $\mathcal{C}$, two of the six tangents to $\mathcal{C}$ issuing from P coincide with the tangent at P).

If the quadruple of tangents issuing from a generic point M of $\mathcal{C}$ to touch the cubic elsewhere is harmonic or equianharmonic (that is, if the quadruple of points cut out by the four tangents on a line not passing through M is harmonic or equianharmonic, cf. 1.1.1), we say that $\mathcal{C}$ is a *harmonic* or respectively *equianharmonic* cubic.

5.7.18. Let P be a flex of a cubic X. The polar conic of P splits into the tangent at X and a residual line called the *harmonic polar* of P. Show that the cubic is mapped into itself under the harmonic homology having P as its center and as axis its harmonic polar.

If the tangent at the flex $P = [0, 1, 0]$ is the line $x_0 = 0$ and if the harmonic polar of P is the line $x_1 = 0$ one finds that X has an equation of the form $x_0 x_1^2 - \varphi(x_0, x_2) = 0$, with φ a form of degree 3. One sees immediately that X is mapped into itself by the harmonic homology defined by $[x_0, x_1, x_2] \mapsto [x_0, -x_1, x_2]$ (cf. 1.1.14).

5.7.19. Let X be a non-singular cubic, F a flex of X. Calculate the modulus of the cubic (cf. Exercise 5.7.17).

Let r be the polar harmonic of F (cf. Exercise 5.7.18). Since X is non-singular, the line r meets X in three points A, B, C of contact of the tangents issuing from F and distinct from the tangent in F (by the Reciprocity Theorem 5.4.6). To calculate the modulus of the cubic it then suffices to calculate the cross ratio $\mathcal{R}(A, B, C, D)$ where D is the intersection of the line r with the tangent at F. If the tangent at the flex $F = [0, 1, 0]$ is the line $x_0 = 0$ and if the harmonic polar of F is $x_1 = 0$, the equation of X has the form $x_0 x_1^2 - \varphi(x_0, x_2) = 0$, with φ a form of degree 3.

The points A, B, C, the intersections of X with the line r, are given by $x_1 = \varphi(x_0, x_2) = 0$ and so in the induced coordinate system on the line r : $x_1 = 0$ one has $A = [a, 1]$, $B = [b, 1]$, $C = [c, 1]$, $D = [0, 1]$, where a, b, c are the roots (surely distinct by the hypothesis that X is non-singular) of the cubic equation $\varphi(x, 1) = 0$, $x = x_0/x_1$. Then

$$\mathcal{R}(A, B, C, D) = \frac{\begin{vmatrix} c & 1 \\ a & 1 \end{vmatrix}}{\begin{vmatrix} c & 1 \\ b & 1 \end{vmatrix}} : \frac{\begin{vmatrix} 0 & 1 \\ a & 1 \end{vmatrix}}{\begin{vmatrix} 0 & 1 \\ b & 1 \end{vmatrix}} = \frac{b(c-a)}{a(c-b)}.$$

5.7.20. Let O be a fixed point on a non-singular cubic X. Given two points P, Q on X we define the sum $P + Q$ to be the point that one obtains on X by projecting the remaining intersection of X with the line r_{PQ} from O. Show that in this way one obtains the structure of an abelian group on X with O as the neutral element.

The only property which is not obvious is the associative law. In the Figure 5.4 the points $P + Q$ and $R + Q$ are shown. In order to prove that $(P + Q) + R = P + (Q + R)$ it suffices to show that the three points P, Z, and $Q + R$ are collinear, where Z is the remaining intersection of the cubic with the line joining R with $P + Q$. To this end, consider the triples of points $P + Q, N, O$ and R, Q, M. The lines $r_{P+Q,R}, r_{NQ}$ and r_{OM} intersect X in the three points $Z, P, Q + R$, which are therefore collinear by Exercise 5.7.15 (2).

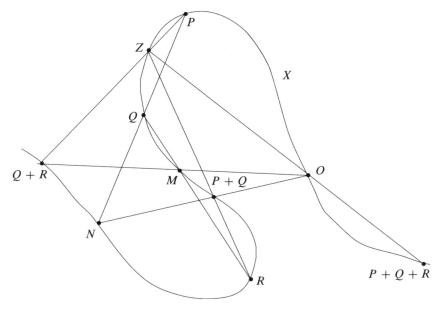

Figure 5.4

5.8 Surfaces in $\mathbb{P}^3$

The remarks and exercises which follow serve to illustrate the theory developed in this chapter with regard to some of the properties of surfaces in $\mathbb{P}^3$; see also [92, Chapters IX, XIII]. We assume that $K = \mathbb{C}$.

5.8.1 Normal singularities of a surface. One says that a surface X in $\mathbb{P}^3$ has *normal* (or *ordinary*) *singularities* if its singularities are (at most) the following:

(1) A *double nodal* curve $\mathcal{L}$, that is, such that in each generic point of $\mathcal{L}$ the tangent cone to X is composed of a pair of distinct planes; and in this case the point is said to be a *double biplanar* point for X.

(2) On the curve $\mathcal{L}$ a finite number of double points (for X) with coincident tangent planes; such points are called *cuspidal points* (or *pinch-points*, or also *uniplanar double points*).

(3) On the curve $\mathcal{L}$ a finite number of *triplanar triple points* (that is, triple points of X at which the tangent cone to X splits into three distinct planes) and which are also triple points for the curve $\mathcal{L}$.

If each point of the curve $\mathcal{L}$ is a cuspidal point, $\mathcal{L}$ is said to be a *cuspidal double curve* (and in that case X does not have normal singularities).

Normal singularities are the only singularities that the generic projection in $\mathbb{P}^3$ of a non-singular surface $\mathcal{F}$ embedded in $\mathbb{P}^5$ can possess [37]. The interest of the notion of normal singularities depends on the following fact:

- Every algebraic surface S has a non-singular birational model (that is, birationally isomorphic to it) embedded in $\mathbb{P}^5$.

This important theorem was discovered by various authors. The first rigorous proof is due to Levi [64]; for more recent proofs see, for example, Walker [112] and Zariski [116], [117] (see also Exercise 13.1.21 for a discussion of the analogous result for algebraic curves).

The extension of Levi's theorem to three dimensional varieties is due to Zariski [118]; finally, in 1964, Hironaka [51] proved the fundamental result that every irreducible algebraic variety over a field K of characteristic zero possesses non-singular birational models.

With regard to the generic projection X in $\mathbb{P}^3$ of a non-singular algebraic surface $\mathcal{F}$ in $\mathbb{P}^5$, Franchetta [39] has proved that the double curve of X is irreducible with a unique exceptional case when $\mathcal{F}$ is the Veronese surface, whose generic projection is in fact the Steiner surface (cf. Exercise 10.5.6) whose nodal double curve is a triple of lines issuing from a point.

5.8.2. Consider a surface X in $\mathbb{P}^3$, of order r, and having a double point P. Study the behavior at P of the first polar $X_1(Q)$ of a point $Q \neq P$.

We take P and Q to be the points $A_0 = [1, 0, 0, 0]$ and $A_1 = [0, 1, 0, 0]$, so that

$$X: x_0^{r-2}\varphi_2(x_1, x_2, x_3) + x_0^{r-3}\varphi_3(x_1, x_2, x_3) + \cdots = 0,$$

$$X_1(Q): x_0^{r-2}\frac{\partial \varphi_2}{\partial x_1} + x_0^{r-3}\frac{\partial \varphi_3}{\partial x_1} + \cdots = 0.$$

Initially we suppose that the cone Γ (with equation $\varphi_2 = 0$) tangent to X at P is not a pair of planes both of which pass through Q. This is equivalent to supposing that the polynomial $\frac{\partial \varphi_2}{\partial x_1}$ not be null. In that case $X_1(Q)$ passes simply through P and the plane π (with equation $\frac{\partial \varphi_2}{\partial x_1} = 0$) tangent to $X_1(Q)$ at P is the polar plane of Q with respect to Γ.

If Q does not belong to Γ, then π is the locus of the harmonic conjugates of P with respect to the pairs of points of Γ collinear with Q. (In particular, if Γ is a pair of distinct planes α, β neither of which passes through Q, then π is the harmonic conjugate with respect to α and β, in the pencil of planes with axis the line $r = \alpha \cap \beta$, of the plane that joins Q with r.) The point P is a node for the line $\mathcal{L} = X \cap X_1(Q)$ and $\Gamma \cap \pi$ is the pair of tangents of $\mathcal{L}$ at P.

If, however, Q belongs to Γ, π is the tangent plane to Γ at Q and P is (in general) a double point with coincident tangents for $\mathcal{L}$.

In the exceptional case that $\frac{\partial \varphi_2}{\partial x_1}$ is the null polynomial (which happens if and only if Γ is a pair of planes both of which contain Q) the point P is (at least) double for $X_1(Q)$.

152   Chapter 5. Hypersurfaces in Projective Space

5.8.3. Prove that if two surfaces F, G in $\mathbb{P}^3$ are mutually tangent at a point P (simple for both of them), that is, if at P they have the same tangent plane, P is at least double for the curve $\mathcal{L} = F \cap G$.

It suffices to observe that if P is the origin of a system of affine coordinates x, y, z and if
$$f = z + \varphi_2 + \varphi_3 + \cdots = 0; \quad g = z + \theta_2 + \theta_3 + \cdots = 0$$
are the equations of F and G, the curve $\mathcal{L}$ can be represented by the system of equations $f = f - g = 0$ and is therefore the intersection of F with the surface of equation $\varphi_2 - \theta_2 + \varphi_3 - \theta_3 + \cdots = 0$ which passes doubly through P.

5.8.4 (Tangent cone in a point of a double curve). Let $\mathcal{L}$ be a curve in $\mathbb{P}^3$ and X a surface passing doubly through $\mathcal{L}$. Show that if P is a simple point of $\mathcal{L}$ the tangent cone to X at P is a pair of planes passing through the tangent of $\mathcal{L}$ at P.

The question is of local nature, and so we may suppose that $\mathcal{L}$ is the complete intersection of two surfaces with equations $f = 0, g = 0$ both passing simply through P and not mutually tangent there.

It is known (cf. the next note 5.8.5) that if $F = 0$ is the equation of the surface X passing doubly through $\mathcal{L}$, the polynomial F belongs to the ideal $(f, g)^2 = (f^2, fg, g^2)$. Let then
$$F = Af^2 + 2Bfg + Cg^2, \quad A, B, C \in \mathbb{C}[x_0, x_1, x_2, x_3].$$

We have,
$$F_{ij} := \frac{\partial^2 F}{\partial x_i \partial x_j} = 2A \frac{\partial f}{\partial x_i} \frac{\partial f}{\partial x_j} + 2B \left(\frac{\partial f}{\partial x_i} \frac{\partial g}{\partial x_j} + \frac{\partial f}{\partial x_j} \frac{\partial g}{\partial x_i} \right) + 2C \frac{\partial g}{\partial x_i} \frac{\partial g}{\partial x_j} + \cdots.$$

Thus, on putting $f_i = \frac{\partial f}{\partial x_i}$, $g_i = \frac{\partial g}{\partial x_i}$, $i = 0, 1, 2, 3$, we have
$$\frac{\partial^2 F}{\partial x_i \partial x_j} - 2(Af_i f_j + B(f_i g_j + f_j g_i) + Cg_i g_j) \in (f, g).$$

Hence,
$$\left(\frac{\partial^2 F}{\partial x_i \partial x_j} \right)_P = 2[A(P)f_i(P)f_j(P) + B(P)(f_i(P)g_j(P)$$
$$+ f_j(P)g_i(P)) + C(P)g_i(P)g_j(P)]$$

and the tangent cone to X at P (namely, the cone of equation $F_{00}(P)x_0^2 + 2F_{01}(P)x_0 x_1 + \cdots = 0$) is
$$A(P) \sum_{i,j} f_i(P)f_j(P)x_i x_j + B(P) \sum_{i,j} [f_i(P)g_j(P) + f_j(P)g_i(P)]x_i x_j$$
$$+ C(P) \sum_{i,j} g_i(P)g_j(P)x_i x_j = 0,$$

which is to say

$$A(P)\left(\sum_i f_i(P)x_i\right)^2 + 2B(P)\left(\sum_i f_i(P)x_i\right)\left(\sum_i g_i(P)x_i\right)$$
$$+ C(P)\left(\sum_i g_i(P)x_i\right)^2 = 0,$$

and thus it is a pair of planes passing through the line $\sum_i f_i(P)x_i = \sum_i g_i(P)x_i = 0$ tangent to $\mathcal{L}$ at P.

If $B(P)^2 - A(P)C(P) = 0$ the two planes coincide and P is a cuspidal point (or pinch-point) of X.

In the special case that $B^2 - AC \in (f, g)$ each point of $\mathcal{L}$ is a cuspidal point (of the surface) and $\mathcal{L}$ is a cuspidal double curve.

If $\mathcal{L}$ is a nodal double curve it can have only finitely many cuspidal points, namely the solutions of the system $f = g = B^2 - AC = 0$.

To obtain the tangent cone to X at a generic point $P \in \mathcal{L}$ it suffices to intersect X with an arbitrary plane π passing through P (but not through the tangent of $\mathcal{L}$ at P) and then to take the two planes that join the tangent of $\mathcal{L}$ at P with the tangents at P to the curve $X \cap \pi$.

5.8.5. Note. If $\mathfrak{p}$ is a prime ideal of the ring A (commutative and with identity), the symbol $\mathfrak{p}^{(s)}$ denotes its sth *symbolic power*, that is, the set (which one immediately sees to be an ideal) of elements $x \in A$ such that there is a $y \notin \mathfrak{p}$ for which one has $xy \in \mathfrak{p}^s$.

If $A = K[y_1, \ldots, y_n]$ one has $f \in \mathfrak{p}^{(s)}$ if and only if the affine hypersurface $f = 0$ passes s-fold through $V(\mathfrak{p})$. It is then obvious that $\mathfrak{p}^s \subset \mathfrak{p}^{(s)}$. It follows that $\mathfrak{p}^s = \mathfrak{p}^{(s)}$ if and only if $\mathfrak{p}$ is an ideal of principal class (and thus, in particular, if $\mathfrak{p}$ is generated by only two elements); cf. [17], [16].

An interesting example of a prime ideal $\mathfrak{p} \subset K[x, y, z]$ such that $\mathfrak{p}^2 \neq \mathfrak{p}^{(2)}$ is the ideal $\mathfrak{p} = (xz - y^2, x^3 - yz, z^2 - x^2y)$ already encountered in Exercise 3.4.11 (cf. [71, Chapter I]); indeed, the polynomial $f = x^5 - 3x^2yz + xy^3 + z^3$ does not belong to $\mathfrak{p}^2$ inasmuch as a polynomial of $\mathfrak{p}^2$ can not contain the monomial z^3, but it belongs to $\mathfrak{p}^{(2)}$ because one has

$$xf = (x^3 - yz)^2 + (xz - y^2)(z^2 - x^2y) \in \mathfrak{p}^2.$$

5.8.6 Tangent cone at a point of an s-fold variety. Let V_k be an s-fold variety for the hypersurface X in $\mathbb{P}^n$ defined by the equation $F = 0$. Moreover, let $P = [1, 0, \ldots, 0]$ be a simple point of V_k and suppose that $x_1 = \cdots = x_{n-k} = 0$ is the tangent space S_k at P to V_k. Since V_k is locally (near P) a complete intersection, we can suppose that V_k is the locus of zeros of a homogeneous prime ideal $\mathfrak{p} = (f_1, \ldots, f_{n-k})$, where

$$f_j = x_0^{d_j - 1} x_j + \cdots, \quad j = 1, \ldots, n - k; \quad d_j = \deg f_j.$$

Chapter 5. Hypersurfaces in Projective Space

Since $F \in \mathfrak{p}^s$ we can write

$$F = \sum_i A_i \theta_i + H,$$

where $\theta_i \in \mathfrak{p}^s$, $H \in \mathfrak{p}^{s+1}$, and the A_i are polynomials not all of which vanish at P.

One sees immediately that the coefficient of the highest power of x_0 in F is a form of degree s in the indeterminates $x_1, \ldots, x_{n-k}$ with constant coefficients.

Therefore, *the tangent cone to X at P has as its vertex the tangent space to V_k at P.* In particular, if $\mathcal{L}$ is an s-fold curve of a surface X in $\mathbb{P}^3$, the tangent cone to X at a simple point P of $\mathcal{L}$ consists of s planes passing through the tangent to $\mathcal{L}$ at P.

For our later purposes it is useful to give more detail in the case of a hypersurface passing through a linear space which we may assume to be defined by the equations $x_0 = \cdots = x_{n-k-1} = 0$. Let $F \in (x_0, \ldots, x_{n-k-1})^2$ so that

$$F = L_{00}x_0^2 + 2L_{01}x_0x_1 + \cdots + L_{n-k-1\,n-k-1}x_{n-k-1}^2 + G(x_0, \ldots, x_{n-k-1}),$$

where the coefficients $L_{ij} = L_{ij}(x_{n-k}, x_{n-k+1}, \ldots, x_n)$ are homogeneous polynomials of degree $r-2$ in the indeterminates $x_{n-k}, x_{n-k+1}, \ldots, x_n$ and G belongs to $(x_0, \ldots, x_{n-k-1})^3$.

At a generic point $P = [0, \ldots, 0, a_{n-k}, a_{n-k+1}, \ldots, a_n]$ of S_k we have

$$\frac{1}{2}\left(\frac{\partial^2 F}{\partial x_i \partial x_j}\right)_P = L_{ij}(a_{n-k}, a_{n-k+1}, \ldots, a_n),$$

and thus the tangent cone to X at P has equation:

$$L_{00}(a_{n-k}, a_{n-k+1}, \ldots, a_n)x_0^2 + 2L_{01}(a_{n-k}, a_{n-k+1}, \ldots, a_n)x_0x_1 + \cdots$$
$$\cdots + L_{n-k-1\,n-k-1}(a_{n-k}, a_{n-k+1}, \ldots, a_n)x_{n-k-1}^2 = 0.$$

This equation represents a quadric cone that has as vertex the space S_k.

The points of S_k belonging to the hypersurface $\mathcal{D}$ (in S_k) with equation

$$\Delta = \det(L_{ij}(a_{n-k}, a_{n-k+1}, \ldots, a_n)) = 0$$

are exceptional. If P is a point of $\mathcal{D}$ for which the matrix

$$(L_{ij}(a_{n-k}, a_{n-k+1}, \ldots, a_n))$$

has rank $k + 1 - \rho$, the tangent cone at P has a double space of dimension $k + \rho$ (even though P is in general only a double point for F). It is not impossible that Δ is the null polynomial.

In particular, if $k = n - 2$ and thus

$$F = L_{00}x_0^2 + 2L_{01}x_0x_1 + L_{11}x_1^2 + G(x_0, x_1),$$

the tangent cone at P consists of the pair of hyperplanes

$$L_{00}(a_2,\ldots,a_n)x_0^2 + 2L_{01}(a_2,\ldots,a_n)x_0x_1 + L_{11}(a_2,\ldots,a_n)x_1^2 = 0.$$

We will say that P is a *bihyperplanar double point* if the two hyperplanes are distinct, and a *unihyperplanar double point* if the two hyperplanes coincide. The hypersurface $\mathcal{D}$ of the space S_{n-2} which is the locus of the unihyperplanar double points is the intersection of S_{n-2} with the hypersurface having equation

$$L_{01}(x_2,\ldots,x_n)^2 - L_{00}(x_2,\ldots,x_n)L_{11}(x_2,\ldots,x_n) = 0,$$

and so has order $2(r-2)$. On it there may very well be points of multiplicity > 2, and so on.

In the case in which X is a surface in $\mathbb{P}^3$ and $k = 1$, the locus $\mathcal{D}$ is in general composed of a finite number of points: the uniplanar points (or pinch-points) and possible points of multiplicity > 2 that X possesses on the double line. The most general case is that in which one has $2(r-2)$ pinch-points. But it can happen that all the points of the double line are uniplanar, and in that case the line is said to be *cuspidal double*.

The extension to the case in which the double variety is not a linear space is only formally more complex. An example of a surface in $\mathbb{P}^3$ having a cuspidal double curve is given by the surface spanned by the tangents of the space curve itself.

5.8.7 (Hypersurfaces in $\mathbb{P}^4$ with a double line). In general a hypersurface X of $\mathbb{P}^4$ with a double line r has at each point of r a tangent cone whose vertex is just the line r.

Consider, for example, the cubic hypersurface X with equation

$$x_0x_2^2 + x_1(x_3^2 + ax_4^2) + \varphi_3(x_2, x_3, x_4) = 0,$$

where φ_3 is a form of degree 3 and $a \in \mathbb{C}$, which passes doubly through the line $r : x_2 = x_3 = x_4 = 0$. The tangent cone at the generic point $P = [\lambda, \mu, 0, 0, 0]$ of r is

$$\lambda x_2^2 + \mu(x_3^2 + ax_4^2) = 0.$$

In each generic point of r the tangent cone is irreducible, and is the cone that projects the conic $\gamma : x_0 = x_1 = \lambda x_2^2 + \mu(x_3^2 + ax_4^2) = 0$ from r. This cone is a pair of hyperplanes if γ is degenerate, that is, if $\lambda\mu^2 = 0$. For $\lambda = 0$ one has two distinct hyperplanes $x_3^2 + ax_4^2 = 0$; while $\mu = 0$ gives the double hyperplane $x_2^2 = 0$.

In particular, if $a = 0$, the line r is a particular double line because in each of its points the tangent cone is composed of the pair of hyperplanes $\lambda x_2^2 + \mu x_3^2 = 0$ (which coincide if $\lambda\mu = 0$).

In this case we say that r is a *bihyperplanar double line* (cf. Exercise 13.1.34).

5.8.8. Find all the cubic surfaces passing doubly through the line $x_2 = x_3 = 0$ and for each of them find the pinch-points on that line (i.e., the uniplanar points).

The generic cubic surface X passing doubly through the line $x_2 = x_3 = 0$ has equation of the type
$$x_2^2 L + 2x_2 x_3 M + x_3^2 N = 0,$$
where L, M, N are linear forms in x_0, x_1, x_2, x_3. The tangent cone to X at the generic point $P = [a, b, 0, 0]$ of the double line has equation (cf. §5.8.6)
$$x_2^2 L(a, b, 0, 0) + 2x_2 x_3 M(a, b, 0, 0) + x_3^2 N(a, b, 0, 0) = 0,$$
and so it splits into two coincident planes if $M(P)^2 - L(P)N(P) = 0$. The pinch-points are thus the points of intersection of the line $x_2 = x_3 = 0$ with the quadric having equation $M^2 - LN = 0$.

5.8.9 Apparent boundary. The *apparent boundary of a surface* $X \subset \mathbb{P}^3$ *from a point O* (or *with respect to O*) is the closure Γ of the locus of points $P \in X$ such that the tangent plane at P to X passes through O. It is nothing more than the curve of intersection of X with the first polar $X_1(O)$ of O.

Therefore (cf. Lemma 5.6.2) an s-fold point of X is (in general) an $s(s-1)$-fold point for Γ. It is easy to see that *a point A of Γ that is simple for X is in general also simple for Γ*. Indeed, let $A = [1, 0, 0, 0]$, $O = [0, 1, 0, 0]$ and assume that the tangent plane to X at A is $x_2 = 0$, so that the equation of X may be written in the form
$$f := x_0^{n-1} x_2 + x_0^{n-2}(ax_1^2 + \cdots) + \cdots = 0,$$
where n is the order of X. One then has
$$\frac{\partial f}{\partial x_1} = x_0^{n-2}(2ax_1 + \cdots) + \cdots = 0$$
and in general the two planes $x_2 = 0$, $2ax_1 + \cdots = 0$ are distinct. (The two planes coincide only if the two principal tangents of X at A coincide with the line r_{AO}).

The apparent boundary *carried by O over a plane π*, is the line Γ' which is the intersection of π with the apparent boundary of X from O, that is, the projection of Γ from O onto π. If $\mathcal{L}$ is a curve traced on X and passing through a point A of Γ, the tangents to $\mathcal{L}$ and to Γ at A are both contained in the tangent plane σ to X at A; since this plane passes through O, *the projection $\mathcal{L}'$ of $\mathcal{L}$ from O onto π is tangent to Γ' at the point A', the projection of A*. This fact is known as the "theorem of the apparent boundary" (Figure 5.5).

5.8.10. Prove that an algebraic surface of order 3 contains a line, and that a surface X^r of order $r > 3$ does not (at least in general) contain any line.

The necessary and sufficient condition for the surface X^r with equation $f(x, y, z) = 0$ in $\mathbb{A}^3$ to contain the line of equations $x - az - b = y - cz - d = 0$ is that a, b, c, d annull the $r + 1$ coefficients of the polynomial $f(az + b, cz + d, z) \in \mathbb{C}[z]$. One finds a system of $r + 1$ equations in the four unknowns a, b, c, d. If $r + 1 \leq 4$, that is, if $r \leq 3$, the system

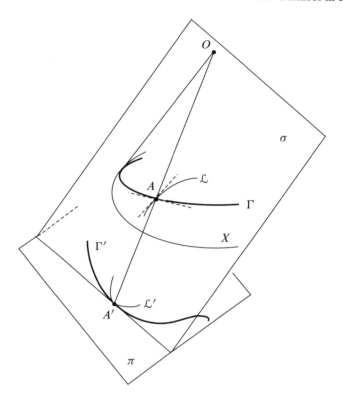

Figure 5.5.

has solutions and thus X^r contains lines (infinitely many lines if $r \leq 2$, a finite number, in general, if $r = 3$). If, however, $r \geq 4$ the system, in general, has no solutions.

5.8.11. Let X be a surface (non-singular or with at most only nodes of the most general type, cf. §5.6.5) of $\mathbb{P}^3$. Then, its class is $\nu(X) = 2$ if and only if X is a quadric.

That a quadric of $\mathbb{P}^3$ has class equal to two is obvious. The converse follows by duality from Exercise 5.3.4.

5.8.12. Consider a cubic surface X of general type, and choose as the point $[1, 0, 0, 0]$ one of its non-singular points P, so that the equation of X has the form

$$f = x_0^2 \varphi_1(x_1, x_2, x_3) + 2x_0 \varphi_2(x_1, x_2, x_3) + \varphi_3(x_1, x_2, x_3) = 0,$$

where $\varphi_j(x_1, x_2, x_3)$ is a homogeneous polynomial of degree j, $j = 1, 2, 3$. Put

158   Chapter 5. Hypersurfaces in Projective Space

$\Delta := \varphi_1 \varphi_3 - \varphi_2^2$, and let V be the cone with equation $\Delta = 0$. Verify the identity

$$\varphi_1 f = \left(\frac{1}{2}\frac{\partial f}{\partial x_0}\right)^2 + \Delta,$$

and use it to prove the following facts:

(1) The tangent planes to V are precisely the tangent planes to X that contain P.

(2) Every double generator of the cone V which does not belong to the plane with equation $\varphi_1 = 0$ contains a double point of X, and conversely each double point of X belongs to a double generator of V.

The first polar $X_1(P)$ of P with respect to X has equation $\frac{\partial f}{\partial x_0} = 0$ and, by the Reciprocity Theorem (Theorem 5.4.6), the curve $C := X \cap X_1(P)$ is the locus of the points of contact of X with the tangent planes containing P. On the other hand, Δ belongs to the ideal generated by f and by $\frac{\partial f}{\partial x_0}$, and so V contains the curve C; since the variable x_0 does not appear in Δ, we conclude that V is the cone that projects C from P (cf. §3.4.5). From this (1) follows.

Each double point of X is obviously also double for the surface of equation $\left(\frac{\partial f}{\partial x_0}\right)^2 = 0$ and hence is double for the cone V; therefore belongs to a double generator of V. Moreover, a double generator of V that does not belong to the plane $\varphi_1 = 0$ meets the surface $X_1(P)$: $\frac{\partial f}{\partial x_0} = 0$ in a double point for the surface with equation $\varphi_1 f = 0$, and hence a double point for X. This proves (2).

5.8.13. Show that a non-singular cubic surface X (of general type) contains twenty-seven lines.

In the notation of 5.8.12, let $P = [1, 0, 0, 0]$ be a non-singular point of X, and

$$f = x_0^2 \varphi_1(x_1, x_2, x_3) + 2x_0 \varphi_2(x_1, x_2, x_3) + \varphi_3(x_1, x_2, x_3) = 0$$

the equation of X. Let V be the cone, with equation $\Delta := \varphi_1 \varphi_3 - \varphi_2^2 = 0$, that projects the curve $C = X \cap X_1(P)$ from P. A *bitangent plane* of the cone V (that is, a plane tangent to V along two distinct generators) is a bitangent plane of X that passes through P. It intersects X in a cubic with two double points; by Theorem 4.2.1 this cubic splits and contains the line ℓ joining the two double points. Hence, every bitangent plane of the cone V contains a line lying in X.

Conversely, the plane joining P with a line of X meets X in a (reducible) cubic with two double points, and so is bitangent to X. By the generality hypothesis made on X, the polynomials $\varphi_1, \varphi_2, \varphi_3$ are such that the plane quartic C with equation $\varphi_1 \varphi_3 - \varphi_2^2 = x_0 = 0$ is non-singular (equivalently, V is a quartic cone of general type). Using Plücker's formulas proved in 5.7.9, we have that the number of flexes ρ and the class ν of C are $\rho = 24$ and $\nu = 12$, so that the number δ of bitangent lines to C is $\delta = 28$. So V has twenty-eight bitangent planes. One of these is the plane $\varphi_1 = 0$ that is a generic tangent plane (i.e., non-bitangent) to X, and which therefore does not contain lines of the surface, since otherwise its intersection with X would be reducible. The other twenty-seven bitangent planes each contain a line of X.

For a description of the configuration of the lines of X see for instance [74, III, §7] or also [15, §11].

We recall a few definitions. We say that a surface in $\mathbb{P}^3$ (or more generally in $\mathbb{P}^r$) is *ruled* if it is the locus of ∞^1 lines, called *generators*. A curve of the surface that meets each generator in only one point is said to be a *directrix*. A generator g of a ruled surface X is said to be *simple* if the generic point of g is non-singular for X. On the other hand, g is said to be *multiple* (double, triple, ...) if each of its points is multiple (double, triple, ...) for X. A generator g is said to be *singular* if g is simple for X and the tangent plane to X in the generic points $P \in g$ is fixed as P varies in g. A ruled surface is said to be *developable* if the simple generators are all singular.

5.8.14 Criterion for developability of a ruled surface.
In a system of affine coordinates in $\mathbb{A}^3$, let us represent a ruled surface X in the form

$$\begin{cases} x = \alpha(u) + tl(u), \\ y = \beta(u) + tm(u), \\ z = t, \end{cases} \quad (5.38)$$

where $\alpha(u), l(u), \beta(u), m(u)$ are twice continuously differentiable functions of the parameter u, and suppose that the generator $g(u)$ corresponding to the value u of the parameter is singular, that is, that the tangent plane to X along $g(u)$ is fixed.

We consider two plane sections of X, for example the two sections $\mathcal{L}_0$, $\mathcal{L}_1$ that one has for $t = 0$ and $t = 1$, and the points A, B where they meet $g(u)$. If the generator is singular the tangent planes to X at A and B coincide and so the tangent to $\mathcal{L}_0$ at A and the tangent to $\mathcal{L}_1$ at B are coplanar. Thus, since they are contained in the two parallel planes $z = 0, z = 1$, they must be parallel. It follows that their direction vectors $(\alpha'(u), \beta'(u), 0)$ and $(\alpha'(u) + l'(u), \beta'(u) + m'(u), 0)$, must be parallel, where the prime indicates the derivative with respect to u. Hence $\alpha'(u)(\beta'(u) + m'(u)) - \beta'(u)(\alpha'(u) + l'(u)) = 0$, which is to say

$$\alpha'(u)m'(u) - \beta'(u)l'(u) = 0. \quad (5.39)$$

This is the necessary and sufficient condition in order for $g(u)$ to be singular. If the ruled surface is developable, that is, if all of its generators are singular, this condition is verified for all values of the parameter u. Condition (5.39) is obviously satisfied if $l'(u) = m'(u) = 0$, that is, if $l(u)$ and $m(u)$ are constants (in which case X is a cylinder).

If $l'(u), m'(u)$ are not both zero, condition (5.39) implies that there exists a function $\sigma(u)$ such that

$$\alpha'(u) = \sigma(u)l'(u), \quad \beta'(u) = \sigma(u)m'(u). \quad (5.40)$$

On every generator we then take the point $P(u)$ obtained for $t = -\sigma(u)$. Supposing that this point is not independent of u [if $P(u)$ were independent of u, the ruled surface X would be a cone with vertex $(\alpha(u), \beta(u), 0)$] one obtains the curve Γ on X defined by parametric equations

$$\begin{cases} x = \alpha(u) - \sigma(u)l(u), \\ y = \beta(u) - \sigma(u)m(u), \\ z = -\sigma(u). \end{cases}$$

The tangent line to Γ at its generic point $P(u)$ joins the point $P(u)$ (which belongs to $g(u)$) to the improper point

$$[\alpha'(u) - \sigma(u)l'(u) - \sigma'(u)l(u), \beta'(u) - \sigma(u)m'(u) - \sigma'(u)m(u), -\sigma'(u), 0]$$
$$= [-\sigma'(u)l(u), -\sigma'(u)m(u), -\sigma'(u), 0],$$

that is, to the improper point $[l(u), m(u), 1, 0]$ of $g(u)$. The generator $g(u)$ is then the tangent to Γ at $P(u)$ and X is the surface spanned by the tangents of Γ.

Bearing in mind (5.40), the osculating plane of Γ at $P(u)$ has equation

$$\begin{vmatrix} x - \alpha(u) + \sigma(u)l(u) & y - \beta(u) + \sigma(u)m(u) & z + \sigma(u) \\ \sigma'(u)l(u) & \sigma'(u)m(u) & \sigma'(u) \\ \sigma''(u)l(u) + \sigma'(u)l'(u) & \sigma''(u)m(u) + \sigma'(u)m'(u) & \sigma''(u) \end{vmatrix} = 0,$$

that is,

$$\begin{vmatrix} x - \alpha(u) & y - \beta(u) & z \\ l(u) & m(u) & 1 \\ l'(u) & m'(u) & 0 \end{vmatrix} = 0.$$

Thus Γ coincides with the tangent plane to X at the point of the generator $g(u)$ that comes from the value $t = 0$ of the parameter, that is, with the tangent plane to the ruled surface along the generator.

Note in addition that, again recalling (5.40), criterion (5.39) is the necessary and sufficient condition in order that the tangent plane to X at the point $P(u)$ coming from the value $t = -\sigma(u)$ of the parameter t should be indeterminate.

In conclusion, for developable ruled surfaces the following property holds: *On every generator of a developable ruled surface X there is a singular point, at which the ruled surface does not have a well-defined tangent plane. If this point is fixed (that is, does not depend on u) X is a cone with vertex in that point. Otherwise, the locus of the singular points $P(u)$ of the various generators is a curve Γ, called the* **regression edge**, *having as tangent and as osculating plane at $P(u)$ the generator $g(u)$ and the tangent plane to X along $g(u)$.*

As an example, we consider the affine cubic surface $\mathcal{F} : z^2 y - x^2 = 0$. The line $x = z = 0$ is double for $\mathcal{F}$. Every plane of the pencil with axis r therefore meets the surface in a line, and thus $\mathcal{F}$ is ruled. For $\mathcal{F}$ we consider the parametric representation

$$\begin{cases} x = ut, \\ y = u^2, \\ z = t. \end{cases}$$

For each fixed u the generator $g(u)$ is the line passing through $(\alpha(u), \beta(u), 0) = (0, u^2, 0)$ and having direction vector $(l(u), m(u), n(u)) = (u, 0, 1)$. Applying the preceding criterion, in order for $g(u)$ to be a singular generator it is necessary and sufficient that

$$\beta'(u)l'(u) = 2u = 0,$$

which means that $u = 0$; thus one has the singular generator $x = y = 0$ (along which the fixed tangent plane is $x = 0$).

5.8.15. Let $\mathcal{F}$ be the surface with equation $x_0^2 x_1 - x_2^2 x_3 = 0$. Noting that $\mathcal{F}$ has a double line and then observing that it is a ruled surface, find the singular generators and the pinch-points on the double line.

Since $x_0^2 x_1 - x_2^2 x_3 \in (x_0, x_2)^2 \cap (x_1, x_3)$, $\mathcal{F}$ contains the two lines $r : x_0 = x_2 = 0$, $s : x_1 = x_3 = 0$ and r is a double line. A generic plane passing through r meets $\mathcal{F}$ in another line. Therefore $\mathcal{F}$ is a ruled surface.

A plane passing through s meets $\mathcal{F}$ in a conic that has a double point P on r and which thus splits into two lines g, g'. The two planes $\langle r, g \rangle$, $\langle r, g' \rangle$ form the tangent cone to $\mathcal{F}$ at P. If $g = g'$, P is a uniplanar point (or pinch-point). On r there are two pinch-points. Indeed, the section with the plane $\lambda x_1 + \mu x_3 = 0$ (containing s) is

$$\begin{cases} \lambda x_1 + \mu x_3 = 0, \\ x_3(\mu x_0^2 + \lambda x_2^2) = 0 \end{cases}$$

and it consists of the line s and the two lines g, g' given by the system $\lambda x_1 + \mu x_3 = \mu x_0^2 + \lambda x_2^2 = 0$, which coincide if $\lambda \mu = 0$. The two pinch-points (intersections of r with the two planes $x_1 = 0$, $x_3 = 0$) are $P_1 = [0, 0, 0, 1]$ and $P_2 = [0, 1, 0, 0]$. The singular generators are $p_1 : x_1 = x_2 = 0$ and $p_2 : x_0 = x_3 = 0$; along each of them the tangent plane to $\mathcal{F}$ is fixed (the plane $x_1 = 0$ and the plane $x_3 = 0$ respectively).

5.8.16. Let X be a surface in $\mathbb{P}^3$ having at most nodes of the most general type and let d be the number of its nodes. Let r be the order of X and suppose that X is not ruled. Prove that $d \leq 4$ if $r = 3$ and $d \leq 16$ if $r = 4$.

Then write the equation of a cubic surface with four nodes.

By 5.8.11 we know that the class $\nu(X)$ of a ruled surface X of order > 2 is at least 3. Thus we have (cf. §5.6.5)

$$\nu(X) = r(r-1)^2 - 2d \geq 3.$$

Since d is an integer by parity one obtains $r(r-1)^2 - 2d \geq 4$; and thus $d \leq 4$ if $r = 3$ and $d \leq 16$ if $r = 4$.

A cubic that has four nodes in the vertices of the fundamental tetrahedron has equation
$$ax_1x_2x_3 + bx_2x_3x_0 + cx_3x_0x_1 + dx_0x_1x_2 = 0, \quad a,b,c,d \in \mathbb{C}, \ abcd \neq 0.$$

5.8.17 (The Kummer surface). Let X be the quartic hypersurface defined by the equation
$$f = x_0^2 \varphi_2(x_1, x_2, x_3) + 2x_0\varphi_3(x_1, x_2, x_3) + \varphi_4(x_1, x_2, x_3) = 0,$$
where $\varphi_j(x_1, x_2, x_3)$ are homogeneous polynomials of degree j, $j = 2, 3, 4$, such that $\Delta = \varphi_2\varphi_4 - \varphi_3^2$ is the product of six linear forms $\ell_i \in \mathbb{C}[x_1, x_2, x_3]$ (cf. 5.7.12) Using the identity (immediately verified)
$$\varphi_2 f = \left(\frac{1}{2}\frac{\partial f}{\partial x_0}\right)^2 + \Delta, \tag{5.41}$$
prove that X has sixteen double points.

Let $P_{ij} = [0, a, b, c]$ be a double point of the curve with equation $\Delta = x_0 = 0$, namely one of the fifteen points defined by $\ell_i = \ell_j = x_0 = 0$. The tangent cone $\varphi_2 = 0$ to X at its double point $[1, 0, 0, 0]$ meets the plane $x_0 = 0$ in a conic γ that is tangent to the six lines $\ell_i = x_0 = 0$. Since $P_{ij} \notin \gamma$ one has $\varphi_2(a, b, c) \neq 0$. Consider the point $Q_{ij} = \left[-\frac{\varphi_3(a,b,c)}{\varphi_2(a,b,c)}, a, b, c\right]$ (of which P_{ij} is the projection on the plane $x_0 = 0$). The coordinates of Q_{ij} annull the partial derivative $\frac{\partial f}{\partial x_0} = 2(x_0\varphi_2 + \varphi_3)$; and so Q_{ij} is double for the surface $\left(\frac{1}{2}\frac{\partial f}{\partial x_0}\right)^2 = 0$. Furthermore, Q_{ij} is double for the cone with equation $\Delta = 0$ and so, bearing in mind that $\varphi_2(a, b, c) \neq 0$, we obtain by (5.41) that Q_{ij} is a double point for X. Thus X has sixteen double points; the fifteen points Q_{ij} and in addition the point $[1, 0, 0, 0]$.

Historical note. The difficult problem of determining the maximum number $\mu(r)$ of isolated double points (i.e., not belonging to multiple lines) that a surface of order r in $\mathbb{P}^3$ can have has been resolved only for $r \leq 6$. Besides the results $\mu(2) = 1$, $\mu(3) = 4$, $\mu(4) = 16$ one has:

$\mu(5) = 31$ (Togliatti [107]: $\mu(5) \geq 31$, Beauville [7]: $\mu(5) \leq 31$);
$\mu(6) = 65$ (Jaffe and Rubermann [56]: $\mu(6) \leq 65$, Barth [5]: $\mu(6) \geq 65$).

5.8.18. Let X be a ruled surface in $\mathbb{P}^3$ and let g be a simple generator of X. Prove the following:

(1) If g is a singular generator of X then g contains a point that is multiple for X (which is said *singular point of the singular generator*).

(2) (Chasles' theorem) If g is non-singular, associating to each point of g the tangent plane to X in that point one obtains a projectivity between the pointed line g and the pencil of planes having g as axis.

In the affine space $\mathbb{A}^3$ let X be the ruled surface which is the locus of $P(u, t) = (a(u) + tl(u), b(u) + tm(u), c(u) + tn(u))$, where $a(u), b(u), c(u), l(u), m(u), n(u)$ are functions of a parameter u.

It suffices to observe that the tangent plane at the point $P(u, t)$ of the generator $g = g(u)$ is $\pi_t : L + tM = 0$, where $L = 0$ and $M = 0$ are the equations of the tangent planes to X in the points $P(u, 0)$ and $P(u, t_\infty)$, and t_∞ denotes the improper point on the t-axis in the (u, t)-plane.

If $g(u)$ is singular, and hence $L = kM$, $k \in \mathbb{C}$, the tangent plane at $P(u, t)$ has equation $(k + t)M = 0$ and so on $g(u)$ there is the singular point $P(u, -k)$, at which the tangent plane does not exist.

5.8.19. Show that the normals to a ruled surface $\mathcal{F} \subset \mathbb{A}^3(\mathbb{R})$ in the points of a non-singular generator g span a hyperbolic paraboloid.

The normal n_P at a point P to g is the perpendicular at P to the tangent plane π_P to $\mathcal{F}$ at P. Its improper point N_P thus belongs to the improper line r_∞ of the planes perpendicular to g.

The correspondence $\pi_P \mapsto N_P$ between the pencil of planes with axis g and the line r_∞ is algebraic and bijective, and hence projective.

By Chasles' theorem, see 5.8.18(2), the points P and N_P thus correspond under a projectivity between the two lines g, r_∞.

A simple check shows that the locus of the lines that join points corresponding under a projectivity between two skew lines is a non-singular quadric (cf. 13.1.41). Then the lines n_P are the generators of a quadric that is a hyperbolic paraboloid because it contains the improper line r_∞.

5.8.20. If $X \subset \mathbb{P}^3$ is a (non-developable) ruled surface of order r, its class is $\nu(X) = r$.

A generic line ℓ meets X in r distinct points $P_1, \ldots, P_r$; from each of these points P_i there issues a generator g_i of the ruled surface. It then suffices to observe that the tangent planes to X (at the points P_i) containing the line ℓ are the r planes that join ℓ with the single generators g_i, $i = 1, \ldots, r$, cf. 5.8.18(2).

5.8.21. Let X be a non-developable ruled surface of order $r > 2$. Show that X contains a multiple curve $\mathcal{L}$, which in general is a double curve, that meets every non-singular generator in $r - 2$ points. The curve $\mathcal{L}$ is also called a *double directrix*.

Let g be a simple generator. A generic plane π passing through g meets X in a curve Γ of order r which splits into g and a residual curve γ of order $r - 1$. The curve γ intersects g in $r - 1$ double points of Γ. By 5.8.18(2) the plane π is tangent to X at only one of these points; the other $r - 2$ points are double for X (cf. Exercise 5.2.6). As π varies in the pencil with axis g, these $r - 2$ points describe a double curve $\mathcal{L}$.

5.8.22. The double curve of a non-developable ruled cubic X is necessarily a line $\mathfrak{d}$ and the tangent cone at a point $P \in \mathfrak{d}$ splits into a pair of planes passing through $\mathfrak{d}$.

164   Chapter 5. Hypersurfaces in Projective Space

The ruled surface X is said to be *general* (or *of general type*) if the tangent cones in the points P of $\mathfrak{d}$ are pairs of planes both of which vary as P varies on $\mathfrak{d}$; and then they are the pairs of corresponding elements in an involution ω of the pencil of planes having $\mathfrak{d}$ as axis (cf. [52, Vol. 1, Chapter VIII, §2]). The two points of $\mathfrak{d}$ corresponding to the two fixed planes of ω are the cuspidal points (or pinch-points) of X; at them the tangent cone consists of a pair of coinciding planes. Note that if X is a general ruled cubic, from each point $P \in \mathfrak{d}$, which is not cuspidal, there issue two distinct generators of the ruled surface: the two lines that when joined with $\mathfrak{d}$ define the two planes into which the tangent cone to X at P splits.

A general ruled cubic X also has a rectilinear directrix which is skew to the double line.

Prove these results and use them to find a simple analytic representation for a general ruled cubic.

The double curve $\mathcal{L}$ of a ruled cubic is necessarily a straight line, which we will denote by $\mathfrak{d}$. Indeed, two arbitrary points A, B of $\mathcal{L}$ are joined by a line lying on X, having at least four intersections with it (two in A, two in B). Thus, $\mathcal{L}$ can not be a plane curve of order ≥ 2 because otherwise X would contain the plane of $\mathcal{L}$.

But neither can it be a space curve, because the chords of a space curve span all the space (cf. Exercise 7.5.2).

Let P be a generic point of the double line $\mathfrak{d}$ and let α, β be the two tangent planes to X at P. The three intersections of a line passing through P and contained in one of these two planes coincide with P. Therefore, each of the two planes meets the ruled cubic in a curve of third order having P as triple point, and thus that curve must be composed of the double line $\mathfrak{d}$ counted twice and of another line containing P. Therefore from each point P of $\mathfrak{d}$ there issue two generators a, b of X (which coincide only if P is one of the pinch-points). The plane of these two lines intersects X in another line r which is skew to $\mathfrak{d}$. A plane passing through r meets the ruled cubic in a curve consisting of r and a pair of lines issuing from the point in which the plane intersects $\mathfrak{d}$. All the generators of X are thus supported by r, which is therefore a simple directrix.

Take the cuspidal points to be $A_3 = [0, 0, 0, 1]$ and $A_2 = [0, 0, 1, 0]$ (so that the double line $\mathfrak{d}$ has equations $x_0 = x_1 = 0$) and take as points $A_0 = [1, 0, 0, 0]$ and $A_1 = [0, 1, 0, 0]$ the intersections of r with the planes (of equations $x_1 = 0$ and $x_0 = 0$) which when doubly counted give the tangent cones in A_3 and A_2 respectively, and finally take the unit point $[1, 1, 1, 1]$ to be a point of the surface. One finds that the ruled surface then has the simple equation $x_0^2 x_2 - x_1^2 x_3 = 0$. (Note that the planes $x_1 = 0$ and $x_0 = 0$ meet X, besides in the double line $\mathfrak{d}$ counted twice, respectively in the lines $x_1 = x_2 = 0$ and $x_0 = x_3 = 0$.)

5.8.23 (Cayley's ruled cubic, cf. 10.5.17). Represent analytically a ruled cubic X which is not a cone, and such that the tangent cones in the points of the double directrix are pairs of planes, one of which is fixed.

We have seen in 5.8.22 that the double directrix is a line $\mathfrak{d}$. A cubic hypersurface passing doubly through the line with equations $x_0 = x_1 = 0$ has an equation of the form

$$x_2 \varphi_2(x_0, x_1) + x_3 \varphi_2'(x_0, x_1) + \theta_3(x_0, x_1) = 0,$$

where φ_2, φ_2' are quadratic forms and θ_3 is a cubic form. The tangent cone in the generic point $P = [0, 0, h, k]$ of $\mathfrak{b}$ has equation $h\varphi_2(x_0, x_1) + k\varphi_2'(x_0, x_1) = 0$. If the fixed plane is $x_1 = 0$ one then finds an equation for X of the form

$$x_1 x_2 (ax_0 + bx_1) + x_1 x_3 (cx_0 + dx_1) + mx_0^3 + nx_0^2 x_1 + px_0 x_1^2 + qx_1^3 = 0.$$

Note that $m \neq 0$, for otherwise X would be reducible. One must also have $ad - bc \neq 0$. Indeed, if $ad = bc$, the surface X would have $Q = [0, 0, -c, a] = [0, 0, -d, b]$ as a triple point and thus would be a cone with vertex Q.

The equation of X may be rewritten as

$$mx_0^3 + x_1^2(px_0 + qx_1 + bx_2 + dx_3) + x_0 x_1 (nx_0 + ax_2 + cx_3) = 0.$$

We can then effect a change of coordinates such that X receives the equation

$$x_0^3 + x_1^2 x_2 + x_0 x_1 x_3 = 0.$$

Note that the fixed plane $x_1 = 0$ osculates the ruled surface along the double line $\mathfrak{b}$ (which means that it intersects X along the line $\mathfrak{b}$ counted three times) and from each point P of $\mathfrak{b}$ there issues a single generator g_P (and not two, as in the case of the general ruled surface in 5.8.22). More precisely, the tangent cone to X at the point $P = [0, 0, h, k]$ of $\mathfrak{b}$ has equation $hx_1^2 + kx_0 x_1 = x_1(hx_1 + kx_0) = 0$ and the plane $hx_1 + kx_0 = 0$ meets X in the line $\mathfrak{b}$ counted twice and in the further generating line g_P. The generator contained in the plane $x_1 = 0$ is $\mathfrak{b}$ which is thus simultaneously a double directrix and a generator.

Chapter 6
Linear Systems

The notion of "linear system of divisors" on an algebraic variety X plays a crucial role in algebraic geometry. The present chapter is dedicated to that concept in the case of linear systems of projective hypersurfaces, that is, in the case $X = \mathbb{P}^n$. The topics discussed here constitute an indispensable prerequisite for the reading of Chapters 9 and 10.

In Sections 6.1, 6.2 we will give the general definitions, in particular that of the dimension of a linear system, and we consider the hypersurfaces of a linear system that satisfy specific conditions.

In Section 6.3 we study the base locus of a linear system Σ, that is, the locus of those points common to all the hypersurfaces which make up the system Σ. In this regard, Bertini's first theorem (Theorem 6.3.11), is one of the fundamental theorems of algebraic geometry. It assures us that the generic hypersurface of Σ does not have singularities outside of its base subvariety.

Some properties of the Jacobian variety of a linear system Σ, that is, the projective variety which is the locus of the zeros of the ideal generated by the minors of maximal order of the Jacobian matrix associated to Σ, are discussed in Section 6.4.

In Section 6.5 we consider the notions of simple and composite linear systems, and we state Bertini's second theorem (Theorem 6.5.2), which describes the structure of reducible linear systems, that is, those consisting entirely of reducible hypersurfaces.

In Section 6.6 we study the notion, fundamental in algebraic geometry, of the projective image of a linear system and describe unirational and rational varieties in terms of projective images of linear systems. In this regard, we state Lüroth's theorem (Theorem 6.6.2), proved in Section 7.4, and Castelnuovo's theorem (Theorem 6.6.3), which concern the cases of curves and surfaces respectively.

Section 6.7 is dedicated to the Veronese varieties, that is, to the varieties $V_{n,d}$ which are the projective images of the linear systems $\Sigma_{n,d}$ of all the hypersurfaces of a suitable order d in $\mathbb{P}^n$. The Veronese varieties constitute a very interesting example of rational varieties; we shall dedicate ample space to Veronese surfaces in the course of Chapter 10.

Finally, in Section 6.8 we mention a class of rational transformations widely used in algebraic geometry: blow-ups of a variety X along a given subvariety B. We examine the interesting particular cases in which B is a point or a linear space $\mathbb{P}^b$. The case of blowing up a plane at a point will be reconsidered in §9.2.7 within the context of the study of quadratic transformations between planes.

6.1 Linear systems of hypersurfaces

Let $f_j(T_0, \ldots, T_n) = 0$, $j = 0, 1, \ldots, h$, be algebraic hypersurfaces in $\mathbb{P}^n := \mathbb{P}^n(K)$ all of the same order r. The totality of the hypersurfaces with equations of the form

$$\lambda_0 f_0 + \cdots + \lambda_h f_h = 0, \tag{6.1}$$

where $\lambda_0, \ldots, \lambda_h$ are elements of the base field K not all equal to zero, is said to be a *linear system* of hypersurfaces of order r or, more briefly, a linear system of order r.

Since the K-vector space V of homogeneous polynomials of degree r in $n+1$ indeterminates is generated by the $\binom{n+r}{r}$ monomials $T_0^{\alpha_0} T_1^{\alpha_1} \ldots T_n^{\alpha_n}$, where $0 \le \alpha_i \le r$, $\alpha_0 + \cdots + \alpha_n = r$, which are linearly independent, all the hypersurfaces of a given order r in $\mathbb{P}^n$ constitute a linear system which can be viewed as the projective space $\mathbb{P}(V) = \mathbb{P}^{N(r)}$, where

$$N(r) = \binom{n+r}{n} - 1.$$

The linear system Σ represented by the equation (6.1) is the subspace S of $\mathbb{P}^{N(r)}$ generated by the $h+1$ "points" given by the hypersurfaces with equation $f_j = 0$ (see also Section 6.6).

By the *dimension* of Σ we mean its dimension as a subspace of $\mathbb{P}^{N(r)}$. If $\dim \Sigma (\le h)$ is the dimension of the system Σ with equation (6.1), the same system can be obtained by taking linear combinations of any $\dim \Sigma + 1$ of its linearly independent hypersurfaces.

A linear system Σ of dimension 1 will be called a *pencil*; if $\dim \Sigma = 2$ we will say that Σ is a *net* of hypersurfaces.

From now on we will assume, as we may without loss of generality, that the hypersurfaces $f_j = 0$ in (6.1) are linearly independent, and thus that $h = \dim \Sigma$. We will also write that Σ is a system ∞^h.

In the case $n = 1$ linear systems are customarily referred to as *linear series* (of groups of points of $\mathbb{P}^1$). A linear series of dimension h and order r is denoted by the symbol g_r^h. Its elements are groups of r points (not necessarily all distinct) of $\mathbb{P}^1$.

All statements regarding the space $\mathbb{P}(V) = \mathbb{P}^{N(r)}$ and its subspaces can be referred to the totality of the hypersurfaces $X_{n-1}^r \subset \mathbb{P}^n$ of order r and to linear systems of hypersurfaces. Thus, for example, if Σ_1 and Σ_2 are two linear systems of hypersurfaces of order r in $\mathbb{P}^n$, then so too are their intersection $\Sigma_1 \cap \Sigma_2$ and their join $\Sigma_1 + \Sigma_2$ (i.e., the set of hypersurfaces having equation $f + g = 0$ with $f \in \Sigma_1, g \in \Sigma_2$) and one has

$$\dim \Sigma_1 + \dim \Sigma_2 = \dim(\Sigma_1 + \Sigma_2) + \dim(\Sigma_1 \cap \Sigma_2).$$

The linear systems of hypersurfaces $X_{n-1}^r \subset \mathbb{P}^n$ are particular *algebraic systems*, where by algebraic system we mean an algebraic variety X in $\mathbb{P}^{N(r)}$. The dimension of X is, by definition, the *dimension* of the algebraic system. An algebraic system is said to be *irreducible, reduced, pure*, ... according to whether or not the variety X that represents it is irreducible, reduced, pure, ... (cf. Chapter 3).

6.2 Hypersurfaces of a linear system that satisfy given conditions

We will say that a condition $\mathcal{K}$ imposed on the hypersurfaces X_{n-1}^r of $\mathbb{P}^n$ is *linear* if it translates into a system of linear equations among the coefficients of the equation of X_{n-1}^r. The set of all the hypersurfaces of a given order that satisfy a linear condition constitutes a linear system.

We will say that $\mathcal{K}$ is a *linear condition of dimension d* if it translates into d independent linear equations, so that the hypersurfaces of given order r that satisfy the condition constitute a linear system of dimension $N(r) - d$.

Similarly, we say that $\mathcal{K}$ is an *algebraic condition of dimension d* if it translates into polynomial equations that define a variety of codimension d in $\mathbb{P}^{N(r)}$.

The hypersurfaces in $\mathbb{P}^n$ of a given order r that satisfy a linear condition are the points of a subspace of $\mathbb{P}^{N(r)}$; those that satisfy an algebraic condition are the points of a closed algebraic subset of $\mathbb{P}^{N(r)}$.

An important example of a linear condition is that of passage through a *given point P with assigned multiplicity*. In order that the hypersurface with equation $f = 0$ have a point P with multiplicity s it is necessary and sufficient that in that point all the partial derivatives of order $s - 1$ of f should vanish. The number of such derivatives is equal to the number of combinations with repetitions of $s - 1$ objects chosen from a class of $n + 1$ objects, namely $\binom{n+s-1}{s-1} = \binom{n+s-1}{n}$.

It is useful to perform this calculation in the following alternative fashion. We may suppose that P is the origin of a system of affine coordinates. Then in the equation of a hypersurface having P as an s-fold point, the constant term, the $n = \binom{n}{1} = \binom{n}{n-1}$ coefficients of the linear terms, the $\binom{n+1}{2} = \binom{n+1}{n-1}$ coefficients of the terms of degree two, ..., the $\binom{n+s-2}{s-1} = \binom{n+s-2}{n-1}$ coefficients of the terms of degree $s - 1$ must be zero. Moreover, we have

$$1 + \binom{n}{n-1} + \binom{n+1}{n-1} + \cdots + \binom{n+s-2}{n-1} = \binom{n+s-1}{n}.$$

This argument assures us that these are indeed independent linear conditions, and therefore s-fold passage through a *given* point P is indeed a linear condition of dimension $\binom{n+s-1}{n}$.

6.2. Hypersurfaces of a linear system that satisfy given conditions

Example 6.2.1 (The case of plane curves). The algebraic plane curves of order r that pass with multiplicity $s\,(\le r)$ through a given point P constitute a linear system whose dimension is

$$\binom{r+2}{2} - 1 - \binom{s+1}{2} = \frac{r(r+3)}{2} - \frac{s(s+1)}{2}.$$

Example–Definition 6.2.2 (Regular systems). Let $P_1, P_2, \ldots, P_q$ be q points of $\mathbb{P}^n$. Passage through any one of them imposes a linear condition on the hypersurfaces X_{n-1}^r of degree r in $\mathbb{P}^n$. For particular choices of q, n, or r these q conditions can fail to be independent.

A trivial example is the following: we take three collinear points in the plane and impose on the lines of the plane to contain these points. One obtains three linear conditions which are manifestly not independent inasmuch as passing through two of these points implies passage also through the third.

A more interesting example is the following. Once again in the plane, we consider two curves of order 3 and the nine points that they have in common. Each of these points imposes a linear condition on the cubics in the plane. But the nine conditions are not independent because otherwise the curve of order 3 containing these points would be unique.

In any case, one does have that if r is sufficiently large with respect to q (for example if $r > q-1$), then the conditions imposed by the q points are independent. Indeed, one immediately finds curves of order $> q-1$ (having $q-1$ lines as components) that pass through $q-1$ chosen arbitrarily among the q assigned points but not through the remaining point.

The algebraic plane curves of order r that pass with multiplicities $s_1, s_2, \ldots, s_t$ through t distinct points $P_1, P_2, \ldots, P_t$ constitute a linear system Σ of dimension

$$\dim \Sigma \ge \frac{r(r+3)}{2} - \sum_{i=1}^{t} \frac{s_i(s_i+1)}{2}. \tag{6.2}$$

Note that the right-hand side of (6.2) may well be negative; even in that case the system Σ can still be non-empty. For example, on imposing that a quartic have five given double points, the right-hand side of (6.2) yields -1, but the system Σ of such quartics is non-empty since it clearly contains the square of the conic through the five points.

The non-negative integer

$$\sigma := \dim \Sigma - \frac{r(r+3)}{2} + \sum_{i=1}^{t} \frac{s_i(s_i+1)}{2}$$

is called the *superabundance* of Σ. If $\sigma = 0$ the system is said to be *regular*; in that case the $\sum_{i=1}^{t} \binom{s_i+1}{2} = \sum_{i=1}^{t} \frac{s_i(s_i+1)}{2}$ conditions imposed by the s_i-fold points P_i, $i = 1, \ldots, t$, are independent. As we shall see in Lemma 7.2.14 this is always the case if the curves are rational. If $\sigma > 0$ one says that Σ is *superabundant*.

Another example of a linear condition on the hyperfurfaces X_{n-1}^r in $\mathbb{P}^n$ is that of possessing a given component with an assigned multiplicity s. This amounts to considering hypersurfaces with equations of the form $p^s f$ where p is a fixed and f an arbitrary homogeneous polynomial.

An important example of an algebraic but non-linear condition is that of possessing a *not specified* point of a given multiplicity $s \geq 2$. It is a condition of dimension

$$\binom{n+s-1}{n} - n.$$

In particular, when $s = 2$ one has a condition of dimension 1. In this case the condition corresponds to requiring the existence of a non-trivial solution to the system of equations arising from the vanishing of the $n+1$ first partial derivatives of the polynomial defining the hypersurface.

Consider for example the case of the quadrics in $\mathbb{P}^n$, that is, the case $r = 2$. In order that the quadric with equation $\sum a_{ij} x_i x_j = 0$ should have a double point it is necessary and sufficient that there be proper solutions to the system of $n+1$ linear homogeneous equations in $n+1$ indeterminates having (a_{ij}) as its matrix of coefficients. This means that the point of $\mathbb{P}^{N(r)}$ having the coefficients a_{ij} as coordinates must belong to the algebraic hypersurface of order $n+1$ represented in $\mathbb{P}^{N(r)}$ by the equation $\det(a_{ij}) = 0$.

Furthermore it can be proved that the algebraic plane curves of order r with d nodes constitute an algebraic system of dimension $\frac{r(r+3)}{2} - d$. The classical proof is due to Severi [101] (cf. also [36, Vol. III, pp. 386–387]). For a modern treatment of this topic we refer the reader, for example, to Sernesi's book [94, IV.7].

6.3 Base points of a linear system

A point x that belongs to all the hypersurfaces having equations $f_0 = 0$, $f_1 = 0$, ..., $f_h = 0$ which determine a linear system Σ of equation (6.1) evidently belongs to all the hypersurfaces of the system. We will say that x is a *base point* of Σ. If the generic hypersurface of Σ has x as a simple point, we will say that x is a *simple base point*.

More generally, suppose that the point x has, for each of the hypersurfaces $f_j = 0$, a multiplicity at least as great as s. Then the same fact will hold for any other hypersurface of Σ. If in addition x has multiplicity exactly s for at least one of the $f_j = 0$, the same will hold for the generic hypersurface of Σ and we will say that x is an *s-fold base point*. One sees immediately that if x is an s-fold base point, the system Σ can be obtained by taking a linear combination of hypersurfaces all having the same multiplicity s in that point.

The locus of base points of Σ is the *base variety*. It is the algebraic variety $B = V(f_0, f_1, \ldots, f_h)$ associated to the homogeneous ideal $(f_0, f_1, \ldots, f_h)$. One says that B, or one of its components, is an *s-fold base variety* of Σ if each of its generic points is an s-fold base point.

We define the *degree* deg Σ of a linear system Σ of hypersurfaces in $\mathbb{P}^n$ to be the number of points, not belonging to the base variety, which are common to n generic hypersurfaces of Σ.

One says that P is an *isolated base point* if it does not belong to any irreducible base variety of dimension > 0. Similarly, an irreducible base variety of dimension k is said to be an *isolated base variety* if it does not belong to any irreducible base variety having dimension $> k$.

An s-fold isolated base point is said to be *ordinary* if the tangent cone there to the generic hypersurface of the system does not have any multiple generator and it varies with the hypersurface.

An s-fold isolated base variety B_k of dimension k is an *s-fold ordinary base variety* if the tangent cone to the generic hypersurface of Σ at the generic point of B_k does not have any multiple generator S_{k+1}, and that cone varies with the hypersurface.

It is not excluded that on the ordinary s-fold base variety there are subvarieties that are base varieties of multiplicity $s' > s$ for Σ. For example, the surfaces in $\mathbb{P}^3$ that pass simply through a curve having a triple point x with *non* coplanar tangents are constrained to have x as at least a double point.

One can say something more regarding the s-fold base points, not limiting ourselves to their multiplicities, but also considering the possibility of there being fixed tangents. For example, if at an s-fold base point x of the linear system (6.1) the hypersurfaces that define it have a common tangent (that is, a line common to their tangent cones at x), then all the hypersurfaces of the system will have that same common tangent at x.

We now return to the case in which the hypersurfaces of the h-dimensional linear system Σ are merely required to pass (simply) through the various given points. Passage through any one of these points is equivalent to one condition, so that giving a number q of points, with $q \leq h$, one will have, *in general*, a linear system of dimension $h - q$.

However, we have seen that the dimension will rise whenever imposing the passage through those q points does not constitute imposition of independent conditions.

In particular, for $q = h$, one has the following:

Theorem 6.3.1. *In a linear system ∞^h of hypersurfaces there is, in general, only one hypersurface that passes through h given points (that is, the system includes only one hypersurface that passes through h generic points, namely points that represent h independent conditions).*

Remark 6.3.2 ("Generic" points with respect to a linear system). In regard to the statement of Theorem 6.3.1, one does well to look a bit more deeply at the meaning of the term "generic". To start with, one takes an arbitrary first point P_1, provided only that it not belong to the base variety of the given linear system Σ of dimension h. Then the hypersurfaces of Σ that pass through P_1 do not exhaust the entire system, but rather constitute a certain linear system ∞^{h-1}. This new system will contain as base points all those of the previous system, together with the point P_1 as well, and possibly other new points too. One then takes a point P_2 different from all these base points. The hypersurfaces of the second linear system which pass through P_2 do not exhaust it, but rather constitute a certain linear system ∞^{h-2}, which will have as base points all those already encountered, together with the point P_2 and possibly other new points as well. In analogous fashion one then takes a point P_3, and so on, until one has h points, through which will really pass one and only one hypersurface of Σ.

The fact that the preceding theorem has a converse is important. More precisely, one has the following result which we will only state here (and for whose proof one may consult, for example [14, Chapter 10]).

Theorem 6.3.3. *Let Λ be an algebraic system ∞^h of hypersurfaces in $\mathbb{P}^n$ all having the same order, and with generic member non multiple. If only one hypersurface of Λ passes through h generic points of $\mathbb{P}^n$, then Λ is a linear system.*

We observe that the hypothesis in Theorem 6.3.3 that the generic hypersurface of Λ should not be multiple is essential. It suffices to consider the algebraic system Λ of dimension 2 consisting of the double degenerate conics of a plane (that is, constituted by a double line). Through two generic points of the plane there passes one and only one conic of Λ, and yet Λ is not a linear system.

The reasoning by which we arrived at a set of h points through which there passes only one hypersurface of the linear system Σ with equation (6.1) can be repeated when, instead of a linear system Σ, one has an algebraic system Λ of dimension h consisting of algebraic hypersurfaces. Rather than a single hypersurface one now arrives at a 0-dimensional algebraic system of hypersurfaces, that is, at a finite number $i(\Lambda)$ of hypersurfaces of Λ.

The number $i(\Lambda)$, namely, the number of hypersurfaces of Λ that pass through h generic points of the space, is called the *index* of the algebraic system Λ. If Λ is linear, then $i(\Lambda) = 1$. The following fact is easily seen:

- If Λ is a pure and reduced algebraic system of hypersurfaces of order r with the dimension of Λ being $h > 0$, and if the generic hypersurface of Λ is without multiple subvarieties that vary with it, then the index of Λ is equal to the order of the variety V_h that represents Λ in $\mathbb{P}^{N(r)}$.

Indeed, the hypersurfaces of Λ that pass through a point P are the points of a hyperplane section of V_h, and thus the index of Λ is the number of points that V_h has in common with the space in which h generic hyperplanes of $\mathbb{P}^{N(r)}$ meet.

6.3. Base points of a linear system   173

We also note that the hypothesis that the generic hypersurface of Λ be without multiple subvarieties that vary with it is also essential. For example, the system Λ of double lines in the plane has index $i(\Lambda) = 1$ since through two points there passes only one doubly degenerate conic, but it is represented in $\mathbb{P}^5$ by a surface of fourth order (the Veronese surface, which we will study in Example 10.2.1 and Section 10.4).

6.3.4 Section of a linear system by a subspace. Given a linear system Σ of hypersurfaces of order r in $\mathbb{P}^n$, we consider a subspace S not contained in the base variety of Σ and the linear system Σ_0 of the hypersurfaces of Σ that contain it. The intersections with S of the hypersurfaces of Σ constitute a linear system Σ' of hypersurfaces of S (cf. Section 5.1) and it is easy to see that

$$\dim \Sigma' = \dim \Sigma - \dim \Sigma_0 - 1.$$

Indeed, we assume that the reference system is chosen so that the equations of S ($= S_t$) are $x_{t+1} = x_{t+2} = \cdots = x_n = 0$. Put $h = \dim \Sigma$ and $h_0 = \dim \Sigma_0$, and choose $h+1$ hypersurfaces $\varphi_0 = 0, \varphi_1 = 0, \ldots, \varphi_h = 0$ in Σ among which there are $h_0 + 1$, for example those with equations $\varphi_0 = 0, \varphi_1 = 0, \ldots, \varphi_{h_0} = 0$, contained in Σ_0 (and thus the forms $\varphi_0, \varphi_1, \ldots, \varphi_{h_0}$ all belong to the ideal generated by $x_{t+1}, x_{t+2}, \ldots, x_n$, while none of the remaining forms $\varphi_{h_0+1}, \ldots, \varphi_h$ belongs to that ideal). The system of equations

$$\sum_{i=0}^{h} \lambda_i \varphi_i = x_{t+1} = \cdots = x_n = 0$$

is equivalent to the system

$$\sum_{i=h_0+1}^{h} \lambda_i \varphi_i(x_0, \ldots, x_t, 0, \ldots, 0) = x_{t+1} = \cdots = x_n = 0.$$

Therefore $\dim \Sigma' \leq h - h_0 - 1$; and in fact the equality $\dim \Sigma' = h - h_0 - 1$ holds in view of the hypothesis that no hypersurface having as equation a linear combination of $\varphi_{h_0+1}, \ldots, \varphi_h$ contains S. So we have the following

Theorem 6.3.5. *A linear system Σ of hypersurfaces cuts out on a subspace S a linear system whose dimension is equal to that of Σ diminished by the maximum number of linearly independent forms belonging to Σ and passing through S.*

Exercise 6.3.6. In $\mathbb{P}^3$ we consider a plane π and in π we take a linear system Σ of algebraic curves of order r. Supposing that $\dim \Sigma = h \ (\leq \frac{r(r+3)}{2})$, determine the maximal dimension δ that a linear system of surfaces of order r in $\mathbb{P}^3$ that cuts Σ in π can have. Similarly, given in a subspace S_t of $\mathbb{P}^n$ a linear system Σ of hypersurfaces of order r, determine the maximal dimension of the linear systems of hypersurfaces of a fixed order r that can cut Σ in S_t.

6.3.7 Section of a linear system by an irreducible subvariety. We consider a linear system Σ of hypersurfaces in $\mathbb{P}^n$ and an irreducible and reduced variety X. The hypersurfaces of Σ meet X in a totality $\mathcal{D}$ of divisors (that is, of subvarieties of codimension 1) which we will call a *linear system* or *linear series of divisors of* X. If it happens that there is a one-to-one correspondence between the elements of $\mathcal{D}$ and the hypersurfaces of Σ we will say that $\mathcal{D}$ has *dimension* equal to that of Σ. As we have seen in §6.3.4 in the case in which X was a linear space, one can take in Σ a linear system Σ' complementary to the system Σ_0 of hypersurfaces of Σ passing through X and one sees that there is a one-to-one correspondence between hypersurfaces of Σ' and divisors of $\mathcal{D}$. Thus the relation

$$\dim \mathcal{D} = \dim \Sigma - \dim \Sigma_0 - 1$$

holds. In most cases it is convenient to consider, rather than $\mathcal{D}$, the totality (which we will continue to call $\mathcal{D}$), of the variable parts of the divisors of $\mathcal{D}$. In other words, one can dispense with possible components common to all the divisors of $\mathcal{D}$. By contrast, it can sometimes be opportune to add a fixed divisor to all the elements of a linear system.

6.3.8 Tangent cones at a base point. It is easy to see that the tangent cones to the hypersurfaces X_{n-1}^r of a linear system Σ of dimension h at an s-fold base point x constitute a linear system of dimension $h - h_0$, where h_0 is the maximum number of linearly independent hypersurfaces that can be found in Σ all having x as a point of multiplicity $> s$.

To prove this it suffices to assume that x is the point $[1, 0, \ldots, 0]$ and to choose a system of generators $g_0, g_1, \ldots, g_{h_0-1}, f_{h_0}, \ldots, f_h$ in Σ that are linearly independent and such that the h_0 hypersurfaces $g_j = 0$ have x as at least an $(s+1)$-fold point:

$$g_j = x_0^{r-s-1} \varphi_{s+1}^{(j)} + x_0^{r-s-2} \varphi_{s+2}^{(j)} + \cdots, \quad j = 0, \ldots, h_0 - 1;$$
$$f_i = x_0^{r-s} \varphi_s^{(i)} + x_0^{r-s-1} \varphi_{s+1}^{(i)} + \cdots, \quad i = h_0, \ldots, h.$$

The generic hypersurface of Σ has equation $\sum_i \lambda_i f_i + \sum_j \mu_j g_j = 0$, that is

$$x_0^{r-s}(\lambda_{h_0}\varphi_s^{(h_0)} + \cdots + \lambda_h\varphi_s^{(h)}) + x_0^{r-s-1}(\cdots) + \cdots = 0.$$

As the parameters λ vary, the tangent cone to that hypersurface at x runs over the linear system

$$\lambda_{h_0}\varphi_s^{(h_0)} + \cdots + \lambda_h\varphi_s^{(h)} = 0,$$

generated by the $h - h_0 + 1$ cones with equations $\varphi_s^{(h_0)} = 0, \ldots, \varphi_s^{(h)} = 0$. These equations are linearly independent. Indeed, if there were $h - h_0 + 1$ elements $a_{h_0}, \ldots, a_h$ not all zero in K such that

$$a_{h_0}\varphi_s^{(h_0)} + \cdots + a_h\varphi_s^{(h)} = 0,$$

the hypersurface $a_{h_0}f_{h_0} + \cdots + a_h f_h = 0$ would have x as at least an $(s+1)$-fold point and so its equation would be a linear combination of $g_0, g_1, \ldots, g_{h_0-1}$ and the polynomials $g_0, g_1, \ldots, g_{h_0-1}, f_{h_0}, \ldots, f_h$ would not be linearly independent.

Remark 6.3.9 (Tangent hyperplanes to the hypersurfaces of a pencil at a base point). Let P be an s-fold base point of a pencil Σ of hypersurfaces. One observes that if in the pencil there is a hypersurface having P as at least an $(s+1)$-fold point, then all the other hypersurfaces of the pencil have at P the same tangent cone; and conversely, if two generic hypersurfaces of the pencil have the same tangent cone at P, in the pencil there is a hypersurface having multiplicity at least $s+1$ at P. In particular, two hypersurfaces have at the point P, simple for both of them, the same tangent hyperplane if and only if in the pencil that they determine there is an (obviously unique) hypersurface having P as singular point.

6.3.10 Bertini's first theorem. An important theorem, known as "Bertini's first theorem", states that over a base field of characteristic zero the generic hypersurface of a linear system Σ without fixed components does not have singularities outside the base variety. For a modern proof we refer to [48, Lecture 17] and also to [50, Chapter III, §10].

Note that the hypothesis requiring the base field K to have characteristic zero is essential, since there are counterexamples in characteristic $p > 0$. The modern proof of Bertini's theorem uses "generic smoothness" type results in an essential way, while the classical proof uses some delicate analytical arguments. For a complete panorama of Bertini type theorems we also refer to [57].

Here we offer an elementary proof, along the lines of the classical approach, and also inspired by the argument given in [92, Chapter VI, §1].

Theorem 6.3.11 (Bertini's first theorem). *Let K be a field of characteristic zero, and let Σ be a linear system without fixed components of hypersurfaces of order r in $\mathbb{P}^n(K)$. If the generic hypersurface of Σ has a (variable) s-fold point, with $s \geq 2$, the locus of such points is a base variety that is at least $(s-1)$-fold for Σ.*

Proof. Let

$$\Sigma: \lambda_0 f_0(x_1, \ldots, x_n) + \lambda_1 f_1(x_1, \ldots, x_n) + \cdots + \lambda_h f_h(x_1, \ldots, x_n) = 0,$$

where $f_j(x_0, \ldots, x_n)$ are linearly independent homogeneous polynomials of degree r, without common factors, $h = \dim \Sigma$.

If $f \in \mathbb{C}[x_0, \ldots, x_n]$, we will use f^* to denote an arbitrary $(s-2)$-nd derivative of f.

Fix an index $j \in \{1, \ldots, h\}$. We suppose that the generic hypersurface of the pencil

$$\Phi_j: f_0(x_0, x_1, \ldots, x_n) + t f_j(x_0, x_1, \ldots, x_n) = 0,$$

contained in Σ, has an s-fold point P which *varies with the parameter t*. Thus the point P satisfies the equations

$$\frac{\partial}{\partial x_i}(f_0^* + tf_j^*) = \frac{\partial f_0^*}{\partial x_i} + t\frac{\partial f_j^*}{\partial x_i} = 0, \quad i = 0, \ldots, n. \tag{6.3}$$

By Euler's theorem on homogeneous functions (cf. Exercise 3.1.18) we have

$$(r - s + 2)(f_0^* + tf_j^*) = \sum_{i=0}^{n} x_i \frac{\partial}{\partial x_i}(f_0^* + tf_j^*) = 0.$$

Arguing by contradiction, suppose that $f_j^*(P) \neq 0$ (or, equivalently, $f_0^*(P) \neq 0$) which means that P is not an $(s-1)$-fold point of the base variety of the pencil Φ_j. Substituting the value

$$t = -\frac{f_0^*}{f_j^*}$$

in (6.3) we find

$$\varphi_i(x_0, \ldots, x_n) := \frac{\partial f_0^*}{\partial x_i} - \frac{f_0^*}{f_j^*} \frac{\partial f_j^*}{\partial x_i} = 0, \quad i = 0, \ldots, n,$$

as well as

$$\frac{\partial}{\partial x_i}\left(\frac{f_0^*}{f_j^*}\right) = \frac{1}{f_j^*}\left(\frac{\partial f_0^*}{\partial x_i} - \frac{f_0^*}{f_j^*} \frac{\partial f_j^*}{\partial x_i}\right) = 0, \quad i = 0, \ldots, n.$$

This implies that the point P does not depend on t (since P satisfies the equation $\varphi_i(x_0, \ldots, x_n) = 0$, where the rational function φ_i is independent of t, $i = 0, \ldots, n$), and so contradicts the assumption made on P. Thus one must have $f_j^*(P) = f_0^*(P) = 0$, which is to say that P is an $(s-1)$-fold point for the base variety of the pencil Φ_j.

Since the same reasoning applies to each of the pencils Φ_j as j varies in $\{1, \ldots, h\}$, this establishes the desired conclusion.

Let us finally note a crucial point: $\varphi_i(x_0, x_1, \ldots, x_n)$ can not be identically zero for every $i = 0, \ldots, n$. Otherwise

$$\frac{\partial}{\partial x_i}\left(\frac{f_0^*}{f_j^*}\right)$$

would be identically zero for every i, and so one would have $f_0^* = kf_j^*$ for some $k \in \mathbb{C}$. Therefore, for each $i = 0, \ldots n$, it would follow that

$$\frac{\partial}{\partial x_i}(f_0^*) = k\frac{\partial}{\partial x_i}(f_j^*),$$

and so (6.3) would be equivalent to $\frac{\partial}{\partial x_i}(f_j^*) = 0$, $i = 0, \ldots, n$. Thus there would not exist variable s-fold points for the generic hypersurface of Σ. $\square$

Example 6.3.12. The quadric $\mathcal{F}_\lambda \subset \mathbb{P}^3$ with equation

$$x_1^2 + x_0(x_2 - \lambda x_3) = 0$$

is a cone with vertex $[0, 0, \lambda, 1]$. As λ varies it describes a pencil Σ of quadrics whose generic element has a double point. The locus of the double points $[0, 0, \lambda, 1]$ of the quadrics of Σ is the line $x_0 = x_1 = 0$ which is a (simple) base line for Σ.

6.4 Jacobian loci

We consider the linear system Σ of hypersurfaces $X_{n-1}^r \subset \mathbb{P}^n$ of equation

$$\lambda_0 f_0 + \cdots + \lambda_h f_h = 0,$$

where f_j are linearly independent homogeneous polynomials of degree r, so that $\dim \Sigma = h$.

Theorem 6.3.11 assures us that the generic hypersurface of Σ does not have singular points outside the base variety. There may however be particular hypersurfaces in Σ having a multiple point that is not a base point of the system.

Let $P = [y_0, y_1, \ldots, y_n]$ be a point of $\mathbb{P}^n$ that is multiple for some hypersurface of the system. Then there exist $h + 1$ elements $\lambda_0, \ldots, \lambda_h$ (not all of which are zero) from the field K such that

$$\lambda_0 \frac{\partial f_0}{\partial y_i} + \lambda_1 \frac{\partial f_1}{\partial y_i} + \cdots + \lambda_h \frac{\partial f_h}{\partial y_i} = 0, \quad i = 0, \ldots, n, \tag{6.4}$$

where we have written $\frac{\partial}{\partial y_i}$ rather than $\left(\frac{\partial}{\partial x_i}\right)(y_0, y_1, \ldots, y_n)$. The linear homogeneous system (6.4) thus admits non-trivial solutions and so the Jacobian matrix $\left(\frac{\partial f_j}{\partial x_i}\right)_{\substack{j=0,\ldots,h \\ i=0,\ldots,n}}$ (with $n + 1$ rows and $h + 1$ columns) has rank $< h + 1$ at P, that is

$$\varrho\left(\frac{\partial f_j}{\partial y_i}\right) < h + 1. \tag{6.5}$$

Conversely, if this condition holds the system (6.4) has non-trivial solutions and so there exist hypersurfaces of Σ having P as at least a double point.

The projective variety which is the locus of the zeros of the ideal generated by the minors of order $h + 1$ of the Jacobian matrix is called the *Jacobian variety* of Σ.

If $h > n$, condition (6.5) is always satisfied (indeed, the Jacobian matrix then has rank $\leq n + 1 < h + 1$). It follows that if Σ has dimension $h > n$, every point of $\mathbb{P}^n$ is multiple for some hypersurface of the system (this means that the Jacobian variety coincides with all of $\mathbb{P}^n$). In fact we know that the hypersurfaces of Σ that have a given point P (which is not a base point) as double point constitute a linear system whose dimension is $h - n - 1$. (One really does have a linear condition

178   Chapter 6. Linear Systems

of dimension $n + 1$, inasmuch as it is expressed by the $n + 1$ independent linear conditions obtained by setting the first partial derivatives of the equation of the generic hypersurface of Σ equal to zero.)

If $h = n$ the condition (6.5) reduces to $\det\left(\frac{\partial f_j}{\partial y_i}\right) = 0$. The Jacobian variety is then the hypersurface with equation

$$\det\left(\frac{\partial f_j}{\partial x_i}\right) = 0;$$

and its order is $(n + 1)(r - 1)$.

One can prove that when $h \leq n$, the Jacobian variety generally has dimension $h - 1$ and order (cf. [89, n. 5] and also [14, Chapter 10, n. 11])

$$\binom{n+1}{h}(r-1)^{n-h+1}.$$

As for the dimension, if one desires that the Jacobian matrix have rank h (as happens in the general case), it suffices to annul all the minors of order $h + 1$ that contain a non-zero minor of order h. The number of such minors is $n + 1 - h$. Thus one obtains $n+1-h$ equations that define a variety of dimension $n-(n+1-h) = h-1$.

6.4.1 The Jacobian group of a series g_r^1. Consider $\mathbb{P}^1$ with homogeneous coordinates x_0, x_1 and the linear series g_r^1 on $\mathbb{P}^1$ represented by

$$\lambda_0 \varphi_0(x_0, x_1) + \lambda_1 \varphi_1(x_0, x_1) = 0,$$

where φ_0, φ_1 are forms of degree r.

We impose the passage through a point $P \in \mathbb{P}^1$ and we let $\{P, P_2, \ldots, P_r\}$ be the group of r points of the line that constitutes the group of points of g_r^1 containing P. Thus we obtain an involution of order r on the line, that is, an algebraic totality of groups of r points that has the property that its groups are in algebraic one-to-one correspondence with the values of a parameter, and that every point of the line determines the group containing it (one notes that to each point $P, P_2, \ldots, P_r$ of the group there corresponds the same value of the parameter, cf. §1.1.2). Making each point P of the line correspond to the $r - 1$ points $P_2, \ldots, P_r$ (which are, in general, distinct from P) one obtains an algebraic correspondence $\omega \colon \mathbb{P}^1 \to \mathbb{P}^1$ of indices $(r - 1, r - 1)$, endowed with $2(r - 1)$ fixed points (cf. §1.1.3). These $2(r - 1)$ points are the double points of the involution and constitute the *Jacobian group* of g_r^1.

The Jacobian group consists of the *s-fold points of the series* g_r^1, $s \geq 2$. These are the points $P \in \mathbb{P}^1$ for which the group $\{P, P_2, \ldots, P_r\}$ of the g_r^1 containing P is such that $s - 1$ of its remaining points $P_2, \ldots, P_r$ coincide with P. Hence an *s-fold point of the series is counted $s - 1$ times in the Jacobian group*.

This fact is immediate from an analytic point of view. If, for example, the s-fold point is $A_1 = [0, 1]$ and $\varphi_1(x_0, x_1) = 0$ is the equation of the group containing A_1, one clearly has $\varphi_1(x_0, x_1) = x_0^s f_{r-s}(x_0, x_1)$, where f_{r-s} is a form of degree $r - s$. It then suffices to calculate the Jacobian matrix of φ_0, φ_1 and observe that its determinant is divisible by x_0^{s-1}.

Example 6.4.2. In $\mathbb{P}^1$ consider the linear series g_3^1 represented by

$$\lambda_0(x_0 - x_1)x_0^2 + \lambda_1 x_1^3 = 0.$$

The group $\{P, P_2, P_3\}$ of g_3^1 that contains the point $P = [a_0, a_1]$ corresponds to the value of the parameter $\lambda_0 : \lambda_1$ such that $\lambda_0(a_0 - a_1)a_0^2 + \lambda_1 a_1^3 = 0$, that is

$$\lambda_0 : \lambda_1 = a_1^3 : a_0^2(a_1 - a_0).$$

So one finds that group by solving the equation $a_1^3(x_0-x_1)x_0^2 + a_0^2(a_1-a_0)x_1^3 = 0$, or

$$(a_1 x_0 - a_0 x_1)[(a_1 x_0 - a_0 x_1)^2 + 3a_0 a_1 x_0 x_1 - a_1 x_1(a_1 x_0 + a_0 x_1)] = 0.$$

The point P is (at least) double for g_3^1 if $a_1 x_0 - a_0 x_1$ is a divisor of the polynomial $(a_1 x_0 - a_0 x_1)^2 + 3a_0 a_1 x_0 x_1 - a_1 x_1(a_1 x_0 + a_0 x_1)$; that is, if $a_0 a_1 = 0$. The two points $A_0 = [1, 0]$ and $A_1 = [0, 1]$ are both triple for g_3^1.

6.4.3 Exercises. Some further properties of the Jacobian variety are described in the exercises proposed here.

(1) *What is the Jacobian variety of a generic pencil of quadrics in $\mathbb{P}^n$?*

The Jacobian variety of a pencil Σ of quadrics is constituted by the vertices of the quadric cones belonging to Σ. If the pencil is generic the Jacobian variety is a finite set with $n + 1$ points.

For the first few values of n this fact is easily verified in the following way. If $n = 2$, let

$$\lambda_0 f_0(x_0, x_1, x_2) + \lambda_1 f_1(x_0, x_1, x_2) = 0$$

be the equation of the pencil of conics. The Jacobian matrix is

$$\begin{pmatrix} \dfrac{\partial f_0}{\partial x_0} & \dfrac{\partial f_0}{\partial x_1} & \dfrac{\partial f_0}{\partial x_2} \\ \dfrac{\partial f_1}{\partial x_0} & \dfrac{\partial f_1}{\partial x_1} & \dfrac{\partial f_1}{\partial x_2} \end{pmatrix}.$$

The Jacobian variety then consists of the three points belonging to the intersection of the two conics of equation

$$\frac{\partial f_0}{\partial x_0}\frac{\partial f_1}{\partial x_1} - \frac{\partial f_0}{\partial x_1}\frac{\partial f_1}{\partial x_0} = 0; \quad \frac{\partial f_0}{\partial x_0}\frac{\partial f_1}{\partial x_2} - \frac{\partial f_0}{\partial x_2}\frac{\partial f_1}{\partial x_0} = 0$$

180 Chapter 6. Linear Systems

and distinct from the point in which the lines $\frac{\partial f_0}{\partial x_0} = 0$, $\frac{\partial f_1}{\partial x_0} = 0$ meet.

In the case of a pencil of quadrics

$$\lambda_0 f_0(x_0, x_1, x_2, x_3) + \lambda_1 f_1(x_0, x_1, x_2, x_3) = 0$$

in $\mathbb{P}^3$, the Jacobian matrix is

$$\begin{pmatrix} \frac{\partial f_0}{\partial x_0} & \frac{\partial f_0}{\partial x_1} & \frac{\partial f_0}{\partial x_2} & \frac{\partial f_0}{\partial x_3} \\ \frac{\partial f_1}{\partial x_0} & \frac{\partial f_1}{\partial x_1} & \frac{\partial f_1}{\partial x_2} & \frac{\partial f_1}{\partial x_3} \end{pmatrix}.$$

Consider the quadrics with equation

$$\frac{\partial f_0}{\partial x_0}\frac{\partial f_1}{\partial x_1} - \frac{\partial f_0}{\partial x_1}\frac{\partial f_1}{\partial x_0} = 0, \qquad \frac{\partial f_0}{\partial x_0}\frac{\partial f_1}{\partial x_2} - \frac{\partial f_0}{\partial x_2}\frac{\partial f_1}{\partial x_0} = 0$$

and

$$\frac{\partial f_0}{\partial x_0}\frac{\partial f_1}{\partial x_3} - \frac{\partial f_0}{\partial x_3}\frac{\partial f_1}{\partial x_0} = 0.$$

The three quadrics contain the line r with equation $\frac{\partial f_0}{\partial x_0} = \frac{\partial f_1}{\partial x_0} = 0$. Two of the three quadrics intersect the third, outside of r, in two cubics belonging to the same family, that is, both are curves of type $(2, 1)$ (while r is a curve of type $(0, 1)$). Thus they meet in $(2, 1)(2, 1) = 4$ points which make up the Jacobian variety of the pencil (cf. Section 7.3).

(2) *Let $C \subset \mathbb{P}^n$ be a rational curve of order r (cf. Section 7.4). How many of the hyperplanes that pass through a generic S_{n-2} are tangent to C?*

The pencil of hyperplanes with center the given S_{n-2} cuts out a linear series g_r^1 on the curve C ($\cong \mathbb{P}^1$) where r is the order of C. The hyperplanes of the pencil that are tangent to C are in one-to-one correspondence with the $2(r-1)$ points of the Jacobian group of g_r^1, cf. §6.4.1.

(3) *Prove that the Jacobian curve of a net of algebraic plane curves having an s-fold base point P passes through P with multiplicity $3s - 1$.*

Let r be the order of the curves of the net, and let $P = [1, 0, 0]$. Further, let

$$f_i = x_0^{r-s}\varphi_s^{(i)} + \cdots, \quad i = 1, 2, 3,$$

with $\varphi_s^{(i)}$ a homogeneous polynomial of degree s in x_1, x_2, be three independent curves of the net. By Euler's formula for homogeneous functions (Exercise 3.1.18), we have

$$s\varphi_s^{(i)} = x_1\frac{\partial \varphi_s^{(i)}}{\partial x_1} + x_2\frac{\partial \varphi_s^{(i)}}{\partial x_2}, \quad i = 1, 2, 3,$$

and so

$$\det \begin{pmatrix} \dfrac{\partial \varphi_s^{(1)}}{\partial x_1} & \dfrac{\partial \varphi_s^{(2)}}{\partial x_1} & \dfrac{\partial \varphi_s^{(3)}}{\partial x_1} \\ \dfrac{\partial \varphi_s^{(1)}}{\partial x_2} & \dfrac{\partial \varphi_s^{(2)}}{\partial x_2} & \dfrac{\partial \varphi_s^{(3)}}{\partial x_2} \\ \varphi_s^{(1)} & \varphi_s^{(2)} & \varphi_s^{(3)} \end{pmatrix} = 0.$$

From this it follows easily that in the homogeneous polynomial of degree $3(r-1)$ given by

$$\det \begin{pmatrix} \dfrac{\partial f_1}{\partial x_0} & \dfrac{\partial f_2}{\partial x_0} & \dfrac{\partial f_3}{\partial x_0} \\ \dfrac{\partial f_1}{\partial x_1} & \dfrac{\partial f_2}{\partial x_1} & \dfrac{\partial f_3}{\partial x_1} \\ \dfrac{\partial f_1}{\partial x_2} & \dfrac{\partial f_2}{\partial x_2} & \dfrac{\partial f_3}{\partial x_2} \end{pmatrix}$$

the indeterminate x_0 appears at most to degree $3r - 3s - 2$. The multiplicity of P for the Jacobian curve is thus $3(r-1) - (3r - 3s - 2) = 3s - 1$.

(4) *Show that the Jacobian curve of a net of algebraic plane curves of order r can also be defined as the locus of the points in the plane that are contact points between curves of the net.*

Suppose that the point x (not a base point) is at least double for a curve C of the net. Then the curves of the net passing through x constitute a pencil of curves all (except for the curve C) having the same tangent at the point x (cf. Remark 6.3.9). Indeed, if $x = [1, 0, 0]$, the curve C has equation of the type $x_0^{r-2} \varphi_2(x_1, x_2) + \cdots = 0$, φ_2 a form of degree 2. Hence the curves of Σ that pass through x describe a pencil (defined by the curve C and by a curve of Σ passing simply through x, with equation $x_0^{r-1} \varphi_1(x_1, x_2) + \cdots = 0$, φ_1 a linear form) that has an equation of the form

$$\mu_1 x_0^{r-1} \varphi_1(x_1, x_2) + \mu_2 x_0^{r-2} \varphi_2(x_1, x_2) + \cdots = 0.$$

Therefore (if $\mu_1 \neq 0$) they all have as their tangent at x the line with equation $\varphi_1(x_1, x_2) = 0$. Hence x is a point of contact of the two curves.

Conversely, suppose we have a point x in the plane that is a contact point for two curves of the net. These two curves will determine a pencil of curves (of the net) all tangent to one another at the point x. Therefore, there will be a curve in the pencil having x as a double point. Thus x is a point of the Jacobian variety. This reasoning is no longer valid if x is a base point; nevertheless, any possible base points also belong to the Jacobian variety.

(5) *Prove that the Jacobian surface of a linear system ∞^3 of surfaces in $\mathbb{P}^3$ may be regarded as the locus of the points of contact of two surfaces of the system, or also as the locus of the points x such that the surfaces of the net of surfaces that pass through x have a common tangent at x.*

The reasoning is analogous to that of the preceding exercise.

(6) *In a pencil of algebraic plane curves of order r there are, in general, $3(r-1)^2$ curves endowed with a double point.*

Indeed, let f, g be two forms of degree r in x_0, x_1, x_2. The points where the Jacobian matrix

$$\begin{pmatrix} \dfrac{\partial f}{\partial x_0} & \dfrac{\partial f}{\partial x_1} & \dfrac{\partial f}{\partial x_2} \\ \dfrac{\partial g}{\partial x_0} & \dfrac{\partial g}{\partial x_1} & \dfrac{\partial g}{\partial x_2} \end{pmatrix}$$

has rank 1 are the points that annull the two minors of order 2 which contain the first column, except for the points that annull $\frac{\partial f}{\partial x_0}$ and $\frac{\partial g}{\partial x_0}$ (where the matrix has, in general, rank 2). Thus one finds the $4(r-1)^2 - (r-1)^2 = 3(r-1)^2$ points of the Jacobian group of the pencil. This number coincides with that of the curves of the pencil which are endowed with a double point.

6.5 Simple, composite, and reducible linear systems

Let Σ be a linear system of hypersurfaces X_{n-1}^r in $\mathbb{P}^n$. We say that Σ is *simple* if the hypersurfaces of Σ that contain a point P are not required to contain a variety W, properly containing P (and depending on P); in the contrary case we say that Σ is *composed with the congruence Γ of the variety W*, or simply that Σ is *composite*. If the varieties W are 0-dimensional, that is, groups of points, one says that Σ is *composed with the involution Γ*.

As is clear from the analysis carried out in Section 6.6, the following is a characteristic property of the congruence Γ: every point of the space (which is not a base point of Σ) belongs to one and only one variety of Γ, and every variety W of Γ is determined by each of its points (the case of a composite linear system Σ corresponds to the case in which the closure of the projective image of Σ has dimension $< n$, or has dimension n and the generic fiber of the associated morphism φ consists of a finite number $t > 1$ of points; cf. Section 6.6).

Obviously every linear system Σ of dimension $h < n$ is composite. Indeed, the hypersurfaces of such a system which pass through a generic point P are those of a linear system Σ' of dimension $h-1$ (cf. Section 6.3), which certainly has a base variety W passing through P and hence not contained in the base variety of Σ. The variety W is the intersection of h hypersurfaces that define Σ'. In particular, a pencil

6.5. Simple, composite, and reducible linear systems

Σ of hypersurfaces is a composite system and the varieties W are the hypersurfaces of Σ.

A linear system of cones having the same vertex is composite, because the cones that pass through a generic point P have in common a linear space that contains P.

Example 6.5.1 (Geiser's involution). A non-trivial example of a composite linear system is the linear system Σ of plane cubic curves that pass through seven arbitrary points of the plane. Indeed, the cubics of Σ that contain an eighth point P form a pencil Φ and thus all of them also pass through the ninth base point P' of Φ. Similarly, the cubics of the system that pass through P' also pass through P. The pairs (P, P') are the elements of a plane involution Γ which is called *Geiser's involution* and Σ is composed with Γ.

The same discussion could be made for plane curves of order $r \geq 4$ which contain $r^2 - 2$ points; but one must note that for $r^2 - 2$ generic points with $r > 4$ such a linear system does not exist, while for $r = 4$ it consists of a single curve, on which one can not impose an additional condition of passing through another arbitrary point.

A linear system Σ of hypersurfaces X_{n-1}^r in $\mathbb{P}^n$ is said to be *irreducible* if its generic hypersurface is irreducible. One says that Σ is *reducible* if its hypersurfaces are all reducible. Note that in the case of an algebraic system of hypersurfaces X_{n-1}^r we have said that the system is irreducible or reducible according to the irreducibility or reducibility of the variety that represents the system in $\mathbb{P}^{N(r)}$. A linear system, which is represented by a linear space, obviously is always an irreducible algebraic system. Thus, to call a linear system of reducible hypersurfaces a reducible system should not cause any confusion.

The following are examples of reducible linear systems.

- *Linear system with a fixed component*, that is, a linear system defined by

$$\theta(\lambda_0 f_0 + \lambda_1 f_1 + \cdots + \lambda_h f_h) = 0, \qquad (6.6)$$

whose hypersurfaces have a *fixed component* (or *part*), namely, the common component with equation $\theta = 0$.

- *Linear system composed with a pencil*, that is, of the type

$$\lambda_0 \varphi^h + \lambda_1 \varphi^{h-1} \psi + \lambda_2 \varphi^{h-2} \psi^2 + \cdots + \lambda_h \psi^h = 0, \qquad (6.7)$$

where φ, ψ are forms of the same degree. The left-hand side of (6.7) is a binary form in φ, ψ and so is the product of h factors of the type $\lambda \varphi + \mu \psi$. Every hypersurface of Σ is thus split into h hypersurfaces of the pencil Φ with equation $\lambda \varphi + \mu \psi = 0$: the hypersurfaces of Σ that pass through a point P all have the hypersurface of Φ that passes through that point as a component.

184   Chapter 6. Linear Systems

In regard to reducible linear systems, there is the following important theorem, known as "Bertini's second theorem", for the proof of which we refer the reader to [14, Chapter 10, n. 13] and [100, p. 45]).

Theorem 6.5.2 (Bertini's second theorem). *A reducible linear system of hypersurfaces in $\mathbb{P}^n$ ($n > 1$) either has a fixed component, or is composed with a pencil, or satisfies both these conditions.*

The two foregoing examples thus exhaust all possible cases for reducible linear systems.

Exercise 6.5.3. Let Σ be a linear system of hypersurfaces in $\mathbb{P}^n$, of dimension $h \leq n$, and with equation

$$\theta(\lambda_0 f_0 + \lambda_1 f_1 + \cdots + \lambda_{h-1} f_{h-1}) + \lambda_h f_h = 0.$$

Then the hypersurface Θ with equation $\theta = 0$ belongs to the Jacobian variety of Σ. In particular, the fixed component of a linear system with equation (6.6) is contained in the Jacobian variety of Σ.

This is geometrically evident. Indeed, if P is a generic point of Θ every hypersurface F_P of Σ that one obtains for values of the parameters λ such that $\lambda_0 f_0(P) + \lambda_1 f_1(P) + \cdots + \lambda_{h-1} f_{h-1}(P) = \lambda_h = 0$ passes through P and contains Θ. The residual component meets Θ in a locus of singular points of F_P. Thus P belongs to the Jacobian variety of Σ.

This fact can also be seen analytically as follows. The Jacobian matrix of the polynomials θf_j and f_h is

$$J = \begin{pmatrix} \frac{\partial \theta}{\partial x_0} f_0 + \theta \frac{\partial f_0}{\partial x_0} & \frac{\partial \theta}{\partial x_0} f_1 + \theta \frac{\partial f_1}{\partial x_0} & \cdots & \frac{\partial f_h}{\partial x_0} \\ \frac{\partial \theta}{\partial x_1} f_0 + \theta \frac{\partial f_0}{\partial x_1} & \frac{\partial \theta}{\partial x_1} f_1 + \theta \frac{\partial f_1}{\partial x_1} & \cdots & \frac{\partial f_h}{\partial x_1} \\ \cdots & \cdots & \cdots & \cdots \\ \frac{\partial \theta}{\partial x_n} f_0 + \theta \frac{\partial f_0}{\partial x_n} & \frac{\partial \theta}{\partial x_n} f_1 + \theta \frac{\partial f_1}{\partial x_n} & \cdots & \frac{\partial f_h}{\partial x_n} \end{pmatrix}$$

and may be written as a sum $J = A + \theta B$, where

$$A = \begin{pmatrix} \frac{\partial \theta}{\partial x_0} f_0 & \frac{\partial \theta}{\partial x_0} f_1 & \cdots & \frac{\partial f_h}{\partial x_0} \\ \frac{\partial \theta}{\partial x_1} f_0 & \frac{\partial \theta}{\partial x_1} f_1 & \cdots & \frac{\partial f_h}{\partial x_1} \\ \cdots & \cdots & \cdots & \cdots \\ \frac{\partial \theta}{\partial x_n} f_0 & \frac{\partial \theta}{\partial x_n} f_1 & \cdots & \frac{\partial f_h}{\partial x_n} \end{pmatrix} \quad \text{and} \quad B = \begin{pmatrix} \frac{\partial f_0}{\partial x_0} & \cdots & \frac{\partial f_{h-1}}{\partial x_0} & 0 \\ \frac{\partial f_0}{\partial x_1} & \cdots & \frac{\partial f_{h-1}}{\partial x_1} & 0 \\ \cdots & \cdots & \cdots & \cdots \\ \frac{\partial f_0}{\partial x_n} & \cdots & \frac{\partial f_{h-1}}{\partial x_n} & 0 \end{pmatrix}.$$

The minors of order $h+1$ of the matrix J that are not divisible by θ are all contained in A, and they are all zero since in the matrix A all the columns except the last are proportional; therefore the Jacobian variety contains the hypersurface Θ.

6.6 Rational mappings

In the projective space $\mathbb{P}^n = \mathbb{P}^n(K)$, with $x_0, x_1, \ldots, x_n$ as homogeneous projective coordinates, let

$$\Sigma: \lambda_0 f_0 + \lambda_1 f_1 + \cdots + \lambda_h f_h = 0, \tag{6.8}$$

with $\lambda_0, \ldots, \lambda_h$ elements of the field K which are not all zero, be a linear system of dimension h spanned by the hypersurfaces of order r with equations $f_j = 0$. The system Σ forms an h-dimensional projective space, and $\lambda_0, \ldots, \lambda_h$ are the projective coordinates there with respect to a reference system having $\{f_0, \ldots, f_h\}$ as its fundamental $(h+1)$-hedron and $\sum_{j=0}^{h} f_j$ as unit element.

Now let $\mathbb{P}^h$ be a projective space of dimension h over the field K, with homogeneous coordinates $X_0, X_1, \ldots, X_h$, let $\mathbb{P}^{h*}$ be the dual space of $\mathbb{P}^h$, that is, the space whose points are the hyperplanes of $\mathbb{P}^h$, and let $\varphi: \Sigma \to \mathbb{P}^{h*}$ be a (non-degenerate) projectivity.

We may assume that we have chosen the reference system of $\mathbb{P}^h$ so that a hypersurface of Σ and the hyperplane corresponding to it under φ have the same coordinates, and so that to the hypersurface

$$\lambda_0 f_0 + \lambda_1 f_1 + \cdots + \lambda_h f_h = 0 \tag{6.9}$$

there corresponds the hyperplane with equation

$$\lambda_0 X_0 + \lambda_1 X_1 + \cdots + \lambda_h X_h = 0. \tag{6.10}$$

We consider a linear system Σ_1, of dimension $h-1$, contained in Σ. The hyperplanes of $\mathbb{P}^h$ which are images under the projectivity φ of the hypersurfaces of Σ_1 are those of an $(h-1)$-dimensional star, namely, they are the hyperplanes that pass through a point of $\mathbb{P}^h$.

Let $P = [x_0, x_1, \ldots, x_n]$ be a point of $\mathbb{P}^n$ not belonging to the base variety of Σ, that is, to the variety B which is the locus of the zeros of the ideal $(f_0, f_1, \ldots, f_h)$. One may then associate to P the center $P' = [X_0, X_1, \ldots, X_h]$ of the star of hyperplanes of $\mathbb{P}^h$ that φ associates to the linear system Σ_1 of those hypersurfaces of Σ that pass through P.

Thus one has a rational map $\varphi^*: \mathbb{P}^n \to \mathbb{P}^h$, defined on the open set U of $\mathbb{P}^n$ which is the complement of B. To obtain the analytic representation of this map it suffices to observe that the hypersurfaces of Σ that pass through P, in other words the hypersurfaces that one obtains by imposing the condition

$$\lambda_0 f_0(x_0, x_1, \ldots, x_n) + \cdots + \lambda_h f_h(x_0, x_1, \ldots, x_n) = 0$$

on the parameters λ, correspond to the hyperplanes (6.10) with coefficients satisfying this same condition. In other words, these are the hyperplanes that pass through the point P' with coordinates

$$f_0(x_0, x_1, \ldots, x_n), f_1(x_0, x_1, \ldots, x_n), \ldots, f_h(x_0, x_1, \ldots, x_n).$$

Here then are the equations of φ^*:

$$X_j = f_j(x_0, x_1, \ldots, x_n), \quad j = 0, 1, \ldots, h. \tag{6.11}$$

The closure V (in the Zariski topology) of the image $\varphi^*(U)$ is called the *projective image of the linear system* Σ. If one replaces the basis $\{f_0, f_1, \ldots, f_h\}$ chosen to define Σ with another basis, one finds as projective image of Σ a variety that is the transform of V under a projectivity of $\mathbb{P}^h$. Therefore, the projective image of a linear system of hypersurfaces is defined by the system only up to a projectivity.

A variety of this type, namely a projective variety that can be represented in the form (6.11), is called a *unirational variety*; and (6.11) is its parametric representation. In other words:

- A unirational variety is the projective image of a linear system of hypersurfaces of a projective space (cf. 2.7.22).

Remark 6.6.1. Assume that Σ has an equation of the form

$$\theta(\lambda_0 f_0 + \lambda_1 f_1 + \cdots + \lambda_h f_h) = 0,$$

where $\theta = 0$ is the equation of the fixed component of Σ, common to all the hypersurfaces of Σ. It is then evident that the projective image V of Σ coincides with the projective image of the system Σ' having equation $\lambda_0 f_0 + \lambda_1 f_1 + \cdots + \lambda_h f_h = 0$, and of the same dimension h as Σ.

Hence by considering a projective image of a linear system Σ we can always assume that Σ has no fixed components.

Let φ_U^* be the restriction of φ^* to U. If $P' = [a_0, a_1, \ldots, a_h]$ is a point of $\varphi^*(U)$, the subvariety of $\mathbb{P}^n$ which is the locus of points which bestow rank one on the matrix

$$\begin{pmatrix} f_0(x_0, x_1, \ldots, x_n) & f_1(x_0, x_1, \ldots, x_n) & \cdots & f_h(x_0, x_1, \ldots, x_n) \\ a_0 & a_1 & \cdots & a_h \end{pmatrix}$$

is the fiber $\varphi^{*-1}(P')$ of φ^* over P'. That locus contains the base variety B and is the union of B and the fiber $\varphi_U^{*-1}(P')$ of φ_U^* over P', that is, the set of points P of U such that $\varphi_U^*(P) = P'$ (cf. §2.6.9).

The dimension of V coincides with the rank of the Jacobian matrix associated to the rational map (6.11) (cf. Section 3.2). In any case it is obvious that the dimension of V can not exceed n; and the necessary and sufficient condition for $\dim(V) = n$ is that for every generic point $P' \in \varphi^*(U)$, the fiber $\varphi_U^{*-1}(P')$ be a 0-dimensional set, that is, consist of a finite number t of points.

If $t = 1$, that is, if every generic point of $\varphi^*(U)$ comes from a single point of U, the variety V is rational and $\varphi^* \colon \mathbb{P}^n \to V \subset \mathbb{P}^h$ is a birational isomorphism (cf. Section 2.6). We note explicitly that

- the projective image of a simple linear system of hypersurfaces of a projective space is a rational variety.

If $t > 1$ the system Σ is composed with the involution Γ (of order t) consisting of the groups of t points that are inverse images of generic points of $\varphi^*(U)$ (cf. Section 6.5).

If, on the other hand, $\dim(V) < n$, the system Σ is composed with the congruence of the fibers $\varphi^{*-1}(P')$ of φ^* over points $P' \in \varphi^*(U)$. Note that, in view of equations (6.11), the fiber $\varphi^{*-1}(P')$ is an algebraic set of $\mathbb{P}^n$; if $\varphi^{*-1}(P')$ has dimension q for a generic point P', one has $\dim(V) = n - q$.

In the case in which Σ is not simple, it is possible that the same variety is also the projective image of a simple linear system Σ'. This is always true in the two cases $n = 1$ (curves) and $n = 2$ (surfaces), which will be further studied in Chapter 7, Section 7.4 and in Chapter 10, by virtue of two classical theorems that we here only state. For the case of surfaces we refer the reader to [23] and also to [6, Chapter V] and [4, Chapter VI, §2] (for the proof of Lüroth's theorem see also Exercise 2.7.31 and Section 7.4).

Theorem 6.6.2 (Lüroth's theorem). *Every unirational curve is rational.*

Theorem 6.6.3 (Castelnuovo's theorem [23]). *Every unirational surface is rational.*

However, examples are known of varieties of dimension ≥ 3 which are unirational but not rational; for example, the general cubic hypersurface X_3^3 in $\mathbb{P}^4$ (in this regard see [48, Lecture 18, Example 18.19] and [26]).

6.6.4. We note explicitly that the $h + 1$ hypersurfaces with equations $f_j = 0$ in (6.8) are linearly independent (that is, $\dim \Sigma = h$) if and only if the variety V, the projective image of Σ, is embedded in $\mathbb{P}^h$ and not in a space of lower dimension. Indeed, the condition in order that all the points

$$P' = [f_0(x_0, x_1, \ldots, x_n), f_1(x_0, x_1, \ldots, x_n), \ldots, f_h(x_0, x_1, \ldots, x_n)]$$

belong to the hyperplane with equation $\sum_{j=0}^{h} u_j X_j = 0$ is that one have, for each choice of the variables x_i, $i = 0, \ldots, n$,

$$\sum_{j=0}^{h} u_j f_j(x_0, x_1, \ldots, x_n) = 0;$$

and this means the linear dependence of the f_j's.

6.6.5 (Order of the projective image of a linear system). Let Σ be a linear system of dimension h and degree D of hypersurfaces of $\mathbb{P}^n$, and let V be its projective image. If Σ is composed with an involution Γ of degree t (in particular $\dim(V) = n$ and so $h \geq n$) the *order of V is* $\frac{D}{t}$.

This fact can be seen more precisely as follows. Let B be the base variety of Σ and let $\varphi^* \colon \mathbb{P}^n \to \mathbb{P}^h$ be the rational map defined on the open subset $U \subset \mathbb{P}^n$ which is the complement of B. We can suppose that the system Σ is irreducible, so that it does not contain any fixed component, and consequently that B has codimension at least 2 in $\mathbb{P}^n$.

We consider the graph $\mathcal{G}$ of φ^* in $U \times V$, i.e., $\mathcal{G} \ (\cong U) := \{(P, \varphi^*(P)) \mid P \in U\}$, and let Z be the closure of $\mathcal{G}$ in $\mathbb{P}^n \times V$. Let $p_1 \colon \mathbb{P}^n \times V \to \mathbb{P}^n$, and $p_2 \colon \mathbb{P}^n \times V \to V$ be the projections on the factors, and let $\pi = p_{1|Z} \colon Z \to \mathbb{P}^n$ and $f = p_{2|Z} \colon Z \to V$ be the restrictions to Z. One obtains a commutative diagram (the elementary resolution of the locus of indeterminacy of φ^*)

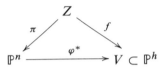

with π and f surjective morphisms. Moreover π is a birational morphism, and an isomorphism outside of B. Since Σ is composed with an involution of order t, the morphism φ^* (and so also f) is generically finite, and of degree t.

Let $H_1, \ldots, H_n$ be n (generic) hyperplane of $\mathbb{P}^h$ whose intersection with V is therefore a finite set of points, in number equal to the order $\deg(V)$ of V. To these points there correspond n divisors $f^*H_1, \ldots, f^*H_n$ of Z and the intersection

$$f^*H_1 \cap \cdots \cap f^*H_n = t(H_1 \cap \cdots \cap H_n \cap V)$$

consists of $D = t \deg(V)$ points. This number D depends only on Σ and coincides with the number of points that n generic hypersurfaces of Σ have in common outside the base locus B, that is, with the degree of Σ. In particular, if $t = 1$, the degree of Σ coincides with the order of V. Since $B \neq \mathbb{P}^n$ the (possibly reducible) variety $Y := f(\pi^{-1}(B))$ is properly contained in V, for purely dimensional reasons. Hence it is always possible to choose hyperplane $H_1, \ldots, H_n$ in such a way that $H_1 \cap \cdots \cap H_n \cap Y = \emptyset$. Indeed, if $m := \deg(Y) < n$, we can choose $H_1, \ldots, H_m$ so that $H_1 \cap \cdots \cap H_m \cap Y$ is a finite number of points, and thus $H_1 \cap \cdots \cap H_n \cap Y = \emptyset$.

It follows that the n (generic) hypersurfaces $F_1, \ldots, F_n$ of Σ that correspond bijectively to $H_1, \ldots, H_n$ meet outside of B in only D points.

Now suppose that $\dim(V) < n$, so that the system Σ is composed with the congruence of the fibers of φ^*. The reasoning given then clearly shows that in this case n generic hyperplane sections of V do not have any point in common, and so the corresponding (generic) hypersurfaces $F_1, \ldots, F_n$ of Σ do not meet outside of B. In this case Σ has degree $D = 0$.

For example, if Σ is a pencil of lines in $\mathbb{P}^2$ with center a point B, the variety Z is the blow-up, $\pi : Z \to \mathbb{P}^2$, of $\mathbb{P}^2$ at the point B, $\pi^{-1}(B)$ is the exceptional line E (see Section 6.8), and the morphism $f : Z \to \mathbb{P}^1$ is that contracting all the proper transforms of the lines of the pencil. In this case one has $Y = \mathbb{P}^1$ and obviously $D = 0$.

6.6.6 (Projective image of a composite linear system). With the notations as in Section 6.5, let Σ be a linear system of hypersurfaces in $\mathbb{P}^n$.

First suppose that the system Σ is composed with a pencil $\lambda \varphi + \mu \psi = 0$, that is, that it has an equation of the type

$$\lambda_0 \varphi^h + \lambda_1 \varphi^{h-1} \psi + \lambda_2 \varphi^{h-2} \psi^2 + \cdots + \lambda_h \psi^h = 0, \quad \lambda_0 \lambda_h \neq 0.$$

Set $t = \frac{\varphi}{\psi}$. One then has a parametric representation for the projective image V of Σ given by

$$X_0 : X_1 : \cdots : X_h = t^h : t^{h-1} : \cdots : 1,$$

so that V is a rational curve in $\mathbb{P}^h$ (cf. Section 7.4).

More generally, consider a linear system Σ of hypersurfaces of $\mathbb{P}^n$ composed with the congruence Γ of the varieties that are the loci of zeros of the minors of the matrix

$$\begin{pmatrix} \varphi_0 & \varphi_1 & \cdots & \varphi_\rho \\ a_0 & a_1 & \cdots & a_\rho \end{pmatrix},$$

with $\varphi_j \in K[x_0, x_1, \ldots, x_n]$ homogeneous polynomials of the same degree, $a_j \in K$, $j = 0, \ldots, \rho$, and with $\rho \leq n$. Thus, we are considering a linear system Σ having equation

$$\lambda_0 F_0(\varphi_0, \ldots, \varphi_\rho) + \lambda_1 F_1(\varphi_0, \ldots, \varphi_\rho) + \cdots + \lambda_h F_h(\varphi_0, \ldots, \varphi_\rho) = 0,$$

with F_i forms in the $\varphi_0, \ldots, \varphi_\rho$ of the same degree. The rational transformation $\phi : \mathbb{P}^n \to \mathbb{P}^h$, associated to Σ, where $h = \dim \Sigma$, has as fibers the subvarieties (of dimension $\geq n - \rho$) that constitute Γ, and so the projective image V of Σ has dimension $\dim(V) \leq n - (n - \rho) = \rho$.

In the preceding case in which Σ was composed with a pencil, the congruence Γ was constituted by the hypersurfaces of the pencil $\lambda \varphi + \mu \psi = 0$.

6.7 Projections and Veronese varieties

We return to consideration of the system Σ of equation (6.8) and we suppose that Σ is contained in a linear system Σ_0, of dimension $h_0 > h$, of hypersurfaces in $\mathbb{P}^n$ all of the same order d. We denote by B_0 the base variety of Σ_0, by U_0 the open complement of B_0 in $\mathbb{P}^n$, and we do for Σ_0 what we have done in Section 6.6 for Σ.

We use $\varphi_0: \Sigma_0 \to (\mathbb{P}^{h_0})^*$ to denote the projectivity between Σ_0 and the space of all the hyperplanes of $\mathbb{P}^{h_0}$, and V_0 to indicate the projective image of Σ_0.

Consider the hyperplanes of $\mathbb{P}^{h_0}$ that are images under φ_0 of the hypersurfaces of $\Sigma \subset \Sigma_0$. They constitute an h-dimensional star. Indeed, they are the hyperplanes of $\mathbb{P}^{h_0}$ containing a fixed linear space $\mathcal{O}$ of dimension $h_0 - h - 1$. These are the hyperplanes $S_{h_0-1} = J(\mathcal{O}, S_{h-1})$ that project the subspaces S_{h-1} of a space S_h skew to $\mathcal{O}$ from the vertex $\mathcal{O}$. Thus, on restricting φ_0 to Σ, one gets a projectivity $\varphi: \Sigma \to S_h^*$ between Σ and the space of hyperplanes of S_h. In this way one obtains a variety V in S_h which is the projective image of Σ.

If U is the open complement in $\mathbb{P}^n$ of the base variety B of Σ, one has $U \subset U_0$ (since $B_0 \subset B$).

Let $\varphi_0^*: U_0 \to \mathbb{P}^{h_0}$ be the morphism associated to the projectivity φ_0, which is thus defined on U. To the hypersurfaces of Σ that pass through a point $x \in U$ the projectivity φ_0 associates the hyperplanes of $\mathbb{P}^{h_0}$ that pass through $\mathcal{O}$ and through the point $\varphi_0^*(x)$. These hyperplanes intersect S_h in the spaces S_{h-1} that pass through the projection $\varphi_0^*(x)$ from $\mathcal{O}$ onto S_h. It follows that the variety V, the projective image of Σ, is the projection of V_0 from $\mathcal{O}$ onto S_h; and one has a commutative diagram

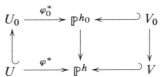

where the vertical arrows indicate the projection from $\mathcal{O}$ onto S_h.

The preceding can also be seen immediately in an analytic fashion by choosing a basis $\{f_0, f_1, \ldots, f_h, \ldots, f_{h_0}\}$ of Σ_0 that contains the basis $\{f_0, f_1, \ldots, f_h\}$ of Σ. For V and V_0 one finds the following parametric representations:

$$V: \begin{cases} X_0 = f_0(x_0, x_1, \ldots, x_n), \\ \vdots \\ X_h = f_h(x_0, x_1, \ldots, x_n), \end{cases} \quad \text{and} \quad V_0: \begin{cases} X_0 = f_0(x_0, x_1, \ldots, x_n), \\ \vdots \\ X_h = f_h(x_0, x_1, \ldots, x_n), \\ \vdots \\ X_{h_0} = f_{h_0}(x_0, x_1, \ldots, x_n). \end{cases}$$

As a consequence of this fact one has the following important result.

Proposition–Definition 6.7.1. *Every unirational variety V, of dimension n, in $\mathbb{P}^h$ is the projection of the variety $V_{n,d}$, the projective image of the linear system $\Sigma_{n,d}$ of all the hypersurfaces of any sufficiently large order d in $\mathbb{P}^n$. The variety $V_{n,d}$ is called the **Veronese variety of indices** (n, d).*

6.7. Projections and Veronese varieties

Proof. Assume that the variety V is the projective image of a linear system Σ of hypersurfaces of a given order d in $\mathbb{P}^n$. It then suffices to apply the reasoning used above to the system $\Sigma_0 = \Sigma_{n,d}$. □

With regard to Veronese varieties $V_{n,d}$ we note the following facts.

(1) The variety $V_{n,d}$ is embedded in a space $\mathbb{P}^{N(d)}$ of dimension

$$N(d) = \binom{n+d}{d} - 1$$

but not in any space of lower dimension.

(2) The system $\Sigma_{n,d}$ is without base points, and thus determine a morphism $\varphi_{n,d} : \mathbb{P}^n \to V_{n,d} \subset \mathbb{P}^{N(d)}$.

(3) The system $\Sigma_{n,d}$ is obviously simple, because the hypersurfaces of a given order that contain a given point P need not necessarily pass through other points determined by P: therefore the morphism $\varphi_{n,d} : \mathbb{P}^n \to V_{n,d}$ is an isomorphism.

(4) The order of $V_{n,d}$ is d^n. Indeed, the order of an n-dimensional variety is the number of points in which it is met by a general linear subspace of codimension n; that is, by the space in which n general hyperplanes intersect. Such hyperplanes correspond to n general hypersurfaces of $\Sigma_{n,d}$, which meet in d^n points (note that d^n is the degree of $\Sigma_{n,d}$, cf. §6.6.5).

(5) The equation of an arbitrary hypersurface of order d in $\mathbb{P}^n$ can be written in the form

$$\sum a_{i_1 \ldots i_d} x_{i_1} \ldots x_{i_d} = 0,$$

where the sum is extended over all the $N(d)+1$ combinations with repetitions of class d of the $n+1$ integers $0, 1, 2, \ldots, n$. Therefore, denoting by $X_{i_1 \ldots i_d}$ the projective coordinates in $\mathbb{P}^{N(d)}$, one has the parametric representation for $V_{n,d}$:

$$X_{i_1 \ldots i_d} = x_{i_1} \ldots x_{i_d}. \tag{6.12}$$

Equation (6.12) is the analytic expression of the morphism $\varphi_{n,d} : \mathbb{P}^n \to \mathbb{P}^{N(d)}$ (whose image is $V_{n,d}$) which is called the *Veronese morphism*, or *Veronese immersion of $\mathbb{P}^n$ realized by the hypersurfaces of order d*. To each subvariety W of $\mathbb{P}^n$, the Veronese morphism $\varphi_{n,d}$ associates a subvariety $\varphi_{n,d}(W) \subset \mathbb{P}^{N(d)}$, an isomorphic transform of W, called *Veronese image* of W.

(6) One sees immediately that $V_{n,d}$ is the locus of the zeros of the quadratic forms

$$X_{\alpha_1 \ldots \alpha_d} X_{\beta_1 \ldots \beta_d} = X_{\gamma_1 \ldots \gamma_d} X_{\delta_1 \ldots \delta_d}, \tag{6.13}$$

where $(\alpha_1,\ldots,\alpha_d,\beta_1,\ldots,\beta_d)$ and $(\gamma_1,\ldots,\gamma_d,\delta_1,\ldots,\delta_d)$ are the same combination with repetition of class $2d$ of the $n+1$ integers $0,1,\ldots,n$ (so that $\alpha_1,\ldots,\alpha_d,\beta_1,\ldots,\beta_d$ and $\gamma_1,\ldots,\gamma_d,\delta_1,\ldots,\delta_d$ differ only in the order of the terms).

One can prove that the quadratic forms (6.13) are the generators of the ideal $I(V_{n,d})$ of polynomials of $K[X_{i_1\ldots i_d}]$ which vanish on $V_{n,d}$ (for this see [48, pp. 24 and 51] and also [20]).

(7) If X is a projective variety in $\mathbb{P}^n$ and F is a hypersurface of order n that intersects on X the subvariety Y, then the Veronese image $\varphi_{n,d}(Y)$ is a hyperplane section of $\varphi_{n,d}(X)$.

Examples 6.7.2. Among the Veronese varieties there are the rational normal curves C^d in $\mathbb{P}^d$. They are projective images of the linear system $\Sigma_{1,d}$ of all the groups of d points of $\mathbb{P}^1$.

If $\Sigma = \Sigma_{2,d}$ is the linear system of all the algebraic plane curves of a given order d one has the *Veronese surface* $V_{2,2} \subset \mathbb{P}^5$ (of order 4) in the case $d=2$, and the *Del Pezzo surface* $V_{2,3} \subset \mathbb{P}^9$ (of order 9) in the case $d=3$.

These special varieties and their projections will be more extensively studied in the sequel: see in particular Section 7.4, Example 10.2.1, Section 10.4 and Exercise 10.5.7.

6.8 Blow-ups

In the projective space $\mathbb{P}^n$ we consider an irreducible, reduced and non-singular variety V of dimension $d < n-1$, the locus of the zeros of a homogeneous ideal $\mathfrak{a}$, and a simple linear system Σ (and so of dimension $h > n$) of hypersurfaces

$$\sum_{i=0}^{h} \lambda_i f_i(x_0,\ldots,x_n) = 0,$$

with $f_i \in \mathfrak{a}$ linearly independent forms of the same degree, and without base points outside of V.

Since Σ is simple, the projective image of Σ is a variety W birationally isomorphic to $\mathbb{P}^n$, and embedded in $\mathbb{P}^h$.

We suppose moreover that in the open set $U = \mathbb{P}^n \setminus V$ there does not exist any pair of points such that the hypersurfaces of Σ passing through one of them all pass through the other as well. This implies that if $\phi \colon \mathbb{P}^n \to W$ is the birational map determined by Σ, the restriction $\phi_U \colon U \to \phi(U)$ is an isomorphism.

All these hypotheses hold, for example, if Σ is the linear system of all the hypersurfaces passing through V and having order m which is sufficiently large, for example such that there exist hypersurfaces of order $m-1$ passing through V.

6.8. Blow-ups

Indeed, if P and Q are two arbitrary points of U it suffices to add a hyperplane that contains P but not Q to a hypersurface of order $m-1$ passing through V in order to obtain a hypersurface of Σ that passes through only one of the two points.

In the sequel we will suppose that Σ is the linear system of all the hypersurfaces of order $m \gg 0$, although, as we shall see in some examples, a linear system contained in Σ may suffice provided that it has sufficiently high dimension.

Now let P be a point of V and Φ_P the star ∞^{n-d-1} of the linear spaces S_{d+1} in $\mathbb{P}^n$ that pass through the space T_P tangent to P at V. The hypersurfaces of Σ whose tangent hyperplane at P contains a given linear space S_{d+1} passing through T_P form a linear system of dimension $h-1$ (inasmuch as they are obtained by imposing on the tangent hyperplanes at P, which already pass through T_P, the passage through a point of S_{d+1} not belonging to T_P) to which there corresponds a star Σ_1, of dimension $h-1$, of hyperplanes of $\mathbb{P}^h$. As S_{d+1} varies in Φ_P the center P' of Σ_1 describes (under the present hypotheses) a linear space L_P of dimension $n-d-1$.

Analytically we may obtain L_P as follows. To the generic point $[\lambda a_0 + \mu b_0, \ldots, \lambda a_n + \mu b_n]$ of a line r issuing from $P = [a_0, \ldots, a_n]$ there corresponds on W the point with coordinates

$$X_i = f_i(\lambda a_0 + \mu b_0, \ldots, \lambda a_n + \mu b_n)$$
$$= \lambda^n f_i(a_0, \ldots, a_n) + \lambda^{n-1}\mu \sum_{k=0}^{n}\left(\frac{\partial f_i}{\partial x_k}\right)_P b_k + \cdots \quad (i = 0, \ldots, h). \quad (6.14)$$

If $P \in V$ (and so $f_i(a_0, \ldots, a_n) = 0$) the right-hand terms of (6.14) are divisible by μ. On suppressing the common factor μ and then setting $\mu = 0$ one sees that to the point P (which lies on r for $\mu = 0$) there is associated the point with coordinates

$$X_i = \sum_{k=0}^{n}\left(\frac{\partial f_i}{\partial x_k}\right)_P b_k, \quad i = 0, \ldots, h, \quad (6.15)$$

which runs over the linear space $L_P = J(P^{(0)}, P^{(1)}, \ldots, P^{(h)})$, where

$$P^{(i)} = \left[\frac{\partial f_i}{\partial x_0}, \frac{\partial f_i}{\partial x_1}, \ldots, \frac{\partial f_i}{\partial x_n}\right], \quad i = 0, \ldots, h,$$

as the line r varies in the star with center P (that is, as the parameters b_k vary). The matrix whose rows are formed by the coordinates of these points has rank $d+1$. (Indeed, it is the matrix of coefficients of the equations of the hyperplanes tangent at P to V, hyperplanes which meet in the space T_P whose dimension is d.)

For each point $P \in V$ one thus has a linear space L_P of dimension $n-d-1$. The locus $E = \bigcup_{P \in V} L_P$ of these linear spaces is a hypersurface of W (in fact $\dim(E) = (n-d-1)+d = n-1$) and is represented parametrically by the equations (6.15) in which not only the parameters $b_0, \ldots, b_n$ are considered as variable, but also the point P, that is, the parameters $a_0, \ldots, a_n$. The correspondence $\sigma : W \to \mathbb{P}^n$ defined by setting $\sigma_{W \setminus E} = \phi_U^{-1}$ is a rational map of varieties. It is called the *blow-up of* $\mathbb{P}^n$ *with center* V or also the *dilatation of* V *in* $\mathbb{P}^n$. The hypersurface E is called the *exceptional divisor* of the blow-up. The effect of σ is that of replacing $\mathbb{P}^n$ by the variety W and the variety $V \subset \mathbb{P}^n$ by a divisor of W.

We observe that in order to obtain the blow-up of the variety V which is the locus of the zeros of the ideal generated by the homogeneous polynomials $f_0, f_1, \ldots, f_t$ it suffices to take as Σ the linear system of all hypersurfaces of order $m = 1 + \max_{i=1,\ldots,t} \deg f_i$ (cf. paragraph 11.3.4). For example, if V is a linear space it suffices to take the linear system consisting of the quadrics that contain it.

Example 6.8.1 (Blowing up $\mathbb{P}^n$ at a point, cf. §9.2.7). When one wishes to study a question that regards the neighborhood of a point P of $\mathbb{P}^n$ (for example, to examine the behavior of a hypersurface at the point P) it is convenient to use not the blow-up W of $\mathbb{P}^n$ with center P, obtained by way of the linear system Σ (of dimension $h = \binom{n+2}{2} - 2$) of the quadrics passing through P, but rather the variety obtained from W by projection from a suitable space Π of $\mathbb{P}^h$ onto a linear space skew to Π. This is equivalent to replacing Σ with a linear system contained in Σ (cf. Section 6.7).

Suppose that P is the point $A_0 = [1, 0, \ldots, 0]$ and consider one of the linear systems Σ_i with equation

$$\lambda_0 x_i^2 + x_0(\lambda_1 x_1 + \lambda_2 x_2 + \cdots + \lambda_n x_n) = 0, \quad i = 1, 2, \ldots, n,$$

whose base variety is the union of the point P and the space S_{n-2} with equation $x_0 = x_i = 0$.

Since $\dim \Sigma_i = n$, the projective image W^* of Σ_i is a projective space S_n. The rational transformation $\phi_i : \mathbb{P}^n \to S_n$ represented by the equations

$$\phi(:= \phi_i): \begin{cases} X_0 = x_i^2, \\ X_1 = x_0 x_1, \\ \vdots \\ X_n = x_0 x_n, \end{cases} \tag{6.16}$$

which may be inverted rationally by

$$\sigma : \begin{cases} x_0 = X_0 X_i, \\ x_1 = X_1 X_i, \\ \vdots \\ x_i = X_0^2, \\ \vdots \\ x_n = X_n X_i, \end{cases}$$

is a birational isomorphism.

The exceptions to the bijectivity of the correspondence are: in $\mathbb{P}^n$ the point A_0 and the space $\Pi: x_0 = x_i = 0$, and in S_n the point A_i^* and the space $\Pi^*: X_0 = X_i = 0$.

In the affine chart $U_0 = \{x_0 \neq 0\} \subset \mathbb{P}^n$ the system Σ_i has P as its unique base point.

We cover U_0 with the open sets U_{0i} where $x_0 x_i \neq 0$. Putting $x_i = 1$ in (6.16) one finds that σ makes the point $[x_0, x_1, \ldots, x_{i-1}, 1, x_{i+1}, \ldots, x_n]$ of U_{0i} correspond to the point of S_n with coordinates

$$X_0 : X_1 : \cdots : X_n = 1 : x_0 x_1 : \cdots : x_0 x_{i-1} : x_0 : x_0 x_{i+1} \cdots : x_0 x_n,$$

which belongs to the open set $U'_{0i} = \{X_0 X_i \neq 0\}$ of S_n.

Therefore, if $x_0, x_1, \ldots, x_{i-1}, x_{i+1}, \ldots, x_n$ are non-homogeneous coordinates in U_{0i} and $X_1, X_2, \ldots, X_n$ are non-homogeneous coordinates in $U'_0 = \{X_0 \neq 0\}$, the restrictions $\phi_{U_{0i}}$ and $\sigma_{U'_{0i}}$ have the following equations:

$$\phi_{U_{0i}}: \begin{cases} X_1 = x_0 x_1, \\ X_2 = x_0 x_2, \\ \vdots \\ X_{i-1} = x_0 x_{i-1}, \\ X_i = x_0, \\ X_{i+1} = x_0 x_{i+1}, \\ \vdots \\ X_n = x_0 x_n, \end{cases} \quad \text{and} \quad \sigma_{U'_{0i}}: \begin{cases} x_0 = X_i, \\ x_1 = \dfrac{X_1}{X_i}, \\ x_2 = \dfrac{X_2}{X_i}, \\ \vdots \\ x_{i-1} = \dfrac{X_{i-1}}{X_i}, \\ x_{i+1} = \dfrac{X_{i+1}}{X_i}, \\ \vdots \\ x_n = \dfrac{X_n}{X_i}. \end{cases}$$

Example 6.8.2 (Blow-up of $\mathbb{P}^n$ along a line, cf. Exercise 9.5.8). In $\mathbb{P}^3$ we consider the line ℓ with equations $x_0 = x_1 = 0$. The projective image of the linear system Σ of dimension 6 of all the quadrics passing through ℓ, with equation

$$\Sigma: x_0(\lambda_0 x_0 + \lambda_1 x_1 + \lambda_2 x_2 + \lambda_3 x_3) + x_1(\lambda_4 x_1 + \lambda_5 x_2 + \lambda_6 x_3) = 0,$$

is the variety W_3^4 locus of the point $[x_0^2, x_0 x_1, x_0 x_2, x_0 x_3, x_1^2, x_1 x_2, x_1 x_3]$. The exceptional surface is the locus of the point $[0, 0, a_0 b_2, a_0 b_3, 0, a_1 b_2, a_1 b_3]$ and thus is the quadric with equation $X_0 = X_1 = X_4 = X_2X_6 - X_3X_5 = 0$. The linear spaces L_P, $P \in \ell$ (each of which is obtained by fixing the ratio $b_2 : b_3$) are the generators of an array.

If instead of taking the system Σ of all quadrics passing through ℓ one takes a system contained in Σ (but satisfying the conditions imposed at the beginning of this section) one obtains another model W' of the blow-up, an isomorphic projection of W_3^4.

Example 6.8.3. Suppose now that B is a linear space S_b in $\mathbb{P}^n$ and that Σ is the linear system of all the hypersurfaces of order r having B as s-fold subvariety. If we assume that B has equations $x_{b+1} = x_{b+2} = \cdots = x_n = 0$ then for Σ one has the equation

$$\sum \varphi_{\alpha_1\alpha_2\ldots\alpha_{n-b}}(x_0,\ldots,x_n) x_{b+1}^{\alpha_1} x_{b+2}^{\alpha_2} \cdots x_n^{\alpha_{n-b}} = 0,$$

where $0 \le \alpha_j \le s$, $\sum_j \alpha_j = s$ and $\varphi_{\alpha_1\alpha_2\ldots\alpha_{n-b}}(x_0,\ldots,x_n)$ are homogeneous polynomials of degree $r - s$.

Bearing in mind that the maximal number of hypersurfaces of order r in $\mathbb{P}^{n-b-1}$ which are linearly independent is $\binom{n-b-1+r}{r}$, the projective image of Σ is the closure of the image of the rational transformation $\varphi: \mathbb{P}^n \setminus B \to \mathbb{P}^N$ (where $N = \binom{n-b-1+r}{r} - 1$) given by the equations

$$X_{\alpha_1\alpha_2\ldots\alpha_{n-b}} = \varphi_{\alpha_1\alpha_2\ldots\alpha_{n-b}}(x_0,\ldots,x_n) x_{b+1}^{\alpha_1} x_{b+2}^{\alpha_2} \cdots x_n^{\alpha_{n-b}}.$$

Let $\xi = [\xi_0, \xi_1, \ldots, \xi_n]$ be a point of $\mathbb{P}^n$ not belonging to B and $T = J(\xi, S_b)$ the space S_{b+1} that joins it to B. The image $\varphi(P)$ of the generic point

$$P = [\lambda_0, \ldots, \lambda_b, \mu\xi_{b+1}, \mu\xi_{b+2}, \ldots, \mu\xi_n], \quad \lambda_0, \ldots, \lambda_b, \mu \in K,$$

of T has coordinates

$$X_{\alpha_1\alpha_2\ldots\alpha_{n-b}} = \varphi_{\alpha_1\alpha_2\ldots\alpha_{n-b}}(\lambda_0,\ldots,\lambda_b,\mu\xi_{b+1},\ldots,\mu\xi_n)\mu^s \xi_{b+1}^{\alpha_1} \cdots \xi_n^{\alpha_{n-b}},$$

that is,

$$X_{\alpha_1\alpha_2\ldots\alpha_{n-b}} = \varphi_{\alpha_1\alpha_2\ldots\alpha_{n-b}}(\lambda_0,\ldots,\lambda_b,\mu\xi_{b+1},\ldots,\mu\xi_n)\xi_{b+1}^{\alpha_1} \cdots \xi_n^{\alpha_{n-b}}. \quad (6.17)$$

Putting $\mu = 0$ in (6.17), one finds the following representation for the exceptional variety E of the blow-up of $\mathbb{P}^n$ along B:

$$X_{\alpha_1\alpha_2\ldots\alpha_{n-b}} = \varphi_{\alpha_1\alpha_2\ldots\alpha_{n-b}}(\lambda_0,\ldots,\lambda_b,0,\ldots,0)\xi_{b+1}^{\alpha_1} \cdots \xi_n^{\alpha_{n-b}}.$$

From this it is apparent that E is the locus of ∞^b Veronese varieties $V_{n-b-1,s}$.

In particular, in the case $s = 1$, E is the locus of linear spaces of dimension $n - b - 1$; if moreover B is an s-fold point, one finds a Veronese variety $V_{n-1,s}$.

Chapter 7
Algebraic Curves

We have already dedicated ample attention to algebraic plane curves in Section 5.7 (see also [42] and [92, Chapter II]). In this chapter we consider algebraic curves embedded in projective spaces of arbitrary dimension.

In Section 7.1 we recall some general properties of projective curves. In Section 7.2 we use an approach which may be historically attributed to Riemann, cf. Remark 7.2.15, to introduce the notion of the genus of an algebraic curve. This is an important birational invariant which plays a key role in the study and classification of algebraic curves. Several exercises, proposed at the close of the section, furnish a useful complement to the general theory; among other things, rational curves are characterized as those of genus zero.

Section 7.3 is dedicated to the study of an important class of algebraic curves which is particularly rich in geometric properties, namely those contained in a quadric in $\mathbb{P}^3$.

In Section 7.4 we introduce rational normal curves, that is, the curves of order n embedded in a projective space $\mathbb{P}^n$. These curves constitute the 1-dimensional case of the Veronese varieties introduced in Section 6.7; thus they have a simple matricial algebraic representation. We mention, among other results, the elementary proof of Lüroth's theorem (Theorem 7.4.1), which establishes that unirational curves are rational, and the analysis of the projective generation of a rational normal curve, discussed in paragraph 7.4.5, which extends Steiner's theorem for conics.

Finally Section 7.5 contains a collection of completely solved exercises, and constitutes an essential complement to the theory developed in the chapter.

7.1 Generalities

In Chapter 3, we defined algebraic projective curves as algebraic projective varieties of dimension 1. Among them there are in particular the hypersurfaces of $\mathbb{P}^2$, that is, the algebraic plane curves about which we have written at length in Section 5.7.

For an arbitrary projective variety V, we have also defined its order, the multiplicity of a point (simple or singular) of V, and the concepts of tangent space and tangent cone at a point of V. We assume that $K = \mathbb{C}$.

7.1.1 Algebraic curves in $\mathbb{P}^r$. In $\mathbb{P}^r$ we consider an irreducible algebraic curve C of order n, which we will suppose not to be contained in spaces of lower dimension. In this case we will say that C is *embedded* (or *non-degenerate*) in $\mathbb{P}^r$. We recall the following facts.

(1) C is birationally isomorphic to a plane algebraic curve C' and one obtains a generically bijective and algebraic map $C \to C'$ via projection from a generic S_{r-3}.

(2) The order of C, equal to that of its generic plane projection, is the number of points that it has in common with a generic hyperplane; we will also write C^n to indicate that C is a curve of order n.

(3) The multiplicity $m_P(C)$ of a multiple point P of C is the minimal intersection multiplicity at P of C with hyperplanes passing through P, so that the generic hyperplane passing through P has $n - m_P(C)$ points different from P in common with C.

(4) If P is a simple point of C the tangent line at P to C is the intersection of the tangent hyperplanes to C at P, that is, the tangent hyperplanes at P to the hypersurfaces containing C and passing simply through P. The tangent line to C at one of its simple points is thus the center of a star Σ_{r-2}, of dimension $r-2$, of hyperplanes tangent there to C. If H is the generic such hyperplane, the intersection multiplicity $m_P(C)$ at P with C is 2, and the hyperplanes of Σ_{r-2} for which one has $m_P(C) > 2$ are those of a star having as its center the *osculating plane* of C at P.

The most general situation, namely when P is a generic point of the curve, is that there are $r-1$ stars Σ_j, of dimensions $j = r-2, r-3, \ldots, 1, 0$, each contained in the preceding, and such that the hyperplanes of Σ_j have at P intersection multiplicity $r-j$ with C. The center of Σ_j is the space S_{r-j-1} *osculating* of C at P. In particular, the star Σ_0 consists of a single hyperplane, the *osculating hyperplane*; it is the unique hyperplane having (at least) r intersections with the curve C all absorbed by P.

(5) If a hyperplane S_{r-1} contains $n+1$ points of a curve C^n, then it contains infinitely many; therefore, if C^n is irreducible, S_{r-1} contains the entire curve. The following corollary is important (a special case of Proposition 4.5.6).

- If an irreducible C^n is embedded in $\mathbb{P}^r$, then $n \geq r$. (Indeed, the hyperplane defined by r points of C^n contains at least r points of C^n.)

The definition of an irreducible algebraic curve as the locus of the zeros of a prime polynomial ideal implies that every algebraic curve is the intersection of algebraic hypersurfaces, and we know that the intersection of $r-1$ algebraic hypersurfaces in $\mathbb{P}^r$ is, in general, an algebraic curve. Be aware, however, that an algebraic curve $C \subset \mathbb{P}^r$ is not, in general, a *complete intersection*, in the sense that it can be obtained as the intersection of $r-1$ algebraic hypersurfaces that intersect along it transversally, that is, of $r-1$ hypersurfaces that have independent tangent hyperplanes in almost all of its points (i.e., in all of the points of C except for at most a finite number of them). Examples of curves in $\mathbb{P}^r$ which are not complete intersections are, for $r > 2$, the curves of order r, about which we shall speak in Section 7.4.

Exercise 7.1.2 (More on the order of a curve). We again take up the discussion of Section 3.4, and we show the constancy of the order of the cone that simply projects a curve C from a space S_{r-3} contained in a space S_{r-2} that does not intersect C in any point (by Definition 3.3.1 we know that a generic S_{r-2} is disjoint from C). This constant value is the order of C.

Indeed, let α and β be two S_{r-3} contained in the same S_{r-2} which does not meet C in any point. If n is the order of the cone X_{r-1}^n projecting C from α, a hyperplane passing through S_{r-2} contains n spaces S_{r-2} generators of that cone (those that join α with the n intersection points of the cone with a generic line), therefore it meets C in n points which give, when joined with β, n spaces S_{r-2} generators of the cone that projects C from β. Thus the two cones have the same order n.

If α and β are not contained in some S_{r-2} that does not meet C, we insert between them (as it is obviously possible) a finite number of spaces $L_1, L_2, \ldots, L_q$ of dimension $r - 3$ in such a way as to obtain a chain $\alpha, L_1, L_2, \ldots, L_q, \beta$ of S_{r-3} such that for any two consecutive terms there is an S_{r-2} containing them but not meeting C. Thus one returns to the preceding case.

7.1.3 Algebraic curves in $\mathbb{P}^3$. Suppose that the curve C is embedded in $\mathbb{P}^3$. At each of its simple points there is a tangent line, the axis of the pencil Σ_1 of tangent planes at P. The osculating plane is the plane of Σ_1 whose intersection multiplicity with the curve at P is at least 3. Possible points on the curve for which this intersection multiplicity is at least 4 are said to be *stationary points*.

A simple point of the curve at which all the planes of Σ_1 have intersection multiplicity with C at least 3 is a *flex* of C.

The *class* of C is defined to be the number of osculating planes that pass through a generic point of the space. Another projective characteristic of the curve is its *rank*, namely the order of the ruled surface formed by the tangents to the curve, which is given by the number of tangents that meet a generic line.

The figure formed by the points, the tangents and the osculating planes of an algebraic space curve C is self-dual in the given space: to the points of C there correspond by duality the osculating planes, and the tangents correspond to tangents.

By Bézout's theorem two algebraic surfaces in $\mathbb{P}^3$, having orders m_1, m_2 and without common components, meet along an algebraic curve of order $m_1 m_2$, and it is then evident that an algebraic curve C whose order is a prime number and that is not contained in any plane can not be the complete intersection of two algebraic surfaces that meet transversally along C.

We note, however, that when one must study the local properties of a simple point, namely properties that concern the neighborhood of the point (for example questions of intersection multiplicity at the point) one does not lose generality in supposing that the curve C is a complete intersection. Indeed, every simple point of C belongs to some Zariski open set U such that the restriction of C to U is a complete intersection of restrictions to U of two algebraic surfaces. If the curve has no multiple points this holds for each of its points: this fact is expressed by saying that *a non-singular curve is locally a complete intersection*.

7.1.4 Planar projections of an algebraic curve in $\mathbb{P}^3$. If we project a curve $C \subset \mathbb{P}^3$ from a generic point O of the ambient space onto a plane π we obtain a curve C' for which it is easy to find the projective characteristics.

First of all, the order of C' is equal to the order of C.

The class of C' is equal to the rank of C. Indeed, let P be a point of π, t' one of the tangents to C' that pass through P and T' the corresponding point of contact. The tangent t of C that has t' as projection is supported by the line r_{OP} at a point P_0. Thus, the class of C' is equal to the number of tangents of C that meet the line r_{OP}, which is a generic line in $\mathbb{P}^3$. But, the number so defined is precisely the rank of C.

Furthermore, the following properties hold true.

(1) A double point of C' that is not the projection of a double point of C comes from a chord of C issuing from O. Since O is generic (and so does not belong to the ruled surface formed by the tangents to C, to the ruled surface formed by the trisecants of C, or to the ruled surface formed by the lines that join pairs of points of contact of bitangent planes to C), the trace of every chord of C passing through O is a double point with distinct tangents for the curve C'.

(2) Since the chords of C are in number ∞^2 while two conditions are imposed on a line in $\mathbb{P}^3$ by passage through O, one sees that a finite number of chords of C pass through O; hence one can not avoid having chords that pass through the center of projection.

(3) A flex of C' can arise either from a flex of C or from an osculating plane of C that passes through O. Since there are ∞^1 osculating planes of C, the point O belongs to a finite number of osculating planes.

(4) A bitangent of C' that is not the projection of a bitangent of C is the trace in π of a bitangent plane of C that passes through O. We note that an algebraic space curve of order $n \geq 4$ has ∞^1 bitangent planes. Indeed, let t be a tangent to C. We say that two points A and A' of C correspond to one another when they are two of the $n-2$ further intersections of C with a plane passing through t. Thus one obtains a symmetric algebraic correspondence

ω of indices $(n-3, n-3)$ on C. The fixed points of ω are points of contact of C with planes through t and tangent to C in another point. Therefore, if $n \geq 4$, every tangent to C belongs to a finite number of bitangent planes.

Moreover, let us explicitly point out the following facts.

(1) If the center of projection O belongs to the tangent to C at a point P, the projection P' of P is a cusp of C', and the cuspidal tangent is the trace in π of the osculating plane to C at P.

(2) The projected curve C' has a tacnode if O belongs to the line that joins the two points of contact of C with a bitangent plane (cf. §9.2.5).

(3) The projection of C has a triple point if O belongs to a trisecant of C.

In conclusion:

- The *generic* plane projection of a non-singular algebraic curve in $\mathbb{P}^3$ has no other singularities except double points with distinct tangents, flexes and bitangents.

We leave to the reader the task of seeing how all this must be modified in the case in which the curve C is projected into π from one of its points.

We merely observe that if one projects the curve C of order n from one of its generic points (which is thus non-singular) onto a plane π one obtains as its projection a curve of order $n-1$ which passes simply through the trace O' of the tangent to C at O, and has as tangent at O' the trace in π of the osculating plane to C at O.

7.2 The genus of an algebraic curve

We consider an irreducible algebraic curve C in $\mathbb{P}^n = \mathbb{P}^n(\mathbb{C})$, the locus of zeros of a prime ideal $\mathfrak{p}$, and its homogeneous coordinate ring $\mathbb{C}[y_0, \ldots, y_n] = \mathbb{C}[x_0, \ldots, x_n]/\mathfrak{p}$ together with its field of fractions, that is, the field of rational functions on C. We consider a pencil Σ of algebraic hypersurfaces

$$\Sigma: \lambda f(x_0, \ldots, x_n) + \mu h(x_0, \ldots, x_n) = 0, \tag{7.1}$$

and assume that the curve C is not contained in all of the hypersurfaces of Σ.

If all the hypersurfaces of Σ meet C in a single common set of points, that is, if the points common to C and the generic hypersurface of Σ are all contained in the base variety of Σ, then Σ contains one (and only one) hypersurface that passes through C. Indeed, if P is a point of C outside of the base variety of the pencil, then the hypersurface of Σ passing through P contains the entire curve C because it has more intersections with C than are allowed by Bézout's theorem.

Conversely, if Σ contains a hypersurface that passes through C, we may suppose this hypersurface to be one of the hypersurfaces chosen to define the pencil, for example, $h(x_0, \ldots, x_n) = 0$; and one sees immediately that the group of points constituted by the intersection of C with the hypersurfaces (7.1) does not change as λ, μ vary.

We exclude this case, and thus suppose that Σ has no hypersurfaces passing through C and consequently that the hypersurface (7.1) intersects C in a group of points not all of which are fixed as λ, μ vary, so that one has an ∞^1 of groups of points of C. We will say that these are the groups of a *linear series* of *dimension* 1 ($= \dim \Sigma$) on C. Possible fixed points, that is, possible points which are common to all the groups of the series, may at pleasure be understood as included (entirely or in part) in all the groups of the series, or excluded (entirely or in part) from all such groups. If r is the number of points of a generic group of the series (or better, of all groups of the series when the points of a group are counted with suitable multiplicity) we will say that r is the *order* of the linear series; and we will denote this series with the symbol g_r^1 (cf. §6.3.7). We will say that a linear series is *partially contained* in another series if by adding a fixed group to each of its groups one obtains groups of the second series. We refer to Chapter 8 for the general theory of linear series on an algebraic curve C.

We observe that the groups of the series g_r^1 cut out on C by the pencil (7.1) appear as groups of constant value (level) of the rational function

$$\theta = \frac{f(y_0, \ldots, y_n)}{h(y_0, \ldots, y_n)} \in \mathbb{C}(C)$$

(that is, as inverse images $\varphi^{-1}(t)$, for $t \in \mathbb{C} \cup \{\infty\}$, where $\varphi \colon C \to \mathbb{C} \cup \{\infty\}$ is the map associated to θ, defined where $h \neq 0$).

A series g_r^1 on C may thus be defined as the totality of the groups of constant value of a rational function on the curve. We note that possible points of C that are common zeros of the two polynomials f and h are the points of indeterminacy of the function θ, and constitute a fixed divisor that we mean to exclude from all the groups of the series. Two rational functions

$$\frac{f(y_0, \ldots, y_n)}{h(y_0, \ldots, y_n)}, \quad \frac{f'(y_0, \ldots, y_n)}{h'(y_0, \ldots, y_n)} \in \mathbb{C}(C)$$

define the same linear series on C if

$$f(x_0, \ldots, x_n) h'(x_0, \ldots, x_n) - f'(x_0, \ldots, x_n) h(x_0, \ldots, x_n)$$

belongs to the ideal $\mathfrak{p}$ of the polynomials in $\mathbb{C}[x_0, \ldots, x_n]$ that vanish on C.

This way of looking at a linear series g_r^1 makes it clear that the notions of linear series and the order of such a series are *birational invariants*. Indeed, if C' is a curve birationally equivalent to C, which means that there exists a birational isomorphism

7.2. The genus of an algebraic curve

$\sigma: C \to C'$, to each rational function on C and to its groups of constant level, there correspond (under the isomorphism $\mathbb{C}(C) \cong \mathbb{C}(C')$ between the function fields) a rational function of C' and its groups of constant level.

The notion of the Jacobian group of a linear series on $\mathbb{P}^1$ (cf. §6.4.1) is easily extended to the case of a linear series g_r^1 on the curve C. For each point $P \in C$ we consider the fiber (group of constant level)

$$\varphi^{-1}\varphi(P) = \{P, P_2, \ldots, P_r\}$$

of the rational map $\varphi: C \to \mathbb{C} \cup \{\infty\}$. Thus the groups of points of g_r^1 constitute an algebraic totality of groups of r points that have the property that its groups are in algebraic and bijective correspondence with the values of a parameter $t \in \mathbb{C}$ and that each point of the curve C specifies the group to which it belongs. If one makes each point $P \in C$ correspond to the $r-1$ points $P_2, \ldots, P_r$ one obtains an algebraic correspondence $\phi: C \to C$. The *Jacobian group* of g_r^1 may then be defined as the collection of points $P \in C$ such that P coincides with (at least) one of the points $P_2, \ldots, P_r$ that correspond to it, that is, with the set of points $P \in C$ such that the fiber $\varphi^{-1}\varphi(P)$ is made up of $< r$ distinct points.

The Jacobian group consists of the *s-fold points of the series g_r^1*, $s \geq 2$. These are the points $P \in C$ for which the group $\{P, P_2, \ldots, P_r\}$ of the g_r^1 containing P is such that $s-1$ of its remaining points $P_2, \ldots, P_r$ coincide with P. Hence an *s-fold point of the series is counted $s-1$ times in the Jacobian group*, cf. paragraph 6.4.1.

Example 7.2.1. If C is a plane curve, a point P is s-fold for the g_r^1 cut out on C by a pencil of lines through (a point) O if there is a line of the pencil which has s of its intersections with the curve concentrated at P. For example, an ordinary flex P with tangent $\langle O, P \rangle$ is a triple point of the g_r^1 and thus should be counted twice in calculating the order r of the Jacobian group. Moreover, an ordinary multiple point of C belongs to the Jacobian group only if one of the tangents at P passes through O.

We now introduce an important birational invariant of an algebraic curve. We know that every algebraic curve is the birational transformation of a plane curve. Moreover we shall see in Chapter 9, Theorem 9.2.4, that it can be reduced via a suitable birational transformation to a plane curve whose only singularities are ordinary multiple points (see also Exercise 13.1.21). We may therefore limit ourselves to the consideration of plane curves having only multiple points with distinct tangents.

Let C be an algebraic curve and Γ a plane model for C endowed with only ordinary singularities (that is, ordinary multiple points). On Γ we consider two linear series $g := g_m^1$, $g' := g_{m'}^1$ of dimension 1 and orders respectively m and m' (which, for simplicity, we suppose to be without fixed points). Let r be the number of points of Γ that constitute the Jacobian group of g (cf. §6.4.1), and let r' be the analogous object for g'. We note that a point of a group of g that is an ordinary s-fold point of the curve does not belong, in general, to the Jacobian group of g. Indeed, we shall see in Chapter 9, Section 9.2 that by means of a quadratic transformation the ordinary s-fold point can be replaced by s distinct simple points.

Assuming this, we say that two groups of the series g *correspond* when they contain two points of the same group of the series g' (i.e., two groups G_1, G_2 of g correspond when there exists a group of g' that contains a point A of G_1 and a point B of G_2). We obtain in this way an algebraic correspondence ω between the groups of g which is symmetric, and which therefore has equal indices (cf. §1.1.3). If $G = \{P_1, P_2, \ldots, P_m\}$ is a group of g and $\{P_i, H_{i2}, \ldots, H_{im'}\}$ is the group of g' that contains P_i, the $m' - 1$ groups $G'_{i2}, \ldots, G'_{im'}$ of g defined by $H_{i2}, \ldots, H_{im'}$ are all in correspondence with G. Indeed, $P_i \in g$ and $H_{it} \in G'_{it}$ so that P_i and H_{it} are two points of the same group of g' (the one defined by P_i). Thus to the group G there correspond $m(m' - 1)$ groups G'_{it} (Figure 7.1). Therefore $\omega \colon \Gamma \to \Gamma$ is an algebraic correspondence of indices $(m(m' - 1), m(m' - 1))$.

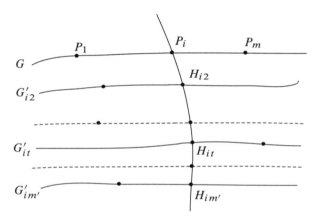

Figure 7.1

We now seek the fixed points of ω, namely the groups of g that coincide with one of their corresponding groups. The coinciding of the two corresponding groups can happen because the two points (belonging to groups of g') by which they are determined coincide, and therefore each of the points of the Jacobian group of g' leads to a fixed point of ω. In this way one finds $\mathfrak{r}'$ fixed points. But the coincidence of the two corresponding groups can also arise from a multiple point of the curve Γ. Indeed, let O be an s-fold point of Γ. The group G of g that contains O is formed by s points $O = O_1, O_2, \ldots, O_s$ coinciding at O (but to be considered distinct) and by $m - s$ further points. The group of g' determined by any one of the points O_j, $j = 1, \ldots, s$, contains, in addition to O, also other $s - 1$ points each of which defines a group of g' which is in correspondence with G and which coincides with G (having the point O in common with G). An s-fold point of the curve leads to

$s(s-1)$ fixed points of the correspondence ω.

Let $\Sigma: \lambda f(x_0, x_1, x_2) + \mu h(x_0, x_1, x_2) = 0$ be a pencil of algebraic curves that meets Γ, outside of possible fixed points, in the series g. The groups of g are in algebraic one-to-one correspondence with the curves of Σ and thus the series g is a projective line (on which λ, μ are homogeneous coordinates). Hence one can apply Chasles' principle of correspondence (cf. §1.1.3) to the correspondence ω to conclude that there are $2m(m'-1)$ fixed points. One then has the relation

$$r' + \sum s(s-1) = 2m(m'-1). \tag{7.2}$$

Similarly, interchanging the roles of the two series g, g', one has

$$r + \sum s(s-1) = 2m'(m-1). \tag{7.3}$$

Thus, on subtracting term by term, we have

$$r - 2m = r' - 2m'.$$

Thus we have arrived at the following important result.

Theorem 7.2.2. *The difference* $r - 2m$ *between the number* r *of points in the Jacobian group and twice the order* m *of a linear series* g_m^1 *(without fixed points) of dimension 1 belonging to an irreducible algebraic curve* C *does not depend on the choice of the series, but rather is a characteristic of the curve which is invariant under birational isomorphisms.*

Equation (7.3) makes it evident that r is an even integer. We can then consider the integer

$$p := \frac{1}{2}r - m + 1 \tag{7.4}$$

which is called the *genus* of the curve C. It is a birational invariant that occupies a fundamental position in the theory of algebraic curves.

Remark 7.2.3. Let C be an irreducible plane curve of order m and genus p, whose multiple points are ordinary singularities. On C, the lines of the pencil having center in a generic point O of the plane cut out a linear series g_m^1. The order of the Jacobian group of g_m^1 coincides with the class, ν, of C (cf. §5.6.5). Therefore the class of C is expressed by the relation

$$\nu = 2(p + m - 1).$$

Remark 7.2.4. If one makes no hypothesis on the singularities of a plane curve C, the evaluation of the order of the Jacobian group of a linear series g_m^1 on C is somewhat delicate and requires, among other things, a detailed study of the behavior of the polars at the multiple points of the curve (see [111]), as well as the theory of branches. For a brief introduction to linear branches and basic references on this topic, see Section 8.1. For more extensive discussions of these matters we refer the reader to the classical treatises, and in particular, for the case in which the curve C has only linear branches, to [86, n. 37], [91, §9, p. 238 ff.], [100, p. 112 ff.], and [36, Vol. I, Chapter I, pp. 174–180; Vol. III, Chapter I, p. 55].

The following result, which also includes the case in which the series g_m^1 has fixed points, is particularly useful: *A point P that is s-fold for a generic group of a linear series g_m^1 and $(s + s')$-fold for a particular group of the series is to be counted $2s + s' - 1$ times in the calculation of the order of the Jacobian group of the series.*

We shall see in 7.2.8 that when the curve C also has k cusps of the first kind, then the class ν of C satisfies

$$\nu = 2m + 2p - 2 - k.$$

If, as Enriques [36, Vol. I, p. 280] writes, one bears in mind that the points of the curve C in a neighborhood of a cusp P belong to a single (non-linear) branch, every line passing through P has two infinitely near intersections there with a branch of the curve, and so can be considered to be tangent to a branch of C. We can then say that the number of tangents to branches of such a curve C passing through a point remains $2m + 2p - 2$. Therefore the relation expressed in Remark 7.2.3 continues to hold (when one replaces ν by $\nu + k$). We also refer forward to paragraph 9.2.5 for details on the notion of successive neighborhoods of a point on a plane curve.

Corollary 7.2.5. *An irreducible algebraic rational curve C, that is, a curve birationally isomorphic to $\mathbb{P}^1$, has genus $p = 0$.*

Proof. It suffices to consider a linear series g_1^1 on $\mathbb{P}^1$; for that series one has $r = 0$ (cf. §6.4.1) and equation (7.4) then gives $p = 0$. □

It is useful to introduce another numerical character for an algebraic plane curve C.

Definition 7.2.6. Let C be an irreducible plane curve of order n whose multiple points $P_1, \ldots, P_t$ have multiplicities $s_1, \ldots, s_t$ ($t \geq 0$). Define the *deficiency* of C to be the integer

$$\delta := \frac{(n-1)(n-2)}{2} - \sum_{i=1}^{t} \frac{s_i(s_i - 1)}{2}.$$

Proposition 7.2.7. *Let C be an irreducible plane curve of order n whose multiple points $P_1, \ldots, P_t$ have multiplicities $s_1, \ldots, s_t$ ($t \geq 0$). Then:*

(1) *If C is irreducible the deficiency is non-negative.*

7.2. The genus of an algebraic curve

(2) *If C is irreducible and its multiple points are all with distinct tangents, the deficiency coincides with the genus.*

Proof. (1) If Q is a generic point of the plane, the first polar $C_1(Q)$ of Q with respect to C is a curve having multiplicity $s_i - 1$ at the points P_i and its intersection multiplicity with C at P_i is $m_{P_i}(C, C_1(Q)) \geq s_i(s_i - 1)$, $i = 1, \ldots, t$ (cf. Lemma 5.6.2 and Exercise 5.7.3). By Bézout's theorem one then has

$$n(n-1) - \sum_{i=1}^{t} s_i(s_i - 1) \geq 0$$

and thus also

$$\frac{n(n-1)}{2} - \sum_{i=1}^{t} \frac{s_i(s_i - 1)}{2} + n - 1 \geq n - 1 > 0,$$

which is to say

$$\frac{(n-1)(n+2)}{2} - \sum_{i=1}^{t} \frac{s_i(s_i - 1)}{2} > 0.$$

Given these facts, we consider the linear system Σ of the curves of order $n - 1$ having multiplicity $s_i - 1$ at the points P_i. We find that, cf. Example–Definition 6.2.2,

$$\dim \Sigma \geq \frac{(n-1)(n+2)}{2} - \sum_{i=1}^{t} \frac{s_i(s_i - 1)}{2} \quad (>0)$$

whence for $\frac{(n-1)(n+2)}{2} - \sum_{i=1}^{t} \frac{s_i(s_i-1)}{2}$ simple points Q_j chosen arbitrarily on C there passes a curve C' of Σ (and actually there pass infinitely many since $\dim \Sigma > 0$).

This C' and the given curve C thus have at least $\sum_{i=1}^{t} s_i(s_i - 1)$ intersections in the points P_i and at least $\frac{(n-1)(n+2)}{2} - \sum_{i=1}^{t} \frac{s_i(s_i-1)}{2}$ intersections in the points Q_j. Thus, by Bézout's theorem,

$$n(n-1) \geq \sum_i m_{P_i}(C, C') + \sum_j m_{Q_j}(C, C')$$

$$\geq \sum_{i=1}^{t} s_i(s_i - 1) + \frac{(n-1)(n+2)}{2} - \sum_{i=1}^{t} \frac{s_i(s_i - 1)}{2},$$

that is

$$\frac{(n-1)(n-2)}{2} - \sum_{i=1}^{t} \frac{s_i(s_i - 1)}{2} \geq 0.$$

(2) Consider the pencil of lines having center in a generic point O of the plane. On C the lines of this pencil cut out a linear series g_n^1 of order n. As noted in Remark 7.2.3, the number of points of the Jacobian group of g_n^1 coincides with the class $v = n(n-1) - \sum_i s_i(s_i - 1)$ of C. Equation (7.4) then establishes the desired result. □

We give, in the form of exercises, some useful complements regarding the notion of the genus of an (irreducible) algebraic curve.

7.2.8. *Let C be an algebraic plane curve of order n having only ordinary multiple points and cusps of the first kind as singular points. Let v be the class of C and k the number of cusps. Prove that the Jacobian group of the linear series g_n^1 cut out on C by a generic pencil Φ of lines consists of $v + k$ points.*

Let a, b be the lines that join a generic point O (not belonging to the plane π that contains C) with an ordinary s-fold point A and with a cusp B respectively. Let Q be a quadric (not a cone) passing through a, b.

In the projection of Q from O onto π (cf. Example 2.6.4 and also Section 7.3), to the curve C there corresponds on Q a curve Γ having s distinct points $A_1, A_2, \ldots$ on the line r_{OA} and which is tangent to the line r_{OB} at a point B' (Figure 7.2).

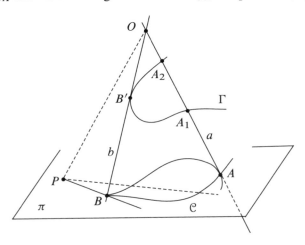

Figure 7.2

The series g_n^1 cut out on C by the lines of π issuing from the center P of Φ has as its correspondent on Γ the series g_n^1 cut out by the planes passing through the line r_{OP}. While the plane $J(A, r_{OP})$ intersects Γ in s distinct points (whence A does not belong to the Jacobian group of g_n^1), the plane $J(B, r_{OP})$ is tangent at B' to Γ and so B is a point of the Jacobian group of g_n^1. Furthermore, from P there issue v tangents to C in distinct points which are points of the Jacobian group of g_n^1.

7.2. The genus of an algebraic curve

7.2.9. *The genus of a plane curve C of order n whose only singularities are d nodes and k ordinary cusps is*

$$p = \frac{1}{2}(n-1)(n-2) - d - k.$$

Consider the pencil of lines having center in a generic point O of the plane. On C the lines of this pencil cut out a linear series g_n^1 of order n. By Plücker's formulas (see 5.7.9), the class of C is $\nu = n(n-1) - 2d - 3k$ and so by Exercise 7.2.8 the number of points of the Jacobian group of g_n^1 is $r = n(n-1) - 2d - 2k$. Equation (7.4) then furnishes the required expression.

7.2.10. *Let C be an irreducible plane algebraic curve of order n and endowed with Plückerian singularities, that is, d nodes, k ordinary cusps, ρ flexes of the first kind, δ bitangents. Define the **genus of the envelope of tangents** of C to be the number*

$$\frac{1}{2}(\nu-1)(\nu-2) - \delta - \rho,$$

where ν is the class of C. Verify that C and the envelope of its tangents have the same genus.

Eliminating d from the first two of Plücker's formulas (cf. 5.7.9) one obtains

$$3\nu - \rho = 3n - k.$$

This relation may also be written in the form

$$n - 2(\nu - 1) - k = \nu - 2(n - 1) - \rho,$$

or also, using the first and third Plücker formulas,

$$\nu(\nu-1) - 2\delta - 3\rho - 2(\nu-1) + \rho = n(n-1) - 2d - 3k - 2(n-1) + k,$$

or

$$(n-1)(n-2) - 2d - 2k = (\nu-1)(\nu-2) - 2d - 2\rho.$$

In conclusion

$$\frac{1}{2}(n-1)(n-2) - d - k = \frac{1}{2}(\nu-1)(\nu-2) - \delta - \rho$$

and thus in view of 7.2.9 one obtains the desired equality.

7.2.11. *Let C be an irreducible algebraic curve. Show that its genus is non-negative.*

Since the genus is invariant under birational isomorphisms, the genus of C coincides with that of its generic plane projection C^*. Moreover, we shall see in Chapter 9, Theorem 9.2.4, that C^* can be reduced via a suitable birational transformation to a plane curve whose only singularities are ordinary multiple points. One then concludes by applying Proposition 7.2.7.

7.2.12. *The maximum number of double points that an irreducible plane algebraic curve of order n can have is*

$$\frac{(n-1)(n-2)}{2}.$$

It suffices to note that for an irreducible curve that has as its only singularities d double points one has, by 7.2.9 and 7.2.11, $\frac{(n-1)(n-2)}{2} - d \geq 0$.

7.2.13. *An irreducible curve C is rational if and only if it has genus zero.*

A rational curve has genus zero, as observed in Corollary 7.2.5.

We prove the converse. We know that two birationally isomorphic algebraic curves have the same genus, and that every algebraic curve is the birational transform of a plane curve. Again using Theorem 9.2.4 we may assume that C is a plane curve of a given order n and genus p with ordinary multiple points of multiplicities s_i, and that

$$p = \frac{(n-1)(n-2)}{2} - \sum_i \frac{s_i(s_i-1)}{2} = 0. \tag{7.5}$$

The dimension of the linear system Σ of the curves of order $n-2$ that pass with multiplicity $s_i - 1$ through each of the s_i-fold points of C and through $n-3$ additional points of C is at least, cf. Example–Definition 6.2.2,

$$\frac{(n-2)(n+1)}{2} - (n-3) - \sum_i \frac{s_i(s_i-1)}{2} = 1, \tag{7.6}$$

where the equality is a consequence of (7.5). Equation (7.6) may be rewritten as

$$n(n-2) = \sum_i s_i(s_i-1) + (n-3) + 1.$$

By Bézout's theorem it follows that the generic curve of a pencil $\Phi \subset \Sigma$ has only one variable point in common with C, and therefore the points of C are in one-to-one algebraic correspondence with the values of one parameter.

From this result one obtains a practical procedure for finding a parametric representation of a curve of genus zero.

In the projective plane consider, for example, the curve C of order $n = 4$ whose Cartesian equation is $(x^2 + y^2)^2 - xy = 0$ (the *lemniscate of Bernoulli*). It has three double points: the origin O and the two circular points, $[1, i, 0]$, $[1, -i, 0]$. Hence $p = 0$. We take a pencil of curves of order $n - 2$ (that is, a pencil of conics) passing through the double points (that is, of circles passing through O) and through an additional simple point P of C ($1 = n - 3$), chosen arbitrarily. For example we can take as P one of the two non-singular points that C possesses in the first order neighborhood of O; that is, we impose on the circles to have tangent at O that coincides with one of the two tangents of C, for example, with the axis x.

Thus we have the pencil $\Phi : x^2 + y^2 = \lambda y$, $\lambda \in \mathbb{C}$. The system of equations

$$\begin{cases} (x^2 + y^2)^2 - xy = 0, \\ x^2 + y^2 = \lambda y \end{cases}$$

is equivalent to the system

$$\begin{cases} \lambda y(x^2 + y^2) - xy = 0, \\ y(\lambda^2 y - x) = 0. \end{cases}$$

The unique solution that depends on λ is the point $(\frac{\lambda^3}{\lambda^4+1}, \frac{\lambda}{\lambda^4+1})$. So we have the parametric equations for the lemniscate:

$$\begin{cases} x = \dfrac{\lambda^3}{\lambda^4 + 1}, \\ y = \dfrac{\lambda}{\lambda^4 + 1}. \end{cases}$$

We now prove a result, which will be useful in the sequel, and which implies, in particular, that a linear system of rational plane curves is regular (cf. Example–Definition 6.2.2).

In this regard, recall that the degree of a linear system Σ of plane curves is defined to be the number $\deg \Sigma$ of intersections of two generic curves of Σ excluding base points.

Lemma 7.2.14. *Let Σ be a linear system of plane curves of order n and genus p, with base points of given multiplicities s_i. Let σ be the superabundance of Σ. Then*

(1) $\deg \Sigma \geq \dim \Sigma - 1$;

(2) $p \geq \sigma$.

Proof. For simplicity we suppose that the base points are ordinary, although the same conclusions hold in general. If $\dim \Sigma = h$, for $h - 1$ points of the plane (distinct from the base points) there pass at least ∞^1 curves of Σ. Two of these curves have in common at least $h - 1$ points and so $\deg \Sigma \geq h - 1$. From the relations (cf. Example–Definition 6.2.2, Proposition 7.2.7 (2))

$$\dim \Sigma = \frac{n(n+3)}{2} - \sum_i \frac{s_i^2 + s_i}{2} + \sigma,$$

and
$$p = \frac{n^2 - 3n + 2}{2} - \sum_i \frac{s_i^2 - s_i}{2}$$

one obtains
$$h + p = n^2 + 1 - \sum_i s_i^2 + \sigma.$$

On the other hand, by Bézout's theorem,
$$\deg \Sigma = n^2 - \sum_i s_i^2.$$

Thus, $h + p = \deg \Sigma + \sigma + 1$ and so
$$\deg \Sigma = h + p - \sigma - 1 \geq h - 1,$$

that is $p \geq \sigma$. □

Remark 7.2.15. The path that we have followed to introduce the genus p of an algebraic curve C (which may be historically attributed to Riemann) is based on the notion of the Jacobian group, and on the construction and study of a suitable algebraic correspondence between groups of points of 1-dimensional linear series on a plane model Γ of C, with Γ having only ordinary singularities. Although this approach involves some rather weighty technical issues which make for difficult reading, it has the advantage of leading to a direct proof of the crucial fact that the genus p is a birational invariant (cf. Theorem 7.2.2).

An alternative approach to the introduction of the genus p of an algebraic curve C originates in the work of Weierstrass. It may be found in Chapter 8, Section 8.5.

7.3 Curves on a quadric

In this section we illustrate some of the properties of the geometry of an important class of algebraic curves in $\mathbb{P}^3$, namely those contained on a quadric surface.

7.3.1 The stereographic projection of a quadric. Let $\mathcal{Q}$ be a non-specialized quadric in $\mathbb{P}^3$, O an arbitrary point of $\mathcal{Q}$ and π a plane not passing through O. We say that a point P of $\mathcal{Q}$ and a point P' of π correspond if P is the second intersection of $\mathcal{Q}$ with the line $r_{OP'}$; this means that P' is the projection of P from O onto π. The correspondence that is thus obtained is clearly a birational isomorphism $\varphi \colon \mathcal{Q} \to \pi$. One usually says that φ is the *stereographic projection* of $\mathcal{Q}$ onto π (cf. Example 2.6.4).

It is easy to see which points are exceptions to the bijectivity of φ.

The unique point of $\mathcal{Q}$ to which φ does not associate a well-defined point is the point O: the points corresponding to O are all the points of the line r in which π

7.3. Curves on a quadric

intersects the plane σ of the two generators a and b of $\mathcal{Q}$ passing through O (i.e., the intersection with the tangent plane to $\mathcal{Q}$ at O). Indeed, if P' is any point of r (not contained in $a \cup b$), the second intersection of $\mathcal{Q}$ with the line $r_{OP'}$ coincides with O. Thus there is a *fundamental point* on $\mathcal{Q}$, the point O, to which there corresponds in π an *exceptional line*, the line r. In the plane π there are instead two fundamental points, the traces A and B of the two generators a and b. All the points of a, except for the point O, are in fact projected from O into the same point A of π, and similarly the points of b other than O all have the same projection B. Since the lines a and b belong to the plane σ, the two points A and B belong to r.

There are no other exceptions, since φ induces an isomorphism between the two open sets $\mathcal{Q} \setminus a \cup b$ and $\pi \setminus r$.

Now consider a line t of σ passing through O and a curve $\mathcal{L}$ belonging to $\mathcal{Q}$ that passes through O and has t as tangent there. The cone Γ that projects $\mathcal{L}$ from O has the line t among its generators and thus the projection $\mathcal{L}'$ of $\mathcal{L}$ from O (that is, the locus of the traces in π of the generators of Γ) passes through the trace T of t in π (note that $T \in r$), and has there as tangent the trace h of the osculating plane of $\mathcal{L}$ at O (cf. §7.1.4). Bearing in mind that σ is the tangent plane to $\mathcal{Q}$ at O (so that every direction of σ issuing from O is that of the tangent to a curve of $\mathcal{Q}$) one therefore has a bijective algebraic correspondence, and hence a projectivity, between the

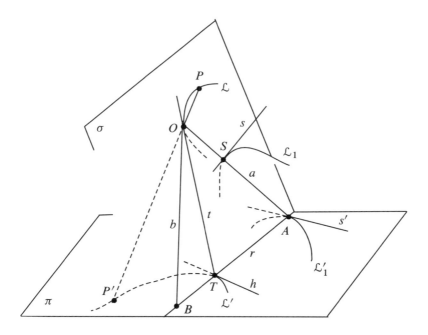

Figure 7.3

directions t issuing from O and belonging to the quadric and the points T of r. Hence the stereographic projection of $\mathfrak{Q}$ has resolved the *first order neighborhood* on $\mathfrak{Q}$ (i.e., the set of those directions) replacing it with a line. One usually says that in the neighborhood of O the quadric has an *infinitesimal line* (namely, the line r).

Note that all the curves $\mathcal{L}$ belonging to the quadric, passing through O and having in common there both the tangent t and the osculating plane, have as projections in π curves having at T the same tangent h as the projection $\mathcal{L}'$ of $\mathcal{L}$. The stereographic projection of the quadric is thus useful also for the study of the second order neighborhood of O on $\mathfrak{Q}$, that is, the set of the first order neighborhoods of the points in the first order neighborhood of O.

One can say something analogous for the two points A and B. There exists a projectivity between the directions of π issuing, for example, from A and the points of the line a. Indeed, if S is a point of a, the tangent plane there to the quadric (contains a and) meets π along a line s' of the pencil with center A, and thus determines a direction of π issuing from A. Conversely, each line of the plane π that passes through A is the image of a plane passing through a, which, in virtue of Chasles' theorem 5.8.18 (2), is tangent to the quadric at a well-defined point S of a. The correspondence that one obtains in this way between the pencil of lines of π with center A and the line a with distinguished point is algebraic and bijective, whence it is a projectivity. All the curves $\mathcal{L}_1$ traced in $\mathfrak{Q}$ and having a given tangent s at a given point S of a have as projections in π curves having the line s' that corresponds to S as tangent at A. This s' depends only on the tangent plane of $\mathfrak{Q}$ at S, which means that s' does not change as s varies in this plane.

Again let $\mathfrak{Q}$ be a non-specialized quadric in $\mathbb{P}^3$, and let a and b the two lines of $\mathfrak{Q}$ issuing from one of its points O. We denote by $\{a\}$ that of the two rulings of $\mathfrak{Q}$ that contains a and by $\{b\}$ the one that contains b.

Let C be an algebraic curve (not necessarily irreducible) of order n traced on $\mathfrak{Q}$. Every plane meets it in n points and this happens, in particular, for the tangent planes of $\mathfrak{Q}$, which meet $\mathfrak{Q}$ along two generators. The n intersections of C with a tangent plane, for example with the plane σ of the two lines a and b, are distributed on the lines a and b and therefore if α is the number of points that C has on a and β is the number of points that C has on b, one has $n = \alpha + \beta$. Any other line $a' \in \{a\}$ is skew to a and supported by b. If, as we suppose, a' is generic, it will not pass through any of the β points in which C meets b, and thus if α' is the number of intersections of a' with C one has $\alpha' + \beta = n$ so that $\alpha = \alpha'$. All the lines of $\{a\}$ thus meet C in the same number of points, and the same holds for the lines of $\{b\}$.

A curve that has α points in common with the lines of a ruling, for example with the lines of $\{a\}$, and β points in common with those of $\{b\}$ will be called a *curve of type* (α, β), or a curve belonging to the *family* (α, β). The algebraic curves of order n traced on $\mathfrak{Q}$ are partitioned into the $n + 1$ families $(0, n), (1, n - 1)$, $(2, n - 2), \ldots, (n - 1, 1), (n, 0)$. In particular, the two rulings $\{a\}, \{b\}$ are the families $(0, 1)$ and $(1, 0)$.

A surface of order n meets every generator of $\mathcal{Q}$ in n points, and thus cuts $\mathcal{Q}$ in a curve of type (n, n).

We shall see later that conversely each curve of type (n, n) on $\mathcal{Q}$ is the complete intersection of $\mathcal{Q}$ with a surface of order n. A curve split into two curves of types (α, β) and (α', β') is a curve of type $(\alpha + \alpha', \beta + \beta')$, and so on.

We return to the projection of $\mathcal{Q}$ from the point O into the plane π. A generic curve C of type (α, β) not passing through O has as projection into π a curve C' of order $n = \alpha + \beta$ which passes through the two points A and B with multiplicities α and β respectively. Conversely, let C' be a curve of π having order $n = \alpha + \beta$ and for which A and B are respectively α-fold and β-fold multiple points. The cone Γ that projects C' from O has the two lines a and b as respectively α-fold and β-fold multiple generators. Hence the cone intersects the quadric in a curve of order $2(\alpha + \beta)$ which includes the two lines a and b counted respectively α and β times. The residual component, of order $\alpha + \beta$, is a curve of type (α, β). Its intersections, for example, with a generic line $q \in \{a\}$, have as projections the intersections of C' with the projection of q, that is, with a line of the pencil of center B. Indeed, since the line q is skew to a and is supported by b, its projection is a line of the pencil of center B which meets C', outside of B, in α points.

Since the curves of type (α, β) have as projections in π the curves of a linear system (the linear system of the curves of order $\alpha + \beta$ with the two multiple points A and B of respective multiplicities α and β), we will say that they form a linear system. Moreover, what has been said for linear systems of algebraic plane curves can now be extended to linear systems of algebraic curves of $\mathcal{Q}$. This will be further clarified by analytic means in Section 11.1.

It is easy to determine the number of common points of two generic curves of types (α, β) and (α', β') belonging to $\mathcal{Q}$: it suffices to count the number of common points of their projections from O into π excluding the intersections that fall at A and B (because the latter are not projections of points common to the two curves). The result is:

$$(\alpha, \beta)(\alpha', \beta') := (\alpha + \beta)(\alpha' + \beta') - \alpha\alpha' - \beta\beta' = \alpha\beta' + \beta\alpha'. \quad (7.7)$$

Since the linear system of the curves of type (α, β) is projected into the linear system of the plane curves of order $\alpha + \beta$ with two points of assigned multiplicities α and β respectively (note that such a system is regular, as can be seen by writing the equation of one of its curves, cf. Example–Definition 6.2.2) its dimension is

$$\frac{(\alpha + \beta)(\alpha + \beta + 3)}{2} - \frac{\alpha(\alpha + 1)}{2} - \frac{\beta(\beta + 1)}{2} = \alpha\beta + \alpha + \beta. \quad (7.8)$$

Since the genus of a space curve is equal to that of its plane projections, the genus of a non-singular curve of type (α, β) is given by

$$p = \frac{(\alpha + \beta - 1)(\alpha + \beta - 2)}{2} - \frac{\alpha(\alpha - 1)}{2} - \frac{\beta(\beta - 1)}{2} = (\alpha - 1)(\beta - 1) \quad (7.9)$$

(see 7.2.9 and note that under the tacit hypothesis that the projection of $\mathcal{Q}$ from O into π is generic one has that the points A and B of multiplicity α and β are ordinary singularities).

In particular, an $(\alpha, 1)$-curve is rational, in agreement with the fact that its points are in bijective correspondence with the lines of one of the two rulings, and similarly for $(1, \beta)$-curves.

In order to prove that every curve of type (n, n) is the complete intersection of the quadric with a surface of order n it suffices to observe that the linear system $\Sigma_\mathcal{Q}$ cut out on $\mathcal{Q}$ by all the surfaces of order n in $\mathbb{P}^3$ has the same dimension $n^2 + 2n$ as that of the linear system of the curves of type (n, n) given by (7.8). Indeed, the linear system Σ of all the surfaces of order n in $\mathbb{P}^3$, whose dimension is $\binom{n+3}{3} - 1$ (cf. Section 6.1), contains the linear system Σ_0 consisting of all the surfaces that split into $\mathcal{Q}$ and a surface of order $n - 2$ and which thus has dimension $\binom{n+1}{3} - 1$. It follows that
$$\dim \Sigma_\mathcal{Q} - \dim \Sigma_0 - 1 = \binom{n+3}{3} - 1 - \binom{n+1}{3} = n^2 + 2n.$$

Examples 7.3.2. Let $\mathcal{Q}$ be a non-specialized quadric. The curves of second order on $\mathcal{Q}$ are those of types $(0, 2)$ and $(2, 0)$, which are the pairs of lines of a given ruling, and the curves of type $(1, 1)$ which are the plane sections of $\mathcal{Q}$.

Let us consider the curves of order 3. Those of types $(0, 3)$ and $(3, 0)$ are the triples of lines of a given ruling. On $\mathcal{Q}$ there are two families of dimension 5 consisting of space cubics which are generically irreducible, namely the curves of type $(1, 2)$ and those of type $(2, 1)$. Through five generic points of $\mathcal{Q}$ there passes one and only one cubic of each family. In view of (7.7), two cubics of the same family have in common $(1, 2)(1, 2) = 4$ points, while two cubics of different families meet in $(1, 2)(2, 1) = 5$ points.

As far as curves of order 4 are concerned, besides the quadruples of lines corresponding to the types $(0, 4)$ and $(4, 0)$, there are two families of rational quartics, the curves of types $(1, 3)$ and $(3, 1)$, and in addition there is a family of curves of type $(2, 2)$ which give quartics of genus 1. These $(2, 2)$-curves are the complete intersections of $\mathcal{Q}$ with the other quadrics in $\mathbb{P}^3$. (One notes that the quadrics of $\mathbb{P}^3$ constitute a system of dimension 9, so that on $\mathcal{Q}$ they cut out a linear system of dimension 8, and 8 is, by (7.8), the dimension of the system of curves $(2, 2)$ on $\mathcal{Q}$.) On the other hand, a curve of type $(1, 3)$, and similarly of type $(3, 1)$, belongs to a single quadric. Indeed, a quadric F that contains a curve C of type $(3, 1)$ must contain every trisecant of C, that is, every line $(0, 1)$. Therefore it coincides with $\mathcal{Q}$. Curves of types $(3, 1)$ and $(1, 3)$ are the residual intersections of $\mathcal{Q}$ with cubic surfaces that contain two lines of $\mathcal{Q}$ belonging to the same ruling.

Remark 7.3.3 (The case of a specialized quadric). One can also do stereographic projection onto a plane π in the case in which $\mathcal{Q}$ is specialized and irreducible, that is, when $\mathcal{Q}$ is a cone. One must take the center of projection O *distinct* from the vertex V of $\mathcal{Q}$. The tangent plane σ to the quadric at O touches the quadric along a generator and so we obtain two coinciding exceptional lines, $a = b$, on $\mathcal{Q}$ and two coinciding fundamental points $A = B$ in π.

A point P of the cone, distinct from O, has a well-defined projection P' in π. On the other hand, to the point O there correspond all the points of the line r common to the two planes σ and π.

The fundamental point A $(A = B)$ comes from every point of the generator a.

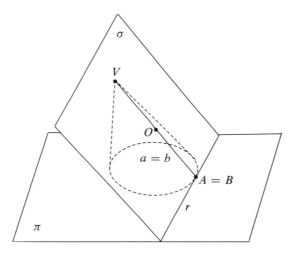

Figure 7.4

The generators of the cone have as their projections the lines of the plane π issuing from the point A: indeed, each generator passes through the vertex of the cone which has the point A as its projection.

A generic plane section has as its projection a conic tangent at A to the line r. Conversely, a conic γ of π tangent to r at A is projected from O onto a quadric cone that has the line a counted twice in common with $\mathcal{Q}$ and then another plane section whose projection is γ.

Let C be an arbitrary algebraic curve traced on the cone. If n is its order, the n intersections with a plane α passing through the vertex are located on the two generators p and q common to the cone and α. If one allows α to run over the pencil of planes that has as axis the line p one sees that all the generators q of the cone must contain the same number of points of C distinct from the vertex V. So if h is the number of points (other than V) common to C and a generic generator of $\mathcal{Q}$ and m is the multiplicity of the vertex V for C, one has $n = 2h + m$. Moreover the curves of the cone may be classified on the basis of the numbers h and m.

The projection of C from the point O is a curve C' of order n with a point of multiplicity $h + m$ at A to which an h-fold point A' is "infinitely near" (that is, A' is a point in the first order neighborhood of A, cf. §9.2.5), arising from the fact that the tangents to C in the h points that it has in common with the line a lie in the plane σ and hence have the same projection r.

It is a simple exercise to prove that the curves having multiplicity m at the vertex of the cone and meeting the generators in h further points constitute a linear system Σ whose dimension is $h^2 + hm + 2h + m$. Their genus (if they do not have other singularities outside the m-fold point at the vertex) is $(h-1)(h+m-1)$. Furthermore, two curves C^{2h+m} and $C^{2h'+m'}$ meet in $2hh' + hm' + h'm$ points outside of the vertex. If one counts the intersections at the vertex too, rather than just the variable intersections, the two curves have $2hh' + hm' + h'm + mm'$ points in common.

7.4 Rational curves

The following result plays a crucial role in the theory of algebraic curves (cf. Exercise 2.7.31 and Theorem 6.6.2).

Theorem 7.4.1 (Lüroth's theorem). *Let C be a curve with parametric equations (in non-homogeneous coordinates)*

$$x_i = \frac{\phi_i(t)}{\phi_0(t)}, \quad \phi_0(t), \phi_i(t) \in \mathbb{C}[t], \ i = 1, \ldots, n.$$

If every point of C comes from $k \geq 1$ values of the parameter t it is possible to find a new parameter τ, a rational function of t, in such a way that there is a bijection between points of C and values of τ.

Proof. We suppose, as is permissible, that the polynomials $\phi_i(t)$, $i = 0, 1, \ldots, n$, do not have common factors of positive degree and are all of the same degree.

Let t_1 be a generic value of t and let $P_1 = \left(\frac{\phi_1(t_1)}{\phi_0(t_1)}, \ldots, \frac{\phi_n(t_1)}{\phi_0(t_1)}\right)$. The values $t_1, \ldots, t_k$ that give the same point are those that give rank 1 to the matrix

$$\begin{pmatrix} \phi_0(t) & \phi_1(t) & \cdots & \phi_n(t) \\ \phi_0(t_1) & \phi_1(t_1) & \cdots & \phi_n(t_1) \end{pmatrix},$$

that is, they are the common zeros of the polynomials

$$\Phi_{ij}(t, t_1) := \phi_i(t_1)\phi_j(t) - \phi_j(t_1)\phi_i(t), \quad i, j = 0, \ldots, n, \ i \neq j.$$

This means that $t_1, \ldots, t_k$ are the zeros of the greatest common divisor, $\Delta(t, t_1)$, of the polynomials Φ_{ij}. Let

$$\Delta(t, t_1) := a_0(t_1)t^k + a_1(t_1)t^{k-1} + \cdots + a_k(t_1).$$

As t_1 varies the coefficients $a_h(t_1)$ can not remain proportional to constants, since otherwise the roots of $\Delta(t, t_1) = 0$ would be independent of t_1, and that is absurd because t_1 is among those roots.

We may suppose that the roots of $\Delta(t, t_1) = 0$ all vary with t_1 because if one of them were to remain fixed we could neglect it and say that every point of C comes from $k - 1$ values of the parameter. In particular if t_1 is generic, $\Delta(t, t_1)$ is a polynomial of degree k in t and so $a_0(t_1) \neq 0$.

It is obvious that we would have arrived at the same equation $\Delta(t, t_1) = 0$ if instead of t_1 we had initially chosen any other of the values $t_2, t_3, \ldots, t_k$ that give the same point $P(t_1)$. Therefore, for at least one index j (let it be, for instance, $j = 1$), the ratio $\frac{a_j(t_1)}{a_0(t_1)}$ really depends on t_1 and is

$$\frac{a_1(t_1)}{a_0(t_1)} = \frac{a_1(t_2)}{a_0(t_2)} = \cdots = \frac{a_1(t_k)}{a_0(t_k)}.$$

If we then set

$$\tau := \frac{a_1(t)}{a_0(t)},$$

we obtain a bijective correspondence between the points of the curve and the values of the parameter τ. Indeed, if $t_1, \ldots, t_k$ are all the values of t giving the same point P, one has

$$\frac{a_1(t_1)}{a_0(t_1)} = \frac{a_1(t_2)}{a_0(t_2)} = \cdots = \frac{a_1(t_k)}{a_0(t_k)}. \tag{7.10}$$

The function $\tau = \frac{a_1(t)}{a_0(t)}$ then assumes the same value for $t = t_1, \ldots, t_k$ and thus to a point P there corresponds a unique value of τ.

Conversely, for each fixed $\tau \in \mathbb{C}$ we consider the equation

$$\tau a_0(t) - a_1(t) = 0, \tag{7.11}$$

which is of degree k and uniquely determined by τ. If t_1 is one of its roots, that is, if $\tau a_0(t_1) - a_1(t_1) = 0$, or equivalently if $\tau = \frac{a_1(t_1)}{a_0(t_1)}$, then $t_2, \ldots, t_k$ are also roots, in view of (7.10). Thus (7.11) furnishes the entire set of values of t corresponding to a common point of the curve.

To get the new parametric representation of C we eliminate t from each of the n systems of equations ($i = 1, \ldots, n$)

$$\begin{cases} \tau a_0(t) - a_1(t) = 0, \\ x_i \phi_0(t) - \phi_i(t) = 0. \end{cases}$$

The elimination (which is done by writing the Euler–Sylvester resultant, cf. Section 4.1) leads to an algebraic equation $g(x_i, \tau) = 0$ which to each value of τ associates a unique value of x_i (the ith coordinate of the point that corresponds to τ). We know that x_i turns out to be a rational function of τ, see Section 2.6. (We observe that the Euler–Sylvester resultant is a polynomial of degree k in the x_i which will be the kth power of a polynomial of degree one with respect to x_i.)

One has infinitely many parameters in bijective algebraic correspondence with the points of the curve: if τ is one of them, every other analogous parameter τ', having to be related to τ by way of the points of the curve via a bijective algebraic correspondence, will necessarily have the form

$$\tau' = \frac{a\tau + b}{c\tau + d} \tag{7.12}$$

with $ad - bc \neq 0$. Thus it is possible (since equation (7.12) is a non-degenerate bilinear relation which thus represents a projectivity) to choose a parameter whose values are in bijective algebraic correspondence with the points of the curve, and in such a way as to assign three arbitrary distinct assigned values to three arbitrarily chosen distinct points of the curve. □

We can define a rational curve in $\mathbb{P}^r$ as the locus C of a point of $\mathbb{P}^r$ whose homogeneous coordinates x_i may be expressed in the form

$$x_i = f_i(\lambda), \quad i = 0, \ldots, r, \tag{7.13}$$

where the $f_i(\lambda)$ are polynomials in a non-homogeneous coordinate λ without common factors. Equations (7.13) in fact define a rational map $\mathbb{P}^1 \to C$ which in view of Lüroth's theorem (cf. Exercise 2.7.31 and also [14, p. 270]) is a birational isomorphism. Hence C is a rational curve (cf. 7.2.13 and Corollary 2.6.6).

It is obvious that the property of being a rational curve is invariant under projectivities.

- *A rational curve C is algebraic.*

Indeed, consider a generic S_{r-3}, which we may assume to have equations $x_0 = x_1 = x_2 = 0$. The equation of the cone that projects C from that space may be obtained immediately by eliminating the parameter λ from the two equations

$$f(\lambda) = f_1(\lambda)x_0 - f_0(\lambda)x_1 = 0,$$
$$g(\lambda) = f_2(\lambda)x_0 - f_0(\lambda)x_2 = 0.$$

We know that the elimination is performed with rational operations (for example by the method of Euler–Sylvester, cf. Section 4.1) and so leads to an algebraic equation $F(x_0, x_1, x_2) = 0$, which is homogeneous inasmuch as the coefficients of $f(\lambda)$ and $g(\lambda)$ are linear and homogeneous in the variables x_0, x_1, x_2. Thus the cone projecting C from a generic S_{r-3} is algebraic, and so C appears as an intersection of algebraic hypersurfaces (the cones projecting it from the various S_{r-3} in $\mathbb{P}^r$). This proves that C is a (1-dimensional) algebraic variety.

7.4. Rational curves

By Lüroth's theorem one may assume that to each λ there corresponds a point of C, and conversely each generic point of C comes from a single value of λ. One then sees immediately that the order of C is the maximal degree n of the polynomials $f_i(\lambda)$.

Writing λ/μ rather than λ and multiplying by μ^n, we obtain homogeneous polynomials of degree n in λ, μ in the second terms of equations (7.13):

$$x_i = a_{i0}\lambda^n + a_{i1}\lambda^{n-1}\mu + \cdots + a_{in}\mu^n, \quad i = 0, \ldots, r. \tag{7.14}$$

The hypothesis that the given curve C^n of order n is embedded in $\mathbb{P}^r$ assures us that these polynomials are linearly independent, which is to say that the matrix of coefficients (a_{ij}) has rank $r + 1$.

If $r < n$ we may adjoin to the equations (7.14) other $n - r$ equations of the same type, chosen so as to have $n + 1$ equations whose second terms are linearly independent (and this may be done in infinitely many ways).

Such equations will give a curve of order n in $\mathbb{P}^n$ of which the given C^n in $\mathbb{P}^r$ is a projection. Thus we have the following result which gives a way to study the rational curves of a given order n belonging to an arbitrary space (of dimension $\leq n$) by deducing them via projection from degree n curves in $\mathbb{P}^n$.

Theorem 7.4.2. *A rational irreducible curve C^n is either a curve C^n in $\mathbb{P}^n$ or the projection of such a curve in a space $\mathbb{P}^r$ with $r < n$.*

We now study some noteworthy properties of irreducible curves C^n in $\mathbb{P}^n$.

Proposition 7.4.3. *An irreducible curve C^n in $\mathbb{P}^n$ is rational.*

Proof. Take $n - 1$ arbitrary distinct points of C^n; they are linearly independent, and therefore they generate a space S_{n-2}. Otherwise, for these points and two further points of C^n there would pass some hyperplane containing the $n + 1$ points of C^n and thus the entire curve. It follows that the points of C^n correspond bijectively to the spaces S_{n-1} of the pencil with axis S_{n-2} and so to the values of a parameter λ. The homogeneous coordinates of the points of the curve may then be written as polynomials in that parameter in the form of equation (7.13). □

Proposition 7.4.4. *An irreducible curve C^n of $\mathbb{P}^n$ has no multiple points.*

Proof. Indeed, the order of an irreducible curve $\mathcal{L}$ embedded in $\mathbb{P}^n$ and having a multiple point P is certainly $> n$, as one sees immediately by observing that a hyperplane containing P and other $n - 1$ points of the curve $\mathcal{L}$ must have at least $n + 1$ intersections with $\mathcal{L}$. (For the notion of multiple point of a curve in a higher dimensional spaces see Section 3.4.) □

We will also say that a curve C^n in $\mathbb{P}^n$ is a *rational normal curve* (of order n).

7.4.5 Projective generation of a curve C^n in $\mathbb{P}^n$. Let $\Phi_1, \Phi_2, \ldots, \Phi_n$ be pencils of hyperplanes. We will say that they are *projectively referred* if there exist projectivities $\omega_{ij} : \Phi_i \to \Phi_j$ such that $\omega_{ij} \circ \omega_{jk} = \omega_{ik}$, $\omega_{ii} = \text{id}_{\Phi_i}$.

Assuming this, we now prove the following theorem which extends Steiner's theorem on the projective generation of a conic to the case of a curve C^n in $\mathbb{P}^n$, [31, Chapter XIV].

Theorem 7.4.6 (Projective generation of a curve C^n in $\mathbb{P}^n$, I). *An irreducible algebraic curve C^n of order n in $\mathbb{P}^n$ is the locus of the points common to n corresponding hyperplanes H_i of n projectively referred pencils (that is, such that $\omega_{ij}(H_i) = H_j$).*

Conversely, let $S_{n-2}^{(1)}, S_{n-2}^{(2)}, \ldots, S_{n-2}^{(n)}$ be the centers of n projectively referred pencils of hyperplanes $\Phi_1, \Phi_2, \ldots, \Phi_n$, and suppose that if $S_{n-1}^{(i)} \in \Phi_i$, $i = 1, \ldots, n$, are corresponding hyperplanes then their intersection consists of a single point (this happens in general). Then the locus of the points common to n corresponding hyperplanes is a curve C^n in $\mathbb{P}^n$.

Proof. Consider n spaces $S_{n-2}^{(1)}, S_{n-2}^{(2)}, \ldots, S_{n-2}^{(n)}$ that are $(n-1)$-secants of C^n and the pencils $\Phi_1, \Phi_2, \ldots, \Phi_n$ having them as centers.

We say that two hyperplanes H_i and H_j correspond to each other when they contain a common point P of C^n. Thus one has a bijective algebraic correspondence between the two pencils, and hence a projectivity ω_{ij}. The n pencils are thus projectively referred. (We note explicitly that if P is one of the points belonging to the center, for example, of Φ_j then to the hyperplane of Φ_i passing through P there corresponds under ω_{ij} a hyperplane of Φ_j tangent at P to C^n.) This proves the first part of the statement.

To prove the converse we take two hyperplanes $L_1 = 0$ and $M_1 = 0$ in the pencil Φ_1. If $L_i = 0$ and $M_i = 0$ are the hyperplanes of the pencil Φ_i corresponding to them it is permissible to suppose that to the hyperplane $L_1 + \lambda M_1 = 0$ there correspond the hyperplanes $L_i + \lambda M_i = 0$ of Φ_i, $i = 1, \ldots, n$.

Solving the system formed by the equations of the n corresponding hyperplanes,

$$\begin{cases} L_1 + \lambda M_1 = 0, \\ L_2 + \lambda M_2 = 0, \\ \quad \vdots \\ L_n + \lambda M_n = 0, \end{cases} \tag{7.15}$$

one finds the homogeneous coordinates of their common point as a polynomial of degree n in the parameter λ:

$$x_i = a_{i0}\lambda^n + a_{i1}\lambda^{n-1} + \cdots + a_{in}, \quad i = 0, 1, \ldots, n. \tag{7.16}$$

This proves that the locus described by that point as λ varies is a curve C^n contained in $\mathbb{P}^n$. If the n pencils of hyperplanes are chosen generically, C^n turns out to be embedded in $\mathbb{P}^n$, that is, $\det(a_{ij}) \neq 0$: if $\det(a_{ij}) = 0$ one would in fact have a linear relation between the variables x_i and so C^n would belong to some subspace S_{n-1}. (We note explicitly that what we have seen holds in general: it is not excluded that by choosing the n pencils in some particular manner, the n corresponding hyperplanes might have in common not merely a point, but, for example, a line; and then rather than a curve one would find a ruled surface.) □

Since, as it has been noted, we may suppose that $\det(a_{ij}) \neq 0$ in (7.16), it follows that the given curve C^n is projectively identical to the curve C'^n given by

$$x_0 : x_1 : \cdots : x_{n-1} : x_n = \lambda^n : \lambda^{n-1} : \cdots : \lambda : 1. \tag{7.17}$$

This means that $C'^n = \varphi(C)$, where φ is the projectivity of $\mathbb{P}^n$ associated to the matrix $A = (a_{ij})$. So we may draw the following conclusion:

- All the irreducible curves C^n in $\mathbb{P}^n$ are projectively identical and are parametrically representable in the form (7.17).

System (7.17) shows that the points of C^n are those that give rank 1 to the matrix

$$\begin{pmatrix} L_1 & L_2 & \cdots & L_n \\ M_1 & M_2 & \cdots & M_n \end{pmatrix} \tag{7.18}$$

and therefore C^n is the base curve of a linear system of quadrics.

For example, if $n = 3$, one finds that the space curve of order 3 in $\mathbb{P}^3$ is the base curve of the net of quadrics

$$\lambda_1(L_2 M_3 - M_2 L_3) + \lambda_2(L_1 M_3 - M_1 L_3) + \lambda_3(L_1 M_2 - M_1 L_2) = 0.$$

For $n = 2$ one finds the equation of a conic in the form $L_1 M_2 - M_1 L_2 = 0$ which places its projective generation in evidence.

The consideration of the matrix (7.18) leads to another way of projectively generating the curve C^n. Consider the two stars Σ and Σ' of hyperplanes having as centers the points $L_1 = \cdots = L_n = 0$, $M_1 = \cdots = M_n = 0$ and a projectivity between them. We may assume that the two hyperplanes

$$\mu_1 L_1 + \mu_2 L_2 + \cdots + \mu_n L_n = 0, \quad \mu_1 M_1 + \mu_2 M_2 + \cdots + \mu_n M_n = 0 \tag{7.19}$$

that one has for the same value of the parameters $\mu_1, \ldots, \mu_n$ correspond in the projectivity.

The coordinates of an arbitrary point P of C^n render the L_i proportional to the M_i; therefore, if we substitute the aforesaid coordinates in (7.19) we find two linear equations in the μ_i such that each system of values of the μ_i that satisfies the first also satisfies the other, and conversely, $i = 1, \ldots, n$. This means that each hyperplane of the first star passing through P has as its corresponding hyperplane a hyperplane that also passes through P; that is, P lies on two corresponding and incident lines of the two stars (each line is the intersection of $n-1$ hyperplanes of the star). It is clear that the two centers are arbitrary points of C^n.

In conclusion, in addition to the statement of Theorem 7.4.6 we also have the following.

- (Projective generation of a C^n in $\mathbb{P}^n$, II) *An irreducible curve C^n in $\mathbb{P}^n$ is the locus of the points common to two corresponding and incident lines under a projectivity between two stars of lines (having two points of C^n as centers).*

Conversely, two stars of projective lines generate (in general) a curve of order n, as the locus of points of incidence of corresponding lines. Indeed, intersecting the two stars with a hyperplane one finds a projectivity between two superposed hyperplanes; and the n fixed points of the projectivity are the points of the hyperplane that belong to the two corresponding and incident lines.

An immediate consequence of the foregoing is the following result that extends the case $n = 2$.

Proposition 7.4.7. *For $n + 3$ generic points $P_1, P_2, \ldots, P_{n+3}$ of $\mathbb{P}^n$ there passes one and only one curve C^n of order n.*

Proof. By Theorem 1.1.12 we know that there is a single projectivity between the stars of lines having as centers P_{n+2} and P_{n+3} and under which the $n+1$ pairs of lines $(r_{P_{n+2}P_i}, r_{P_{n+3}P_i}), i = 1, 2, \ldots, n, n+1$, correspond. $\square$

7.4.8 Osculating hyperplanes to a curve C^n in $\mathbb{P}^n$. We consider a curve C^n in $\mathbb{P}^n$ and denote by $u_0, \ldots, u_n$ the coordinates of the osculating hyperplane Π to C^n at one of its points P, that is, of the unique hyperplane whose n intersections with the curve are all absorbed by P (cf. §7.1.1). If C^n is represented in the form (7.17) and $P = [x_0, \ldots, x_n] = [\lambda_0^n, \ldots, 1]$, the equation

$$u_0 \lambda^n + u_1 \lambda^{n-1} + \cdots + u_n = 0,$$

which furnishes the intersections of C^n with Π, must be equivalent to the equation

$$(\lambda - \lambda_0)^n = \lambda^n - n\lambda_0 \lambda^{n-1} + \binom{n}{2} \lambda_0^2 \lambda^{n-2} - \cdots + (-1)^n \lambda_0^n = 0.$$

Thus,

$$u_0 = 1, \quad u_1 = -n\lambda_0, \quad u_2 = \binom{n}{2} \lambda_0^2, \ldots, \quad u_n = (-1)^n \lambda_0^n.$$

Hence one has the relations

$$u_0 = x_n, \quad u_1 = -nx_{n-1}, \quad u_2 = \binom{n}{2} x_{n-2}, \ldots, \quad u_n = (-1)^n x_0$$

among the coordinates x_i of P and the u_i. Thus we have proved the following

Theorem–Definition 7.4.9 (Clifford). *The points of a curve C^n in $\mathbb{P}^n$ correspond to its osculating hyperplanes under the reciprocity φ associated to the matrix*

$$A = \begin{pmatrix} 0 & \cdots & 0 & 0 & 1 \\ 0 & \cdots & 0 & -n & 0 \\ 0 & \cdots & \binom{n}{2} & 0 & 0 \\ \cdots & \cdots & \cdots & \cdots & \cdots \\ (-1)^n & \cdots & 0 & 0 & 0 \end{pmatrix}.$$

If n is even, the matrix A is symmetric and φ is a polarity with respect to a quadric; if n is odd the matrix A is antisymmetric and φ is a null polarity.

*In the first case C^n is contained in the quadric which is the locus of the self-conjugate points, called the **Clifford quadric** (and coincides with it if $n = 2$). In the second case every point of $\mathbb{P}^n$ is self-conjugate and the spaces S_{n-1} osculating C^n at its intersections with a given hyperplane H meet in a point of H.*

7.5 Exercises on rational curves

7.5.1. Determine the singularities of the rational plane cubic $\mathcal{C}^3$ with parametric equations:

$$x_i = a_i t^3 + b_i t^2 + c_i t + d_i, \quad i = 1, 2, 3, \ a_i, b_i, c_i, d_i \in \mathbb{C}.$$

The possible singularity of $\mathcal{C}^3$ can only be a double point (cf. 7.2.12). To calculate its coordinates we first procure the condition for three points of the cubic to be collinear.

Let t_1, t_2, t_3 be the parameters of three collinear points $P_i = P(t_i), i = 1, 2, 3$. This is equivalent to saying that there exist $u_1, u_2, u_3 \in \mathbb{C}$, not all zero, such that the line $u_1 x_1 + u_2 x_2 + u_3 x_3 = 0$ meets $\mathcal{C}^3$ in these three points. Thus t_1, t_2, t_3 are the roots of the equation $\sum_{i=1}^{3} u_i(a_i t^3 + b_i t^2 + c_i t + d_i) = 0$, which means

$$\left(\sum_{i=1}^{3} a_i u_i\right) t^3 + \left(\sum_{i=1}^{3} b_i u_i\right) t^2 + \left(\sum_{i=1}^{3} c_i u_i\right) t + \sum_{i=1}^{3} d_i u_i = 0.$$

If we set $\sigma_1 = t_1 + t_2 + t_3, \sigma_2 = t_1 t_2 + t_2 t_3 + t_3 t_1, \sigma_3 = t_1 t_2 t_3$ then we have

$$\left(\sum_{i=1}^{3} a_i u_i\right)\sigma_1 + \sum_{i=1}^{3} b_i u_i = 0, \quad \left(\sum_{i=1}^{3} a_i u_i\right)\sigma_2 - \sum_{i=1}^{3} c_i u_i = 0,$$

$$\left(\sum_{i=1}^{3} a_i u_i\right)\sigma_3 + \sum_{i=1}^{3} d_i u_i = 0,$$

that is,

$$\begin{cases} \sum_{i=1}^{3} u_i(a_i\sigma_1 + b_i) = 0, \\ \sum_{i=1}^{3} u_i(a_i\sigma_2 - c_i) = 0, \\ \sum_{i=1}^{3} u_i(a_i\sigma_3 + d_i) = 0. \end{cases}$$

The necessary and sufficient condition for the existence of such u_1, u_2, u_3 is then

$$\begin{vmatrix} a_1\sigma_1 + b_1 & a_2\sigma_1 + b_2 & a_3\sigma_1 + b_3 \\ a_1\sigma_2 - c_1 & a_2\sigma_2 - c_2 & a_3\sigma_2 - c_3 \\ a_1\sigma_3 + d_1 & a_2\sigma_3 + d_2 & a_3\sigma_3 + d_3 \end{vmatrix} = 0. \tag{7.20}$$

Subtracting the first column multiplied by $\frac{a_2}{a_1}$ from the second column and then the first column multiplied by $\frac{a_3}{a_1}$ from the third column one sees that (7.20) may be rewritten in the form:

$$At_1t_2t_3 + B(t_1t_2 + t_2t_3 + t_3t_1) + C(t_1 + t_2 + t_3) + D = 0, \quad A, B, C, D \in \mathbb{C}. \tag{7.21}$$

Equation (7.21) is the necessary and sufficient condition in order for the points coming from the values t_1, t_2, t_3 of the parameter to be collinear.

With this in hand, we order (7.21) with respect to t_3,

$$t_3(At_1t_2 + B(t_1 + t_2) + C) + Bt_1t_2 + C(t_1 + t_2) + D = 0, \tag{7.22}$$

and we resolve the symmetric system

$$\begin{cases} At_1t_2 + B(t_1 + t_2) + C = 0, \\ Bt_1t_2 + C(t_1 + t_2) + D = 0. \end{cases}$$

We find a pair of numbers t_1, t_2 which together with any t_3 satisfy equation (7.22). Since $P(t_1), P(t_2), P(t_3)$ are collinear no matter how t_3 is chosen (and they exhaust the intersection of $\mathcal{C}^3$ with a line) we must have that $P(t_1) = P(t_2)$ is a double point of $\mathcal{C}^3$ (a node if $t_1 \neq t_2$; and a cusp if $t_1 = t_2$, which means that the two polynomials $At^2 + 2Bt + C$ and $Bt^2 + 2Ct + D$ have a common zero).

7.5.2. Let C be an irreducible non-planar cubic in $\mathbb{P}^3$.

(1) *Show both synthetically and analytically that for each point of $\mathbb{P}^3$ there passes one and only one chord (or a tangent) of C, and write the equation for the ruled surface of the tangents to C.*

(2) *Study the singular points and flexes of the plane projections of C.*

(1) For a point not belonging to a cubic C there can not pass more than one chord of the curve, because otherwise C would lie in the plane of two concurrent chords (since that plane has at least four intersections with C). The chords of C (which are in number ∞^2) can not fill merely a surface, because in that case through each point of that surface there would have to pass infinitely many (i.e., the ∞^1 determined by passage through the point). On the other hand, a surface that contains all the chords of a curve would also have to contain all the cones that project the curve from one of its points, and the curve would be planar.

This fact can be derived analytically in the following fashion. Let C be the locus of the point $P(t) = [t^3, t^2, t, 1]$ and $O = [a_0, a_1, a_2, a_3]$ a point not belonging to C. If $P(\lambda)$ and $P(\mu)$ are two points of the curve C which are collinear with O, the matrix

$$\begin{pmatrix} a_0 & a_1 & a_2 & a_3 \\ \lambda^3 & \lambda^2 & \lambda & 1 \\ \mu^3 & \mu^2 & \mu & 1 \end{pmatrix}$$

has rank two. Hence

$$\begin{vmatrix} a_0 & a_1 & a_2 \\ \lambda^3 & \lambda^2 & \lambda \\ \mu^3 & \mu^2 & \mu \end{vmatrix} = \begin{vmatrix} a_1 & a_2 & a_3 \\ \lambda^2 & \lambda & 1 \\ \mu^2 & \mu & 1 \end{vmatrix} = 0,$$

that is,

$$\begin{cases} a_0 - a_1(\lambda + \mu) + a_2 \lambda\mu = 0, \\ a_1 - a_2(\lambda + \mu) + a_3 \lambda\mu = 0. \end{cases}$$

This implies that λ and μ are the roots of the quadratic equation

$$\begin{vmatrix} x^2 & x & 1 \\ a_0 & a_1 & a_2 \\ a_1 & a_2 & a_3 \end{vmatrix} = 0. \tag{7.23}$$

Thus for each point O not belonging to C there passes one and only one chord of C.

The trace P' of this chord upon a plane π not passing through O is the double point of the cubic Γ, the projection of C from O onto π.

If the roots of (7.23) coincide, that is, if O is a point of the quartic surface with equation

$$(x_0 x_3 - x_1 x_2)^2 - 4(x_0 x_2 - x_1^2)(x_1 x_3 - x_2^2) = 0, \tag{7.24}$$

one has a tangent rather than a chord of C (at the point $P(\lambda) = P(\mu)$) and P' is a cusp of Γ.

Equation (7.24) is the equation of the ruled surface $\mathcal{F}$ of the tangents of C. From it one sees that a space cubic has rank 4, which means that four is the number of

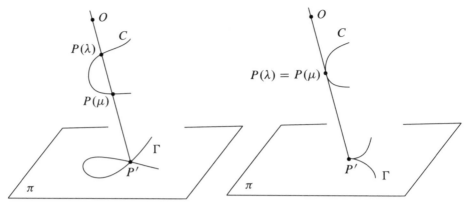

Figure 7.5

tangents of the cubic supported by a generic line (in the four points in which the line intersects the quartic $\mathcal{F}$).

(2) Suppose now that the osculating plane α to C in one of its points P passes through O. Since by Clifford's theorem (Theorem 7.4.9), α is the polar plane of P under the null polarity determined by C, the polar plane of O passes through P. Conversely, if the polar plane of O passes through P, the osculating plane of P passes through O. Hence the points of C with osculating plane that passes through O are the intersections of C with the plane corresponding to O under the null polarity.

Since under a null polarity every point belongs to its polar hyperplane (inasmuch as each point of the space is self-conjugate) we may conclude as follows.

i) If O does not belong to the quartic $\mathcal{F}$, that is, if through O there passes no tangent of C, the curve has a node and hence three flexes in view of (5.35), which belong to the line $\alpha \cap \pi$.

ii) If O belongs to $\mathcal{F}$ (but not to C), let P be the point of contact of C with the tangent t of C passing through O. The projection Γ of C from O onto the plane π has a cusp (in the point $P' = t \cap \pi$) and thus only one flex. The osculating plane of C at P passes through O and does not meet C outside of P. Hence, the cuspidal tangent of Γ is the trace in π of the osculating plane of C at P.

7.5.3. *The twisted cubic C^3 in $\mathbb{P}^3$ is the base curve of a net Σ of quadrics. Two quadrics of the net meet also in a line, which is a chord of C^3. The net Σ contains infinitely many quadric cones having vertex on C^3, and C^3 may be defined as the further intersection of two quadric cones having a common generator (but not tangent along that line).*

Exploit this fact to find the parametric representation $P(t) = [t^3, t^2, t, 1]$ of C^3.

7.5. Exercises on rational curves

We have seen in Section 7.4 that C^3 is the base curve of a net of quadrics. Two quadrics of the net meet in a curve C^4 which is composed of C^3 and a further line r, which is a chord of C^3. Indeed, the complete intersection of a quadric Q with another quadric is a quartic curve C^4 of type $(2,2)$. If C^3 is on Q a curve of type $(1,2)$, then the residual component is a line $r = (1,0)$ and meets C^3 in $(1,2)(1,0) = 2$ points (cf. Section 7.3).

Let $r = r_{VV'}$ be a chord of C^3, with $V = A_0 = [1,0,0,0]$ and $V' = A_3 = [0,0,0,1]$, and let F and F' be the quadric cones projecting C^3 from V and V' and thus having in common the generator r. We take the point $A_2 = [0,0,1,0]$ in the tangent plane to F along r, and the point $A_1 = [0,1,0,0]$ in the plane tangent to F' along the same line r.

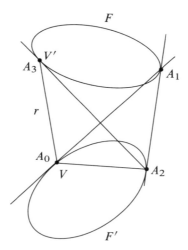

Figure 7.6

The equation of F (in which the variable x_0 does not appear since F is a cone with vertex A_0) represents in the plane $x_0 = 0$ the intersection of F with that plane, that is a conic tangent at A_3 to the line $x_1 = 0$. Thus the equation of F is

$$x_1(x_3 + ax_1 + bx_2) + hx_2^2 = 0.$$

Similarly F' has equation

$$x_2(x_0 + a'x_1 + b'x_2) + h'x_1^2 = 0.$$

The equations

$$X_0 = x_0 + a'x_1 + b'x_2, \quad X_1 = x_1, \quad X_2 = x_2, \quad X_3 = x_3 + ax_1 + bx_2$$

represent a projectivity (non-degenerate) that fixes the two points V and V' and transforms the two cones into the surfaces with equations

$$X_1 X_3 + h X_2^2 = 0; \quad X_2 X_0 + h' X_1^2 = 0.$$

It then suffices to suppose that the unit point belongs to these two cones in order to have $h = h' = -1$ and thus the parametric representation $P(t) = [t^3, t^2, t, 1]$ for the residual C^3 that is their intersection outside of the line r (see also Example 11.1.1).

7.5.4. *Consider a curve C^4 in $\mathbb{P}^4$. Determine the number of osculating hyperplanes and the number of trisecant planes of C^4 that pass through a generic point of $\mathbb{P}^4$.*

For each generic point P of $\mathbb{P}^4$ there pass four osculating hyperplanes of C^4. Indeed, by Theorem 7.4.9, the osculating hyperplanes of C^4 correspond to the points of osculation under the polarity with respect to a quadric $\mathcal{Q}$. The osculating hyperplanes that pass through P are then the osculating hyperplanes at the points in which C^4 is met by the polar hyperplane of P with respect to $\mathcal{Q}$.

We now seek the trisecant planes of C^4 which issue from a generic point A not on C^4.

We project C^4 from one of its points O onto $S_3 = \pi$ and we let γ be the projected cubic. For the point A', the projection of A from O onto S_3, there passes a chord $\langle B', C' \rangle$ of γ. Moreover, the plane determined by the points B', C', O is a trisecant plane of our C^4 passing through A because the lines (coplanar) $r_{OA'}$, $r_{OB'}$, $r_{OC'}$ each contain a point of the quartic. As O varies on C^4 one obtains ∞^1 trisecant planes of C^4 passing through A.

One can also obtain this result in the following way.

Let C^4 be the locus of the point $P(t) = [t^4, t^3, t^2, t, 1]$ and let $P_i = P(t_i)$, $i = 1, 2, 3$, be three points of C^4 contained in a plane α passing through A. Each point Q of α lies in the hyperplane π defined by the points P_1, P_2, P_3 and by an arbitrary $P_4 = P(t_4)$ on C^4. On the other hand, if

$$u_0 x_0 + u_1 x_1 + u_2 x_2 + u_3 x_3 + u_4 x_4 = 0$$

is the equation of π, the four numbers t_1, t_2, t_3, t_4 are the solutions of the equation

$$u_0 t^4 + u_1 t^3 + u_2 t^2 + u_3 t + u_4 = 0$$

whence

$$u_0 : u_1 : u_2 : u_3 : u_4 = 1 : -\sigma_1 : \sigma_2 : -\sigma_3 : \sigma_4$$

where

$$\sigma_1 = \sum_i t_i, \quad \sigma_2 = \sum_{ij} t_i t_j, \quad \sigma_3 = \sum_{ijk} t_i t_j t_k, \quad \sigma_4 = t_1 t_2 t_3 t_4.$$

7.5. Exercises on rational curves   231

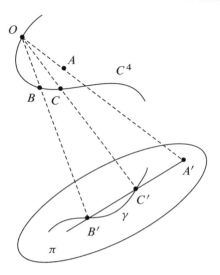

Figure 7.7

Therefore, the equation of π is

$$x_0 - \sigma_1 x_1 + \sigma_2 x_2 - \sigma_3 x_3 + \sigma_4 x_4 = 0;$$

and if the coordinates of A are a_0, a_1, a_2, a_3, a_4 we will have

$$a_0 - \sigma_1 a_1 + \sigma_2 a_2 - \sigma_3 a_3 + \sigma_4 a_4 = 0 \qquad (7.25)$$

for any choice of t_4. But (7.25) is a linear equation in t_4 of the type

$$t_4 f(t_1, t_2, t_3) + g(t_1, t_2, t_3) = 0.$$

Hence, the equations of the trisecant plane are

$$f(t_1, t_2, t_3) = g(t_1, t_2, t_3) = 0, \qquad (7.26)$$

where

$$f(t_1, t_2, t_3) = x_1 - (t_1 + t_2 + t_3)x_2 + (t_1 t_2 + t_2 t_3 + t_3 t_1)x_3 - t_1 t_2 t_3 x_4,$$
$$g(t_1, t_2, t_3) = x_0 - (t_1 + t_2 + t_3)x_1 + (t_1 t_2 + t_2 t_3 + t_3 t_1)x_2 - t_1 t_2 t_3 x_3.$$

Since A does not belong to C^4 the matrix

$$\begin{pmatrix} x_0 & x_1 & x_2 & x_3 \\ x_1 & x_2 & x_3 & x_4 \end{pmatrix}$$

has rank 2 (since the Cartesian equations of C^4 are obtained by setting the second order minors of the matrix equal to zero, cf. Exercise 7.5.10) and therefore equations (7.26) are two independent linear equations in the three unknowns $t_1 + t_2 + t_3$, $t_1t_2 + t_2t_3 + t_3t_1$, $t_1t_2t_3$. This suffices to conclude that there exist ∞^1 trisecant planes of C^4 passing through A.

7.5.5. Write the equation of the variety of the chords of the curve C^4 in $\mathbb{P}^4$ which is the locus of the point $P(t) = [t^4, t^3, t^2, t, 1]$.

Let r be the chord that joins the two points $P_1 = P(t_1)$ and $P_2 = P(t_2)$ of C^4 in $\mathbb{P}^4$, and let $x = [x_0, x_1, x_2, x_3, x_4]$ be any one of its points. If $P_3 = P(t_3)$ and $P_4 = P(t_4)$ are two points of C^4, the fact that x, P_1 and P_2 are collinear implies that the point x belongs to the hyperplane $J(P_1, P_2, P_3, P_4)$ for any choice of t_3, t_4. Thus the polynomial, in t_3, t_4,

$$x_0 - \sigma_1 x_1 + \sigma_2 x_2 - \sigma_3 x_3 + \sigma_4 x_4 = 0, \tag{7.27}$$

where

$$\sigma_1 = \sum_i t_i, \quad \sigma_2 = \sum_{ij} t_i t_j, \quad \sigma_3 = \sum_{ijk} t_i t_j t_k, \quad \sigma_4 = t_1 t_2 t_3 t_4,$$

must be the null polynomial. On the other hand (7.27) can be written in the form

$$\begin{aligned}x_0 - (t_1 + t_2)x_1 + t_1 t_2 x_2 - (t_3 + t_4)(x_1 - (t_1 + t_2)x_2 + t_1 t_2 x_3) \\ + t_3 t_4 (x_2 - (t_1 + t_2)x_3 + t_1 t_2 x_4) = 0,\end{aligned} \tag{7.28}$$

and hence must be

$$\begin{cases} x_0 - (t_1 + t_2)x_1 + t_1 t_2 x_2 = 0, \\ x_1 - (t_1 + t_2)x_2 + t_1 t_2 x_3 = 0, \\ x_2 - (t_1 + t_2)x_3 + t_1 t_2 x_4 = 0. \end{cases} \tag{7.29}$$

So the coordinates of the point x satisfy the equation

$$\begin{vmatrix} x_0 & x_1 & x_2 \\ x_1 & x_2 & x_3 \\ x_2 & x_3 & x_4 \end{vmatrix} = 0 \tag{7.30}$$

(otherwise the system (7.29) in the two unknowns $t_1 + t_2$, $t_1 t_2$ would have no solution). The variety swept out by the chords of C^4 is thus a cubic hypersurface of $\mathbb{P}^4$, of equation (7.30).

7.5.6. Prove that for any C^4 in $\mathbb{P}^4$ the ruled surface formed by its tangents is a surface of order 6.

The order of the ruled surface of the tangents of C^4 is the number of the tangents of C^4 that meet a plane π. Each of these tangents gives, when joined with π, a tangent hyperplane to C^4. Conversely, a tangent hyperplane to C^4 in a point P passes through the tangent to C^4 at P, which meets every plane π contained in that hyperplane. We then consider the pencil of hyperplanes having π as axis and we say that two points of C^4 correspond when they belong to the same hyperplane in the pencil. On the curve C^4, the locus of the point $P(t) = [t^4, t^3, t^2, t, 1]$ (or, if one prefers, on the line on which the parameter t varies) one obtains a symmetric algebraic correspondence of indices $(3, 3)$, cf. §1.1.3. The six fixed points of that correspondence are the points of contact of the tangents of C^4 that meet π. The order of the ruled surface is therefore 6.

7.5.7. *Let Γ be a rational quartic curve in $\mathbb{P}^3$. A point P of Γ is stationary if the osculating plane to Γ at P is hyperosculating (which means that the four intersections with Γ are all absorbed by P). Show that Γ possesses four stationary points and ∞^1 trisecants, lies on a unique quadric Q and is met by the generators of one of the two rulings of Q in three points, and by those of the other ruling in a point (that is, Γ is a curve of type $(3, 1)$ on Q, cf. Section 7.3)*.

The curve Γ is the projection of a curve C^4 in $\mathbb{P}^4$ from a point A onto a hyperplane S_3 of $\mathbb{P}^4$, cf. Theorem 7.4.2. If A is a generic point of $\mathbb{P}^4$, then Γ is a rational quartic of general type in $\mathbb{P}^3$. It possesses four stationary points coming from the four osculating planes of C^4 that pass through A; and it has ∞^1 trisecant lines (which sweep out a surface S) which are the traces on S_3 of the ∞^1 trisecant planes passing through A (cf. Exercise 7.5.4).

For nine arbitrary points of Γ there passes a quadric $Q \subset \mathbb{P}^3$ containing Γ. This is the unique quadric that contains Γ because every quadric passing through Γ must contain all the trisecant lines of Γ (since it has at least three intersections with those trisecants).

Thus a rational quartic in $\mathbb{P}^3$ lies on a unique quadric Q, the locus of the trisecants, and it is met by the lines of one of the two rulings of Q in three points, and in only one point by the lines of the other ruling, which is to say that it is a curve of type $(3, 1)$ on Q.

7.5.8. *Study the quartic curve Γ in $\mathbb{P}^3$ that is obtained as the projection of a rational normal curve C^4 from a generic point of the hypersurface X_3^3 of the chords of C^4.*

Suppose that A is a point lying on the variety $W = X_3^3$ of the chords, but not on the ruled surface X_2^6 of the tangent lines to C^4 (cf. Exercises 7.5.5, 7.5.6). Since C^4 has no quadrisecant planes, through A there passes a single chord r of C^4 (if one had a quadrisecant plane, the space S_3 joining the plane with a point P of C^4 not contained in the plane would have five intersections with C^4). The trace O of the line r on the hyperplane $\pi = \mathbb{P}^3$ is the unique double point of Γ.

If M and N are the points at which r meets C^4, the ∞^1 trisecant planes that pass through A all contain the line that joins the points A, M, and N (cf. Exercise 7.5.4) and thus their trace in π passes through O. Three of these lines can not lie in a common plane of π passing through O. Indeed, each of them meets Γ, outside of O, in a further point, and therefore a plane of π passing through O and containing three of those lines would give, when joined to A, a 5-secant hyperplane (in the points M, N, H, P, Q; see Figure 7.8). Therefore the trisecant planes of C^4 that pass through A meet π along the generators of a quadric cone with vertex O. Those generators are unisecants of Γ (without counting the intersections at the vertex O).

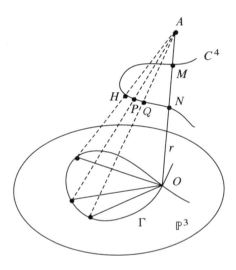

Figure 7.8

In this case the quartic Γ (that has O as a double point) lies on ∞^1 quadrics passing through O; and all these quadrics contain Γ. Hence Γ is a base curve of a pencil of quadrics having a fixed tangent plane at O (for this, see Remark 6.3.9).

7.5.9. *Study the rational plane quartics as projections of a curve C^4 of $\mathbb{P}^4$.*

The rational plane quartics are obtained by projecting the curve C^4 from a line r not meeting C^4 onto a plane π skew to r. The various cases arise in correspondence with the various positions of r with respect to the variety $W = X_3^3$ of the chords of C^4 (cf. Exercise 7.5.5). If r is a generic line which therefore meets W in three distinct points, the projection is a quartic Γ that has as double points the traces on π of the three chords of C^4 that meet r. If r belongs to a trisecant plane of C^4 the quartic Γ has a triple point.

Since the order of the ruled surface X_2^6 of the tangents to C^4 is 6 (cf. Exercise 7.5.6), the plane that joins r with a point R of π meets six tangents of C^4 (each in one of the six points in which the plane meets the surface X_2^6). Each of these tangents has as its projection onto π a tangent of Γ issuing from R. Hence, a plane rational quartic is a curve having class $\nu = 6$.

7.5.10. Prove that the homogeneous ideal of the polynomials in $\mathbb{C}[x_0,\ldots,x_n]$ vanishing on the curve C^n in $\mathbb{P}^n$ which is the locus of the point $P(t)=[t^n, t^{n-1},\ldots,t,1]$ is generated by the second order minors of the matrix

$$\begin{pmatrix} x_0 & x_1 & \cdots & x_{n-1} \\ x_1 & x_2 & \cdots & x_n \end{pmatrix}.$$

Let $I(n)$ be the homogeneous ideal of the polynomials vanishing on C^n and let I_k be the ideal generated by the minors of the matrix

$$\begin{pmatrix} x_k & x_{k+1} & \cdots & x_{n-1} \\ x_{k+1} & x_{k+2} & \cdots & x_n \end{pmatrix}.$$

We prove that $I_0 = I(n)$. The assertion certainly holds for $n = 2$.

Since I_0 is generated by elements of the type $x_i x_{j+1} - x_{i+1} x_j$, $i, j = 0,\ldots, n-1$, which vanish on the points of C^n, one has $I_0 \subseteq I(n)$.

To prove that $I(n) \subseteq I_0$ we may proceed by induction on n assuming that $I(n-1) \subseteq I_0$.

Let $f_m \in I(n)$ be a form of degree m. Each monomial in which the variable x_0 appears may be modified modulo I_0 bearing in mind that $x_0 x_i - x_1 x_{i-1} \in I_0$ and so $x_0 x_i = x_1 x_{i-1}$ modulo I_0. Via successive modifications one arrives at a form $g_{m-1}(x_0, x_1)$ and at a form $g'_m(x_1, x_2, \ldots, x_n)$ such that

$$f_m - x_0 g_{m-1}(x_0, x_1) - g'_m(x_1, x_2, \ldots, x_n) \in I_0. \tag{7.31}$$

But $I_0 \subseteq I(n)$. Hence $x_0 g_{m-1}(x_0, x_1) + g'_m(x_1, x_2, \ldots, x_n) \in I(n)$, that is,

$$t^n g_{m-1}(t^n, t^{n-1}) + g'_m(t^{n-1},\ldots,1) = 0 \quad \text{for all } t. \tag{7.32}$$

Now, $t^n g_{m-1}(t^n, t^{n-1})$ contains only terms that have degree $\geq n+(m-1)(n-1) = mn - m + 1$ with respect to t; while the terms of $g'_m(t^{n-1},\ldots,1)$ are all of degree $\leq m(n-1) = mn - m$. Thus from (7.32) one obtains

$$t^n g_{m-1}(t^n, t^{n-1}) = 0 \quad \text{for all } t,$$

and also

$$g'_m(t^{n-1},\ldots,1) = 0 \quad \text{for all } t.$$

From the first condition it follows that $g_{m-1}(x_0, x_1)$ is the null polynomial, and from the second, on using the inductive hypothesis, one has $g'_m(x_1, x_2, \ldots, x_n) \in I(n-1) \subseteq I_0$. Equation (7.31) then implies that $f_m \in I_0$ and therefore $I(n) \subseteq I_0$, which was to be proved.

7.5.11 (Space curves in $\mathbb{A}^3$ as the set-theoretic intersection of three surfaces). Let $\mathcal{C}$ be an algebraic curve in $\mathbb{A}^3$, where x, y, and z are the coordinates. Suppose that $\mathcal{C}$ does not contain (as a component) any line parallel to the z-axis.

We observe first that the contracted ideal $I(\mathcal{C}) \cap \mathbb{C}[x, y]$ is principal (cf. Section 4.4). Indeed, let f_1, f_2 be polynomials in $\mathbb{C}[x, y] \cap I(\mathcal{C})$. One has $(f_1, f_2) \subset I(\mathcal{C})$ and so $V(f_1, f_2) \supset \mathcal{C}$. Then f_1, f_2 are both multiples of the same polynomial, since otherwise the equations $f_1 = f_2 = 0$ would give a set of lines parallel to the z-axis which contains $\mathcal{C}$ (and hence $\mathcal{C}$ would split in some of them).

Hence $I(\mathcal{C}) \cap \mathbb{C}[x, y] = (g)$ and $g = 0$ is the equation of a curve in the $\langle x, y \rangle$-plane which is the closure of the projection of $\mathcal{C}$ in that plane with axis of projection the z-axis (note that under the projection some points may be missing, for example those corresponding to vertical asymptotes).

Now let $\mathcal{M}$ be the set of polynomials of $I(\mathcal{C})$ having positive degree with respect to z and let $h = az^n + bz^{n-1} + \cdots + c \in \mathcal{M}$ be a polynomial of minimal degree $n(> 0)$ in z, with $a, b \ldots, c \in \mathbb{C}[x, y]$. Let $f = Az^m + Bz^{m-1} + \cdots + C$ be a polynomial of $I(\mathcal{C})$, with $A, B, \ldots, C \in \mathbb{C}[x, y]$. If in $\mathbb{C}[x, y][az]$ we divide the polynomial

$$fa^m = A(az)^m + Ba(az)^{m-1} + \cdots + a^m C$$

by the polynomial

$$ha^{n-1} = (az)^n + b(az)^{n-1} + \cdots + ca^{n-1},$$

we get a relation of the form

$$fa^m = ha^{n-1}q + r, \tag{7.33}$$

where the degree of r with respect to z is less then the degree n of h. By the minimality of n this degree is zero which is to say that $r = r(x, y) \in \mathbb{C}[x, y]$. So we have $r \in (h, f) \cap \mathbb{C}[x, y]$ whence $r \in (g) = I(\mathcal{C}) \cap \mathbb{C}[x, y]$. On the other hand $h \in I(\mathcal{C})$. Thus the two summands on the right-hand side of equation (7.33) are zero on $\mathcal{C}$. A common point of the surfaces $g = 0$ and $h = 0$ belongs to the surface with equation $a^m f = 0$; and if that point is not a zero of f it must be a zero of a. In conclusion

$$V(g, h) = \mathcal{C} \cup \{\text{lines parallel to the } z\text{-axis}\}.$$

It follows that $\mathcal{C}$ is the (set-theoretic) intersection of three surfaces: it suffices to take a third surface that contains $\mathcal{C}$ but no lines parallel to the z-axis (for example, $\mathcal{C} = V(g, h, h')$ where $h' = 0$ is the equation of the cylinder that projects $\mathcal{C}$ parallel to the axis x).

The same result holds also in the projective case. It suffices to choose the reference system such that the plane at infinity does not contain components of the curve $\mathcal{C}$ in order to reduce to the affine case.

Note. It is not known whether a curve in $\mathbb{A}^3$ or $\mathbb{P}^3$ is always the set theoretic intersection of two surfaces. In the special case of a non-singular (or, more generally, locally complete intersection) curve $\mathcal{C}$ in $\mathbb{A}^3$, it is nowadays a classical fact (see [105], and also [60]) that $\mathcal{C}$ is a set theoretic intersection of two surfaces.

Chapter 8
Linear Series on Algebraic Curves

This chapter is the natural development and completion of Chapter 7, and makes use of some results from Section 9.2. Its aim is to furnish an introduction to the study of the geometry on an algebraic curve in complex projective space.

In Section 8.1 we give some remarks on the theory of branches for an algebraic plane curve endowed with ordinary singularities. Moreover, we prove that an arbitrary irreducible algebraic curve has birational models whose only singularities are multiple points with distinct and coplanar tangents. These considerations allow us to interpret a curve as a collection of linear branches rather than as a set of points. That interpretation permits us to introduce the language of divisors on a curve in an elementary way, without requiring any knowledge of local algebra.

The language of divisors is useful in the study of the geometry on an algebraic curve, and in particular for the introduction of linear series, to which Section 8.2 is dedicated. We define the order and the dimension of a linear series and discuss some of their general properties.

In Section 8.3 we define linear equivalence between divisors, which leads to the notion of a complete linear series. Brill and Noether's Restsatz plays a central role and permits us to introduce the notion of the difference between two linear series.

In Section 8.4 we study the projective image of a linear series in terms of simple series and series composed with an involution, and we give a geometric characterization of complete linear series. The material in this section is closely related to the part of the theory developed in Chapter 6, taken up again in Chapter 10 for the case of surfaces.

In Section 8.5 we introduce the genus of a curve according to Weierstrass, an interpretation that offers notable advantages with respect to the definition used in Section 7.2, which dates back to Riemann, and which made use only of the notion of 1-dimensional linear series on a suitable plane model of the given curve. In particular, the alternative interpretation of the genus that we present here allows quick proofs for several key points in the theory (see, for example, the proof of Proposition 8.5.2).

Moreover, in that section we also discuss one of the central problems in the theory of linear series, related to the notion and interpretation of the index of speciality of a divisor. Furthermore, we interpret the non-singular spatial models and the planar models with only nodes of a given curve in $\mathbb{P}^n$ in terms of the projective images of linear series on the curve, a subject that we shall discuss in Section 9.2.

In Section 8.6 the canonical series of a curve is defined and studied, and its crucial birational invariance is proved. The problem cited above (Problem–Definition 8.5.3) finds an answer in the fundamental theorem of Riemann–Roch, which is extensively discussed in this section for the case of a non-singular curve. This theorem is based on a result due to M. Noether, the Noether reduction theorem (Proposition 8.6.10).

In Section 8.7 we discuss some further properties of the canonical series and its projective image, thereby obtaining a crude classification of the non-singular curves in $\mathbb{P}^n$.

Section 8.8 is a brief account of the algebraic correspondences between two curves. After having stated (without proof) Zeuthen's formulas and the Cayley–Brill correspondence principle, we offer some of their applications and consequences in the form of exercises.

In Section 8.9 we give a brief and elementary introduction to moduli varieties for algebraic curves. We illustrate a heuristic method for calculating the number of classes of birational equivalence for curves of a given genus p (≥ 2). The hyperelliptic case, which presents some particularly notable aspects, is based upon the classification theorem 8.9.2.

Finally, various examples and exercises are collected in Section 8.10, where we also state Halphen–Castelnuovo's bound for the genus of an irreducible curve embedded in $\mathbb{P}^n$.

Excellent expositions of the subjects treated in this chapter may be found in the text [1], and the two memoirs [86], [99]. Severi's text [100] has been a particular font of inspiration. It develops the so-called "quick method" for the study of the geometry of algebraic curves. For a modern and completely general discussion of these matters, including a discussion of the Riemann–Roch problem in the singular case, the reader may consult Serre's text [95] (see also [61]).

8.1 Divisors on an algebraic curve with ordinary singularities

For our purposes in the discussion that follows, it is useful to introduce the notion of a (linear) branch of a given algebraic curve. Here we limit ourselves to the consideration of plane curves, and refer the reader to Severi's text [97, pp. 65–72] for the general case (see also [114, Chapter 2]).

8.1.1 The local study of a plane algebraic curve. If one wishes to study an irreducible plane curve C of order d in a neighborhood of its point O, it is convenient to make use of affine coordinates x, y with origin O. The equation of the curve C, ordered according to increasing powers of x, y, then assumes the form:

$$f_1(x, y) + f_2(x, y) + f_3(x, y) + \cdots = 0, \tag{8.1}$$

where $f_1, f_2, f_3, \ldots$ are homogeneous polynomials in x, y of degrees $1, 2, 3, \ldots$.

It is important to show how the given curve C can be approximated near the point O by simpler algebraic curves.

We begin by supposing that O is a simple point. In this case, to simplify the calculations, we choose the x-axis to coincide with the line whose equation is $f_1(x, y) = 0$, namely the tangent to C at O. The equation (8.1) then assumes the form

$$y + f_2(x, y) + f_3(x, y) + \cdots = 0. \tag{8.2}$$

Since the left-hand side of (8.2) vanishes for $x = y = 0$ but its first partial derivative with respect to y is non-zero at $O = (0, 0)$, equation (8.2) defines, in a neighborhood of $x = 0$, an implicit function $y(x)$ which admits a power series representation of the type

$$y = m_1 x + m_2 x^2 + m_3 x^3 + \cdots \tag{8.3}$$

in that neighborhood.

The coefficients $m_1, m_2, m_3, \ldots$ may be calculated by observing that, on substituting in (8.2) the power series (8.3) in place of y, one must obtain an identity with respect to x. Let

$$f_2(x, y) = b_0 x^2 + b_1 xy + b_2 y^2,$$
$$f_3(x, y) = c_0 x^2 + c_1 x^2 y + c_2 xy^2 + c_3 y^3,$$
$$f_4(x, y) = d_0 x^4 + d_1 x^3 y + \cdots.$$

Then setting the successive coefficients of $x, x^2, x^3, \ldots$ equal to zero one obtains the following equations for the $m_1, m_2, m_3, \ldots$:

$$m_1 = 0, \quad m_2 + b_0 = 0, \quad m_3 + b_1 m_2 + c_0 = 0,$$
$$m_4 + b_1 m_3 + (b_2 m_2 + c_1) m_2 + d_0 = 0, \ldots.$$

These equations allow one to calculate the successive coefficients m_i. Thus one finds

$$y = -b_0 x^2 + (b_0 b_1 - c_0) x^3 + \cdots. \tag{8.4}$$

Carrying out the calculation up to the coefficient m_u of x^u, where u is an arbitrary integer ≥ 2, equation (8.4) gives a *parabola of order* u which, in a sufficiently small neighborhood of O, coincides with the given curve up to infinitesimals of order $u + 1$. At O the two curves have a *contact of order* u; we agree to say that the point O counts $u + 1$ times as a point common to both the curves.

We further remark that if the tangent $\ell : y = 0$ to C at O has intersection multiplicity $m_O(C, \ell) = q + 1$ with C there (cf. Section 4.5), then the polynomials $f_2, f_3, \ldots, f_q$ will all be divisible by y; thus in (8.4) the coefficients of the terms of degree $\leq q$ will be zero. In that case O is a flex of order $q - 1$ ($q \geq 2$) having as its tangent the x-axis and the curve C may be approximated in the neighborhood of O by a parabola of the type

$$y = m_{q+1} x^{q+1} + \cdots + m_u x^u$$

8.1. Divisors on an algebraic curve with ordinary singularities 241

of an arbitrary order $u \geq q + 1$ (and having intersection multiplicity u with C at O).

The points of C in the neighborhood of O are said to form a *linear branch with origin O*.

Suppose now that O is a multiple point for the curve C. Here we consider only the case of a singular point with distinct tangents. The study of the general case is much more complicated (beside [114, Chapter 2], see, for example, [10, §8.2] for indications regarding the case of a double point). On the other hand, as we shall show in §8.1.2, for our purposes it is sufficient to reduce to the case of ordinary singularities.

Let O then be an s-fold point with distinct tangents for the curve C. We will suppose that one of the s tangents at O coincides with the x-axis. The equation of C may then be written in the form

$$y\varphi_{s-1}(x, y) + f_{s+1}(x, y) + f_{s+2}(x, y) + \cdots + f_d(x, y) = 0,$$

where $\varphi_{s-1}, f_{s+1}, \ldots, f_d$ are homogeneous polynomials of degrees equal to their indices; more explicitly, we have

$$y(a_0 x^{s-1} + \cdots + a_{s-1} y^{s-1}) + (b_0 x^{s+1} + \cdots) + (c_0 x^{s+2} + \cdots) + \cdots = 0,$$

for suitable constants $a_0, \ldots, b_0, \ldots, c_0, \ldots$, but where, however, we must have $a_0 \neq 0$.

Consider then a parabola, of arbitrarily large order u, with equation of the type

$$y = m_2 x^2 + m_3 x^3 + \cdots + m_u x^u. \tag{8.5}$$

On seeking the intersections at O of this parabola with the curve C one finds the following equation in x:

$$(m_2 x^2 + m_3 x^3 + \cdots)[a_0 x^{s-1} + a_1 x^{s-2}(m_2 x^2 + \cdots) + \cdots]$$
$$+ [b_0 x^{s+1} + b_1 x^s (m_2 x^2 + \cdots) + \cdots] + [c_0 x^{s+2} + \cdots] = 0.$$

That is,

$$(a_0 m_2 + b_0) x^{s+1} + (a_0 m_3 + a_1 m_2^2 + b_1 m_2 + c_0) x^{s+2} + \cdots = 0,$$

which always has $x = 0$ as $(s + 1)$-fold root. The conditions in order for the root $x = 0$ to have multiplicity $s + 2$, or $s + 3$, or $s + 4, \ldots$, are successively

$$a_0 m_2 + b_0 = 0, \quad a_0 m_3 + a_1 m_2^2 + b_1 m_2 + c_0 = 0, \ldots.$$

Since $a_0 \neq 0$, these conditions permit one to calculate successively all the m_i without limitations, that is, no matter how large u may be.

Thus there exists a well-defined parabola of order $u \geq 2$ and arbitrarily large, having contact of order $s + u - 1$ with C at O; we will say that there exists a linear branch of C with origin O which is *tangent* at O to the x-axis. The simplest of these parabolas is that having equation

$$y = -\frac{b_0}{a_0}x^2.$$

Since O is an ordinary s-fold point, one can repeat the preceding considerations for each of the s tangents of C at O and therefore:

- *The neighborhood of an s-fold point O with distinct tangents $\ell_1, \ell_2, \ldots, \ell_s$ is composed of s linear branches with origin O and tangent at O respectively to the lines $\ell_1, \ell_2, \ldots, \ell_s$.*

In general, if P has coordinates (x_0, y_0), for the branch that has origin at P and with tangent the line $y - y_0 = m(x - x_0)$ one finds a power series in $(x - x_0)$ of the type

$$y = y_0 + m(x - x_0) + a(x - x_0)^2 + \cdots, \tag{8.6}$$

for suitable constants $a, \ldots$. To each point P of the curve there are thereby associated the branches that have P as origin and conversely a branch belonging to C, that is represented by a power series like (8.6) for which one has the identity $f(x, y_0 + m(x - x_0) + a(x - x_0)^2 + \cdots) = 0$, has as its origin the point (x_0, y_0) of the curve. In conclusion:

- *We may conceive of the curve (which by hypothesis is endowed only with ordinary singularities) as a set of linear branches rather than as a set of points.*

8.1.2 (Ordinary models of curves). In Section 9.2 we shall prove that a plane algebraic curve may be transformed by a sequence of quadratic transformations into a plane curve having only ordinary singularities, that is, multiple points with distinct tangents.

Later in this chapter (see Theorem 8.5.7) we shall prove that every irreducible algebraic curve has non-singular models in any projective space of dimension $r \geq 3$, as well as plane models with only nodes (that is, double points with distinct tangents). For the present, we establish the following preliminary result.

Proposition–Definition 8.1.3. *For every $r \geq 2$, an irreducible algebraic curve $\mathcal{C}$ has birational models X embedded in $\mathbb{P}^r$ and having as singularities only multiple points with distinct and coplanar tangents. We will say that the curve X is an* **ordinary model** *of the original curve $\mathcal{C}$.*

Proof. Suppose that $\mathcal{C}$ is embedded in a space S_r with $r > 2$. Let $\mathcal{C}'$ be its projection from a generic S_{r-3} onto a plane and let $\mathcal{C}_0$ be a curve endowed with multiple points having distinct tangents obtained from $\mathcal{C}'$ by way of repeated quadratic transformations. The curve $\mathcal{C}_0$ is a birational model for $\mathcal{C}$.

To have a model embedded in $\mathbb{P}^3$ and with its multiple points having distinct and coplanar tangents it suffices to take

i) a line ℓ in the plane π of $\mathcal{C}_0$ whose intersections with $\mathcal{C}_0$ are all distinct, and two distinct points A, B on ℓ which do not belong to $\mathcal{C}_0$;

ii) a point O not belonging to the plane and a non-degenerate quadric $\mathcal{Q}$ passing through the two lines r_{OA}, r_{OB}.

Consider the stereographic projection from the point O, $\varphi \colon \mathcal{Q} \to \pi$, described in §7.3.1. It sends the quadric $\mathcal{Q}$ onto the plane π. The inverse image $\Gamma := \varphi^{-1}(\mathcal{C}_0)$ of the curve $\mathcal{C}_0$ is a birational model of $\mathcal{C}$ and is endowed with multiple points having distinct coplanar tangents (since at each of the multiple points the tangents of Γ there belong to the corresponding tangent plane to $\mathcal{Q}$).

To obtain a birational model X of $\mathcal{C}$ endowed with singular points having distinct and coplanar tangents and embedded in $\mathbb{P}^r$ it suffices to consider a rational normal ruled surface $\mathcal{F}^{r-1} \subset \mathbb{P}^r$ of order $r-1$, and to project onto it the plane curve $\mathcal{C}_0$ from the space S_{r-3} joining $r-2$ generic points of $\mathcal{F}^{r-1}$. For the details we refer to the exposition given in Section 10.3. □

Thus in the sequel when we wish to study the geometry of an irreducible algebraic curve, to which end it is useful to introduce the language of divisors, we may always refer to a non-singular curve or to its ordinary model X, endowed with only singularities having distinct (coplanar) tangents.

If the curve X is non-singular, a *divisor* is defined as an element of the *free abelian group*, $\mathrm{Div}(X)$, *generated by the points of* X. This group is called the *group of Weil divisors* of X.

The question is not so easy when the curve is allowed to have arbitrary singularities. In this regard a complete discussion may be found in Chapter IV of Serre's text [95].

However, in the case of curves whose multiple points all have distinct tangents (so that an s-fold point may be regarded as s superimposed simple points) we can introduce divisors in an elementary fashion, without recourse to any knowledge of local algebra.

8.1.4 (Divisors: the case of plane curves). Let X be a plane curve with ordinary singularities. An *effective divisor* is defined as a formal sum of type

$$D = n_1 \mathcal{R}_1 + n_2 \mathcal{R}_2 + \cdots,$$

where $\mathcal{R}_1, \mathcal{R}_2, \ldots$ are branches belonging to X and the $n_1, n_2, \ldots$ are non-negative integers. By a *divisor* we mean an element of the free abelian group $\mathbf{D}(X)$ generated by the effective divisors. Since the sum of effective divisors is an effective divisor, every divisor turns out to be the difference of two effective divisors. Thus, a divisor D is a formal sum of the type $D = \sum_i n_i \mathcal{R}_i$ with $n_i \in \mathbb{Z}$. We define the *degree* of D to be the integer $\deg(D) = \sum_i n_i$. The *support* $\mathrm{Supp}(D)$ of the divisor D is the set of points of X which are origins of its branches $\mathcal{R}_1, \mathcal{R}_2, \ldots$.

According to the usual modern terminology, we say that an effective divisor D is an *effective Cartier divisor* if, for each point $P \in \mathrm{Supp}(D)$, there exists a regular (polynomial) function φ_P that locally defines D at P. The necessary and sufficient condition in order for this to occur is that for each $P \in \mathrm{Supp}(D)$ all the branches of the curve X with origin P appear in D with positive coefficient.

In general, a divisor is a *Cartier divisor* if for each point $P \in \mathrm{Supp}(D)$ there exists a rational function φ_P that locally defines D at P. The set of Cartier divisors of X constitutes a group, called the *group of Cartier divisors* of X and denoted by $\mathrm{Cart}(X)$. The degree of a divisor defines a homomorphism $\deg \colon \mathrm{Cart}(X) \to \mathbb{Z}$.

If the curve X is non-singular, the Weil group and the Cartier group coincide, and, more precisely, both coincide with the group $\mathbf{D}(X)$ defined above. In fact in such a case each linear branch is identified with the point of X which is the origin of it. In the singular case, the group $\mathbf{D}(X)$ can be interpreted as the group of (Weil or Cartier) divisors of a suitable non-singular birational model of the curve X.

8.1.5 (Intersection multiplicity of a curve with a linear branch). In the same plane as the curve X let us consider another irreducible algebraic curve Γ. Let P be a common point of the two curves and let $\mathcal{R}$ be a branch belonging to X. We define the *intersection multiplicity* $m_P(\Gamma, \mathcal{R})$ of Γ with a linear branch $\mathcal{R}$ of the curve X having P as origin in the following way.

We assume that the coordinate system is chosen so that P is the coordinate origin and so that the tangent to $\mathcal{R}$ at P has the equation $y = 0$, and we suppose that the branch $\mathcal{R}$ is then represented by the power series

$$y = ax^2 + bx^3 + cx^4 + \cdots,$$

with $a, b, c, \ldots$ suitable constants, and moreover that the curve Γ has equation

$$g(x, y) = g_t(x, y) + g_{t+1}(x, y) + \cdots,$$

where $g_j(x, y)$ is a homogeneous polynomial of degree j (so that P is a point of multiplicity t for Γ). If, for some integer ε and suitable constants $k_0, k_1, k_2, \ldots$ with $k_0 \neq 0$, one has

$$g(x, ax^2 + bx^3 + cx^4 + \cdots) = x^{t+\varepsilon}(k_0 + k_1 x + k_2 x^2 + \cdots),$$

then we put

$$m_P(\Gamma, \mathcal{R}) := t + \varepsilon.$$

If $\mathcal{R}_1, \mathcal{R}_2, \ldots, \mathcal{R}_s$ are the s branches of X that have as origin an s-fold point P of the curve X, the divisor

$$D = m_1 \mathcal{R}_1 + m_2 \mathcal{R}_2 + \cdots + m_s \mathcal{R}_s \quad (m_j > 0)$$

can be represented locally by a polynomial $\varphi_P(x, y)$ such that, for each $i = 1, 2, \ldots, s$, one has that m_i is the intersection multiplicity at P of the branch $\mathcal{R}_i$ with the curve having equation $\varphi_P(x, y) = 0$. The intersection multiplicity of X in P with the curve $\Phi_P := \varphi_P(x, y) = 0$ is the sum of the intersection multiplicities of Φ_P with the individual branches $\mathcal{R}_i$ having origin P; that is,

$$m_P(X, \Phi_P) = \sum_{i=1}^{s} m_P(\Phi_P, \mathcal{R}_i).$$

8.1.6 (The case of a space curve). The extension to the case in which X is not a plane curve does not present any substantial new difficulties, once one has a suitable representation for a linear branch $\mathcal{R}$ belonging to X and with origin at a given s-fold point P.

We limit ourselves to some brief remarks, considering for example the case in which X is embedded in $\mathbb{P}^3$. By what has been seen in §8.1.2, we may suppose that X belongs to a non-singular quadric $\mathcal{Q}$ and that it has only multiple points with distinct coplanar tangents. We take a system of affine coordinates x, y, z such that

i) the point P is the coordinate origin;

ii) the tangent to the branch $\mathcal{R}$ at P has equations $x = y = 0$;

iii) the s tangents to X at P are contained in the plane $z = 0$.

If the quadric $\mathcal{Q}$ has equation $z = xy$ and if

$$\begin{cases} z = 0, \\ y = ax^2 + bx^3 + cx^4 + \cdots \end{cases}$$

is the projection of $\mathcal{R}$ in the plane $z = 0$ from the point at infinity on the z-axis, then the branch $\mathcal{R}$ has the representation:

$$\begin{cases} y = ax^2 + bx^3 + cx^4 + \cdots, \\ z = x(ax^2 + bx^3 + cx^4 + \cdots). \end{cases}$$

If $g(x, y, z) = 0$ is the equation of a surface $\mathcal{G}$ having P as t-fold point, and if, for some choice of constants $\varepsilon \geq 1$ and $k_0 \neq 0$,

$$g(x, ax^2 + bx^3 + cx^4 + \cdots, x(ax^2 + bx^3 + cx^4 + \cdots)) \\ = x^{t+\varepsilon}(k_0 + k_1 x + k_2 x^2 + \cdots),$$

then the intersection multiplicity at P of the surface $\mathcal{G}$ with the linear branch $\mathcal{R}$ is

$$m_P(\mathcal{G}, \mathcal{R}) := t + \varepsilon.$$

8.2 Linear series

Let X be an ordinary model, in the sense specified in Proposition 8.1.3, of a given irreducible algebraic curve. We assume that X is embedded in a projective space $\mathbb{P}^n$.

We introduce the notion of linear series by extending the definition of 1-dimensional linear series introduced in Section 7.2 (see also §6.3.7).

Thus we consider an arbitrary linear system Σ of algebraic hypersurfaces

$$\Sigma : \lambda_0 \varphi_0(x_0, \ldots, x_n) + \lambda_1 \varphi_1(x_0, \ldots, x_n) + \cdots + \lambda_t \varphi_t(x_0, \ldots, x_n) = 0,$$

and we suppose that the curve X is not contained in the base variety of Σ. Every hypersurface F of Σ not passing through X intersects X properly and cuts out on X (counting the intersection multiplicities) a divisor $X \cap F = \sum_{i=1}^{k} n_i P_i$ of degree $d := \sum_{i=1}^{k} n_i$ ($= \rho \deg(X)$, where ρ is the degree of the hypersurfaces of Σ). Moreover we suppose that the hypersurfaces of Σ meet X in points not all of which are fixed as $\lambda_0, \ldots, \lambda_t$ vary (that is, they are not all contained in the base variety of Σ) so that one has ∞^r divisors on X, where r is the number of parameters on which the determination of one of its divisors depends. We will say that these are the divisors of a *linear series* g_d^r, of *dimension* r and *order* d. We shall also say that this g_d^r is the *series cut out on X by the linear system* Σ (and, obviously, in this case, $d > 0$). The notation g_d^r was introduced by M. Noether [70].

A point belonging to the support of each of the divisors of such a g_d^r is called a *fixed point* of g_d^r. In most cases one neglects possible fixed points or at least some of them. The fixed points, if there are any, can then be considered in all or in part as making up part of every divisor of the series. Hence, in general, if $D = \sum_i q_i Q_i + \sum_j m_j M_j$ is a divisor of the series, and the M_j are the fixed points, we will also say that the system Σ cuts out (away from the fixed points) a linear series g_q^r of order $q = \sum_i q_i$.

If $\dim \Sigma = t$, one obviously has $r \leq t$; but one can not always affirm that $r = t$. This equality certainly holds in the case when every divisor of the g_d^r belongs to only one hypersurface of Σ so that one has a one-to-one correspondence between the hypersurfaces of Σ and the divisors of the g_d^r.

If, however, two different hypersurfaces F_1, F_2 of Σ cut out the same divisor D of the g_d^r on X, the hypersurface of the pencil $\lambda F_1 + \mu F_2 = 0$ containing a generic point P of X will also contain the whole curve X since it has more intersections with X than allowed by Bézout's theorem. Therefore, the situation in which the correspondence between the hypersurfaces of Σ and the divisors of the g_d^r is not bijective can occur when Σ contains some hypersurface containing X.

Let $\mathbb{H}$ be the linear system of hypersurfaces of Σ that contain X and let h (≥ -1) be its dimension. Note that imposing on a hypersurface the condition that it pass through a curve X requires imposing a certain number of linear conditions. Therefore the hypersurfaces of a linear system that pass through a curve themselves form a linear system. If there are no such hypersurfaces, we will have a linear system of dimension -1.

For a divisor D of g_d^r there pass ∞^{h+1} hypersurfaces of Σ. They constitute the linear system joining $\mathbb{H}$ with a hypersurface of Σ having equation $\psi = 0$ and which cuts out the divisor D on X. Indeed, if $\varphi_0, \ldots, \varphi_h$ is a base of $\mathbb{H}$, the hypersurfaces $\lambda_0 \varphi_0 + \cdots + \lambda_h \varphi_h + \lambda_{h+1} \psi = 0$ all intersect the curve X in the divisor D cut out by the hypersurface $\psi = 0$. Moreover, there can not pass through D more that ∞^{h+1} hypersurfaces of Σ. Otherwise, they would form a linear system $\mathbb{H}_0$ of dimension at least $h + 2$ and the system of dimension at least $h + 1$ consisting of the hypersurfaces of $\mathbb{H}_0$ passing through a generically fixed point of X would be contained in $\mathbb{H}$.

Bearing this in mind, we may suppose that of the $t + 1$ linearly independent forms $\varphi_0, \ldots, \varphi_t$ that generate Σ there are $h + 1$ in $\mathbb{H}$. The $t - h$ remaining forms define a linear system $\mathbb{L}$, of dimension $t - h - 1$, which can not have forms in common with $\mathbb{H}$, since otherwise, the systems $\mathbb{H}$ and $\mathbb{L}$, would, by Grassmann's formula, both belong to a linear system of dimension at most $t - 1$ and containing Σ. The forms of $\mathbb{L}$, none of which pass through the curve X, meet X in the divisors of the series g_d^r cut out by Σ, and there is a bijection between the forms of $\mathbb{L}$ and the divisors they cut out on X. In particular, $r = \dim \mathbb{L}$.

Thus we have proved the following result.

Theorem 8.2.1. *Let X be a curve of $\mathbb{P}^n$, Σ a linear system that cuts out on X a linear series g_d^r, and let $\mathbb{H}$ be the linear system of dimension h consisting of the forms of Σ passing through X. It is always possible to obtain g_d^r on X by way of a linear system $\mathbb{L}$ which is contained in Σ and whose hypersurfaces are in bijective correspondence with the divisors of the given g_d^r. Moreover, one also has the relation*
$$r = \dim \Sigma - h - 1.$$

The fact that a series g_d^r can always be thought of as cut out on X by a linear system $\mathbb{L}$ of dimension r has as a consequence the following facts.

(1) For r generic points of X there passes one and only one divisor of the given g_d^r.

(2) The divisors of g_d^r that pass through a *generic* point P of X form a linear series g_d^{r-1} having P as fixed point. On removing P one has a series g_{d-1}^{r-1} which is called the *residual series of P with respect to g_d^r*.

  More generally, the divisors of a series g_d^r passing through $s \le r$ generic points of X form a linear series g_d^{r-s}. On removing the s points (that are fixed points for this series which may, however, have other fixed points), one obtains the series g_{d-s}^{r-s} residual to these s points with respect to the given g_d^r.

From this it obviously follows that
$$r \le d. \tag{8.7}$$

One then sees immediately that $r = d$ if and only if X is a rational curve. Indeed, the residual series of a divisor made up of $d-1$ generic points with respect to a series g_d^d is a series g_1^1 which, as we have seen above, can be cut out on X by the hypersurfaces of a pencil.

8.3 Linear equivalence

The theory of linear series on a curve is based on the notion of linear equivalence between divisors which we now introduce.

In the course of this section X will denote a curve in $\mathbb{P}^n$, which we suppose to be an ordinary model of a given irreducible algebraic curve. On X we consider two (possibly coinciding) effective divisors A and B both having the same degree d. We will say that A and B are *linearly equivalent*, and we will write $A \equiv B$, if there exists a divisor M on X and two (possibly coinciding) hypersurfaces of the same order ρ that intersect X respectively in the divisors $A+M$ and $B+M$. This is equivalent to saying that the two divisors A and B belong to a common r-dimensional linear series g_d^r on X, $r \geq 1$. For example, one may take the series cut out on X (away from the fixed divisor M) by the linear system Σ of all hypersurfaces of order ρ passing through M. Thus, A and B are linearly equivalent if and only if they are both level divisors of the same rational function on X, namely the rational function arising as quotient of the corresponding two elements of Σ (cf. Section 7.2).

The relation just introduced is clearly reflexive and symmetric. One also sees immediately that if $A \equiv B$ and $B \equiv C$ one also has $A \equiv C$. Indeed, suppose that $f_1 = 0$ and $f_2 = 0$ are two hypersurfaces of the same order that meet X in the two divisors $A + M$ and $B + M$; and suppose that $g_1 = 0$ and $g_2 = 0$ are two hypersurfaces, these too also having the same order, that meet X in the two divisors $B + L$ and $C + L$. Consider the three hypersurfaces of equations $f_1 g_1 = 0$, $f_2 g_1 = 0$, and $f_2 g_2 = 0$. They have the same order, and intersect X respectively in the divisors

$$A + (B + M + L), \quad B + (B + M + L), \quad C + (B + M + L).$$

Thus, $A \equiv C$; moreover, the divisors A, B, and C can be cut out on X, apart from a fixed divisor, by three hypersurfaces of the same order.

In the set of divisors of a given order d, we have thus defined an equivalence relation. The equivalence class of an effective divisor A of degree d is called a *complete series of order d* and is denoted by $|A|$. Moreover, one sees that the (effective) divisors of $|A|$ can be obtained by intersecting the curve X, apart from a fixed divisor, with a linear system of hypersurfaces.

By the above it follows that

- two effective divisors of the same degree d are linearly equivalent if and only if they both belong to a common linear series of order d.

8.3. Linear equivalence

Hence, in particular, a series g_d^r is entirely contained in a single complete one. More precisely, we have:

Lemma 8.3.1. *Let g_d^r be a linear series on the curve X. If g_d^r is not a complete series, then it is contained in a series g_d^{r+1}. In particular, a linear series is contained in a complete series. Moreover a complete series of given order is a linear series.*

Proof. If g_d^r is not a complete series, then there exists a divisor B (of order d) outside of g_d^r which is equivalent to all the divisors of g_d^r. Suppose that g_d^r is cut out by the linear system of hypersurfaces

$$\lambda_0 \varphi_0 + \lambda_1 \varphi_1 + \cdots + \lambda_r \varphi_r = 0,$$

and that $\varphi_0 = 0, \varphi_1 = 0, \ldots, \varphi_r = 0$, respectively, cut out on X the divisors $A_0 + L, A_1 + L, \ldots, A_r + L$. Since $A_0 \equiv B$ there exist two forms of the same order, of equations $\alpha = 0$ and $\beta = 0$, which cut out on X the divisors $A_0 + M$ and $B + M$. For $j = 1, 2, \ldots, r$, the forms with equations $\alpha \varphi_0 = 0$, $\beta \varphi_0 = 0$, and $\alpha \varphi_j = 0$ all have the same order. On X they cut out, respectively, the divisors $A_0 + (A_0 + M + L)$, $B + (A_0 + M + L)$, $A_j + (A_0 + M + L)$, $j = 1, 2, \ldots, r$. Therefore the $(r+1)$-dimensional linear series g_d^{r+1} cut out on X by the linear system

$$\Sigma := \mu \varphi_0 \beta + \lambda_0 \alpha \varphi_0 + \cdots + \lambda_r \alpha \varphi_r = 0$$

contains (outside of the divisor $A_0 + M + L$) the series g_d^r.

Iterating this argument, one sees that a linear series is contained in a complete series (which, in turn, is a linear series).

Assuming this, let $|A|$ be a complete series of order d, and let r be the maximum dimension of the linear series of order d that are contained in $|A|$. One sees immediately that there exists one and only one g_d^r contained in $|A|$, and that it coincides with $|A|$. Indeed, there can be no divisor of degree d in $|A|$ which does not belong to g_d^r because such a divisor would, as an element of $|A|$, be equivalent to all the divisors of g_d^r, but would also be contained in a larger linear series of order d and dimension $r + 1$ still contained in $|A|$. $\square$

From Lemma 8.3.1 we deduce the following statements:

(1) The complete linear series that contains a given linear series is unique.

(2) An arbitrary divisor A of X determines the complete linear series $|A|$ that contains it. In particular, if on the curve X there exist no divisors equivalent to A, then this divisor by itself constitutes a complete linear series, $|A| = g_d^0$, of *dimension zero*.

8.3.2 Sum and difference of linear series. We now introduce the notions of the sum and difference of two complete linear series. In this regard, let A, B, C, D be effective divisors (possibly non-distinct) on the curve X. The notion of linear equivalence leads us to conclude that if $A \equiv B$ and $C \equiv D$ then one has $A + C \equiv B + D$. Indeed, if $A \equiv B$ one has $A + C \equiv B + C$. To see this it suffices to note that if $f_1 = 0$, $f_2 = 0$ are two hypersurfaces that cut out on X (outside of a possible fixed divisor L) the divisors A and B respectively, and if the hypersurface $g = 0$ cuts out the divisor $C + M$, then the two hypersurfaces $f_1 g = 0$ and $f_2 g = 0$ give respectively the divisors $A + C + (L + M)$ and $B + C + (L + M)$. Similarly, from $C \equiv D$ it follows that $B + C \equiv B + D$.

This leads us to the notion of the *sum series* of two complete linear series $|A|$, $|B|$. It is the complete series defined by a divisor of the form $A + B$, with $A \in |A|$, $B \in |B|$, and is denoted by $|A + B|$.

We now prove that subtracting equivalent divisors from equivalent divisors one obtains equivalent divisors. In symbols:

$$\text{if } A + B \equiv C + D \text{ and } A \equiv C \text{ then } B \equiv D. \tag{8.8}$$

Indeed, if $f_0 = 0$ and $f_1 = 0$ cut out $A + B + M$ and $C + D + M$ while $g_0 = 0$ and $g_1 = 0$ cut out $A + L$ and $C + L$, the hypersurfaces $f_0 g_1 = 0$ and $f_1 g_0 = 0$ respectively cut out the divisors $A + B + M + C + L$ and $C + D + M + A + L$, that is, $B + (A + M + C + L)$ and $D + (A + M + C + L)$. Thus $B \equiv D$.

Let $|A|$ be a complete series on the curve X and consider (if there are any) all the divisors of $|A|$ that contain a given divisor B (of degree not larger than that of A). Removing B from these divisors one finds equivalent divisors which constitute a linear series. By Theorem 8.2.1, such a residual series is of dimension $\geq \dim |A| - \deg(B)$, with equality if and only if the passage through B imposes independent conditions on the hypersurfaces of the linear system which cuts out $|A|$ on X. This series is called the *residual series of B with respect to $|A|$*.

This residual series turns out to be complete. Indeed, a possible divisor C equivalent to all the divisors of $|A|$ that contain B, but distinct from each of them, when added to B would give a divisor equivalent to A but not contained in $|A|$, and that is not possible because $|A|$ is complete. Hence the residual series is complete. Let then $B_1 \equiv B$ and suppose that there exist divisors of $|A|$ containing B and divisors of $|A|$ containing B_1; then there exist two divisors R, R_1 on X such that $B + R \equiv A \equiv B_1 + R_1$. By (8.8) one has $R \equiv R_1$ and so $|R| = |R_1|$.

Thus we can state the following proposition known as the "Restsatz" of Brill and Noether (cf. [100, p. 105]).

Proposition 8.3.3 (Restsatz). *Let $|A|$ be a complete series on the curve X. The residual divisors R of a given divisor B with respect to the complete series $|A|$ constitute a complete series, $|R|$. Moreover the residual divisors R are also residual divisors with respect to the same series of every other divisor equivalent to B.*

The series $|R|$ in Proposition 8.3.3 is also called the *difference series* of the two series $|A|$ and $|B|$, and one writes $|R| := |A - B|$.

The definition of the difference series $|A - B|$ may be extended also to the case in which a divisor B on the curve X is not contained in some divisor $A \in |A|$. More precisely, for arbitrary effective divisors A, A_1, B, B_1, we decree that

$$A - B \equiv A_1 - B_1 \iff A + B_1 \equiv A_1 + B.$$

In particular, if $A \equiv B$, the divisor $A - B$ is the *zero divisor* and $|A - B|$ is the *zero series*. Note that if A and B are two divisors of the same degree, but are not equivalent, then the series $|A - B|$ has order zero but is not the zero series.

Linear equivalence extends to divisors and the divisor operations pass to the quotient. The quotient group $\operatorname{Pic}(X) := \operatorname{Cart}(X)/\equiv$ is called *group of divisor classes* of X, or the *Picard group* of X.

The difference series of two divisors A and B may then be defined in an equivalent manner as

$$|A - B| := \{D \text{ effective divisor} \mid A \equiv D + B\}.$$

In fact, the set of all *effective* divisors linearly equivalent to a given divisor A (not necessarily effective) is called the *complete linear series defined by A* and is again denoted by $|A|$. In particular, if A is not effective and if there exist no effective divisors equivalent to A, the series $|A|$ reduces to the empty set and has dimension -1.

One notes that the series $|A - B|$ depends only on the complete series $|A|$ and $|B|$ and not on the particular divisors $A \in |A|$ and $B \in |B|$. We say that a linear series g_d^r on the curve X is *complete* if $g_d^r = |A|$ for some divisor A on X.

We note also the following elementary but useful fact.

Lemma 8.3.4. *Let A be a divisor on the curve X. If $\dim |A| \geq 0$, one has $\deg(A) \geq 0$. Moreover, if $\dim |A| \geq 0$ and $\deg(A) = 0$, then A is a divisor linearly equivalent to zero.*

Proof. If $\dim |A| \geq 0$, the complete linear series $|A|$ is not the empty set. Hence A is linearly equivalent to some effective divisor A'. Since linearly equivalent divisors have the same degree, one has $\deg(A) = \deg(A') \geq 0$. If moreover $\deg(A) = 0$, then A' is an effective divisor of degree zero. But the only effective divisor of degree zero is, up to linear equivalence, the null divisor. $\square$

8.4 Projective image of linear series

In the sequel we consider a curve X lying in a projective space $\mathbb{P}^n$, with homogeneous coordinates $x_0, \ldots, x_n$, and assume that X is an ordinary model for an irreducible algebraic curve. On the curve X we consider the linear series g_d^r cut out by the hypersurfaces of the linear system of dimension r, cf. Theorem 8.2.1,

$$\Sigma: \lambda_0 f_0(x_0, \ldots, x_n) + \cdots + \lambda_r f_r(x_0, \ldots, x_n) = 0.$$

If $y_0, \ldots, y_r$ are homogeneous coordinates in a projective space $\mathbb{P}^r$, a rational map $\varphi: \mathbb{P}^n \to \mathbb{P}^r$ is thereby defined and represented by the equations

$$y_i = f_i(x_0, \ldots, x_n), \quad i = 0, \ldots, r. \tag{8.9}$$

It associates to each point P of $\mathbb{P}^n$, not belonging to the base variety of Σ, a well-defined point $\varphi(P)$ of $\mathbb{P}^r$.

We projectively refer the hypersurfaces of Σ to the hyperplanes of $\mathbb{P}^r$, saying that the hypersurface $\lambda_0 f_0 + \cdots + \lambda_r f_r = 0$ corresponds to the hyperplane $\lambda_0 y_0 + \cdots + \lambda_r y_r = 0$. To the hypersurfaces of Σ that pass through a point $\bar{x} \in \mathbb{P}^n$, that is, to the hypersurfaces that one obtains by imposing the relation $\lambda_0 f_0(\bar{x}) + \cdots + \lambda_r f_r(\bar{x}) = 0$ among the parameters, there correspond the hyperplanes of $\mathbb{P}^r$ whose coefficients are bound by the same relation, namely, the hyperplanes that pass through the point $\bar{y} \in \mathbb{P}^r$ with coordinates $(f_0(\bar{x}), \ldots, f_r(\bar{x}))$.

If P describes the curve X, the point $\varphi(P)$ (if it varies) describes in $\mathbb{P}^r$ an irreducible algebraic curve $\mathcal{X}$ which is called the *projective image* of g_d^r. A parametric representation of $\mathcal{X}$ is given by the equations (8.9), where one understands that the variables $x_0, x_1, \ldots, x_n$ are constrained to satisfy the equations of the curve X. More precisely, if $I(X) \subset K[x_0, \ldots, x_n]$ is the homogeneous ideal of polynomials vanishing on X and $\{p_1, \ldots, p_m\}$ is a system of homogeneous generators of $I(X)$, then the curve $\mathcal{X} \subset \mathbb{P}^r$ is represented by the equations

$$\mathcal{X} := \begin{cases} y_i = f_i(x_0, \ldots, x_n), & i = 0, \ldots, r, \\ p_j(x_0, \ldots, x_n) = 0, & j = 1, \ldots, m. \end{cases}$$

Suppose that in Σ there are no forms containing X. This means that there do not exist non-null $(r+1)$-uples $(\lambda_0, \ldots, \lambda_r)$ for which the equation $\lambda_0 y_0 + \cdots + \lambda_r y_r = 0$ is satisfied by all the points of $\mathcal{X}$. From this one concludes that the space of minimal dimension that contains the curve $\mathcal{X}$ has dimension r. Let us recall that if we replace a basis chosen to define g_d^r with another basis, we find as projective image of g_d^r a curve that is the transform of $\mathcal{X}$ under a projectivity of the ambient space $\mathbb{P}^r$ (cf. Section 6.6).

We observe explicitly that two points $P, P' \in X$, not belonging to the base variety of Σ, have the same image $\varphi(P) = \varphi(P')$ in $\mathcal{X}$ if and only if the $(r+1)$-uples $(f_0(P), \ldots, f_r(P))$, $(f_0(P'), \ldots, f_r(P'))$ are proportional; that is, if and only if the divisors of our g_d^r that pass through P also contain P'.

Two cases are possible, according to whether or not the restriction $\varphi_{|X}: X \to \mathcal{X}$ is generically bijective.

a) A generic point of $\mathcal{X}$ comes from a unique point of X.

In this case the two curves X, $\mathcal{X}$ are birationally equivalent, which means that the restriction $\varphi_{|X}: X \to \mathcal{X}$ is a birational map. Then the divisors of g_d^r containing a generic point of X (varying in the open set on which the map is an isomorphism) do not have in common other points of the curve. In this case we will say that g_d^r is a *simple linear series*. One has:

- The curve $\mathcal{X}$ has order d.

Indeed, the series g_d^r corresponds on $\mathcal{X}$ to the series $\mathcal{H}_{\mathcal{X}}$ cut out by the hyperplanes. But the order of $\mathcal{X}$ is the number of intersections of $\mathcal{X}$ with a generic hyperplane $\sum_{i=0}^r \lambda_i y_i = 0$ of S_r and such points correspond bijectively to the points of the divisor of g_d^r cut out on X by the hypersurface of equation $\sum_{i=0}^r \lambda_i f_i = 0$.

As to the exceptions to bijectivity, several points $P_1, \ldots, P_\mu$ of X, not belonging to the base variety of Σ (and not necessarily distinct), produce the same point of $\mathcal{X}$ if and only if the $(r+1)$-uples

$$(f_0(P_1), \ldots, f_r(P_1)), \ldots, (f_0(P_\mu), \ldots, f_r(P_\mu))$$

are proportional. This means that all the divisors of g_d^r that pass through any one of the points $P_1, \ldots, P_\mu$ must also contain all the others. One then says that $H = \{P_1, \ldots, P_\mu\}$ is a *neutral divisor* with respect to the series g_d^r. The divisor H imposes a single condition on the divisors of g_d^r which are to contain it. All of its points have as corresponding point in $\mathcal{X}$ a single point $P' = \varphi(P_j)$, $j = 0, \ldots, \mu$, which will be an μ-fold point of $\mathcal{X}$.

b) Every point of $\mathcal{X}$ corresponds to $\mu \geq 2$ points of X.

The divisors of order μ, each of which corresponds to a point of $\mathcal{X}$, form an algebraic series σ_μ^1 (in general *not* linear), that is, they constitute a simply infinite totality of divisors of degree $\mu \geq 2$. The series σ_μ^1 is such that every point of X belongs to one and only one of the divisors constituting σ_μ^1. Moreover, the divisor of g_d^r that contains a point P contains the whole divisor of σ_μ^1 to which P belongs. One says in this case that g_d^r is *composed with the involution* σ_μ^1. Every divisor of g_d^r is then made up of $\frac{d}{\mu}$ divisors of the series σ_μ^1, to each of which there corresponds a single point of $\mathcal{X}$. Therefore:

- The curve $\mathcal{X}$ has order $\frac{d}{\mu}$.

The map $\varphi_{|X}: X \to \mathcal{X}$ is only rational and is an μ-fold covering of the curve $\mathcal{X}$.

Thus we have proved the following:

Proposition 8.4.1. *Let g_d^r be a linear series on the curve $X \subset \mathbb{P}^n$ and let $\mathcal{X}$ be the non-degenerate curve in $\mathbb{P}^r$ that is the projective image of g_d^r. If g_d^r is simple, the curve $\mathcal{X}$ is a birational model of X and has order d. Furthermore, the series corresponding to g_d^r is the series $\mathcal{H}_{\mathcal{X}}$ cut out by the hyperplanes of $\mathbb{P}^r$.*

The completeness of a linear series is expressed in geometric terms by the following proposition.

Proposition 8.4.2. *Let g_d^r be a simple linear series on the curve $X \subset \mathbb{P}^n$, and let $\mathcal{X} \subset \mathbb{P}^r$ be the projective image curve of g_d^r. The following are equivalent:*

(1) *The series g_d^r is complete.*

(2) *The curve $\mathcal{X}$ can not be obtained as the projection of a curve of the same order embedded in a projective space of dimension $> r$.*

Proof. If g_d^r is not complete, consider the linear series g_d^m, $m > r$, in which it is (totally) contained. Let $\mathcal{X}^* \subset \mathbb{P}^m$ be the curve, of order $d = \deg(\mathcal{X})$, which is the projective image of g_d^m. On $\mathcal{X}^*$ the given g_d^r corresponds to the series cut out by the r-dimensional linear system of hyperplanes of $\mathbb{P}^m$ passing through a linear space S_{m-r-1} that does not meet $\mathcal{X}^*$ (since the series g_d^r and g_d^m have the same order). Let Γ be the projection of $\mathcal{X}^*$ from S_{m-r-1} into a projective space S_r skew to S_{m-r-1}. There is a bijective correspondence between the projective image $\mathcal{X}$ of g_d^r and the curve Γ under which the hyperplane sections of one correspond to the hyperplane sections of the other. That correspondence is thus induced by a projectivity between the two spaces $\mathbb{P}^r$ and S_r. The curve $\mathcal{X}$ is thus, like Γ, the projection of a curve of the same order belonging to a higher dimensional projective space.

Conversely, suppose that the projective image $\mathcal{X}$ of the series g_d^r (which has order d and is embedded in $\mathbb{P}^r$) were the projection of a curve $\mathcal{X}^*$ embedded in a space S_m of dimension $m > r$. The series $\mathcal{H}_{\mathcal{X}}$ of hyperplane sections of $\mathcal{X}$ is then the projection of the series g^{*r}_d cut out on $\mathcal{X}^*$ by the hyperplane sections that pass through the center of projection, which is contained in the linear series of all the hyperplane sections of $\mathcal{X}^*$. This means that g^{*r}_d (and so too $\mathcal{H}_{\mathcal{X}}$) is not a complete series on $\mathcal{X}$. Consequently, neither is the linear series g_d^r that corresponds to $\mathcal{H}_{\mathcal{X}}$.

For the proof one could also make use of an analytic argument as in Section 6.7 (where one should consider $\mathcal{X}$ and $\mathcal{X}^*$ in place of V and V_0 and r and m in place of h and h_0 respectively, and where the variables $x_0, \ldots, x_n$ are understood to satisfy the equations defining $\mathcal{X}$). □

We will say that the curve X in $\mathbb{P}^n$ is *linearly normal* if the series $\mathcal{H}_X = g^n_{\deg(X)}$ cut out by the hyperplanes of $\mathbb{P}^n$ is a complete series. Since X obviously coincides with the projective image curve of $\mathcal{H}_X$, Proposition 8.4.2 allows us to conclude that a curve $X \subset \mathbb{P}^n$ is linearly normal if it can not be obtained as the projection of a curve of the same order embedded in a projective space of dimension $> n$.

Remark 8.4.3. A given linear series g_d^r on a curve $X \subset \mathbb{P}^n$ may very well have several projective images, not only because there is freedom in the choice of the system of forms whose linear combinations determine the linear system Σ that cuts out the g_d^r on X, but also because a given linear series can be cut out by different linear systems of hypersurfaces.

A projective model (or image) of a series g_d^r may be constructed geometrically by imposing an arbitrary projective correspondence between the divisors of g_d^r and the hyperplanes of S_r. Two projective models of the same g_d^r belonging to two spaces S_r, S_r' thus correspond under the projectivity which makes two hyperplanes that cut homologous divisors correspond to each other.

8.4.4 Birational correspondences between curves. We now consider two irreducible algebraic curves that stand in birational correspondence. To each *generic* (and so non-singular) point of one there corresponds a unique point of the other, but this need not hold for *every* point. The following argument, taken from [100, pp. 79–81], shows a basic fact, important in itself and for the sequel (see also Exercise 8.10.7 for a reformulation of the proof of the second part of the statement of the proposition).

Proposition 8.4.5. *Let $\varphi \colon X \to X'$ be a birational map between two irreducible algebraic curves. Then the following holds.*

(1) *To each non-singular point P of X there corresponds a uniquely determined s-fold point P' of X' with $s \geq 1$; and similarly on interchanging the role of the two curves.*

(2) *If X and X' are non-singular, the map φ is a morphism, and bijective without exceptions.*

Proof. The second statement is an immediate consequence of the first.

We prove (1). By Proposition 8.1.3, we may surely suppose that X and X' are endowed with only ordinary singularities. Suppose that X' is embedded in a projective space S_n and has order d. The hyperplanes of S_n cut out a linear series $\mathcal{H}_{X'}$ on X' which is without fixed points. To $\mathcal{H}_{X'}$ there corresponds on X a linear series g_d^n.

Let P be a non-singular point of X. The divisors of g_d^n containing P form, for any P, a series g_d^{n-1} with base point P. Suppose that the divisors of this series have $d - s$ variable points ($s \geq 1$). Under φ, to g_d^{n-1} there corresponds on X' a linear series that again has order d, and which is contained in the linear series $\mathcal{H}_{X'}$ of hyperplane sections of X'. The linear series thus induced on X' is therefore cut out by the hyperplanes that pass through a well-defined point P' of S_n. Since the divisors of this series have only $d - s$ variable points, the point P' belongs to X' and is an s-fold point of X'. (Note that we do not assert that P' comes from several distinct points of X. For example, if X' is the plane curve obtained by projection of a space curve X from a point O lying on the tangent to X at one of its non-singular points P, the projection P' of P is a cuspidal double point for X' to which there corresponds only the point P.) □

8.5 Special linear series

The method that we have used to introduce the genus p of a curve X (which may be attributed to the work of Riemann) is based on the notion of Jacobian divisor and on the construction and study of a suitable algebraic correspondence between the divisors of 1-dimensional linear series on a plane model $\mathcal{C}$ of X having only ordinary singularities. It presents technical difficulties of some significance which make Section 7.2 rather difficult reading, but also offers the advantage of a direct proof of the crucial fact of birational invariance of p (cf. Theorem 7.2.2 and (7.4)).

For a quick reconstruction of the geometry on a curve X it is particularly convenient to use an alternative definition of the genus, namely the property expressed by a famous theorem of Weierstrass known as the "Lückensatz" (see [100, p. 164]). In 8.5.10 we will show that the two notions of genus do indeed coincide.

Proposition–Definition 8.5.1 (The genus according to Weierstrass). *Let $\mathcal{C}$ be an irreducible algebraic curve and let X be an ordinary model of $\mathcal{C}$. We define the **genus** of X to be the minimal integer p such that $p + 1$ arbitrary points of X are a divisor of a linear series g_{p+1}^r with $r \geq 1$ (while p generic points of X constitute an **isolated divisor**, i.e., a series g_p^0). In particular, the genus p is a **birational invariant**, which allows one to define the **genus of the curve** $\mathcal{C}$ as the genus p of X.*

Notation. In the remainder of this section we use X to denote an ordinary model of a given irreducible algebraic curve.

The following facts are consequences of the definition of the genus of a curve.

Proposition 8.5.2. *Let X be a curve of $\mathbb{P}^n$ of genus p and let g_d^r be a linear series on X. Then*

(1) *$p \geq 0$, and $p = 0$ if and only if the curve X is rational;*

(2) *if g_d^r is complete, one has $r \geq d - p$;*

(3) *if A is a generic divisor consisting of $d \geq p$ points of X, one has $\dim |A| = d - p$.*

Proof. (1) See the comment regarding the equality in formula (8.7), at the end of Section 8.2.

(2) By the Restsatz (Proposition 8.3.3), the residual series of a divisor consisting of $d - p - 1$ generic points with respect to a complete series g_d^r is a complete series

of dimension $r - (d - p - 1)$ and order $p + 1$. From Proposition–Definition 8.5.1 it then follows that $r - (d - p - 1) \geq 1$, that is, $r \geq d - p$.

(3) Now let A be a generic divisor consisting of $d \geq p$ points of X. Observe that A can be obtained as a sum $H + B$ of two generic divisors consisting respectively of p and $d - p$ points. Since B imposes $d - p$ independent conditions on the divisors of the series $|A|$ which are to contain it, the residual series of B with respect to $|A|$ has dimension $r - (d - p)$. Bearing in mind that the divisor B consists of p generic points, Proposition–Definition 8.5.1 ensures that one then has $\dim |A| - (d - p) = 0$. Thus $\dim |A| = d - p$. □

The important inequality expressed by Proposition 8.5.2(2) leads naturally to the division of linear series into two types, according to whether or not the equality actually holds. In the first case ($r = d - p$) the linear series is said to be *non-special*; if instead $r > d - p$, the series is *special*.

Problem–Definition 8.5.3. The central problem in the theory of linear series is that of determining the dimension

$$r = d - p + i, \quad i \geq 0,$$

of the complete series defined by a divisor A of degree d on a curve of genus p, or equivalently to calculate the difference $i := r - (d - p)$, which is called the *index of speciality of the series* $|A|$, or also of the *divisor* A (which is said to be *special* if $i > 0$).

We shall return to this problem later (see Theorem 8.6.12).

Two conditions, both of which are sufficient (but *not necessary*) in order for a series g_d^r to be non-special, are given by the following theorem (compare with Proposition 8.6.15).

Theorem 8.5.4. *Let X be a curve of $\mathbb{P}^n$ with genus p and let g_d^r be a linear series on X. Then:*

(1) *If $d > 2p - 2$ the series is non-special.*

(2) *If $r > p - 1$ the series is non-special (and has order $d \geq 2p$).*

Proof. We may surely suppose that the series g_d^r is complete.

(1) Let $d > 2p-2$. Suppose, by way of contradiction, that one has $r \geq d-p+1$. The residual series of a divisor A consisting of p generic points of the curve X would be a complete series g_{d-p}^{r-p} (partially contained in g_d^r), and the residual series of a divisor of g_{d-p}^{r-p} with respect to g_d^r would be a series of order $d - (d - p) = p$ and dimension $\geq r - (d - p) \geq 1$. By the definition of the genus, see Proposition–Definition 8.5.1, this is not possible because in that series there is the divisor A consisting of p generic points.

(2) Suppose now that $r > p - 1$, and put $r = p + e$ ($e \geq 0$). Then, by (8.7), we have $d \geq p + e$. The residual series of a divisor A consisting of p generic points with respect to g_d^r is a complete series g_{d-p}^e (partially contained in g_d^r), and the residual series of a divisor of g_{d-p}^e with respect to g_d^r would be a series of order p and dimension $\geq r - (d - p)$ containing A. Thus, again in view of Proposition–Definition 8.5.1, we have $r - (d - p) = 0$, that is, $r = d - p$. It then follows that $d = r + p > p - 1 + p$, and so $d \geq 2p$. □

One also has:

Theorem 8.5.5. *Let X be a curve in $\mathbb{P}^n$ of genus p and let g_d^r be a linear series on X. If $d > 2p$, the series does not have neutral divisors (and so is simple) and does not have fixed points.*

Proof. By Theorem 8.5.4, the series g_d^r is non-special, and so one has $r = d - p$.

Let G be a neutral divisor consisting of $\mu \geq 2$ points. Since G imposes a single condition on the divisors of g_d^r which contain it, the residual series of G with respect to g_d^r would be a series $g_{d-\mu}^{r-1}$. Such a series would be simultaneously special inasmuch as

$$r - 1 = d - p - 1 > d - p - \mu = (d - \mu) - p,$$

and non-special inasmuch as

$$r - 1 = d - p - 1 > p - 1.$$

Similarly, the residual series g_{d-1}^r of a fixed point with respect to g_d^r would be special because $r = d - p > (d - 1) - p$, and non-special because $r = d - p > p - 1$. □

8.5.6 Non-singular spatial models and plane models with nodes. Consider a complete linear series g_d^r on a curve X of genus p in $\mathbb{P}^n$ and suppose that $d > 2p$. Then by Theorems 8.5.4 and 8.5.5 it is non-special (i.e., $r = d - p$), and does not have neutral divisors. Therefore its projective image $\mathcal{X}$ is a linearly normal curve of order d in S_r and is birationally isomorphic to X. The fact that g_d^r is simple and without neutral divisors ensures moreover that $\mathcal{X}$ does not have multiple points. Indeed, if P were an s-fold point of $\mathcal{X}$ with $s \geq 2$ the residual series of P with respect to the series g_d^r (which is cut out on $\mathcal{X}$ by the hyperplanes of S_r) would be a series g_{d-s}^{r-1} cut out on $\mathcal{X}$ by the hyperplanes of S_r passing through P. This series g_{d-s}^{r-1} would be, simultaneously, special because $r - 1 = d - p - 1 > (d - s) - p$ and non-special because $r - 1 = d - p - 1 > p - 1$.

- Thus it is always possible to transform a curve X birationally into a non-singular curve $\mathcal{X}$ belonging to a projective space of sufficiently large dimension r.

If $r > 3$, a generic S_{r-4} does not meet the three dimensional variety V_3 of the chords of $\mathcal{X}$ (see, in this regard, [48, (11.24), (11.25)]) and so projecting $\mathcal{X}$ from S_{r-4} into a subspace S_3 skew to it, one thereby has a non-singular projected curve $\mathcal{X}'$.

If one then considers a generic point O in S_3, and projects $\mathcal{X}'$ from O onto a generic plane π, one obtains a plane curve $\mathcal{X}_0$ which has as singularities only double points with distinct tangents, called *nodes*. More precisely, the point O should be chosen outside of the ruled surface consisting of the trisecants of $\mathcal{X}'$, outside of the ruled surface consisting of the tangents of $\mathcal{X}'$, and outside of the ruled surface consisting of the lines that join pairs of points of contact of $\mathcal{X}'$ with its bitangent planes. The nodal singularities described above can not be avoided because the chords of a space curve belonging to S_3 cover the entire space.

Applying the previous reasoning to an ordinary model X of a given spatial irreducible algebraic curve one thus obtains the following fundamental result (see also Section 9.2 and Exercise 13.1.21).

Theorem 8.5.7. *An irreducible algebraic curve $\mathcal{C}$ has non-singular birational models belonging to every space $\mathbb{P}^r$ of dimension $r \geq 3$. Moreover, it has plane models whose singularities are only nodes.*

Exercise 8.5.8. An irreducible algebraic curve $\mathcal{C}$ of genus p can be transformed birationally into a (linearly normal) non-singular curve of order $d > 2p$ in the projective space $\mathbb{P}^{d-p}$. Indeed, a curve of that order and genus in $\mathbb{P}^{d-p}$ is necessarily non-singular.

The first statement is an obvious consequence of Theorems 8.5.4 and 8.5.5. The second statement follows from the fact that the series g_d^{d-p} of hyperplane sections (of an ordinary model of $\mathcal{C}$) is non-special in view of Theorem 8.5.4 and so is complete, and it has no neutral divisors again by Theorem 8.5.5.

8.5.9. Consider a complete linear series g_d^r on a curve X in $\mathbb{P}^n$, and let $\mathcal{C}$ be a plane model with only nodes for the curve X. We note explicitly that the curve $\mathcal{C}$ has order d and that under the birational transformation $X \to \mathcal{C}$ the initial series g_d^r is transformed into a linear series of the same dimension (inasmuch as the dimension of a linear series is clearly a birational invariant) and of the same order d.

Indeed, the map $X \to \mathcal{C}$ factorizes into the birational map $\varphi \colon X \to \mathcal{X} \subset \mathbb{P}^r$ and the two successive projections $\mathbb{P}^r \to \mathbb{P}^3 \to \mathbb{P}^2$ with respective centers a generic $S_{r-4} \subset \mathbb{P}^r$ and a generic point $O \in \mathbb{P}^3$. The linear series corresponding to g_d^r in $\mathcal{X}$ is the series $\mathcal{H}_\mathcal{X}$, which obviously has order d. In view of the genericity of the centers of projection, the curve $\mathcal{C} \subset \mathbb{P}^2$ has the same order, d, as $\mathcal{X}$, and to the series $\mathcal{H}_\mathcal{X}$ there corresponds on $\mathcal{C}$ another series still of the same order d (but not, in general, cut out by the lines in the plane).

If the curve X is non-singular, then under the birational transformation $X \to \mathcal{C}$, *every* linear series is transformed in $\mathcal{C}$ into a series of the same order (and same dimension). In fact we know that the birational model $\mathcal{X}$ is non-singular, and so the birational map $\varphi \colon X \to \mathcal{X}$ is a morphism, bijective without exceptions by what we have seen in paragraph 8.4.4. Every given linear series on X is thus transformed in $\mathcal{X}$ into a series of the same order, to which, by the genericity of the centers of projection described above, there corresponds a series of the same order on the plane model $\mathcal{C}$.

8.5.10. Using Theorem 8.5.7, we show here how *the genus introduced in Definition 8.5.1 does indeed coincide with the genus previously introduced in Section 7.2.*

We consider a plane model $\mathcal{C}$ of an algebraic and irreducible curve of genus p according to Definition 8.5.1. Let $\mathcal{C}$ be of order d and let it have $\mathfrak{d}$ nodal double points which make up a divisor D, and let $\mathcal{C}$ have no other singular point. Then let P be a divisor consisting of p generic points of $\mathcal{C}$ and let Γ be an algebraic curve, of arbitrary order N, passing through the points P and the nodes D. We can also suppose that the curves $\mathcal{C}$ and Γ have intersection multiplicity 2 at the double points of $\mathcal{C}$ (cf. Section 4.2). If N is the order of Γ, the curves Γ and $\mathcal{C}$ have in common an additional divisor A consisting of $Nd - 2\mathfrak{d} - p$ points. The linear system Σ_D of the curves of order N that pass through the nodes D has dimension (cf. Example–Definition 6.2.2)

$$\dim \Sigma_D \geq \frac{N(N+3)}{2} - \mathfrak{d}.$$

Moreover, the linear system Σ_0 of the curves of order N that contain $\mathcal{C}$ as component has dimension $\dim \Sigma_0 = \frac{(N-d)(N-d+3)}{2}$. Hence the linear series $\mathcal{D}$ of divisors cut out on $\mathcal{C}$ (outside of the base locus D) by the curves of Σ_D has dimension, cf. 6.3.7,

$$\dim \Sigma_D - \dim \Sigma_0 - 1 \geq \frac{N(N+3)}{2} - \mathfrak{d} - \frac{(N-d)(N-d+3)}{2} - 1$$

$$= Nd - \binom{d-1}{2} - \mathfrak{d}.$$

Among the divisors of this series there is the divisor $A + P$; and the residual series $|R|$ of $|A|$ (with respect to the complete series defined by $\mathcal{D}$) is a series that has among its divisors the divisor P, and which therefore has dimension 0 by Definition 8.5.1. On the other hand, the dimension of $|R|$ is

$$\geq Nd - \binom{d-1}{2} - \mathfrak{d} - \deg(A) = Nd - \binom{d-1}{2} - \mathfrak{d} - (Nd - 2\mathfrak{d} - p)$$

$$= \mathfrak{d} + p - \binom{d-1}{2}.$$

Note that if $N \gg 0$ we may assume that the left-hand inequality is in fact an equality. Hence we have $\mathfrak{d} + p - \binom{d-1}{2} = 0$, that is

$$p = \frac{(d-1)(d-2)}{2} - \mathfrak{d}, \qquad (8.10)$$

a relation that coincides with the expression of the genus as defined by Riemann and introduced in Section 7.2; see in particular 7.2.9.

We explicitly note the following immediate consequence of the above remarks.

Corollary 8.5.11. *A non-singular curve X of genus p has plane models of order d with only $\mathfrak{d} = \frac{(d-1)(d-2)}{2} - p$ nodes.*

8.5.12. Let X be an irreducible curve in $\mathbb{P}^n$ of genus p. If $p = 1$, X is called *elliptic*. If $p \geq 2$ and X contains a series g_2^1, the curve X is called *hyperelliptic*. We will discuss some noteworthy properties of elliptic and hyperelliptic curves in Section 8.7.

8.6 Adjoints and the Riemann–Roch theorem

In this section we illustrate one of the fundamental results of the theory of linear series, the Riemann–Roch theorem, and we discuss some of its consequences. The proof of this theorem is based on the notion of adjoint curve to a given algebraic plane curve, and on the derived notion of canonical series. An essential property of the canonical series, that of being a birational invariant, allows one to extend the theory to include also non-singular curves in projective spaces. The first part of this section is dedicated to these preliminary results.

8.6.1 The adjoints and the canonical series of a plane curve. Let $\mathcal{C}$ be a fixed irreducible algebraic plane curve of order n and genus p, and whose multiple points $P_1, \ldots, P_t$ have multiplicities $s_1, \ldots, s_t$ ($t \geq 0$). We suppose for simplicity that the multiple points are all ordinary (i.e., with distinct tangents), although this last hypothesis is not really necessary; for this see Section 9.2.

An *adjoint curve* (or simply an *adjoint*) of $\mathcal{C}$ is a plane algebraic curve that has multiplicity greater than or equal to $s_i - 1$ at the point P_i for each $i = 1, \ldots, t$.

The adjoints of order m of $\mathcal{C}$ constitute a linear system, $\mathcal{A}_m$. Let σ_m be the superabundance of $\mathcal{A}_m$.

8.6.2. With the notation and assumptions as above, let us summarize here some basic properties of adjoints.

(1) *If there exist adjoints of a given order $m - 1$, then the adjoints of order m cut out on $\mathcal{C}$, away from the singular points, a linear series g_d^r whose degree is $d = mn - \sum_i s_i(s_i - 1)$.*

262   Chapter 8. Linear Series on Algebraic Curves

If m is sufficiently large the adjoints of order m have multiplicities $s_i - 1$, $i = 1, \ldots, t$, at the singular points $P_1, \ldots, P_t$ of $\mathcal{C}$, and present the simple case with the curve $\mathcal{C}$ (see Section 4.2). Thus the statement is clear for integers $m \gg 0$. It then suffices to observe that the adjoints of order $m - 1$ together with an arbitrary straight line furnish adjoints of order m and therefore cut out on $\mathcal{C}$, away from the singular points, a linear series of order $mn - \sum_i s_i(s_i - 1) - n = (m-1)n - \sum_i s_i(s_i - 1)$.

(2) If $n \geq 2$, then adjoints of every order $m \geq n - 2$ exist; and if $p \geq 1$ (and so $n \geq 3$) there also exist adjoints of order $n - 3$.[1]

Indeed, the dimension of the linear system of adjoints of order m is

$$\begin{aligned}
\dim \mathcal{A}_m &= \frac{m(m+3)}{2} - \sum_i \frac{s_i(s_i-1)}{2} + \sigma_m \\
&= \frac{m(m+3)}{2} - \frac{(n-1)(n-2)}{2} + p + \sigma_m \quad (8.11) \\
&= \frac{(m+n)(m-n+3)}{2} + p + \sigma_m - 1,
\end{aligned}$$

where the second equality is a consequence of the genus formula $2p = (n-1)(n-2) - \sum_i s_i(s_i-1)$ (cf. Definition 7.2.6, Proposition 7.2.7). Therefore, if $m \geq n - 2$, one has $\dim \mathcal{A}_m \geq n - 2 + p + \sigma_m \geq 0$. If $p \geq 1$, then

$$\dim \mathcal{A}_{n-3} \geq p + \sigma_{n-3} - 1 \geq \sigma_{n-3} \geq 0.$$

(3) If $p \geq 1$, the linear system of adjoints of order $m \geq n - 2$ is regular, and cuts out a complete and non-special linear series g_d^r on $\mathcal{C}$, away from the singular points.

By (2), we know that there exist adjoints of order $\geq n-3$. Hence property (1) gives

$$d \geq n(n-2) - \sum_i s_i(s_i-1) = 2p + n - 2 > 2p - 2.$$

Thus the series g_d^r is non-special by Theorem 8.5.4, and so

$$r \leq d - p \left(= mn - \sum_i s_i(s_i-1) - p\right). \quad (8.12)$$

On the other hand $r = \dim \mathcal{A}_m - \dim \Sigma_0 - 1$, where Σ_0 is the linear system of adjoints of order m that contain the curve $\mathcal{C}$ (cf. 6.3.7). In view of (8.11)

---
[1] With regard to the possible existence of adjoints of order $\leq n - 4$, see Exercise 8.10.16.

8.6. Adjoints and the Riemann–Roch theorem

it follows that

$$r \geq \frac{(m+n)(m-n+3)}{2} + p$$
$$+ \sigma_m - 1 - \frac{(m-n)(m-n+3)}{2} - 1 \qquad (8.13)$$
$$= mn - p - \sum_i s_i(s_i - 1) + \sigma_m.$$

Therefore

$$mn - p - \sum_i s_i(s_i - 1) + \sigma_m \leq mn - p - \sum_i s_i(s_i - 1),$$

and so $\sigma_m = 0$. In particular (8.13) gives the equality in (8.12), which implies the completeness of the linear series g_d^r.

(4) *If $p \geq 1$ (and so $n \geq 3$) the linear system of adjoints of order $n - 3$ is regular and cuts out on $\mathcal{C}$, away from the singular points, a complete linear series g_{2p-2}^{p-1} with index of speciality $i(g_{2p-2}^{p-1}) = 1$.*

From (2) and (3) we know that $\dim \mathcal{A}_{n-2} = n + p - 2 \geq n - 1$. Thus we can impose on the curves of $\mathcal{A}_{n-2}$ the condition that they pass through $n - 1$ collinear but otherwise arbitrary points. The adjoints that one obtains all split into the line containing the points and an adjoint curve of order $n - 3$. Conversely, every adjoint of order $n - 3$ taken together with a line forms an adjoint of order $n - 2$.

Hence the adjoints of order $n - 3$ cut out on $\mathcal{C}$ a linear series g_d^r where

$$d = n(n-2) - \sum_i s_i(s_i - 1) - n = 2p - 2$$

and, since $r = \dim \mathcal{A}_{n-2} - (n-1) + \sigma_{n-3}$,

$$r = n + p - 2 - (n - 1) + \sigma_{n-3} = p - 1 + \sigma_{n-3}.$$

On adjoining a fixed point to this series, one obtains a series $g_{2p-1}^{p-1+\sigma_{n-3}}$ which, if $\sigma_{n-3} > 0$, would have dimension $> p - 1 = (2p - 1) - p$, and thus would be special; however, its degree being $2p - 1 > 2p - 2$, any such $g_{2p-1}^{p-1+\sigma_{n-3}}$ is also non-special by Theorem 8.5.4 (1). In conclusion, $\sigma_{n-3} = 0$.

The linear series g_{2p-2}^{p-1} is complete because there exists no g_{2p-2}^p that contains it. Otherwise, such a g_{2p-2}^p would be simultaneously non-special by Theorem 8.5.4 (2) (having dimension $> p - 1$) and also special (having dimension $p > (2p - 2) - p$).

Then, by definition of the index of speciality, one has $i(g_{2p-2}^{p-1}) = 1$.

The adjoints of order $n - 3$ are called *canonical adjoints* of $\mathcal{C}$.

The linear series cut out on the curve $\mathcal{C}$ (away from its multiple points) by the canonical adjoints will be called the *canonical series of* $\mathcal{C}$ and will be denoted by $|K_{\mathcal{C}}|$. By a *canonical divisor* $K_{\mathcal{C}}$ we mean a divisor belonging to the canonical series.

In view of property (4) above, the equality (8.11) and the genus formula (8.10), we know that

$$\deg(K_{\mathcal{C}}) = 2p - 2 = n(n - 3) - \sum_i s_i(s_i - 1),$$

as well as

$$\dim \mathcal{A}_{n-3} = \dim |K_{\mathcal{C}}| = p - 1. \tag{8.14}$$

8.6.3 Projective Restsatz. Let $\mathcal{C}$ be an irreducible algebraic plane curve of order n and genus p as above.

We say that two divisors A, B of points of $\mathcal{C}$ are *mutually residual with respect to the adjoints of a given order m* if together they constitute the complete intersection of $\mathcal{C}$ with an adjoint of order m. We also say that B is the *residual divisor* of A with respect to the relevant adjoint.

Two divisors A, A' of $\mathcal{C}$ are said to be *coresidual* if they are both mutually residual (with respect to adjoint curves) to a given divisor B (possibly null), that is, if there exist two adjoints (not necessarily of the same order) whose intersections with $\mathcal{C}$ are the divisors $A + B$ and $A' + B$.

Lemma 8.6.4. *Two linearly equivalent divisors A, A' of $\mathcal{C}$ are coresidual.*

Proof. Let $\mathbb{L} = \lambda_0 f_0 + \cdots + \lambda_r f_r$ be the linear system of curves that cut out on $\mathcal{C}$ the linear series g_d^r to which the two divisors belong (cf. Theorem 8.2.1 and Section 8.3). If the curves of $\mathbb{L}$ are not already adjoints, we consider a fixed adjoint $\mathcal{A}$, with equation $g = 0$. All the curves of the linear system $\mathbb{L}' = \lambda_0 g f_0 + \cdots + \lambda_r g f_r$ then are adjoints of $\mathcal{C}$ and cut $\mathcal{C}$ in the fixed divisor $B = \mathcal{C} \cap \mathcal{A}$ and in the divisors of g_d^r. Thus the two divisors A, A' are both mutually residual to the divisor B and so are coresidual. If the curves of $\mathbb{L}$ are already adjoints, then B is the null divisor. □

8.6.5. *Completeness of the linear series cut out by the adjoints of a given order: an alternative proof.* Let us propose the following different argument (cf. 8.6.2 (3) and (4)).

On the curve $\mathcal{C}$ we consider two equivalent divisors A, A' which in view of the preceding lemma are coresidual. Consider an adjoint of order m containing A (that is, m is an arbitrary integer for which the linear system $\mathcal{A}_m$ of adjoints of order m containing A has dimension ≥ 0). It cuts the curve $\mathcal{C}$, away from multiple points and away from A, in a certain divisor H (of degree ≥ 0) which will therefore be a residual divisor for A. By the Restsatz (Proposition 8.3.3), the divisor H is also a residual divisor for A'.

Since A and A' are divisors of the same order, also the two divisors $A + H$ and $A' + H$ are divisors of the same order. Thus, the adjoint that passes through H and which must give A' as further intersection with $\mathcal{C}$ will also be of order m. Hence A', which is an arbitrary divisor equivalent to A, can be obtained as the intersection of $\mathcal{C}$ with adjoints of order m passing through H. Thus one has that the adjoints of a given order m that pass through H cut out the complete series $|A|$ on $\mathcal{C}$. In particular, if $\deg(H) = 0$, the adjoints of a given order m cut out the complete series $|A|$ on $\mathcal{C}$. From this the desired result follows. □

The argument given above allows us to state the Restsatz in the following form, called the "Projective Restsatz" (cf. [100, p. 153]).

Proposition 8.6.6 (Projective Restsatz). *Every residual divisor of a given divisor D on $\mathcal{C}$ with respect to the adjoints of a given order is the residual divisor with respect to the same adjoints of an arbitrary divisor of $|D|$.*

Remark 8.6.7. The preceding discussion furnishes a method for effectively constructing the complete series defined by a given divisor on an irreducible curve X of $\mathbb{P}^N$.

We refer to a plane model $\mathcal{C}$ of X and consider an adjoint $\mathcal{A}_m$ of sufficiently large order m so that $\mathcal{A}_m$ passes through a given divisor D. If H is the residual divisor, the adjoints of order m that pass through H cut out, away from the singular points of $\mathcal{C}$, the complete series $|D|$. We give some examples to illustrate the procedure.

(1) On a plane quintic $\mathcal{C}$ with three nodes (and so of genus $p = 3$) we consider a divisor $D = \sum_{i=1}^{4} P_i$ of degree 4.

In order to construct the complete series $|D|$ we take an adjoint $\mathcal{A}_3$ of order 3 passing through D. It cuts $\mathcal{C}$, away from the three nodes and the four points P_i, in a divisor H of order 5. The adjoints of order 3 that pass through H cut out on $\mathcal{C}$ (away from the nodes and the divisor H) the complete series $|D|$. They constitute a pencil and so $|D| = g_4^1$.

In particular, in the case in which the four points P_i all belong to a straight line r, the cubic $\mathcal{A}_3$ splits into the line r and a conic γ passing through the three nodes. One of the five points in which $\mathcal{A}_3$ meets $\mathcal{C}$ is the fifth intersection O of r with $\mathcal{C}$; the other four are points of γ. The third order adjoints that pass through these five points are all split into the fixed part γ (because they contain seven points of γ) and a line passing through O. In this case the series $|D|$ is cut out on $\mathcal{C}$ by the straight lines of the pencil with center O (and is again a series g_4^1).

(2) On a plane quintic $\mathcal{C}$ with a double point A, consider three points P, Q, R belonging to a straight line r passing through A and a fourth point O of $\mathcal{C}$ not lying on r. We wish to construct the linear series $|D|$ where $D = O + P + Q + R$.

An adjoint of second order passing through D decomposes into the line r and a second line passing through O. This second line further cuts the quintic in four points collinear with O. The adjoint conics that pass through these four points all then pass through O as well, which is therefore a fixed point of the series $|D|$ (which is the series g_4^1 cut out on $\mathcal{C}$ by the lines passing through O). The residual series of O with respect to this series is the series g_3^1 cut out on $\mathcal{C}$ by the lines of the pencil with center A (see Exercise 13.3.2).

Example 8.6.8. Consider the linear series g_d^r cut out on a plane curve (with ordinary singularities) $\mathcal{C}$ by all the curves of a given order m. One sees immediately that it is complete if $\mathcal{C}$ does not have multiple points, inasmuch as in that case every curve of order m is an adjoint.

If however $\mathcal{C}$ has multiple points, the series g_d^r is not, in general, complete. For example, the lines in the plane cut out on a cubic with a double point a linear series g_3^2 which is not complete. On the other hand, since such a cubic is a rational curve, the complete series that contains its linear sections is a series g_3^3 in agreement with statements (1), (2) of Proposition 8.5.2.

8.6.9 The reduction theorem and the Riemann–Roch theorem.
In view of Theorem 8.5.7, given an irreducible algebraic curve, we may consider a plane model $\mathcal{C}$ for it, and suppose that model to have order n and genus p, and to have δ nodes but no other multiple point. We will assume that $p \geq 1$.

The following result is also known as "Noether's reduction theorem" (see [100, p. 154]).

Proposition 8.6.10 (Noether's reduction theorem). *Let $\mathcal{C}$ be a plane model of an irreducible algebraic curve of $\mathbb{P}^N$. Suppose that a divisor A of the curve $\mathcal{C}$ be such that there exists a canonical adjoint containing its points except for one point, P. Then the complete series $|A|$ has P as fixed point.*

Proof. Consider, as in the Figure 8.1, the curve $\mathcal{C}$, of order n, the divisor A of which P is a part, and a canonical adjoint $\mathcal{A}$ of order $n-3$ that passes through all the points of A except P and cuts out on $\mathcal{C}$ one more divisor B. This adjoint, together with a line ℓ passing through P (and which meets $\mathcal{C}$ in the $n-1$ other points H) forms an adjoint of order $n-2$ which cuts $\mathcal{C}$, away from the multiple points, in the divisor $B + A + H$.

The divisor $B + H$ is thus a residual divisor to A with respect to the adjoints of order $n-2$. Therefore the complete series $|A|$ is cut out on $\mathcal{C}$ by the adjoints of order $n-2$ passing through $B + H$. But these adjoints all contain the line ℓ (inasmuch as they have the $n-1$ points H in common with it). Hence P is a fixed point of $|A|$. □

It is useful to note that the preceding statement may be reformulated in the following way. *If G is a divisor contained in some canonical divisor $K_{\mathcal{C}}$ and P*

8.6. Adjoints and the Riemann–Roch theorem

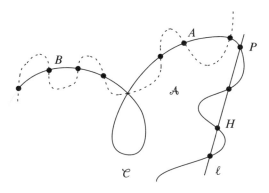

Figure 8.1

is an arbitrary point of the curve $\mathcal{C}$, then one and only one of the following cases holds.

(1) All the canonical divisors that contain G also contain the point P; or

(2) P is a fixed point of the series $|G + P|$.

We can now prove an important property of special linear series.

Proposition 8.6.11. *Let $\mathcal{C}$ be a plane model of an irreducible algebraic curve of $\mathbb{P}^N$. A complete and special linear series on $\mathcal{C}$ is partially contained in the canonical series $|K_\mathcal{C}|$.*

Proof. On the curve $\mathcal{C}$, of genus p, consider a complete and special linear series g_d^r. One then has $r > d - p$. We must prove that every divisor of g_d^r is contained in some canonical divisor.

The statement holds if $r = 0$. Indeed, we know by (8.14) that the canonical series has dimension $p - 1$. On the other hand, g_d^0 is a unique divisor of $d < p$ points. Thus there exists a canonical divisor containing the d points.

We now proceed by induction. Let P be a *generic* point of $\mathcal{C}$ (and thus not fixed for g_d^r) and let G be a divisor residual to P with respect to g_d^r, so that $g_d^r = |G + P|$. The complete series $|G|$ is a special g_{d-1}^{r-1} (because $r - 1 > d - 1 - p$) and so is contained in the canonical series by the induction hypothesis. All the canonical divisors that contain G also contain P, since otherwise in view of Noether's reduction theorem, P would be a fixed point for $|G + P| = g_d^r$ which contradicts the present hypothesis. In conclusion, the series $|G + P|$ is partially contained in $|K_\mathcal{C}|$. □

The following theorem clarifies the issue raised in Problem–Definition 8.5.3, namely to specify the geometric meaning of the index of speciality of a complete linear series. It is a central result in the theory of linear series. In what follows we will make systematic use of Theorem 8.5.7.

Theorem 8.6.12 (Riemann–Roch theorem, I). *Let $\mathcal{C}$ be a plane model, of genus p (≥ 1), of an irreducible algebraic curve in $\mathbb{P}^N$. Let $i(D)$ be the index of speciality of a divisor D of degree d on $\mathcal{C}$. Then $i(D)-1$ is the dimension of the series residual to D with respect to the canonical series $|K_\mathcal{C}|$ of $\mathcal{C}$. (Equivalently, $i(D)$ is the maximum number of linearly independent canonical divisors that contain D, that is, $i(D) - 1 = \dim |K_\mathcal{C} - D|$.)*

Proof. Let $j(D) - 1 := \dim |K_\mathcal{C} - D|$ be the dimension of the residual series of D with respect to the canonical series. It will suffice to prove that

$$\dim |D| = d - p + j(D), \tag{8.15}$$

since by (8.15) and the definition of the index of speciality one then has $j(D) = i(D)$ and so the theorem follows.

Equation (8.15) is obvious if $j(D) = 0$. Indeed, in that case canonical divisors containing D do not exist, so the series $|D|$ is non-special by Proposition 8.6.11 whence $\dim |D| = d - p = d - p + j(D)$.

If $j(D) = 1$ there is a unique canonical divisor $K_\mathcal{C}$ containing D. If then P is a point not belonging to $K_\mathcal{C}$, the divisor $D + P$ does not belong to any canonical divisor (for otherwise a canonical divisor containing $D + P$ would be a canonical divisor $\neq K_\mathcal{C}$ containing D). This means that $j(D + P) = 0$ and so, by the preceding case,

$$\dim |D + P| = (d+1) - p + j(D+P) = d - p + 1 = d - p + j(D).$$

On the other hand, by Noether's reduction theorem (Proposition 8.6.10), the point P is fixed for the series $|D + P|$ and so $\dim |D + P| = \dim |D|$. Thus (8.15) is also true for $j(D) = 1$.

We can now proceed by induction. A point P chosen generically on $\mathcal{C}$ does not belong to the canonical divisors containing D (and does not belong to D) whence

$$\dim |K_\mathcal{C} - D - P| = \dim |K_\mathcal{C} - D| - 1,$$

which is equivalent to $j(D + P) = j(D) - 1$. From the induction hypothesis we have

$$\dim |D+P| = (d+1) - p + j(D+P) = (d+1) - p + j(D) - 1 = d - p + j(D).$$

Since, as already noted, $\dim |D + P| = \dim |D|$, equation (8.15) follows. □

8.6. Adjoints and the Riemann–Roch theorem

As an immediate consequence of Proposition 8.6.11 and Theorem 8.6.12 we obtain that

- *the special linear series on $\mathcal{C}$ are the linear series contained in the canonical series $|K_\mathcal{C}|$.*

In this regard, the following important fact holds.

Remark 8.6.13. With notation as in Theorem 8.6.12, *the canonical series $|K_\mathcal{C}|$ is the unique g_{2p-2}^{p-1} belonging to $\mathcal{C}$.*

In fact, g_{2p-2}^{p-1} is obviously special inasmuch as $p - 1 > (2p - 2) - p$, and is thus contained in $|K_\mathcal{C}|$. Since g_{2p-2}^{p-1} has the same dimension as $|K_\mathcal{C}|$ the two series must coincide.

The above uniqueness property ensures that *the canonical series is a birational invariant*, thus allowing us to define *the canonical series of a non-singular curve X in $\mathbb{P}^N$* of genus p to be the unique (complete) linear series g_{2p-2}^{p-1} belonging to X. More precisely, bearing in mind what was noted in paragraph 8.4.4, it is the series that corresponds on the curve X to the canonical series $|K_\mathcal{C}|$ of the plane model $\mathcal{C}$ of X. We will denote by $|K_X|$ (or simply by $|K|$) the canonical series of X. By a *canonical divisor* K_X (or simply K) on X we mean a divisor belonging to the canonical series. In particular,

$$\deg(K_X) = 2p - 2.$$

We can now reformulate the Riemann–Roch theorem in the non-singular case in the following form.

Theorem 8.6.14 (Riemann–Roch theorem, II). *Let X be a non-singular curve of $\mathbb{P}^N$, of genus p, and let D be a divisor on X of degree d. Then one has*

$$\dim |D| - \dim |K - D| = d + 1 - p.$$

Proof. Let $\mathcal{C}$ be a plane model, endowed with only nodes, of the curve X. The statement then follows from Theorem 8.6.12 and the definition of the index of speciality, once one observes that, under the birational map $X \to \mathcal{C}$ described in §8.5.6, the series $|D|$ is transformed into a linear series g_d^r on $\mathcal{C}$ which still has order $d = \deg(D)$ and dimension $r = \dim |D|$; for this we refer to 8.5.9. □

It is hardly necessary to observe that in the case of a linear series g_d^r on a curve X of genus $p = 0$ the index of speciality is zero, and the Riemann–Roch theorem reduces to the equality $r = d$.

We illustrate a notable consequence of the Riemann–Roch theorem. The following proposition gives a necessary (but *not sufficient*) condition for a linear series to be special (compare with Theorem 8.5.4).

Proposition 8.6.15 (Clifford's theorem). *Let X be a non-singular curve of $\mathbb{P}^N$ and let g_d^r be a linear series on X. If g_d^r is special, then $d \geq 2r$.*

Proof. We can surely suppose that the linear series g_d^r is complete. We take r generic points $P_1, P_2, \ldots, P_r$ on the curve. They impose r independent conditions on the canonical divisors that are to contain them, and so there are at least r conditions imposed on the canonical divisors by the divisor G of g_d^r that contains the points $P_1, P_2, \ldots, P_r$ (and which is determined by them). Hence the dimension of the residual series $|K - G|$ of G with respect to the canonical series is $\leq p - 1 - r$. Thus, by the Riemann–Roch theorem, $i - 1 \leq p - 1 - r$, where i is the index of speciality of G. But we also have $r = d - p + i$, that is, $i - 1 = r + p - d - 1$, and therefore $r + p - d - 1 \leq p - 1 - r$, from which it follows that $2r \leq d$. □

8.7 Properties of the canonical series and canonical curves

In this paragraph we discuss some properties of the canonical series $|K|$ of a non-singular curve X of $\mathbb{P}^n$ and of its projective image, which we shall also call the *canonical model* of X (defined up to a projectivity). These considerations also lead to a "coarse" classification of the curves of $\mathbb{P}^n$, which we summarize for the convenience of the reader in Table 8.1 at the end of this section.

In view of what we have seen in Section 8.6, a rational curve does not have a canonical series in the sense that the canonical divisor K has degree $\deg(K) = -2$ and $|K| = g_{-2}^{-1} = \emptyset$.

If X is an elliptic curve the canonical series $|K|$ is of type g_0^0 and so the divisor K is linearly equivalent to zero (cf. Lemma 8.3.4).

The following results are important in regard to the canonical series of a curve of genus ≥ 2.

Proposition 8.7.1. *Let X be a non-singular curve in $\mathbb{P}^n$ of genus $p \geq 2$ and let $|K|$ be the canonical series. Then:*

(1) *$|K|$ does not have fixed points.*

(2) *If $|K|$ is simple it does not have neutral pairs.*

Proof. (1) If M were a fixed point, then on choosing a point P not belonging to $|K|$, the series $|K - M + P|$ would be a series g_{2p-2}^{p-1} distinct from the canonical series, which contradicts the conclusion of Remark 8.6.13.

(2) If P_1, P_2 were a neutral couple, then on choosing two points A, B not belonging to $|K|$, the series $|K - P_1 - P_2 + A + B|$ would be a series g_{2p-2}^{p-1} distinct from the canonical series, which again contradicts Remark 8.6.13. □

The following two propositions show that the hyperelliptic curves (cf. 8.5.12) are those whose canonical series is composite (with a series g_2^1).

8.7. Properties of the canonical series and canonical curves

Proposition 8.7.2. *Let X be a non-singular curve in $\mathbb{P}^n$ of genus $p \geq 2$. If the canonical series $|K|$ is not simple, it can be composed only with a series g_2^1.*

Proof. Indeed, suppose that $|K|$ is composed with an involution σ_μ^1 of order $\mu \geq 2$. In this case, having chosen $p-1$ arbitrary points of X which pairwise do not belong to a common divisor of σ_μ^1, the canonical divisor that contains them consists of $\mu(p-1)$ points, and so $\mu(p-1) = 2p-2$, which is to say $\mu = 2$.

If then P, Q is a pair of points of σ_2^1, there exist canonical divisors constituted by P, Q and $2p-4$ additional points forming a divisor H consisting of $p-1$ pairs of points of σ_2^1. The divisor H imposes only $p-2$ conditions on the canonical divisors that contain it, and so the residual series of H with respect to $|K|$ is a complete linear series $g_{(2p-2)-(2p-4)}^{(p-1)-(p-2)} = g_2^1$ containing the pair of points P, Q. Every pair of σ_2^1 thus belongs to a linear series g_2^1. But on the curve X there can not be more than one g_2^1, for otherwise X would be birationally isomorphic to a plane cubic (for this, see Exercise 8.10.8); and that is not possible in view of the hypothesis that $p \geq 2$. In conclusion the involution σ_2^1 coincides with g_2^1. □

Proposition 8.7.3. *If a non-singular curve X in $\mathbb{P}^n$ of genus $p \geq 2$ contains a series g_2^1, then the canonical series is composed with that g_2^1.*

Proof. The series g_2^1 is certainly complete (otherwise it would be contained in a series g_2^2 and so the curve would be rational) and therefore, if i is its index of speciality, we have $1 = 2 - p + i$, that is, $i = p - 1$. This means that for every pair of points of the g_2^1 there pass $p-1$ independent canonical divisors, that is, $\dim |K - g_2^1| = \dim |K| - 1$. Thus a given pair of points of g_2^1 impose only one condition on the canonical divisors K which are to contain it; that is, the series $|K|$ is composed with g_2^1. □

Among the hyperelliptic curves, there are in particular all the curves of genus $p = 2$ inasmuch as they have a series g_2^1 as canonical series.

By what we have proved previously, we know that the canonical series g_{2p-2}^{p-1} on a *non-hyperelliptic* curve X, which thus must be of genus $p \geq 3$, is simple, and its projective image is a curve $\mathcal{X}$ of order $2p-2$ in a space S_{p-1}, which is birationally isomorphic to X, and on which the canonical series is cut out by the hyperplanes of S_{p-1}. The curve $\mathcal{X}$ is called *canonical model* of X (and, as we know, it depends on the basis chosen to define g_{2p-2}^{p-1}). In fact, two canonical models of X correspond to each other under a projectivity of their spaces S_{p-1}. One says that $\mathcal{X}$ is a *canonical curve of genus p*. In this regard we note that

- *every non-singular curve of genus p (≥ 3) and order $2p-2$ in S_{p-1} is a canonical curve.*

Indeed, it suffices to observe that the hyperplanes cut out a linear series g_{2p-2}^{p-1} on the given curve.

In the case of a hyperelliptic curve X of genus $p \geq 2$, the image of the canonical series (which is composed with a series g_2^1) is a curve C of order $\frac{2p-2}{2} = p-1$ in S_{p-1}. Hence C is a rational normal curve of order $p-1$ whose points are in bijective correspondence with the pairs of g_2^1. More precisely, one has a double covering $X \to C$ whose fibers are the divisors of g_2^1.

Remark 8.7.4 (Series of type g_2^1 on a curve of genus ≤ 1). Let X be a non-singular curve of genus $p \leq 1$. If $p = 1$, an arbitrary pair of points constitutes a non-special divisor A, and hence by the Riemann–Roch theorem they define a non-special (and complete) linear series $g_2^1 = |A|$. Thus, an elliptic curve contains ∞^1 complete linear series g_2^1.

One may reach the same conclusion in a more elementary fashion by following the procedure described in paragraph 8.5.6. If $p = 1$, the curve X is isomorphic to the projective image of a series g_3^2, which is a non-singular plane cubic $\mathcal{C}$ in view of the inequality $3 > 2p$. If A, B are two points of $\mathcal{C}$, the pencil of lines with center at the third intersection P of $\mathcal{C}$ with the line r_{AB} cuts out on $\mathcal{C}$ a series g_2^1 containing the divisor $A + B$. Thus there are ∞^1 linear series of type g_2^1, just as many as there are points P of $\mathcal{C}$.

If $p = 0$, there are (non-complete) linear series g_2^1 on the rational curve X which are contained in the (complete) series g_2^2 consisting of all the pairs of points of the curve. Such g_2^1 are in number ∞^2, just as many as there are lines in the plane of the projective image conic of g_2^2.

Example 8.7.5 (Canonical curves of genus $p = 3, 4, 5$). The canonical curves of genus $p = 3$ are the non-singular plane quartics. In general, let λ_m be the maximal number of independent hypersurfaces of order $m > 1$ in the space S_{p-1} that contain a canonical curve C^{2p-2}. The dimension r of the linear series $g^r_{m(2p-2)}$ cut out on C^{2p-2} by the linear system of all the hypersurfaces of S_{p-1} of order m is, cf. Theorem 8.2.1,

$$r = \binom{p-1+m}{p-1} - 1 - \lambda_m.$$

On the other hand, this series is non-special because $m > 1$ implies that its order $2m(p-1)$ is greater than $2p-2$. Hence, by the Riemann–Roch theorem,

$$\binom{p-1+m}{p-1} - 1 - \lambda_m \leq 2m(p-1) - p$$

(where the equality holds if the series $g^r_{m(2p-2)}$ is complete), that is,

$$\lambda_m \geq \binom{p-1+m}{p-1} - 1 - 2m(p-1) + p.$$

8.7. Properties of the canonical series and canonical curves

For example, if $p = 4$, we have

$$\lambda_m \geq \binom{m+3}{3} - 6m + 3; \quad \lambda_2 \geq 1; \quad \lambda_3 \geq 5; \quad \ldots.$$

The canonical curve $C^6 \subset \mathbb{P}^3$ of genus $p = 4$ is thus contained in a quadric Q, and really in only one quadric since a space curve of order 6 can not belong to two quadrics. Furthermore, it belongs to all the cubics of a linear system of dimension ≥ 4, the generic member of which does not contain the quadric Q as component, in view of the fact that there are only ∞^3 cubic surfaces in $\mathbb{P}^3$ which split into a given quadric and a residual plane.

Therefore, the canonical curve C^6 of genus $p = 4$ is the complete intersection of a quadric with a cubic surface in $\mathbb{P}^3$. The generators of a system of lines of the quadric (or the generators of the cone if the quadric is a cone) cut out a linear series g_3^1. Thus C^6 is a *trigonal* curve (that is, by definition, 3 is the minimal order of a base point free linear series of dimension ≥ 1 belonging to the curve, cf. [1, Chapter 3, §2]).

If $p = 5$ the canonical curve is a C^8 in $\mathbb{P}^4$. One then has $\lambda_2 \geq 3$ and so C^8 belongs to three linearly independent quadrics, which generate a net. Since $\lambda_3 \geq 15$, C^8 belongs to a linear system of cubic hypersurfaces having dimension ≥ 14. In general, the three independent quadrics passing through C^8 have in common only a curve (as happens for three generic hypersurfaces in $\mathbb{P}^4$), and this curve is the curve C^8, the base locus for the net of quadrics. It can, however, happen that the three independent quadrics contain a common surface $\mathcal{F}$. The quadrics passing through the surface of order 4 common to two quadrics belong to the pencil defined by the latter quadrics. It follows that the order of the surface $\mathcal{F}$ is < 4. Since C^8 does not belong to any S_3, the surface $\mathcal{F}$ has order ≥ 3 (for otherwise $C^8 \subset \mathcal{F} \subset \mathbb{P}^3$). Hence, if the canonical curve C^8 is not the complete intersection of three quadrics, it belongs to a cubic surface $\mathcal{F}^3 \subset \mathbb{P}^4$, the base surface of the net of quadrics passing through C^8.

In the following Table 8.1, X is a non-singular curve in $\mathbb{P}^n$ of genus p, $|K|$ is its canonical series and C is the canonical model of X.

Remark 8.7.6. For an irreducible algebraic curve with arbitrary singularities and lying in a projective space of any given dimension, the canonical series is defined as the image of the canonical series of a non-singular birational model apart from possible fixed points (which may arise from the fact that the two curves are only birationally equivalent).

Table 8.1. "Coarse" classification of the curves of $\mathbb{P}^n$.

p	K	X	C
0	$g_{-2}^{-1} = \emptyset$	rational	
1	g_0^0	elliptic	
2	g_2^1; without fixed points	hyperelliptic	double cover exists $X \to C = \mathbb{P}^1$ whose fibers are the divisors of g_2^1
≥ 3	g_{2p-2}^{p-1}; without fixed points; composed with a series g_2^1	hyperelliptic	$C = C^{p-1} \subset S_{p-1}$ rational normal; double cover exists $X \to C$ whose fibers are the divisors of g_2^1
≥ 3	g_{2p-2}^{p-1}; without fixed points; simple	non-hyperelliptic	$C = C^{2p-2} \subset S_{p-1}$ canonical curve of genus p

8.8 Some results on algebraic correspondences between two curves

In this section we present, in deliberately terse form, some results regarding algebraic correspondences between curves, indicating some applications in the form of exercises. For the proofs and further information one may consult, for example, Severi's text [100].

Possible singularities of the curves under consideration will always be supposed to be ordinary in the sense of Proposition 8.1.3.

By an (m,n) *algebraic correspondence* (or a correspondence with *indices m,n*) between two algebraic curves C, C' we mean a correspondence $\omega: C \to C'$ which associates m points of C' to each point of C and n points of C to each point of C', and such that the coordinates of two associated points are related by algebraic equations.

8.8. Some results on algebraic correspondences between two curves

8.8.1 (Zeuthen's formula). Zeuthen's formula establishes a relation between the genera p, p' of the two curves C, C', the indices of the correspondence, and the numbers δ, δ' of *ramification points* of C and C'. Here by ramification point, for example, of C we mean a point to which there corresponds on C' a group of m points not all of which are distinct. In the case in which the correspondence presents only double points, that is, to each ramification point there correspond only *two* coinciding points, the following formula (due to Zeuthen) holds:

$$\delta - \delta' = 2n(p' - 1) - 2m(p - 1).$$

8.8.2 (Genus of an involution on an algebraic curve). If the divisors of an involution σ_μ^1 on C are birationally referred to the points of an algebraic curve C' of genus p', we will say that p' is the *genus of* σ_μ^1. To give an involution σ_μ^1 is thus equivalent to giving an algebraic correspondence of indices $(1, \mu)$ between the two curves C and C'.

Applying Zeuthen's formula to this correspondence (in which we set $m = 1$ and so $\delta = 0$) one obtains the equality

$$2(p - 1) = 2\mu(p' - 1) + \delta'$$

thus establishing the relation $p' \leq p$. Hence, *on a curve of genus p every involution has genus not greater than p*. In particular (cf. Theorem 6.6.2):

- (Lüroth's theorem) *On a rational curve every involution is rational.*

8.8.3 (The Cayley–Brill correspondence principle). Suppose $C' = C$, that is, we consider an (m, n) algebraic correspondence ω between two superimposed copies of C. Chasles' correspondence principle, according to which an algebraic correspondence of indices m, n on a rational curve possesses $m + n$ fixed points, is a particular case of the *Cayley–Brill correspondence principle*.

Let P be an arbitrary point of C and $\omega(P)$ the group of m points corresponding to it. As P varies on C, the various divisors $\omega(P)$ are not, in general, linearly equivalent. However, it can happen that there exists an integer v such that the divisors $\omega(P) + vP$ are equivalent, with P variable on C. One then says that ω is a *correspondence with valence* and that v is its *valence*.

One proves that on a non-hyperelliptic curve there exist only correspondences with valence.

The Cayley–Brill correspondence principle affirms that *if p is the genus of C then an (m, n) correspondence of valence v has $m + n + 2vp$ fixed points*.

8.8.4 (Number of $(r+1)$-fold points of a series g_d^r on a curve of genus p). For each point P of C let G_P be the divisor of the given g_d^r defined by the point P counted r times. Consider the correspondence $\omega \colon C \to C$ that associates to the point P the remaining $d - r$ points of G_P. The fixed points of ω are the $(r+1)$-fold points of g_d^r. We denote their number by $[d, r + 1]$. We know the first index $m = d - r$ and the valence $v = r$ of ω. It is not difficult to prove that the second index, that is the number n of divisors of the series that contain a point $P' \in C'$ (where now $C' = C$) and which have P as r-fold point, is the number $[d - 1, r]$ of r-fold points of the series g_{d-1}^{r-1} constituting the residual series of P' with respect to g_d^r. By the Cayley–Brill correspondence principle one obtains the recursive formula:

$$[d, r + 1] = d - r + [d - 1, r] + 2rp. \tag{8.16}$$

The number $[d - r + 1, 2]$ of double points of a series g_{d-r+1}^1 on C is also the order $r = 2(r + p - 1)$ of the Jacobian group of g_{d-r+1}^1, cf. (7.4). Thus one obtains:

- The number of $(r + 1)$-fold points of a series g_d^r on a curve C of genus p is

$$[d, r + 1] = (r + 1)(d - r + rp).$$

In particular, the number of *Weierstrass points*, namely the p-fold points for the canonical series of the curve C, is $[2p - 2, p] = p(p^2 - 1)$.

Exercise 8.8.5 (Flexes of an elliptic curve). We say that a point P of an elliptic cubic C corresponds to its tangential P'. Since P' is tangential to four points P (through P' there pass four tangents to C distinct from the tangent at P', cf. Exercise 5.7.17), one thus has an algebraic correspondence $\omega \colon C \to C$ of indices $(1, 4)$ and valence $v = 2$, whose fixed points are the flexes of the cubic. By the Cayley–Brill correspondence principle, their number is then $1 + 4 + (2 \times 2 \times 1) = 9$.

This result can also be obtained immediately by observing that the flexes are triple points of the series g_3^2 cut out on C by the lines of the plane.

Exercise 8.8.6 (Stationary points of a quartic of the first kind). An elliptic space quartic has sixteen stationary points, that is, points at which the osculating plane is hyperosculating. The stationary points are in fact the 4-fold points of the series g_4^3 cut out by plane sections.

Exercise 8.8.7 (Sextatic points of an elliptic cubic). An elliptic cubic C has twenty-seven *sextatic points*, that is, points that absorb the six intersections of C with an irreducible conic. They are the 6-fold points of the series g_6^5 cut out on C by the conics. In view of what was seen in §8.8.4, this series has $6(6 - 5 + 1 \times 5) = 36$ six-fold points. Nine of these are the flexes of the cubic (for which the hyperosculating conic is the flexional tangent counted twice).

Exercise 8.8.8. *Prove that the sextatic points of an elliptic cubic are the tangentials of the flexes.*

Let H be a rectilinear section of the cubic C, and let F and P be a flex and one of its three tangential points (the tangentials of three collinear points of C are collinear, cf. Exercise 5.7.15). We then have $F + 2P \equiv H$ and so $3F + 6P \equiv 3H$. But

$3F \equiv H$, and therefore $6P \equiv 2H$. Hence $6P$ is linearly equivalent to a section by a conic, and is in fact a section by a conic inasmuch as the conics cut out a complete linear series on C.

Exercise 8.8.9. *A non-hyperelliptic curve X can not have infinitely many birational transformations into itself.*

Indeed, a birational transformation of the canonical model C of X sends hyperplane sections into hyperplane sections and so is subordinate to a projectivity of S_{p-1}, where p is the genus of X. But the projectivities of S_{p-1} that send C into itself are finite in number because they must change hyperosculating hyperplanes into hyperosculating hyperplanes, that is, Weierstrass points into Weierstrass points (and the latter are in number $p(p^2 - 1)$).

Exercise 8.8.10. *Let a point O and an irreducible algebraic curve C of order d and genus p be given in the plane. How many normals to C issue from O?*

We say that two points P and P' correspond if the line $r_{PP'}$ and the tangent to C at P are perpendicular.

One thus obtains an algebraic correspondence ω on C whose fixed points are the points of C which are the feet of normals issuing from O.

To a point P of C there correspond the points of contact of C with the tangents issuing from the improper point H of the line orthogonal to the line r_{OP}; their number is the class ν of C. These points are the intersections $P'_1, P'_2, \ldots$ of the curve C with the first polar of H with respect to C. Thus one has an algebraic correspondence ω with first index ν. One sees immediately that the second index is the order d of the curve; indeed, the points corresponding to a point P' are the d intersections of the curve with the perpendicular drawn through O to the tangent at P'.

As P varies, the divisor $\omega(P)$ remains equivalent to itself inasmuch as it consists of the intersections of C with a curve of order $d - 1$ (the first polar of H). This means that the valence of ω is $v = 0$.

By the Cayley–Brill correspondence principle we have that the number of normals issuing from a point P on a plane algebraic curve C of order d and genus p is $d + \nu = 3d + 2p - 2$.

8.9 Some remarks regarding moduli

The classification of algebraic curves is a fundamental problem.

Here we limit ourselves to an elementary introduction outlining a few major points. In addition to the fundamental texts [32] and [67] we refer the reader also to the introductory survey [93] (see also [46, pp. 253–259]).

Let $\mathcal{M}_p$ be the set of birational isomorphism classes of curves of genus p. The problem is to describe $\mathcal{M}_p$. For low values of p the solution is simple.

In view of our discussion on rational curves, the space $\mathcal{M}_0$ reduces to a point.

Consider the case $p = 1$. We begin with the observation that *an elliptic curve X has a non-singular plane cubic as birational model*. Indeed, we consider the complete series defined by a divisor D consisting of three (distinct) points of X. The series $|D|$ is non-special, and so, by the Riemann–Roch theorem, $|D|$ is a series g_3^2. Being of prime order, $|D|$ is surely simple. Thus, its projective image is a non-singular plane cubic $\mathcal{C}$ (compare with 8.10.13). We also refer the reader to [113, Chapter 3, §6] for a description of the further properties of plane cubics.

The results discussed in Section 5.7 (see, in particular, the theorem of G. Salmon, Exercise 5.7.17) then allow one to conclude that the expression

$$J = J(k) := \frac{(k+1)^2(2k-1)^2(k-2)^2}{(k^2-k+1)^3},$$

introduced in 1.1.1, where k denotes the cross ratio of the four tangents drawn to the cubic $\mathcal{C}$ from one of its points, is the *modulus* of the curve X. It depends only on the birational model $\mathcal{C}$ of X and assumes all possible values in $\mathbb{C}$. Thus $\mathcal{M}_1$ may be identified with $\mathbb{C}$ via J.

Deligne and Mumford [32] have shown that $\mathcal{M}_p$ for $p \geq 2$ is an irreducible quasi-projective variety of dimension $3p - 3$ (over any fixed algebraically closed field). In the case $p = 2$, Igusa [55] has given an explicit construction of $\mathcal{M}_2$.

We now illustrate a heuristic method for calculating the number of classes of birational equivalence for curves of a given genus p. By the above remarks we may suppose that $p \geq 2$.

We begin by considering the general case, namely, that of non-hyperelliptic curves. Since all curves of genus 2 are hyperelliptic, we may assume in fact that $p \geq 3$.

If two non-hyperelliptic curves of genus p are birationally equivalent, their canonical curves $C^{2p-2} \subset S_{p-1}$ correspond to each other under a projectivity of their spaces S_{p-1}. In fact, the canonical divisors of the first are mapped to the canonical divisors of the second, and so the hyperplanes of the two spaces S_{p-1} are in correspondence, thus giving rise to a projectivity between the two spaces that maps one curve onto the other.

A C^{2p-2} of S_{p-1} of genus $p \geq 3$ projects from $p - 3$ of its points into a plane π and therein onto a curve $\mathcal{C}_1^{p+1}$ of genus p with $\frac{p(p-1)}{2} - p = \frac{1}{2}(p^2 - 3p)$ double points.

If one changes the centers of projection on C^{2p-2} to $p - 3$ other points, one obtains another curve $\mathcal{C}_2^{p+1}$ as projection in π. For generic choices of the two sets of $p - 3$ points on the curve C^{2p-2}, the corresponding projections in π are not projectively equivalent, that is, they do not correspond to each other under a projectivity of $\mathbb{P}^2$. Indeed, a projectivity of π that sends (collinear) points of $\mathcal{C}_1^{p+1}$ in (collinear) points of $\mathcal{C}_2^{p+1}$ extends to a projectivity of S_{p-1} sending each of the two groups of $p-3$ points on the curve C^{2p-2} into the other, and so would give rise to a birational correspondence of C^{2p-2} with itself. Therefore, the curve C^{2p-2} would have infinitely many automorphisms, in contrast with what was shown in Exercise 8.8.9.

Hence, starting from a given C^{2p-2} one obtains ∞^{p-3} projectively inequivalent plane models (this being the number of spaces S_{p-4} in S_{p-1} generated by groups of $p-3$ points taken as centers of projection). Bearing in mind that there are ∞^8 projectivities of π, one sees that each canonical curve leads to a continuous family of dimension $p-3+8 = p+5$ of curves all of which are birationally equivalent to the initial curve X.

However, the curves of π having order $p+1$ and genus p (and thus endowed with $\frac{1}{2}(p^2-3p)$ double points) constitute an algebraic system of dimension, cf. Example–Definition 6.2.2,

$$\frac{1}{2}(p+1)(p+4) - \frac{1}{2}(p^2-3p) = 4p+2.$$

In conclusion, the number of birational equivalence classes of the curve X is

$$4p+2-(p+5) = 3p-3.$$

In order to treat the hyperelliptic case, which presents some particularly notable aspects, we need the following lemma.

Lemma 8.9.1. *A hyperelliptic curve X of genus $p \geq 2$ is birationally equivalent to a plane curve of order $p+2$ having a point of multiplicity p (and no other singularities).*

Proof. On X take $p+2$ generic points (that is, points belonging to $p+2$ distinct divisors of the g_2^1). They form a non-special divisor D; indeed, since the canonical series is composed with the g_2^1, a canonical divisor that contains D must also contain $p+2$ other points, which is impossible because $2(p+2) > 2p-2$. From this it follows that the complete series $|D|$ is a series g_{p+2}^2.

The series g_{p+2}^2 is simple, that is, without neutral pairs. Indeed, to contain a neutral pair imposes a single condition on the divisors of the series. Thus, the residual series of a neutral pair would be a series g_p^1. But, by definition of the genus, p generic points constitute an isolated divisor. Furthermore, g_{p+2}^2 has no fixed points because the residual series of a fixed point would be a g_{p+1}^2, and the residual series of another (not fixed) point with respect to that g_{p+1}^2 would be a series g_p^1.

In conclusion, the projective image of the series g_{p+2}^2 is a curve $\mathcal{C}^{p+2} \subset \mathbb{P}^2$ birationally equivalent to X. On $\mathcal{C}^{p+2}$, the g_{p+2}^2 is cut out by the lines of the plane.

Let $A+B$ be a divisor of the series g_2^1 on $\mathcal{C}^{p+2}$ and let O be the divisor constituted by the residual intersection of $\mathcal{C}^{p+2}$ with the line $\langle A, B \rangle$. We take another arbitrary divisor $A' + B'$ of our g_2^1. One has the linear equivalences

$$A + B \equiv A' + B' \quad \text{and} \quad A + B + O \equiv A' + B' + O.$$

But $|A + B + O|$ is the complete series defined by the lines of the plane. Therefore, the line $\langle A', B' \rangle$ contains O, no matter how the pair of points A', B' (of the series g_2^1) are chosen. It follows that O is a point of multiplicity p for the curve $\mathcal{C}^{p+2}$. □

Let $\mathcal{C}^{p+2}$ be a plane model of a hyperelliptic curve of genus $p \geq 2$, ant let O be the point of multiplicity p of $\mathcal{C}^{p+2}$. We fix a system of homogeneous coordinates (x_0, x_1, x_2) in the plane in such a way that O is the point $[0, 1, 0]$. For $\mathcal{C}^{p+2}$ we then have an equation of the form, cf. §5.2.1,

$$\mathcal{C}^{p+2} : \varphi_p(x_0, x_2)x_1^2 + 2\varphi_{p+1}(x_0, x_2)x_1 + \varphi_{p+2}(x_0, x_2) = 0,$$

where $\varphi_i(x_0, x_2)$ is a homogeneous polynomial of degree i, $p \leq i \leq p+2$. Passing to affine coordinates $x = \frac{x_0}{x_2}$, $y = \frac{x_1}{x_2}$, O is the point at infinity along the axis y and

$$\mathcal{C}^{p+2} : y^2 \varphi_p(x) + 2y\varphi_{p+1}(x) + \varphi_{p+2}(x) = 0. \tag{8.17}$$

By way of the birational isomorphism

$$\begin{cases} u = x, \\ v = y\varphi_p(x) + \varphi_{p+1}(x), \end{cases} \qquad \begin{cases} x = u, \\ y = \dfrac{v - \varphi_{p+1}(u)}{\varphi_p(u)}, \end{cases}$$

the equation of $\mathcal{C}^{p+2}$ becomes (on eliminating the denominators):

$$v^2 = \varphi_{p+1}^2(u) - \varphi_p(u)\varphi_{p+2}(u).$$

Thus one obtains as birational model for $\mathcal{C}^{p+2}$ a plane curve of order $2p + 2$ with the point O at infinity on the v-axis as a point of multiplicity $2p$, and with equation of the form

$$\mathcal{C}_1^{2p+2} : v^2 = f_{2p+2}(u). \tag{8.18}$$

The polynomial $f_{2p+2}(u)$ may be written in the form

$$f_{2p+2}(u) = \lambda(u - a_1)(u - a_2)\ldots(u - a_{2p+2}), \tag{8.19}$$

where λ is a non-zero constant and the roots $a_i \in \mathbb{C}$ are distinct since the curve $\mathcal{C}_1^{2p+2}$ does not have multiple points on the u-axis. Moreover, the $2p + 2$ double points of the series g_2^1 on $\mathcal{C}_1^{2p+2}$ are projected onto the u-axis by the lines $u = a_1, \ldots, u = a_{2p+2}$ (cf. §8.8.4).

Let ζ denote one of the complex roots of the equation

$$z^2 = (a_1 - a_{2p+2})\ldots(a_{2p+1} - a_{2p+2}),$$

and consider the birational isomorphism

$$\begin{cases} u' = \dfrac{1}{a_{2p+2} - u}, \\ v' = \dfrac{v}{\zeta(a_{2p+2} - u)^{p+1}}, \end{cases}$$

which has as its inverse

$$\begin{cases} u = a_{2p+2} - \dfrac{1}{u'}, \\ v = \zeta \dfrac{v'}{u'^{p+1}}. \end{cases}$$

Under this birational isomorphism, the curve $\mathcal{C}_1^{2p+2}$ is transformed into the curve $\mathcal{C}_2^{2p+1}$ having equation

$$v'^2 = \lambda \frac{((a_{2p+2} - a_1)u' - 1) \ldots ((a_{2p+2} - a_{2p+1})u' - 1)}{(a_{2p+2} - a_1) \ldots (a_{2p+2} - a_{2p+1})},$$

or

$$v'^2 = \lambda \left(u' - \frac{1}{a_{2p+2} - a_1} \right) \cdots \left(u' - \frac{1}{a_{2p+2} - a_{2p+1}} \right). \tag{8.20}$$

In this way one finds a second birational plane model of order $2p + 1$ which has a singular point of multiplicity $2p - 1$ at the point at infinity on the v'-axis. The $2p + 2$ double points of the series g_2^1 are now projected to the u'-axis by the lines $u' = a_1, \ldots, u' = a_{2p+1}$ and by the line at infinity of the plane $\langle u', v' \rangle$.

We can now prove the following classification theorem.

Theorem 8.9.2. *In two affine planes with coordinates x, y and x', y', consider two hyperelliptic curves $\mathcal{C}$ and $\mathcal{C}'$, of degree $2p + 2$ and genus $p \geq 2$, of equations, respectively,*

$$\mathcal{C}: y^2 = f_{2p+2}(x) \quad \text{and} \quad \mathcal{C}': y'^2 = f'_{2p+2}(x').$$

Then a necessary and sufficient condition for $\mathcal{C}$ and $\mathcal{C}'$ to be birationally equivalent is that the $2p+2$ double points of the two series g_2^1, cut out by the two groups of lines $f_{2p+2}(x) = 0$ and $f'_{2p+2}(x') = 0$, respectively, correspond under a projectivity of $\mathbb{P}^1$ (cf. §8.8.4).

Proof. A birational transformation $\varphi \colon \mathcal{C} \to \mathcal{C}'$ changes the series g_2^1 of $\mathcal{C}$ into the series g_2^1 of $\mathcal{C}'$ and subordinates a bijective algebraic correspondence $\omega \colon \Phi \to \Phi'$, where Φ and Φ' are the two pencils of lines cutting out the two g_2^1's on $\mathcal{C}$ and $\mathcal{C}'$, respectively. Note that ω is in fact a projectivity (cf. §1.1.3).

Since φ changes the double points of the series g_2^1 of $\mathcal{C}$ into the double points of the series g_2^1 of $\mathcal{C}'$, the projectivity ω transforms the group of lines $f_{2p+2}(x) = 0$

into the group of lines $f'_{2p+2}(x') = 0$ and viceversa. Thus the condition is necessary.

We now prove the sufficiency of this condition. Write
$$f_{2p+2}(x) = \lambda(x - a_1)(x - a_2)\dots(x - a_{2p+2})$$
and
$$f'_{2p+2}(x') = \lambda'(x' - a'_1)(x' - a'_2)\dots(x' - a'_{2p+2}),$$
as in the form (8.19), with λ, λ' non-zero constants. Suppose that the two groups of points $\{a_1, \dots, a_{2p+2}\}$ and $\{a'_1, \dots, a'_{2p+2}\}$ are transformed into each other by a projectivity of $\mathbb{P}^1$. Let
$$x = \frac{\alpha x' + \beta}{\gamma x' + \delta} \qquad (8.21)$$
be the affine expression of the projectivity with $\alpha, \beta, \delta, \gamma \in \mathbb{C}$. Then
$$f_{2p+2}(x) = f_{2p+2}\left(\frac{\alpha x' + \beta}{\gamma x' + \delta}\right)$$
$$= \frac{\lambda}{(\gamma x' + \delta)^{2p+2}} \prod_{i=1}^{2p+2} (\alpha - \gamma a_i)\left(x' - \frac{\delta a_i - \beta}{\alpha - \gamma a_i}\right).$$

However, in view of (8.21) we also have
$$a'_i = \frac{\delta a_i - \beta}{\alpha - \gamma a_i}, \quad i = 1, \dots, 2p + 2,$$
and so, on setting $\varrho := \frac{\lambda}{\lambda'} \prod_{i=1}^{2p+2}(\alpha - \gamma a_i)$, we find that
$$f_{2p+2}(x) = f_{2p+2}\left(\frac{\alpha x' + \beta}{\gamma x' + \delta}\right) = \varrho \lambda' \frac{\prod_{i=1}^{2p+2}(x' - a'_i)}{(\gamma x' + \delta)^{2p+2}} = \varrho \frac{f'_{2p+2}(x')}{(\gamma x' + \delta)^{2p+2}}. \qquad (8.22)$$

Let t be one of the complex roots of the equation $z^2 - \varrho = 0$, and consider the birational isomorphism
$$\begin{cases} x = \dfrac{\alpha x' + \beta}{\gamma x' + \delta}, \\ y = t \dfrac{y'}{(\gamma x' + \delta)^{p+1}}. \end{cases}$$

In view of (8.22), this birational map transforms the curve $\mathcal{C} : y^2 = f_{2p+2}(x)$ into the curve with equation
$$\varrho \frac{y'^2}{(\gamma x' + \delta)^{2p+2}} = \varrho \frac{f'_{2p+2}(x')}{(\gamma x' + \delta)^{2p+2}},$$
that is, into the curve $\mathcal{C}' : y'^2 = f'_{2p+2}(x')$. $\square$

Remark 8.9.3. Consider a curve $\mathcal{C}^{p+2}$ of order $p+2$ and having an equation of type (8.17). Then $\mathcal{C}^{p+2}$ has the point at infinity O of the y-axis as a p-fold point whose (generally distinct) tangents have equation $\varphi_p(x) = 0$. If $\mathcal{C}^{p+2}$ has no other singularity then it is of genus p. Indeed, the presence of the p-fold point as the only singularity forces the genus to be at most p (as one can readily verify using the genus formula, cf. Definition 7.2.6, Proposition 7.2.7). On the other hand, every plane curve Γ of order $p+2$ having an ordinary p-fold point O is hyperelliptic. Indeed, the pencil of lines with center at O cuts out a linear series g_2^1 on Γ. However, the genus of Γ is not necessarily p because it could very well have other multiple points. The canonical series on Γ is cut out by the adjoints of order $(p+2) - 3 = p - 1$. Since these adjoints pass through O with multiplicity $p-1$ they are split into lines issuing from O. This proves that the canonical series is composed with a g_2^1. If Γ has other multiple points, the lines joining O with the multiple points are fixed components of the adjoint curves of order $p-1$, each line being counted with appropriate multiplicity.

The other two curves $\mathcal{C}_1^{2p+2}$, $\mathcal{C}_2^{2p+1}$ obtained via birational transformations from (8.17) and with equations (8.18) and (8.20) respectively, both have a complicated singularity at the point O at infinity on the y-axis. The tangents at O (which are in number $2p$ or $2p-1$, respectively) all coincide with the line at infinity.

Here we merely note that in the case of an even number of tangents, on resolving the singularity (see also Section 9.2), one sees that O is a $2p$-fold point with p infinitely near double points, while in the odd case O is a $(2p-1)$-fold point with p infinitely near double points. For a detailed analysis of the singularity we refer the reader to [36, Vol. III, pp. 92–94].

8.9.4. Bearing in mind that $\mathcal{M}_p$ is a variety, the condition "X is sufficiently general among the curves of genus p" (or, in the usual terminology, "X has general moduli") may be expressed by saying that X can be chosen in a Zariski open subset of $\mathcal{M}_p$.

It is easy to see that a genus 3 curve X with general moduli has no non-identity birational transformations into itself. Indeed, as seen in Exercise 8.8.9, a birational transformation of the canonical model C^{2p-2} of X is subordinate to a projectivity of S_{p-1} that permutes the set of $p(p^2-1)$ Weierstrass points. If the curve has general moduli, such a projectivity is necessarily the identity.

For particular curves, however, there may exist birational automorphisms that induce a permutation of the set of Weierstrass points. A theorem of Schwarz and Klein (see [100, p. 173]) guarantees that the number of such automorphisms is finite. Here are some famous examples of such particular curves, cf. [36, Vol. III, p. 238].

(1) The *Klein quartic*, with equation

$$x_0^3 x_1 + x_1^3 x_2 + x_2^3 x_0 = 0,$$

has a group of 168 birational automorphisms, called the *Klein group*. In this regard, a classical theorem of Hurwitz [54] states that a smooth curve of genus

$p \geq 2$ (over a field of characteristic 0) has at most $84(p-1)$ automorphisms. Thus, Klein's quartic shows the sharpness of the Hurwitz bound for curves of genus 3.

(2) The *Valentiner sextic*, with equation

$$10x_0^3 x_1^3 + 9x_2(x_0^5 + x_1^5) - 45x_0^2 x_1^2 x_2^2 - 135 x_0 x_1 x_2^4 + 27 x_2^6 = 0,$$

has a group of 360 birational automorphisms, called the *Valentiner group*.

A hyperelliptic curve has at least the involutory automorphism induced by its g_2^1, and in general has no others except for the identity map.

8.10 Complements and exercises

The possible singularities of the curves considered are always supposed to have distinct tangents (cf. Proposition 8.1.3).

We also refer to Section 13.3 for further exercises whose solution makes essential use of the theory of the planar representation of rational surfaces developed in Chapter 10.

8.10.1. *Given four distinct points A, B, C, D on an elliptic cubic X, the divisors $A + B$ and $C + D$ are equivalent if and only if the lines r_{AB} and r_{CD} meet on the curve.*

We note first that an elliptic cubic is a non-singular plane curve. If the lines r_{AB} and r_{CD} meet on X in a point H one then has, by definition, $A + B + H \equiv C + D + H$, so that $A + B \equiv C + D$.

Conversely, assume $A + B \equiv C + D$. If H is the further intersection of X with the line r_{AB}, one then has $A + B + H \equiv C + D + H$. Therefore C, D, H are collinear points, and hence $r_{AB} \cap r_{CD} = H$.

8.10.2. *Two points of an elliptic cubic X are linearly equivalent only if they coincide.*

An arbitrary line passing through a point A of X is a (first order) adjoint that also meets the curve in two other points P, Q. The unique adjoint of first order that contains these two points cuts the curve in the point A and thus it is the unique divisor of the complete series $|A|$. Therefore, there do not exist divisors equivalent to A but distinct from A.

8.10.3. *Prove that three non-collinear points of a non-singular plane quartic form an isolated divisor.*

A conic γ passing through the three points A, B, C (that is an adjoint of order 2) meets the quartic also in five other points. The conics passing through these five points cut out the complete series defined by the divisor $A + B + C$. If among the five points no four are collinear, then there is a unique conic passing through them, namely the conic γ. Consequently, $A + B + C$ does not have divisors to which it is linearly equivalent. If instead four of the five points belong to a line ℓ, then through them there pass ∞^1 conics (among which is γ) all split into ℓ and a variable line in a pencil with center on the quartic.

8.10.4. *Give an example of a linear series with a neutral pair on a curve $X \subset \mathbb{P}^n$.*

It suffices to consider the series cut out on X by the hyperplanes that pass through a generic S_{n-3}. If d and p are the order and genus of X, there are as many neutral pairs as there are chords of X supported by that S_{n-3}. But these are in number $\frac{(d-1)(d-2)}{2} - p$ (the number of double points of a generic plane projection of X).

8.10.5. *Prove that on a plane curve X the linear series cut out by all the curves of a given order m is simple.*

8.10.6 (cf. Proposition 8.7.3). *Give examples of composite linear series on a curve X.*

(1) Let X be a plane curve of order d with a $(d-2)$-fold point O. The curves C of order $d - 3$ (i.e., the canonical adjoints of X) cut out on X a composite series (composed with a series g_2^1), because they are all split into $d - 3$ lines passing through O, and each line has two further points of intersection with X in addition to O. Thus the curves C that pass through a point $P \in X$ contain the line r_{PO}, and consequently pass through P', where P' is the further intersection of C away from P and O.

(2) An example of a linear series composed with an involution is obtained by considering a curve X on a cone $V \subset \mathbb{P}^n$ where X meets the generator spaces in μ points not belonging to the linear space S constituting the vertex of the cone. The hyperplanes passing through S cut out on X a linear series composed with an involution σ_μ^1. Indeed, a hyperplane passing through S and through a generic point P of the curve contains the generator space G joining S with P. Therefore it contains the entire divisor of the intersections of the curve X with G.

In this case σ_μ^1 is said to be an *involution of genus* p (and is said to be *non-rational* if $p > 0$), where p denotes the *sectional genus* of the cone V, that is, the genus of the curve obtained as intersection of V with a generic linear space of dimension $n - \dim(V) + 1$.

In the case, like that of the cone, of a variety V, locus of ∞^1 linear spaces, the sectional genus is also called simply the *genus of V*.

A further (and apparently more general) example is obtained by considering a variety V which is the locus of ∞^1 linear spaces issuing from a linear space S, and on it a curve X that meets each of those linear spaces in μ variable points. The hyperplanes passing through S cut out on X a linear series composed with an involution σ_μ^1 of genus p, where p is the genus of V.

8.10.7 (cf. §8.4.4). *Let $\varphi \colon X \to X'$ be a birational map between two non-singular algebraic curves. Then φ is a morphism and bijective without exceptions.*

Suppose that X is embedded in $\mathbb{P}^n$ and has order d. The linear series g_d^n of the hyperplane sections of X is simple and has no fixed points. Under $\varphi \colon X \to X'$ to g_d^n there corresponds on X' a linear series which is simple and without fixed points. The residual series of a point $P' \in X'$ with respect to that linear series is a g_{d-1}^{n-1}. To this series there again corresponds on X a series g_{d-1}^{n-1} cut out by the hyperplanes passing through a point P (belonging to X). To the point P' there corresponds the point P. Similarly, to each point of X there corresponds a well-defined point of X'.

8.10.8. *An irreducible algebraic curve X on which there are two simply infinite linear series g_d^1 and $g_{d'}^1$, is birationally isomorphic to a plane curve of order $d + d'$ having two multiple points with multiplicities d and d'.*

In particular, if X contains two g_2^1's, then is an elliptic curve.

Consider a plane model $\mathcal{C}$ of the curve X (cf. Theorem 9.2.4). The two linear series can be referred projectively to two coplanar pencils of lines, with centers O and O'. If we consider two divisors of g_d^1 and $g_{d'}^1$ to be in correspondence when they have (at least) a point in common, we obtain an algebraic correspondence ω of indices d', d between the two pencils, and the locus of the points common to corresponding lines is a curve $\mathcal{C}'$ birationally isomorphic to $\mathcal{C}$. Indeed, a generic point P of $\mathcal{C}$ belongs to a divisor of g_d^1 and to a divisor of $g_{d'}^1$, and so determines a pair of corresponding lines in the two pencils, that is a point P' of $\mathcal{C}'$. Conversely, a point P' of $\mathcal{C}'$ belongs to two corresponding lines, and so determines two divisors from the two series which have a point P in common.

To find the order of $\mathcal{C}'$ we must determine the number of its intersections with a generic line ℓ, and these are the fixed points in the (d', d) algebraic correspondence induced by ω on ℓ. By Chasles' correspondence principle, §1.1.3, the order of $\mathcal{C}'$ is $d + d'$. The two series g_d^1 and $g_{d'}^1$ are cut out on $\mathcal{C}'$ by the two pencils of lines having centers O and O' respectively. One then sees immediately that O and O' are multiple points for $\mathcal{C}'$ of multiplicity d and d' respectively (a generic line through O meets $\mathcal{C}'$, away from O, in $(d + d') - d' = d$ points; and similarly for O').

8.10.9. *Verify that the canonical adjoints of an algebraic plane curve are covariant with respect to Cremona transformations, namely that a Cremona transformation of the plane sending a curve X into a curve X' maps the system of canonical adjoints of X into the system of canonical adjoints of X'.*

8.10.10. *Prove that the canonical series on a curve $C(\alpha, \beta)$ lying on a non-singular quadric $\mathcal{Q}$ in $\mathbb{P}^3$ is cut out by the curves of type $(\alpha - 2, \beta - 2)$.*

We make use of the properties of the stereographic projection of a quadric, as discussed in §7.3.1. The projection $\mathcal{C}$ of $C(\alpha, \beta)$ from a point of $\mathcal{Q}$ not belonging to $C(\alpha, \beta)$ onto a plane π has order $\alpha + \beta$ and has (only) two multiple points A and B of multiplicities α and β respectively. The canonical adjoints of $\mathcal{C}$ contain the line r_{AB} as component. The residual components, which are curves of order

$\alpha + \beta - 4$ having multiplicities $\alpha - 2$ in A and $\beta - 2$ in B, are the projections from O of the curves of type $(\alpha - 2, \beta - 2)$ of $\mathcal{Q}$.

Conversely, the curves of type $(\alpha - 2, \beta - 2)$ on $\mathcal{Q}$ constitute a linear system of dimension $\alpha\beta - \alpha - \beta = (\alpha - 1)(\beta - 1) - 1 = p - 1$, where p is the genus of the curve $C(\alpha, \beta)$. On $C(\alpha, \beta)$ that system cuts out a linear series having order $\alpha(\beta - 2) + \beta(\alpha - 2) = 2(\alpha\beta - \alpha - \beta) = 2p - 2$ and having the same dimension $p - 1$.

For example, on a non-singular $(3, 3)$ curve which is the complete intersection of $\mathcal{Q}$ with a cubic surface, the canonical series is given by the $(1, 1)$ curves, that is, by the plane sections of $\mathcal{Q}$. Thus the $(3, 3)$ curve is a canonical projective curve.

8.10.11. *A plane quartic $\mathcal{C}$ with two nodes is the projection of a non-singular quartic in S_3.*

The series $g_4^2 = \mathcal{H}_\mathcal{C}$ of rectilinear sections is non-complete. Indeed, by the Riemann–Roch theorem, the complete series g_4^r that contains it (which is non-special) has dimension $r = 4 - p = 3$, since the genus p of $\mathcal{C}$ is $p = 1$. Hence the quartic is the projection of the curve X which is the projective image of a series g_4^3 (the curve X is the base quartic of a pencil of quadrics).

One can avoid an appeal to the Riemann–Roch theorem by observing that among the second order adjoints of $\mathcal{C}$ (that is, the conics that pass through the two nodes A and B) there are, in particular, those split into the line that joins the two double points and a residual line. Thus the linear series $\mathcal{H}_\mathcal{C}$ is contained in the linear series cut out, away from the two nodes, by the conics passing through A and B. These conics constitute a 3-dimensional linear system. On $\mathcal{C}$ they cut out a complete series g_4^3 containing our g_4^2.

An even more elementary way to reach the same conclusion is the following. Take a point O outside of the plane of $\mathcal{C}$ and consider a quadric $\mathcal{Q}$ passing through the two lines r_{OA} and r_{OB}. The cone that projects $\mathcal{C}$ from O meets $\mathcal{Q}$ in a curve of order 8 that contains the two lines r_{OA}, r_{OB} counted twice; the residual component is a non-singular quartic $C(2, 2)$ and $\mathcal{C}$ is the projection of $C(2, 2)$ from O.

8.10.12. *A plane quartic $\mathcal{C}$ with two nodes is the projection of a non-singular quintic in S_4.*

To a divisor of the g_4^2 of rectilinear sections of $\mathcal{C}$ we adjoin a point (of $\mathcal{C}$), thus obtaining an effective divisor D of degree 5. The genus of $\mathcal{C}$ being $p = 1$, it follows from Theorem 8.5.4 that this divisor is non-special; and so it defines a complete series $|D| = g_5^4$ whose projective image is thus a quintic X in S_4. The series g_4^2 is partially contained in g_5^4 and $\mathcal{C}$ is the projection of X from a point $P \in X$.

8.10.13. *An elliptic curve X of order d is the projection of a curve of order d embedded in $\mathbb{P}^{d-1}$ which is called an **elliptic normal curve** of order d.*

The series $\mathcal{H}$ of the hyperplane sections of X is non-special since X has genus $p = 1$. Hence it is contained in a complete series g_d^{d-1} whose projective image is a curve $\mathcal{X}$ of order d embedded in $\mathbb{P}^{d-1}$. If $\mathcal{H}$ is not already a complete series (that is, if it has dimension $< d - 1$), then X is the projection of $\mathcal{X}$.

8.10.14. *Find the dimension of the linear system Σ_0 of surfaces of order $N(\gg 0)$ in $\mathbb{P}^3$ that pass through a given curve C. In particular, consider the cases in which C is a line, a conic, a cubic, a rational C^4 or an elliptic C^4.*

The linear system Σ_N of the surfaces in $\mathbb{P}^3$ of order N has dimension $t = \binom{N+3}{N} - 1$. Consider the linear series $g^r_{N \deg(C)}$ cut out by the surfaces of Σ_N that do not contain C. By the Riemann–Roch theorem (applied to the complete series defined by the series $g^r_{N \deg(C)}$), we have $r \leq N \deg(C) - p + i$ where p is the genus of C and i is the index of speciality of the series. If $p \leq 1$, then $g^r_{N \deg(C)}$ is non-special by Theorem 8.5.4. Therefore the dimension sought is

$$\dim \Sigma_0 \geq t - r = \binom{N+3}{N} - 1 - N \deg(C) + p.$$

For $N \gg 0$ we may assume that the $g^r_{N \deg(C)}$ is complete and therefore equality holds in the preceding inequality.

8.10.15. *Determine the dimension of the linear system of hypersurfaces of order $N(\gg 0)$ in $\mathbb{P}^n$ that pass through a given curve C. In particular, discuss the cases in which C is a rational normal curve or a elliptic normal curve.*

Consider, for example, the case of an elliptic normal curve in $\mathbb{P}^n$, that is, a non-singular curve of genus $p = 1$ and order $n + 1$. Then the same reasoning given in 8.10.14 shows that the required dimension is

$$\binom{N+n}{N} - N(n+1).$$

8.10.16. *Let $\mathcal{C}$ be a plane algebraic curve of order d and genus p. Find the condition in order that it be the projection of a curve X of the same order embedded in S_R ($R > 2$).*

It is a question of finding the condition for the series g_d^2 cut out on $\mathcal{C}$ by the lines of the plane to be contained in a series g_d^r with $r \geq R$.

The series g_d^2 is contained in a (complete) series of order d and dimension $\geq d - p$, and so for every integer $R \leq d - p$ it is contained in a linear series g_d^R. Therefore, if $R \leq d - p$ (that is, if $p \leq d - R$) the curve $\mathcal{C}$ is the projection of a curve X of order d and contained in S_R.

Suppose then that $p \geq d - R + 1$. If g_d^2 is contained in a series g_d^r with $r \geq d - p + 1 > d - p$, then it must be special, and so, by the Riemann–Roch theorem, $\dim |K_{\mathcal{C}} - g_d^2| \geq 0$. Hence every line ℓ of the plane meets $\mathcal{C}$ in d points which belong to an adjoint of order $d-3$, which splits into the line ℓ and an adjoint of order $d-4$. In particular, if $\mathcal{A}_{d-4}$ denotes the linear system of adjoints of order $d-4$, one has $\dim \mathcal{A}_{d-4} = \dim |K_{\mathcal{C}} - g_d^2|$.

It follows that if $p \geq d - R + 1$ the curve $\mathcal{C}$ can be the projection of a curve of the same order belonging to a space of dimension > 2 only if it possesses adjoints of order $d-4$.

Now on $\mathcal{C}$ the linear system $\mathcal{A}_{d-4}$ of the adjoints of order $d-4$ cuts out a linear series g_m^t which has

- order $m = 2p - 2 - d$, because by adjoining d collinear points to a divisor of the given series one obtains a canonical divisor;

- dimension t greater than or equal to that of $\mathcal{A}_{d-4}$, because no curve of $\mathcal{A}_{d-4}$ can contain $\mathcal{C}$ as component;

- index of speciality $i(g_m^t) \geq 3$, inasmuch as every group of g_m^t belongs at least to the ∞^2 canonical divisors that are obtained by adjoining to it a rectilinear section; and so the g_d^2 is contained in the series $|K_{\mathcal{C}} - g_m^t|$. It follows that $i(g_m^t) - 1 \geq 2$.

Thus by the Riemann–Roch theorem one has

$$t \geq \dim \mathcal{A}_{d-4} \geq (2p - 2 - d) - p + i(g_m^t) \geq p - d + 1.$$

If i is the index of speciality of g_d^r one has $r = d - p + i$, and that is to say that the dimension of the linear system of adjoints of order $d-3$ passing through d collinear points (namely, the dimension of $\mathcal{A}_{d-4}$) is

$$i - 1 = r + p - d - 1.$$

But $r \geq R$, whence

$$\dim \mathcal{A}_{d-4} \geq R + p - d - 1.$$

On the other hand, if $\mathcal{C}$ has s_i-fold points (actual, or in the successive infinitely near neighborhoods) the dimension of $\mathcal{A}_{d-4}$ is

$$\frac{1}{2}(d-4)(d-1) - \frac{1}{2}\sum s_i(s_i - 1) + \sigma$$

$$= \frac{1}{2}(d-2)(d-1) - d + 1 - \frac{1}{2}\sum s_i(s_i - 1) + \sigma = p - d + 1 + \sigma,$$

where σ denotes the superabundance. Thus

$$\dim \mathcal{A}_{d-4} = p - d + 1 + \sigma \geq R + p - d - 1.$$

Therefore
$$\sigma \geq R - 2.$$

Conversely, if $\sigma \geq R-2$, the dimension of the linear system of adjoints of order $d-4$ is $\dim \mathcal{A}_{d-4} \geq p - d + R - 1$. We know that $\dim \mathcal{A}_{d-4} = \dim |K_{\mathcal{C}} - g_d^2|$, and so every rectilinear section of $\mathcal{C}$ belongs to (at least) $p - d + R$ independent canonical divisors, which is to say $i(g_d^2) \geq p - d + R$. Thus, the linear series cut out on $\mathcal{C}$ by the lines is contained in a complete linear series whose dimension is $r \geq d - p + (p - d + R) = R$. In conclusion we have shown:

A necessary and sufficient condition in order for $\mathcal{C}$ to be obtained as a projection of a curve of the same order belonging to a space S_R (but not as a projection of a curve of the same order embedded in S_{R+1}) is that $\mathcal{C}$ admit adjoint curves of order $d - 4$ forming a linear system of superabundance $\geq R - 2$.

8.10.17 (Halphen–Castelnuovo's bound). Let X be a reduced, irreducible curve of order d embedded in $\mathbb{P}^n$. Then the genus p of X is bounded by

$$p \leq \left[\frac{d-2}{n-1}\right]\left(d - n - \left(\left[\frac{d-2}{n-1}\right] - 1\right)\frac{n-1}{2}\right), \tag{8.23}$$

where $[x]$ means the greatest integer $\leq x$.

If $n = 3$, then $\left[\frac{d-2}{2}\right]$ is $\frac{d-2}{2}$ for d even and $\frac{d-3}{2}$ for d odd. Thus for a degree d curve in $\mathbb{P}^3$ one has

$$p \leq \begin{cases} \dfrac{d-2}{2}\left(d - 2 - \dfrac{d-2}{2}\right) = \dfrac{(d-2)^2}{4} & \text{if } d \text{ is even,} \\ \dfrac{d-3}{2}\left(d - 2 - \dfrac{d-3}{2}\right) = \dfrac{(d-1)(d-3)}{4} & \text{if } d \text{ is odd.} \end{cases}$$

Whence, in any case,

$$p \leq \left[\frac{(d-2)^2}{4}\right].$$

This is *Halphen's theorem*, [47]. This theorem was extended to curves in $\mathbb{P}^n$ by Castelnuovo [21], [22], as in (8.23). The maxima assigned by Halphen and Castelnuovo are effectively achieved by certain curves. The curves of maximum genus in $\mathbb{P}^3$ lie on non-singular quadrics, and are the curves of the type (α, α) for $d = 2\alpha$ even, and of the type $(\alpha, \alpha + 1)$ for $d = 2\alpha + 1$ odd.

8.10.18. *Prove that for an algebraic curve X in $\mathbb{P}^3$ having order d and genus $p \geq d - 2$ the chords issuing from a generic point P belong to a cone of order $d - 4$.*

The projection of X from a point P in $\mathbb{P}^3$ is a plane curve $\mathcal{C}$ of order d and genus p which, by what we have seen in 8.10.16, possesses at least one adjoint of order $d - 4$. This means that the chords of X issuing from P belong to a cone of order $d - 4$.

For example, a canonical curve of genus $p = 4$ has order 6 and from every generic point of $\mathbb{P}^3$ there issue six chords (equal in number to the number of double points of its plane model, cf. Corollary 8.5.11) which belong to a quadric cone.

Consider a curve X in $\mathbb{P}^3$ of order $d = 2m$ and genus $p = 2m - 1$. Such curves exist if $2m - 1 \leq (m-1)^2$, that is, $m > 3$ (see §8.10.17). From every generic point of the space there issue $2m^2 - 5m + 2$ chords of X, all of which lie on a cone of order $\deg(X) - 4 = 2m - 4$. This cone also meets the cone projecting X from P in $2m(2m - 4) - 2(2m^2 - 5m + 2) = 2m - 4$ lines.

8.10.19 (Genus formula for a complete intersection curve). Let X be a non-singular curve which is a complete intersection in $\mathbb{P}^n$, that is X is obtained as the intersection of $n - 1$ algebraic hypersurfaces, of degree d_i, $i = 1, \ldots, n-1$, that intersect along it transversally (cf. Section 7.1). The curve X is thus of order $\prod_{i=1}^{n-1} d_i$ by Bézout's theorem (Theorem 4.5.2), and its genus p is expressed by the relation

$$p = 1 + \frac{1}{2} \prod_{i=1}^{n-1} d_i \left(\sum_{i=1}^{n-1} d_i - n - 1 \right).$$

This a consequence of the so called "adjunction formula", which implies that the canonical divisors of $|K_X|$ are cut out on X by the hypersurfaces of degree $\sum_{i=1}^{n-1} d_i - n - 1$. We refer e.g., to [12] for the general theory.

Chapter 9
Cremona Transformations

This chapter is dedicated to the study of generically bijective algebraic correspondences between projective spaces of the same dimension r, with special regard for the case $r = 2$. Such transformations are called "Cremona transformations" in honor of L. Cremona who studied them from the most general point of view, and emphasized their great importance for the development of algebraic geometry.

In Sections 9.1 and 9.2, ample space is given to the quadratic transformations between planes, that is, to the transformations that change the lines of one of the two planes into the conics of a homaloidal net of the other, to some of their important applications, like, for example, the transformation of a given algebraic curve into a plane algebraic curve having as singularities only multiple points with distinct tangents, and to the study of the structure of a singularity of an algebraic curve via the technique of blowing up a point.

In Section 9.3, after having shown how the study of Cremona transformations between planes may be identified with that of homaloidal nets of algebraic plane curves, we study such nets and their fundamental curves. Here the Cremona equations and the Noether–Rosanes inequality (see §9.3.9) are particularly important. They lie at the base of the Noether–Castelnuovo theorem, Theorem 9.3.10, according to which every Cremona transformation between planes is the product of quadratic transformations.

As far as the case $r > 2$ is concerned, we limit ourselves to a few remarks in Section 9.4. In particular, the various possible types of homaloidal linear systems of quadrics in three space are determined. The exercises proposed in Section 9.5, and completely resolved there, constitute an important deepening of the theory developed in this chapter.

For a modern view of the topics discussed in this chapter and also for their applications to advanced problems in algebraic geometry, see the preface of Iskovskikh and Reid in [53].

9.1 Quadratic transformations between planes

A *Cremona transformation* is a birational isomorphism between projective spaces, that is, a dominant rational map between two projective spaces of the same dimension which has a rational inverse mapping (cf. Section 2.6).

The study of these transformations is rather simple in the case in which the spaces in question are two planes.

9.1. Quadratic transformations between planes

In a plane S_2 consider a net of conics, that is, a 2-dimensional linear system Σ of conics (cf. Section 6.1). We will say that it is a *homaloidal net* if it possesses three *base points*, that is, three points through which all the conics of Σ pass. Thus there are three different types of homaloidal nets of conics: the nets $\Sigma = ABC$ of conics with three distinct base points A, B, C; the nets $\Sigma = AAB$ of conics passing through two points A and B and all having at one of them, say A, the same tangent; the nets $\Sigma = AAA$ of conics mutually osculating at a point A (cf. [13, Vol. II, Chapter 16]).

In order to have three conics that generate a homaloidal net Σ it suffices to annihilate the minors of a matrix

$$\begin{pmatrix} L_0 & L_1 & L_2 \\ M_0 & M_1 & M_2 \end{pmatrix}$$

whose elements are linear forms in the indeterminates x_0, x_1, x_2. Indeed, the two conics with equations

$$L_0 M_1 - L_1 M_0 = 0, \quad L_0 M_2 - L_2 M_0 = 0 \tag{9.1}$$

have four points in common, one of which is the point in which the two lines $L_0 = 0$, $M_0 = 0$ intersect. This point does not lie on the third conic which however does pass through the other three because its equation $L_1 M_2 - L_2 M_1 = 0$ is a consequence of (9.1); indeed, it is obtained from (9.1) by eliminating the ratio $L_0 : M_0$. Thus for Σ one has the equation

$$\lambda_0(L_1 M_2 - L_2 M_1) + \lambda_1(L_2 M_0 - L_0 M_2) + \lambda_2(L_0 M_1 - L_1 M_0) = 0,$$

with $\lambda_0, \lambda_1, \lambda_2$ elements of the base field K not all of which are zero.

Assuming this, let x_0, x_1, x_2 be projective coordinates in a plane S_2 and y_0, y_1, y_2 projective coordinates in another plane S_2' that may of course coincide with S_2. Moreover, let θ be the algebraic correspondence between the two planes defined by two algebraic equations

$$\begin{cases} f(x_0, x_1, x_2; y_0, y_1, y_2) = 0, \\ g(x_0, x_1, x_2; y_0, y_1, y_2) = 0. \end{cases} \tag{9.2}$$

We ask what form these two equations must have in order that θ be a birational isomorphism in such a way that every *generic* point of each of the two planes corresponds to a unique point of the other plane.

That certainly occurs if the equations (9.2) are bilinear equations, that is, two equations that can be written in the form

$$\begin{cases} y_0 L_0(x_0, x_1, x_2) + y_1 L_1(x_0, x_1, x_2) + y_2 L_2(x_0, x_1, x_2) = 0, \\ y_0 M_0(x_0, x_1, x_2) + y_1 M_1(x_0, x_1, x_2) + y_2 M_2(x_0, x_1, x_2) = 0, \end{cases} \tag{9.3}$$

and also in the form

$$\begin{cases} x_0 F_0(y_0, y_1, y_2) + x_1 F_1(y_0, y_1, y_2) + x_2 F_2(y_0, y_1, y_2) = 0, \\ x_0 G_0(y_0, y_1, y_2) + x_1 G_1(y_0, y_1, y_2) + x_2 G_2(y_0, y_1, y_2) = 0, \end{cases} \quad (9.4)$$

with L_i, M_i, F_i, G_i linear forms.

We shall see, however, that in this way one obtains only a very particular class of birational isomorphisms between the two planes, namely the *quadratic transformations*. We refer, for example, to (9.3). They immediately give, for some $\rho \in K^*$,

$$\begin{cases} \rho y_0 = L_1 M_2 - L_2 M_1, \\ \rho y_1 = L_2 M_0 - L_0 M_2, \\ \rho y_2 = L_0 M_1 - L_1 M_0, \end{cases}$$

which we will write in the sequel as

$$y_0 : y_1 : y_2 = L_1 M_2 - L_2 M_1 : L_2 M_0 - L_0 M_2 : L_0 M_1 - L_1 M_0. \quad (9.5)$$

In this way y_0, y_1, y_2 can be assumed to be equal to three quadratic forms in the x_i, $i = 0, 1, 2$. Therefore if $P' = P'(y) := [y_0, y_1, y_2]$ belongs to the line $r \subset S_2'$ with equation

$$\lambda_0 y_0 + \lambda_1 y_1 + \lambda_2 y_2 = 0,$$

the point $P = P(x) := [x_1, x_2, x_3]$ corresponding to it via (9.5) belongs to the conic with equation

$$\lambda_0(L_1 M_2 - L_2 M_1) + \lambda_1(L_2 M_0 - L_0 M_2) + \lambda_2(L_0 M_1 - L_1 M_0) = 0.$$

This conic varies in a homaloidal net Σ of conics in S_2 as the line r varies in the net of lines of S_2'. To the lines of a plane there thus correspond the conics of a homaloidal net of the other plane. This correspondence between lines and conics is bijective, algebraic and without exceptions: it is a projectivity between the net of lines of a plane and a homaloidal net of conics in the other plane.

Conversely, let Σ be an arbitrary homaloidal net of conics in S_2 and consider a projectivity ω between the net Σ and the net of lines in S_2'. If P is a point distinct from the base points of Σ, the conics of Σ that pass through P form a pencil that has four base points: the three base points of Σ and the point P. To this pencil there corresponds via ω a pencil of lines in S_2' whose base point P' is determined by P. Moreover, to a point $P' \in S_2'$ there corresponds a pencil of conics in Σ (the image of the pencil of lines with center P') and thus the common point P, the new base point of this pencil different from the base points of the net. Hence the map defined by $P \mapsto P'$ gives rise to a birational isomorphism, that is, to a Cremona transformation $S_2 \to S_2'$ which transforms the conics of Σ into the lines of S_2'.

We may therefore conclude:

- To give a quadratic transformation between two planes is the same as giving a homaloidal net of conics in one of them.

Quadratic transformations are rational maps defined and bijective only on two open sets. In each of the two planes there are in fact points which do not have a well-defined corresponding point. In the plane S_2 the exceptional points are the base points of the net Σ, and analogously in S'_2 the exceptional points are the base points of the homaloidal net Σ' of equation

$$\mu_0(F_1G_2 - F_2G_1) + \mu_1(F_2G_0 - F_0G_2) + \mu_2(F_0G_1 - F_1G_0) = 0,$$

with μ_0, μ_1, μ_2 not all zero in K.

We shall see later that Σ and Σ' are two nets of the same type, and thus one has three projectively distinct types of quadratic transformations between planes.

The base points of each of the two homaloidal nets are called *fundamental points* of the quadratic transformation. To the fundamental points of one of the two planes there is associated a line in the other plane which is called the *exceptional line* corresponding to that fundamental point.

If, for example, at the base point A the conics of Σ are not all tangent to the same line, then to a pencil Φ of conics of Σ having a given tangent t at A there corresponds a pencil of lines in S'_2. The locus described by the center T' of this pencil as t varies is a rational curve E' whose points correspond bijectively to the directions of S_2 issuing from A. One sees immediately that E' is a line. Indeed, a point common to E' and to a generic line r' of S'_2 can correspond only to the unique tangent line at A to the conic of Σ that corresponds to r'. Thus E' is the exceptional curve that corresponds to the fundamental point A.

If the conics of Σ have the same tangent but are not mutually osculating at the base point A, then to define a point T' of E' one can make use of a pencil Φ of conics of Σ that are mutually osculating at A. If the net Σ has a unique base point A (and so the conics of Σ osculate at A) one uses a pencil of hyperosculating conics.

9.1.1 Classification of the quadratic transformations between planes. The study of quadratic transformations between two planes is easily handled with analytic tools by seeking the simplest possible representations for the various types of homaloidal nets Σ of conics in the plane S_2 (in this regard see §9.3.7).

If the three base points of Σ are distinct we may suppose that they coincide with the points $A_0 = [1, 0, 0]$, $A_1 = [0, 1, 0]$, and $A_2 = [0, 0, 1]$ and thus obtain the following equation for the net:

$$\lambda_0 x_1 x_2 + \lambda_1 x_2 x_0 + \lambda_2 x_0 x_1 = 0.$$

If there are only two base points, we may suppose that the conics of Σ are tangent at $A_2 = [0, 0, 1]$ to the line $r_{A_1 A_2} : x_0 = 0$ and furthermore also pass through $A_0 = [1, 0, 0]$. Then for Σ we will have the equation

$$\lambda_0 x_2 x_0 + \lambda_1 x_0 x_1 + \lambda_2 x_1^2 = 0.$$

Finally, it the conics of Σ are mutually osculating at $A_2 = [0, 0, 1]$, we may suppose that Σ contains the two degenerate conics $x_0^2 = 0$ and $x_0 x_1 = 0$. In Σ we take an irreducible conic γ and we assume that the points $A_0 = [1, 0, 0]$ and $A_1 = [0, 1, 0]$ are respectively the further intersection of γ with the line $x_1 = 0$ and the pole of that line with respect to γ. If one chooses as unit point $U = [1, 1, 1]$ a point of γ we find that γ has the equation $x_0 x_2 - x_1^2 = 0$ and so Σ is defined by

$$\lambda_0 (x_0 x_2 - x_1^2) + \lambda_1 x_0^2 + \lambda_2 x_0 x_1 = 0.$$

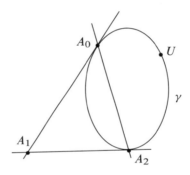

Figure 9.1

Here then are the three types of quadratic transformations.

Type I. $\theta \colon S_2 \to S_2'$ is defined by $y_0 : y_1 : y_2 = x_1 x_2 : x_2 x_0 : x_0 x_1$.

This formula may easily be inverted and gives for θ^{-1} the expression

$$x_0 : x_1 : x_2 = y_1 y_2 : y_2 y_0 : y_0 y_1.$$

Indeed, if $x_0 x_1 x_2 \neq 0$, we have

$$y_0 : y_1 : y_2 = \frac{1}{x_0} : \frac{1}{x_1} : \frac{1}{x_2},$$

or also

$$x_0 : x_1 : x_2 = \frac{1}{y_0} : \frac{1}{y_1} : \frac{1}{y_2}.$$

Thus, multiplying the second term of the last equality by $y_0 y_1 y_2$ one obtains the desired expression for θ^{-1}.

Type II. $\theta: S_2 \to S_2'$ is defined by $y_0 : y_1 : y_2 = x_2 x_0 : x_0 x_1 : x_1^2$.

One has
$$\frac{y_0}{y_1} = \frac{x_2}{x_1} = \frac{y_0 y_2}{y_1 y_2}; \quad \frac{y_1}{y_2} = \frac{x_0}{x_1} = \frac{y_1^2}{y_1 y_2},$$
and then
$$x_0 : x_1 : x_2 = y_1^2 : y_1 y_2 : y_0 y_2$$
is the expression for θ^{-1}.

Type III. $\theta: S_2 \to S_2'$ is defined by $y_0 : y_1 : y_2 = x_0 x_2 - x_1^2 : x_0^2 : x_0 x_1$.

In this case one has
$$\frac{y_0}{y_1} = \frac{x_0 x_2 - x_1^2}{x_0^2} = \frac{x_2}{x_0} - \left(\frac{x_1}{x_0}\right)^2; \quad \frac{y_2}{y_1} = \frac{x_1}{x_0}.$$

Hence
$$\frac{x_2}{x_0} = \frac{y_0}{y_1} + \left(\frac{y_2}{y_1}\right)^2 = \frac{y_0 y_1 + y_2^2}{y_1^2}$$
and then
$$\frac{x_0}{x_1} = \frac{y_1}{y_2} = \frac{y_1^2}{y_1 y_2}; \quad \frac{x_0}{x_2} = \frac{y_1^2}{y_0 y_1 + y_2^2}.$$

Therefore θ^{-1} is given by
$$x_0 : x_1 : x_2 = y_1^2 : y_1 y_2 : y_0 y_1 + y_2^2.$$

In all three cases one sees that the equations of θ and its inverse have the same form (up to a change in the order of the coordinates). Thus it is confirmed that the two homaloidal nets Σ and Σ' are of the same type, and so there exist only three projectively distinct types of quadratic transformations between planes.

9.2 Resolution of the singularities of a plane algebraic curve

In this section we wish to show that by suitably combining quadratic transformations one can transform a plane algebraic curve into an algebraic plane curve endowed with only ordinary singular points.

9.2.1 The transform of a plane algebraic curve by way of quadratic transformations.
We wish to examine the effect of a quadratic transformation θ on a curve C passing through a fundamental point of θ. We limit ourselves to the case in which θ is of type I; it is, however, a useful exercise for the reader to work out the other two cases.

Let $\theta: S_2 \to S_2'$ be a quadratic transformation of type I. In S_2 we consider an algebraic curve C^n of order n passing simply through one of the fundamental points, for example A_0, and tangent there to the line $ax_1 + bx_2 = 0$, $a, b \in K$. The curve C^n, whose equation is written in the form

$$f(x_0, x_1, x_2) = x_0^{n-1}(ax_1 + bx_2) + x_0^{n-2}\varphi_2(x_1, x_2) + \cdots = 0,$$

with $\varphi_2(x_1, x_2)$ form of degree two in x_1, x_2, is transformed by θ into the curve of equation

$$\begin{aligned}f(y_1 y_2, y_2 y_0, y_0 y_1) &= (y_1 y_2)^{n-1} y_0(ay_2 + by_1) \\ &\quad + (y_1 y_2)^{n-2} y_0^2 \varphi_2(y_2, y_1) + \cdots = 0.\end{aligned}$$

Therefore to the curve C^n there corresponds in S_2' a curve that contains the line $y_0 = 0$ as a simple component. That component is the exceptional line associated to the fundamental point A_0. The residual component (which we will call the *proper transform* of C^n, cf. Proposition–Definition 9.2.2), with equation

$$(y_1 y_2)^{n-1}(ay_2 + by_1) + y_0(\cdots) = 0,$$

cuts the line $y_0 = 0$ not only in the fundamental points $B_1 = [0, 1, 0]$, $B_2 = [0, 0, 1]$ in S_2' but also in the point $[0, a, -b]$ which depends only on the tangent line to C^n at A_0. It follows that all the curves through A_0 with a given tangent have as (proper) transforms curves having a common point of intersection with the line $y_0 = 0$. This point is thus associated with the direction issuing from A_0 and belongs to that tangent. Therefore, one introduces the following terminology:

- Given a quadratic transformation of type I, the exceptional line corresponding to a fundamental point represents the first order neighborhood of that point.

More generally, consider a curve C^n having A_0 as s-fold point and let $\varphi_s(x_1, x_2) = 0$ be the equation of the tangent cone at A_0, so that the equation of C^n may be written in the form

$$f(x_0, x_1, x_2) = x_0^{n-s} \varphi_s(x_1, x_2) + x_0^{n-s-1} \varphi_{s+1}(x_1, x_2) + \cdots = 0,$$

with φ_s, φ_{s+1} forms of degree $s, s+1$ respectively. Since

$$\begin{aligned}f(y_1 y_2, y_2 y_0, y_0 y_1) &= (y_1 y_2)^{n-s} \varphi_s(y_2 y_0, y_0 y_1) \\ &\quad + (y_1 y_2)^{n-s-1} \varphi_{s+1}(y_2 y_0, y_0 y_1) + \cdots \\ &= y_0^s ((y_1 y_2)^{n-s} \varphi_s(y_2, y_1) + y_0(\cdots)),\end{aligned}$$

the curve transformed under θ decomposes into the exceptional line $y_0 = 0$ counted s times and a residual curve C' which meets the line $y_0 = 0$ in the fundamental points B_1, B_2 and also in the additional points defined by $y_0 = \varphi_s(y_2, y_1) = 0$ (which may also be, in whole or in part, among the points B_1 and B_2).

9.2. Resolution of the singularities of a plane algebraic curve

Thus there is a bijective correspondence between the set of tangent lines to C^n at A_0 (which correspond to the linear factors of $\varphi_s(x_1, x_2)$) and the intersections of C' with the exceptional line that corresponds to A_0. If $t_1, t_2, \ldots, t_h$ are the tangents of C^n at A_0 and $m_1, m_2, \ldots, m_h$ are their respective multiplicities (that is, the multiplicities of the corresponding roots of the equation $\varphi_s(x_1, x_2) = 0$) we will have h points $T_1, T_2, \ldots, T_h$ on the line $y_0 = 0$ and $m_1, m_2, \ldots, m_h$ will be the intersection multiplicities of C' with the line $y_0 = 0$ in these points. They may be either simple or multiple points for C', of multiplicities $m_i^* \leq m_i, i = 1, \ldots, h$. In particular

$$\sum_{i=1}^{h} m_i^* \leq \sum_{i=1}^{h} m_i = s. \tag{9.6}$$

If C^n has in common with the exceptional line $r_{A_1 A_2}$ of S_2 (that is, with the exceptional line of the transform $\theta^{-1}: S_2' \to S_2$ corresponding to the fundamental point $B_0 = [1, 0, 0]$) the points $L_1, L_2, \ldots, L_k$ different from the points A_1 and A_2, the curve C' will have a corresponding multiple point B_0 with k tangents $\ell_1, \ldots, \ell_k$ (cf. Figure 9.2).

What has been said for the point A_0 may be repeated for A_1 and A_2 and one arrives at the following conclusion:

Proposition–Definition 9.2.2. *Let $\theta: S_2 \to S_2'$ be a quadratic transformation of type I. Let C^n be an algebraic curve in S_2 having multiplicities $s_0, s_1, s_2, s_i \geq 0$, at the points A_0, A_1, A_2 respectively. If the equation of C^n is $f(x_0, x_1, x_2) = 0$, then the equation $f(y_1 y_2, y_2 y_0, y_0 y_1) = 0$ represents a curve in S_2' of order $2n$ which we will call the **total transform** of C^n. The total transform splits and contains the exceptional lines corresponding to A_0, A_1, A_2 counted s_0 times, s_1 times, and s_2 times respectively. The residual curve C', which is irreducible if C^n is irreducible, is the **proper transform** of C^n. Its order is*

$$2n - s_0 - s_1 - s_2.$$

The curve C^n meets the lines $r_{12} = r_{A_1 A_2}$, $r_{02} = r_{A_0 A_2}$, and $r_{01} = r_{A_0 A_1}$, namely the exceptional lines of S_2' corresponding to the fundamental points B_0, B_1, and B_2 of S_2', respectively in $n - s_1 - s_2$ points, in $n - s_0 - s_2$ points, and in $n - s_0 - s_1$ points other than the fundamental points A_0, A_1, and A_2. Therefore the fundamental points B_0, B_1, B_2 are points of multiplicity $n - s_1 - s_2$, $n - s_0 - s_2$, and $n - s_0 - s_1$ for C' (and have all tangents distinct if C^n meets the lines r_{12}, r_{02}, r_{01} in distinct points).

9.2.3 Transformation of an algebraic plane curve into an algebraic plane curve with only multiple points with distinct tangents.

Again let C be an algebraic curve of order n in the plane S_2 and having a non-ordinary s-fold point P and further multiple points Q_j of multiplicities q_j.

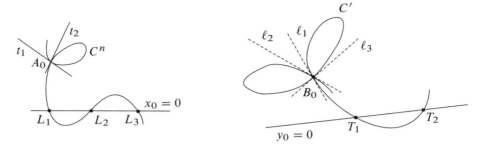

Figure 9.2

We consider two lines, a and b, issuing from P and each having in common with C other $n - s$ distinct points different from P. Let c be a line that meets C in n distinct points not belonging to $a \cup b$.

With this as premise, let Σ be the homaloidal net of conics having as base points P, $M = a \cap c$, $N = b \cap c$. Note that the curve C does not pass through the points M or N. Bearing in mind Proposition–Definition 9.2.2, the quadratic transformation θ (of type I) defined by Σ transforms C into a curve C' of order

$$n' = 2n - s,$$

for which the images Q'_j of the points Q_j have the same multiplicities q_j as the points Q_j for C (because the multiple points Q_j do not belong to the exceptional lines a, b, c and away from them there is an isomorphism which induces projectivities between the pencils of directions issuing from Q_j and from Q'_j, cf. §3.2.1). The point P however will be transformed into a set of points $T_1, \ldots, T_h$ whose multiplicities $s_1, \ldots, s_h$ are such that $\sum_{i=1}^{h} s_i \leq s$ (cf. (9.6)). The curve C' will have moreover three new multiple points B_0, B_1, B_2 each having distinct tangents (since the lines a, b, and c cut the curve C in distinct points), of multiplicities $n - s, n - s$ and n, coming from the lines a, b and c respectively (cf. Proposition–Definition 9.2.2).

If δ and δ' are the deficiencies of C and C', that is

$$\delta := \frac{(n-1)(n-2)}{2} - \frac{s(s-1)}{2} - \sum_j \frac{q_j(q_j-1)}{2},$$

$$\delta' := \frac{(n'-1)(n'-2)}{2} - 2\frac{(n-s)(n-s-1)}{2} - \frac{n(n-1)}{2}$$
$$- \sum_j \frac{q_j(q_j-1)}{2} - \sum_{i=1}^{h} \frac{s_i(s_i-1)}{2},$$

we find that
$$2(\delta - \delta') = \sum_{i=1}^{h} s_i(s_i - 1)$$
and so $\delta' \leq \delta$; and we have $\delta' < \delta$ if $s_i \geq 2$ for at least one index i.

Since the deficiency can not be negative (cf. Proposition 7.2.7), we may conclude that with a finite number of quadratic transformations one arrives at a curve C^* on which in place of P there are simple points while the multiple points that have been adjoined are all ordinary.

If C has another non-ordinary multiple point, the same singularity will be found in a point of C^*. For C^* and this point we do what we have just done for C and P, and so on. Since the number of singular points of C is finite, this procedure terminates after a finite number of steps, and we obtain the following important result (cf. Exercise 13.1.21).

Theorem 9.2.4 (Model of a plane curve with only ordinary singularities). *Let C be an algebraic plane curve. It is always possible to transform C into an algebraic plane curve $\mathcal{C}$ all of whose singularities are ordinary by means of a finite number of quadratic transformations.*

And in fact every linear system Σ of algebraic plane curves may be transformed via a finite number of quadratic transformations into a linear system Σ' with all base points being ordinary points.

9.2.5 Structure of a multiple point. Let P be an s-fold point of a plane curve C. We have seen that by successive quadratic transformations one can transform C into a curve $\mathcal{C}$ such that P is replaced by a finite number of points on $\mathcal{C}$, all of which are simple.

If the point P is changed into the points $P_1, P_2, \ldots$ of C' by a quadratic transformation having P as one of its base points, so that the respective multiplicities are $s_1, s_2, \ldots$, we will say that the curve C has a point P_1 of multiplicity s_1, an s_2-fold point $P_2, \ldots$ in the *first order neighborhood* of P.

If $s_i \geq 2$ we can do for C' and P_i what we have just done for C and P; we obtain the points $P_{i1}, P_{i2}, \ldots$ with multiplicities $s_{i1}, s_{i2}, \ldots$ in the first order neighborhood of P_i. If these are not all simple points we can continue the procedure. The process comes to an end, and leads, after a finite number of steps, to neighborhoods of P in which C has only simple points. Thus at the s-fold point P there are s_1-fold, s_2-fold, $\ldots$ points in the first order neighborhood of P. To each such "first order" s_i-fold point there are then s_{i1}-fold, s_{i2}-fold, $\ldots$ points in the second order neighborhood of P. Furthermore, each "second order" s_{ij}-fold point has (in the third order neighborhood of P) s_{ij1}-fold, s_{ij2}-fold, $\ldots$ points; and so on.

We say that $(s_1, s_2, \ldots, s_{ij}, \ldots, s_{ijk}, \ldots)$ is the *structure of the point P* (or *structure of the singularity P*) of C and one can show that it does not depend on the particular sequence of quadratic transformations by which it was calculated. To distinguish it from the points belonging to one of its successive neighborhoods, we will also say that P is an *actual* point of the curve C.

To represent the structure of P schematically one can think of a "tree" as indicated in the Figure 9.3.

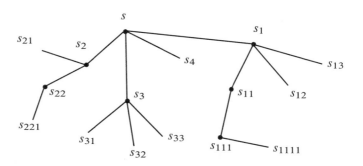

Figure 9.3

Here in the figure that follows are, for example, the "trees" corresponding to a *node* (1), to an *ordinary cusp* (2), to a *tacnode* (3), to a *cusp of the second kind* (4), to an *oscnode* (5), to a *cusp of the third kind* (6) (and these "trees" can be taken as the definitions of the corresponding singularity):

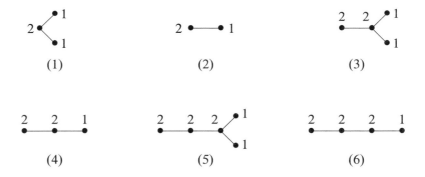

9.2.6 Intersection multiplicity of two plane curves at a point. Let C^n and Γ^m be two coplanar algebraic curves having respective multiplicities r and s at a point P, and let $(r, r_i, r_{ij}, \dots)$ and $(s, s_i, s_{ij}, \dots)$ be the structures of P respectively on C^n and Γ^m. If we agree that the numbers $r_i, r_{ij}, \dots, s_i, s_{ij}, \dots$ can also assume the values zero, we can suppose that the two curves have the same points $P_i, P_{ij}, \dots$ in the successive neighborhoods of P.

9.2. Resolution of the singularities of a plane algebraic curve

The intersection multiplicity at P is $\mu = mn - h$, where h is the number of intersections of the two curves that do not occur at P.

A first quadratic transformation having P as base point and otherwise generic changes C^n and Γ^m into two curves C' and Γ', whose orders are $2n - r$ and $2m - s$, which have multiplicities $r_1, r_2, \ldots$ and $s_1, s_2, \ldots$ respectively at the points $P_1, P_2, \ldots$. Moreover they have three multiple points in common at the base points B_0, B_1, B_2 of the second homaloidal net, having multiplicities respectively $n, n-r, n-r$ for C' and $m, m-s, m-s$ for Γ' (cf. Proposition–Definition 9.2.2); at the points B_0, B_1, B_2 the curves C' and Γ' present the simple case.

The number of intersections of C' and Γ' which do not fall at the points P_i, B_0, B_1, B_2 will be equal to h. If μ_i is the intersection multiplicity of C' and Γ' at P_i we have (cf. Section 4.2)

$$h = (2n - r)(2m - s) - mn - 2(n - r)(m - s) - \sum_i \mu_i = mn - rs - \sum_i \mu_i$$

and so the intersection multiplicity at P is

$$\mu_P(C^n, \Gamma^m) = mn - h = rs + \sum_i \mu_i.$$

In similar fashion we have, with the obvious meaning for the symbols,

$$\mu_i = r_i s_i + \sum_j \mu_{ij}, \quad \mu_{ij} = r_{ij} s_{ij} + \sum_k \mu_{ijk}, \ldots$$

and finally one finds that

$$\mu_P(C^n, \Gamma^m) = rs + \sum_i r_i s_i + \sum_{ij} r_{ij} s_{ij} + \cdots.$$

As far as the number mn of common points of two algebraic curves of orders m and n, it is the sum of the products of the multiplicities of the two curves in all their common points, whether actual or in successive neighborhoods of actual points (cf. Theorem 4.2.1).

9.2.7 Blowing up the plane at a point.
When one wishes to "dilate" a point P of the plane S_2, replacing it with a line on which the first order neighborhood of the point P is extended, one usually prefers to make use of a quadratic transformation θ of type II, assuming P to be a base point with variable tangent of a homaloidal net having only two distinct base points.

If $P = A_0 = [1, 0, 0]$ and if $A_1 = [0, 1, 0]$ is the base point with fixed tangent $r_{A_1 A_2} : x_0 = 0$, one finds that the net Σ has equation (cf. §9.1.1)

$$\lambda_0 x_0 x_2 + \lambda_1 x_0 x_1 + \lambda_2 x_2^2 = 0$$

Chapter 9. Cremona Transformations

and so the quadratic transformation θ will be given by

$$\theta : \begin{cases} y_0 = x_0 x_2, \\ y_1 = x_0 x_1, \\ y_2 = x_2^2. \end{cases}$$

As usual, from

$$\frac{y_0}{y_2} = \frac{x_0}{x_2}, \quad \frac{y_1}{y_0} = \frac{x_1}{x_2}$$

one obtains

$$\frac{x_0}{x_1} = \frac{x_0}{x_2} \frac{x_2}{x_1} = \frac{y_0^2}{y_1 y_2} \quad \text{and} \quad \frac{x_1}{x_2} = \frac{y_1 y_2}{y_0 y_2},$$

whence the inverse quadratic transformation is

$$\theta^{-1} : \begin{cases} x_0 = y_0^2, \\ x_1 = y_1 y_2, \\ x_2 = y_0 y_2. \end{cases}$$

On introducing affine coordinates $x = \frac{x_1}{x_0}$, $y = \frac{x_2}{x_0}$ in S_2, and $x' = \frac{y_1}{y_0}$, $y' = \frac{y_2}{y_0}$ in S_2', we find that the restrictions of θ and θ^{-1} to the affine charts U_0 and U_0' defined by $x_0 \neq 0$ and $y_0 \neq 0$ have equations

$$\theta : \begin{cases} x' = \dfrac{x}{y}, \\ y' = y, \end{cases} \qquad \theta^{-1} : \begin{cases} x = x'y', \\ y = y'. \end{cases}$$

The quadratic transformation θ and its inverse θ^{-1} are also called the *blowing up of the plane S_2 at the point P*. They constitute a particular case of a widely used technique in algebraic geometry for the study of local properties of a given variety (cf. Example 6.8.1).

Now let C be an algebraic curve of S_2 having $P = (0, 0)$ as an s-fold point ($s \geq 1$). In affine coordinates x, y the equation of C is thus of the type

$$f_s(x, y) + f_{s+1}(x, y) + \cdots = 0,$$

with f_j binary forms of degree j. Since

$$f_s(x'y', y') + f_{s+1}(x'y', y') + \cdots = y'^s (f_s(x', 1) + y' f_{s+1}(x', 1) + \cdots), \quad (9.7)$$

the *proper transform C'* of C is the curve with equation

$$f_s(x', 1) + y' f_{s+1}(x', 1) + \cdots = 0$$

and it meets the axis $y' = 0$, that is the exceptional line E that corresponds to P, in the points defined by $f_s(x', 1) = 0$ which correspond to the tangents of C at P.

The curve with equation (9.7) is also called the *total transform* of C (and it splits into the proper transform C' and the exceptional line E counted s times).

If P is an ordinary s-fold point, the roots of the equation $f_s(x', 1) = 0$ are all simple and so on E there are s distinct non-singular points of C' (all with tangents distinct from the line E). In this case the singularity at P is "resolved" and in the first order neighborhood of P the curve C possesses s distinct simple points.

Suppose, in general, that at the s-fold point P the curve C has tangents $t_1, \ldots, t_h$ with multiplicities $m_1, \ldots, m_h$, where $m_i \geq 1$, $\sum_{i=1}^{h} m_i = s$.

If P_i are the points of E that correspond to the tangents t_i, the numbers m_i are the intersection multiplicities of C' with E at P_i and therefore the multiplicity of C' at P_i will be $s_i \leq m_i$, so that $\sum_{i=1}^{h} s_i \leq s$. This gives rise to the following observations:

- If C has at least two distinct tangents at P (i.e., if $h \geq 2$) we have $s_i < s$ for each $i = 1, \ldots, h$. If P is an s-fold point with a single tangent t ($h = 1$), it can happen that C' has multiplicity s in the point of E that corresponds to t.

We can now prove the following important result (for simplicity of notation we will use the same symbol to denote both a curve and its equation).

Theorem 9.2.8 (M. Noether's $Af + B\varphi$ theorem). *In the plane consider two algebraic curves f and φ and an algebraic curve H which has multiplicity $r + s - 1$ at each point P, actual or belonging to a successive neighborhood of an actual point, which is r-fold for f and s-fold for φ. Then there exist two curves A and B having multiplicities $s - 1$ and $r - 1$ respectively at P such that*

$$H = Af + B\varphi.$$

Proof. We set $\deg f = m$, $\deg \varphi = n$, $\deg H = N$ so that we have $\deg A = N - m$, $\deg B = N - n$. Via successive quadratic transformations we can transform the curve with equation $f\varphi = 0$ into a curve having only ordinary multiple points (cf. Theorem 9.2.4). Thus we can reduce to the case in which f and φ have only ordinary multiple points and do not have a common tangent at those points.

First we choose an integer N sufficiently large so that the conditions imposed by the multiplicities $r + s - 1$ on the curves C^N are independent. The dimension of the linear system Σ of such curves C^N will then be (cf. Example 6.2.1)

$$\dim \Sigma = \binom{N+2}{2} - 1 - \sum_P \binom{r+s}{2}.$$

To evaluate the dimension of the linear system Σ^* of the curves having equation of the type $Af + B\varphi = 0$ we see first of all in how many ways a given polynomial H can be written in the form $Af + B\varphi$.

306   Chapter 9. Cremona Transformations

Let $Af + B\varphi = A_1 f + B_1\varphi$, that is, $(A - A_1)f = (B_1 - B)\varphi$. Since f and φ do not have common components and since the ring of polynomial is a unique factorization domain, there must exist a homogeneous polynomial θ of degree $N - m - n$ such that $A - A_1 = \theta\varphi$, $B_1 - B = \theta f$. Conversely, if θ is an arbitrary polynomial of degree $N - m - n$ and $Af + B\varphi = H$, we also have

$$A_1 f + B_1 \varphi = (A - \theta\varphi)f + (B + \theta f)\varphi = H.$$

It follows that a polynomial H such that the curve with equation $H = 0$ belongs to Σ^* can be written in the form $Af + B\varphi$ in ∞^λ ways, where

$$\lambda = \binom{N - m - n + 2}{2}.$$

Hence the dimension of Σ^* is

$$\binom{N - n + 2}{2} - \sum_P \binom{s}{2} + \binom{N - m + 2}{2} - \sum_P \binom{r}{2} - \binom{N - m - n + 2}{2} - 1$$

$$= \binom{N + 2}{2} - 1 - mn - \sum_P \binom{r + s}{2} + \sum_P rs.$$

But $\sum_P rs = mn$; hence

$$\dim \Sigma^* = \binom{N + 2}{2} - 1 - \sum_P \binom{r + s}{2} = \dim \Sigma.$$

Thus if N is sufficiently large we have $\Sigma = \Sigma^*$, which establishes the desired result.

Now we prove that if the theorem is true for a certain integer N, it is also true for $N - 1$.

Let H be a curve of order $N - 1$ through the points P with multiplicities $r + s - 1$ and suppose that the line $x_0 = 0$ does not pass through any of the points P. Then there is an identity of the form

$$x_0 H = A^* f + B^* \varphi, \qquad (9.8)$$

for some homogeneous polynomials $A^*, B^* \in K[x_0, x_1, x_2]$. Setting $x_0 = 0$ here we find that

$$A^*(0, x_1, x_2) f(0, x_1, x_2) + B^*(0, x_1, x_2)\varphi(0, x_1, x_2) = 0$$

and then since $f(0, x_1, x_2)$ and $\varphi(0, x_1, x_2)$ do not have common factors (because none of the points $P \in f \cap \varphi$ are on the line $x_0 = 0$) there exists a form $\theta(x_1, x_2)$ such that

$$A^*(0, x_1, x_2) = \theta(x_1, x_2)\varphi(0, x_1, x_2) \qquad (9.9)$$

whence
$$\theta(x_1, x_2)\varphi(0, x_1, x_2) f(0, x_1, x_2) + B^*(0, x_1, x_2)\varphi(0, x_1, x_2) = 0.$$
Therefore
$$B^*(0, x_1, x_2) = -\theta(x_1, x_2) f(0, x_1, x_2). \tag{9.10}$$
Equations (9.9), (9.10) imply that there are forms $A_0(x_0, x_1, x_2)$, $B_0(x_0, x_1, x_2)$ such that
$$A^*(x_0, x_1, x_2) = \theta(x_1, x_2)\varphi(x_0, x_1, x_2) + x_0 A_0(x_0, x_1, x_2)$$
$$B^*(x_0, x_1, x_2) = -\theta(x_1, x_2) f(x_0, x_1, x_2) + x_0 B_0(x_0, x_1, x_2).$$
Substituting in equation (9.8) and suppressing the factor x_0 one finds that
$$H = A_0 f + B_0 \varphi.$$
Thus the theorem is proved. □

Remark 9.2.9. Notation as in Theorem 9.2.8. We observe that if H is a curve passing through all the points P, a suitable power of H satisfies the hypotheses of Theorem 9.2.8 and so there exists an integer t such that
$$H^t = Af + B\varphi,$$
in agreement with the Hilbert Nullstellensatz (Theorem 2.5.4).

9.3 Cremona transformations between planes

In a plane S_2 consider an irreducible net Σ (that is, consisting of irreducible curves) of order $n \geq 2$. Extending the notion of homaloidal net of conics, we will say that Σ is a *homaloidal net* if it is a net of degree 1, which means that the intersections of any two of its generic curves are all absorbed by the base points with a single exception. If $P_1, P_2, \ldots, P_h$ are the base points and if $s_1, s_2, \ldots, s_h$ are the multiplicities of the curves of Σ at $P_1, P_2, \ldots, P_h$, we will write
$$\Sigma = C^n(P_1^{s_1}, P_2^{s_2}, \ldots, P_h^{s_h}).$$
We will suppose that the base points are all ordinary.

Proposition 9.3.1. *Let* $\Sigma = C^n(P_1^{s_1}, P_2^{s_2}, \ldots, P_h^{s_h})$ *be a homaloidal net. Every irreducible curve of Σ is rational and non-singular outside of the base points $P_1, \ldots, P_h$.*

Proof. One sees immediately that the generic curve $C_0 : \varphi_0 = 0$ of Σ is rational. Indeed, if $C_1 : \varphi_1 = 0$, and $C_2 : \varphi_2 = 0$ are two other curves of the net, both generic, the points of C_0 are in bijective algebraic correspondence with the curves of the pencil $\lambda_1 \varphi_1 + \lambda_2 \varphi_2 = 0$ and thus with the values of the parameter λ_1/λ_2 (cf. Corollary 2.6.6).

By Bertini's first theorem (Theorem 6.3.11), the generic curve of Σ is non-singular outside of the base points. Since it is rational, it has genus $p = 0$ and so (cf. 7.2.13)

$$\frac{1}{2}(n-1)(n-2) - \sum_{i=1}^{h} \frac{s_i(s_i-1)}{2} = 0. \tag{9.11}$$

Let C' be a curve of Σ having a multiple point distinct from the base points; by equation (9.11), the deficiency of C' is then negative, which is not possible if C' is irreducible (cf. Proposition 7.2.7). $\square$

It is easy to prove that the following two relations hold:

$$\sum_{i=1}^{h} s_i^2 = n^2 - 1; \quad \sum_{i=1}^{h} s_i = 3n - 3. \tag{9.12}$$

They are called the *Cremona equations*. The first, which follows from Bézout's theorem (Theorem 4.2.1), expresses the fact that the degree of Σ is 1. From it and relation (9.11), which expresses the rationality of the curves of Σ, the second follows immediately.

Equations (9.12) characterize homaloidal nets in the sense that a linear system Σ of curves of order n with h multiple points P_i of given multiplicities s_i such that equations (9.12) hold can only be a homaloidal net, that is, dim $\Sigma = 2$, deg $\Sigma = 1$. The condition deg $\Sigma = 1$ is a consequence of the first equation in (9.12). From (9.12) one also deduces that

$$\frac{n(n+3)}{2} - \sum_{i=1}^{h} \frac{s_i(s_i+1)}{2} = 2,$$

and therefore dim $\Sigma \geq 2$ (cf. Example–Definition 6.2.2). On the other hand Σ is a regular system, which means that the base points impose independent conditions on its curves (cf. Lemma 7.2.14), and so dim $\Sigma = 2$.

Remark 9.3.2. One says that $G = (n; s_1, \ldots, s_h)$ is an *arithmetic Cremona group* if $(n; s_1, \ldots, s_h)$ is a solution of the Cremona equations (9.12). One then says that G is a *geometric Cremona group* if n is the order of an irreducible homaloidal net Σ and $s_1, \ldots, s_h$ are the multiplicities at the base points of Σ.

9.3. Cremona transformations between planes 309

Clearly a geometric Cremona group is also an arithmetic Cremona group; but the converse does not hold in general. For example, a numeric solution of (9.12) is given by

$$n = 6, \quad h = 9, \quad (s_1, s_2, \ldots, s_9) = (4, 3, 2, 1, 1, 1, 1, 1, 1).$$

To it there corresponds no homaloidal net since an irreducible plane curve of order 6 can not have both a quadruple point and a triple point. Thus $(6; 4, 3, 2, 1, 1, 1, 1, 1, 1)$ is only a Cremona group in the arithmetic sense.

9.3.3. Given a homaloidal net Σ of curves of order n in S_2,

$$\Sigma: \lambda_0 \varphi_0(x_0, x_1, x_2) + \lambda_1 \varphi_1(x_0, x_1, x_2) + \lambda_2 \varphi_2(x_0, x_1, x_2) = 0, \quad (9.13)$$

and having defined a projectivity ω between Σ and the net of lines of a second plane S_2', one obtains a *Cremona transformation* $\theta: S_2 \to S_2'$ between the two planes S_2 and S_2'. We will also say that n is the *order* of the Cremona transformation θ. Geometrically, θ is defined as follows.

- Let P be a point of S_2, distinct from the base points of Σ. Imposing passage through P determines a pencil Φ of curves of Σ, and thus a pencil $\Phi' = \omega(\Phi)$ of lines in S_2'. If P' is the center of Φ', then $\theta(P) := P'$ (since the net is homaloidal the transformation θ thus defined is a birational isomorphism).

If, as we may, we suppose that to the curve in Σ with equation (9.13) there corresponds under ω the line in S_2' defined by $\lambda_0 y_0 + \lambda_1 y_1 + \lambda_2 y_2 = 0$, then one has the following analytic expression for the Cremona transformation θ:

$$\begin{cases} y_0 = \varphi_0(x_0, x_1, x_2), \\ y_1 = \varphi_1(x_0, x_1, x_2), \\ y_2 = \varphi_2(x_0, x_1, x_2). \end{cases} \quad (9.14)$$

From these equations one can recover the x_i which (by reason of the bijectivity of the transformation) are rational functions of the y_i. Prescinding from the technical difficulties, one proceeds as follows. From the two equations

$$y_0 \varphi_1(x_0, x_1, x_2) - y_1 \varphi_0(x_0, x_1, x_2) = 0, \quad y_0 \varphi_2(x_0, x_1, x_2) - y_2 \varphi_0(x_0, x_1, x_2) = 0$$

one eliminates, for example, the variable x_1: this elimination is done via rational operations through calculation of the Euler–Sylvester resultant (cf. Section 4.1). Thus one finds a polynomial equation $R(x_0, x_2) = 0$ (with coefficients homogeneous polynomials in the y_i). On resolving this equation with respect to $\frac{x_0}{x_2}$ one finds only one root that depends on the y_i (since Σ is a homaloidal net); the others are constant and are in fact the ratios of the coordinates x_0, x_2 of the base points of Σ. Thus one obtains for this root an expression of the type $\frac{x_0}{x_2} = f(y_0, y_1, y_2)$, where

310   Chapter 9. Cremona Transformations

f is a rational function of the y_i. Repeat the argument eliminating the variable x_0 to find $\frac{x_1}{x_2}$ as a rational function of y_0, y_1, y_2.

Thus one obtains the inversion formulas for (9.14):

$$\begin{cases} x_0 = \psi_0(y_0, y_1, y_2), \\ x_1 = \psi_1(y_0, y_1, y_2), \\ x_2 = \psi_2(y_0, y_1, y_2), \end{cases}$$

with $\psi_i(y_0, y_1, y_2)$ homogeneous polynomials of the same degree, and

$$\mu_0 \psi_0(y_0, y_1, y_2) + \mu_1 \psi_1(y_0, y_1, y_2) + \mu_2 \psi_2(y_0, y_1, y_2) = 0 \qquad (9.15)$$

represents the homaloidal net Σ' of curves in S_2' that correspond to the lines of S_2.

It is hardly necessary to observe that when $n = 1$, the formulas (9.14) furnish the equations of a projectivity between two planes.

Remark 9.3.4. We note that *the two nets Σ and Σ' have the same order*. Indeed, the intersections of a generic line r' of the plane S_2' with a curve γ' of Σ' correspond bijectively to the intersections of the curve γ in Σ that corresponds to r' with the line in S_2 that corresponds to the curve γ'.

9.3.5 (De Jonquières transformations). Consider the Cremona group $G = (n; n-1, 1, \ldots, 1)$, where 1 appears $2n - 2$ times; it satisfies equations (9.12) and hence is arithmetic. We prove that G is also a geometric Cremona group. The Cremona transformation associated to G, defined by the net $\Sigma = C^n(A^{n-1}, B_1, \ldots, B_{2n-2})$, is called a *De Jonquières transformation*, or *monoidal transformation*.

To write the equations of such transformations θ we observe that there exists a curve (in general unique) Γ of order $n - 1$ with an $(n - 2)$-fold point at A and passing simply through the points B_j. Indeed, one can impose

$$\frac{(n-1)(n-2)}{2} + 2n - 2 = \binom{n+1}{2} - 1$$

linear conditions on the curves of order $n - 1$ which are in number $\infty^{N(n-1)}$ where $N(n-1) = \binom{n+1}{2} - 1$ (cf. Section 6.1).

Let Γ be such a curve and let Φ be the pencil of curves of Σ that have in common with Γ a point M distinct from the points $A, B_1, \ldots, B_{2n-2}$. The number of intersections of Γ with a generic curve of Φ that are absorbed by the points A, B_j, M is at least $(n-1)(n-2) + 2n - 2 + 1 = n(n-1) + 1$, one more than the number implied by Bézout's theorem. It follows that the curves of Φ are all split into Γ and a line of the pencil with center A.

If $A = A_2 = [0, 0, 1]$, Γ has equation of the form

$$x_2 \psi_{n-2}(x_0, x_1) + \psi_{n-1}(x_0, x_1) = 0,$$

9.3. Cremona transformations between planes

with ψ_{n-2}, ψ_{n-1} binary forms of degrees $n-2$ and $n-1$ in x_0, x_1. And for the net Σ, that joins Φ with an arbitrary curve taken from Σ (outside of Φ), one has the equation

$$(x_2\psi_{n-2} + \psi_{n-1})(\lambda_0 x_0 + \lambda_1 x_1) + x_2\varphi_{n-1} + \varphi_n = 0, \qquad (9.16)$$

with φ_{n-1}, φ_n binary forms of degree $n-1$ and n. From equation (9.16) it follows that the De Jonquières transformation may be represented in the form

$$\begin{cases} \rho x_0' = x_0(x_2\psi_{n-2} + \psi_{n-1}), \\ \rho x_1' = x_1(x_2\psi_{n-2} + \psi_{n-1}), \\ \rho x_2' = x_2\varphi_{n-1} + \varphi_n. \end{cases} \qquad (9.17)$$

These formulas are easily inverted. First, one has $\frac{x_0'}{x_1'} = \frac{x_0}{x_1}$ and then

$$x_0 = \sigma x_0', \quad x_1 = \sigma x_1', \qquad (9.18)$$

for some $\sigma \in K^*$. Substituting in the first equation of the system (9.17) one has

$$\rho x_0' = \sigma x_0'(x_2 \sigma^{n-2}\psi_{n-2}(x_0', x_1') + \sigma^{n-1}\psi_{n-1}(x_0', x_1'))$$

whence

$$\rho = \sigma^{n-1}(x_2\psi_{n-2}(x_0', x_1') + \sigma\psi_{n-1}(x_0', x_1')).$$

From the third equation of (9.17) we have

$$\rho x_2' = \sigma^{n-1}(x_2\varphi_{n-1}(x_0', x_1') + \sigma\varphi_n(x_0', x_1'))$$

so that

$$x_2' = \frac{x_2\varphi_{n-1}(x_0', x_1') + \sigma\varphi_n(x_0', x_1')}{x_2\psi_{n-2}(x_0', x_1') + \sigma\psi_{n-1}(x_0', x_1')}.$$

Resolving with respect to x_2 we find

$$x_2'(x_2\psi_{n-2} + \sigma\psi_{n-1}) = x_2\varphi_{n-1} + \sigma\varphi_n,$$

that is,

$$x_2(x_2'\psi_{n-2} - \varphi_{n-1}) + \sigma(x_2'\psi_{n-1} - \varphi_n) = 0.$$

Therefore

$$x_2 = -\frac{\sigma(x_2'\psi_{n-1} - \varphi_n)}{x_2'\psi_{n-2} - \varphi_{n-1}}; \quad \sigma = -\frac{x_2(x_2'\psi_{n-2} - \varphi_{n-1})}{x_2'\psi_{n-1} - \varphi_n}.$$

Bearing in mind (9.18), equations (9.17) are thus inverted as follows:

$$\begin{cases} x_0 = x_0'(x_2'\psi_{n-2}(x_0', x_1') - \varphi_{n-1}(x_0', x_1')), \\ x_1 = x_1'(x_2'\psi_{n-2}(x_0', x_1') - \varphi_{n-1}(x_0', x_1')), \\ x_2 = -x_2'\psi_{n-1}(x_0', x_1') + \varphi_n(x_0', x_1'). \end{cases}$$

One finds formulas of the same type in (9.17), so that the two homaloidal nets Σ, Σ' associated to θ are formed in the same way.

9.3.6 Exceptional curves of a Cremona transformation between planes. Let $\theta: S_2 \to S_2'$ be a Cremona transformation between two planes, and let Σ and Σ' be the two homaloidal nets associated to θ.

A base point of Σ does not have a well-defined corresponding point. To it there is however associated a rational curve. We prove this fact in the simplest case, that in which the base point P does not have fixed tangents: that is, there do not exist lines tangent at P to all the curves of Σ.

The curves of Σ that have a given tangent a at P form a pencil to which there corresponds a pencil of lines in S_2' whose center P_a' will be called *the correspondent of the point infinitely near to P in the direction a*.

Let b be another generic line issuing from P and let P_b' be the correspondent of P in the direction b. We have $P_a' \neq P_b'$. Indeed, if we had $P_a' = P_b'$, the two pencils of lines that have them as centers there would coincide and so too would coincide the two pencils of curves of Σ corresponding to them. Consequently, all the curves of Σ tangent at P to a would also be tangent there to b, and therefore any two of them, having the two points infinitely near to P in the directions of the lines a and b in common (that is, two intersections outside the base points of the net), would be reducible. Since b is a generic line of the pencil with center P, all the curves of Σ would be reducible, and for a homaloidal net that can not happen.

Thus, as the direction of the lines issuing from P varies, one finds ∞^1 points P' that trace out a rational curve α_P (since its points are in bijective algebraic correspondence with the lines of the pencil with center P, cf. 7.2.13). We will say that this is the *exceptional curve of S_2' corresponding to the point P*. It represents the first order neighborhood of P. We will also say that α_P is an *exceptional curve* (or *fundamental curve*) *of the second net Σ'*.

All of this may be repeated interchanging the two planes: to each base point Q' of the homaloidal net Σ' in S_2' there corresponds in S_2 a fundamental curve of the first net Σ.

The Cremona transformation thus has the effect of substituting the first order neighborhoods of the base points of one of the two nets with certain rational curves that are exceptional for the other net.

9.3.7 (Analytic determination of the exceptional lines of a quadratic transformation).

(1) First let $\theta: S_2 \to S_2'$ be a transformation of type I,

$$y_0 : y_1 : y_2 = x_1 x_2 : x_2 x_0 : x_0 x_1; \quad x_0 : x_1 : x_2 = y_1 y_2 : y_2 y_0 : y_0 y_1.$$

The pencil Φ of conics of Σ tangent at $A_0 = [1, 0, 0]$ to the line $ax_1 + bx_2 = 0$, represented by

$$\Phi : \lambda x_1 x_2 + x_0(ax_1 + bx_2) = 0,$$

is transformed into $y_0 y_1 y_2 (\lambda y_0 + a y_2 + b y_1) = 0$. Thus has as its proper transform the pencil of lines $\lambda y_0 + a y_2 + b y_1 = 0$, whose center $[0, a, -b]$

is the point of S'_2 that corresponds to the point of S_2 infinitely close to A_0 in the direction $ax_1 + bx_2 = 0$. It traces out the line $y_0 = 0$ which is therefore the exceptional line corresponding to A_0.

Analogous conclusions also hold for the other two base points.

(2) Now let $\theta : S_2 \to S'_2$ be a transformation of type II given by

$$y_0 : y_1 : y_2 = x_2x_0 : x_0x_1 : x_1^2; \quad x_0 : x_1 : x_2 = y_1^2 : y_1y_2 : y_0y_2.$$

Let A_0 be the fundamental point in which the conics of Σ have variable tangent. To the pencil $\Phi : x_0(ax_1 + bx_2) + \mu x_1^2 = 0$ of conics of Σ tangent at $A_0 = [1, 0, 0]$ to the line $ax_1 + bx_2 = 0$ there corresponds the pencil of lines $ay_1 + by_0 + \mu y_2 = 0$ which has center $[a, -b, 0]$ on the exceptional line $y_2 = 0$.

Now let $A_2 = [0, 0, 1]$ be the point where the conics of Σ have the common tangent $x_0 = 0$. To the pencil $\Phi : x_2x_0 + ax_1^2 + \lambda x_0x_1 = 0$ of conics of Σ that are mutually osculating at A_2 there corresponds the pencil of lines $y_0 + ay_2 + \lambda y_1 = 0$ whose center $[a, 0, -1]$ belongs to the exceptional line $y_1 = 0$ corresponding to A_2.

(3) Finally let $\theta : S_2 \to S'_2$ be a transformation of type III defined by

$$y_0 : y_1 : y_2 = x_0x_2 - x_1^2 : x_0^2 : x_0x_1; \quad x_0 : x_1 : x_2 = y_1^2 : y_1y_2 : y_0y_1 + y_2^2.$$

To the pencil $\Phi : x_0x_2 - x_1^2 + ax_0x_1 + \mu x_0^2 = 0$ (whose four base points coincide in the point A_2) there corresponds the pencil of lines $y_0 + ay_2 + \mu y_1 = 0$ whose center $[a, 0, -1]$ traces the exceptional line $y_1 = 0$ corresponding to $A_2 = [0, 0, 1]$.

9.3.8. We now study some properties of the exceptional curves of a Cremona transformation θ between two planes S_2 and S'_2. Let

$$\Sigma = C^n(A_1^{s_1}, A_2^{s_2}, \ldots, A_h^{s_h}), \quad \Sigma' = C^n(B_1^{r_1}, B_2^{r_2}, \ldots, B_{h'}^{r_{h'}})$$

be the two homaloidal nets associated to θ. Let $\alpha_1, \ldots, \alpha_h$ be the exceptional curves of S'_2 that correspond to the base points $A_1, \ldots, A_h$ of Σ and let t_{ij} be the multiplicity of α_i at B_j. Let $\beta_1, \ldots, \beta_{h'}$ be the exceptional curves of S_2 that correspond to the base points $B_1, \ldots, B_{h'}$ of Σ' and let t'_{ji} be the multiplicity of β_j at A_i.

- To each curve C of Σ there corresponds in S'_2 a line r not passing through any of the points B_j and one has a bijective correspondence between the intersections of r with α_i and the tangents of C at A_i. Thus $\deg \alpha_i = s_i$. Similarly, $\deg \beta_j = r_j$.

- The points of α_i belonging to the neighborhood of B_j (that is, to the exceptional curve of S_2' corresponding to B_j, cf. §9.3.6) correspond bijectively to the points of the neighborhood of A_i belonging to β_j. This means that the multiplicity of α_i at B_j is equal to the multiplicity of β_j at A_i; that is,

$$t'_{ji} = t_{ij}.$$

- The generic line r in S_2 does not pass through A_i; hence the curve $C' = \theta(r)$ (i.e., a generic curve of Σ') does not have points in common with α_i that are not base points of Σ'.

  This implies that if P is a point of α_i distinct from the base points of Σ' the curves of Σ' passing through P form a pencil Φ of curves all of which are split and contain as component α_i. The pencil Φ^* of the residual components corresponds to the pencil of lines in S_2 issuing from A_i. Since the curve α_i is rational one thus has (cf. Proposition 7.2.7)

$$\frac{(s_i - 1)(s_i - 2)}{2} - \sum_{j=1}^{h'} \frac{t_{ij}(t_{ij} - 1)}{2} = 0. \tag{9.19}$$

- If P is a point of α_i and γ is the curve of Φ^* passing through P, then $\alpha_i \cup \gamma$ belongs to Σ' and has P as double point. Therefore each point of α_i is double for some curve of Σ' and therefore belongs to the Jacobian curve J' of Σ' (cf. Section 6.4). This means that the curves α_i are components of J'; and in fact they exhaust J' because the sum of their orders is $\sum_{i=1}^{h} s_i = 3n - 3$, which is equal to the order of J' (cf. Exercise 9.5.13).

  The multiplicity of B_j for the curve J' is $3r_j - 1$, cf. Exercise 6.4.3 (3); on the other hand it is equal to the sum of the multiplicities t_{ij} at B_j of its components $\alpha_1, \ldots, \alpha_h$. Thus, one has the equation

$$\sum_{i=1}^{h} t_{ij} = 3r_j - 1$$

and summing with respect to j gives

$$\sum_{j=1}^{h'} \sum_{i=1}^{h} t_{ij} = 3 \sum_{j=1}^{h'} r_j - h' = 3(3n - 3) - h'. \tag{9.20}$$

Similarly, interchanging the two nets, one has the equations

$$\sum_{j=1}^{h'} t'_{ji} = 3s_i - 1, \tag{9.21}$$

9.3. Cremona transformations between planes 315

$$\sum_{i=1}^{h}\sum_{j=1}^{h'} t'_{ji} = 3\sum_{i=1}^{h} s_i - h = 3(3n-3) - h. \tag{9.22}$$

The left-hand sides of (9.20) and (9.22) are equal, hence

$$h = h'.$$

Thus Σ and Σ' have the same number of base points.

- In virtue of equations (9.19), (9.21) we have (recall that $t'_{ji} = t_{ij}$)

$$\frac{(s_i-1)(s_i-2)}{2} - \sum_{j=1}^{h'}\frac{t_{ij}(t_{ij}-1)}{2} - \left(\sum_{j=1}^{h'} t_{ij} - 3s_i + 1\right)$$

$$= \frac{s_i(s_i+3)}{2} - \sum_{j=1}^{h'}\frac{t_{ij}(t_{ij}+1)}{2} = 0$$

and hence α_i is the unique curve of S'_2 having order s_i and having the points B_j as t_{ij}-fold points (cf. Section 6.2). Similarly β_j is the unique curve of S_2 having order r_j and the multiplicities t_{ij} at the points A_i.

- Finally, we note that two curves α_i, α_j can not meet outside of the base points of Σ' because there is no point of S_2 belonging to the neighborhood of A_i and also to the neighborhood of A_j.

9.3.9 The Noether–Rosanes inequality. Let $\Sigma = C^n(A_1^{s_1}, A_2^{s_2}, \ldots, A_h^{s_h})$ be a homaloidal net of curves of order $n > 1$. If $s_1 \geq s_2 \geq s_3 \geq \cdots \geq s_h$ we have

$$s_1 + s_2 + s_3 \geq n + 1. \tag{9.23}$$

The simple proof that we give here is due to Mlodziejowski (see [66] and [53, p. 10]). Since $\sum_{i=1}^{h} s_i^2 = n^2 - 1$ (and $s_1 \geq s_2 \geq s_3 \geq \cdots \geq s_h$) we have

$$s_1^2 + s_2^2 + s_3(s_3 + s_4 + \cdots + s_h) \geq \sum_{i=1}^{h} s_i^2 = n^2 - 1.$$

On the other hand, by the second equality in (9.12), we have

$$s_3 + s_4 + \cdots + s_h = 3n - 3 - s_1 - s_2.$$

Hence

$$s_1^2 + s_2^2 + s_3(3(n-1) - s_1 - s_2) \geq n^2 - 1,$$

which is to say
$$s_1(s_1 - s_3) + s_2(s_2 - s_3) + 3(n-1)s_3 \geq n^2 - 1.$$
But $s_1 \leq n - 1$, $s_2 \leq n - 1$. Therefore
$$(n-1)(s_1 - s_3) + (n-1)(s_2 - s_3) + 3(n-1)s_3 \geq n^2 - 1$$
and, since $n \neq 1$,
$$s_1 - s_3 + s_2 - s_3 + 3s_3 \geq n + 1,$$
which is (9.23).

Theorem 9.3.10 (Theorem of Noether–Castelnuovo). *A Cremona transformation $\theta \colon S_2 \to S_2'$ between planes can always be obtained as a product of a finite number of quadratic transformations and a projectivity.*

Proof. In the case of Cremona transformations with ordinary base points, that is, with base points in which it is not imposed on the curves to have fixed tangents, the proof is very simple, and was found at the same time by Noether (and by Rosanes and Clifford around 1870; see [77] and also [45, p. 74]). We give the demonstration in that case.

Let $\theta \colon S_2 \to S_2'$ be a Cremona transformation and let $\Sigma = C^n(A_1^{s_1}, \ldots, A_h^{s_h})$, $\Sigma' = C^n(B_1^{r_1}, \ldots, B_h^{r_h})$ be the two homaloidal nets in S_2 and S_2' associated to θ. By 9.3.8 the two nets have the same number of base points.

Let then $\omega \colon S_2' \to S_2''$ be a quadratic transformation between the plane S_2' and a new plane S_2'' and Σ'' the (homaloidal) net of the curves in S_2'' which are transforms of the curves of Σ' by way of ω.

If R is the net of lines in S_2 one has $\theta(R) = \Sigma'$ and thus $\omega(\theta(R)) = \Sigma''$. The product $\phi := \omega \circ \theta$ is a Cremona transformation $\phi \colon S_2 \to S_2''$ which has as its second homaloidal net the net Σ'', the transform of R (this means that Σ, Σ'' are the two homaloidal nets associated to ϕ):

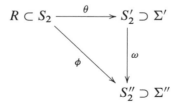

Let B_1, B_2, B_3 be the three points of maximal multiplicity of Σ'. By (9.23) we have
$$r_1 + r_2 + r_3 \geq n + 1;$$
whence the three points are not collinear, and (using here the hypothesis that the three base points are ordinary, cf. Remark 9.3.11) we may choose as ω the quadratic

transformation that has them as its fundamental points. In particular ω is a quadratic transformation of type I and so, bearing in mind Proposition–Definition 9.2.2, the curves of Σ' will be transformed by ω into curves of Σ'' of order $n_1 := 2n - r_1 - r_2 - r_3 < n$. Therefore $\theta = \omega^{-1} \circ \phi$ is the product of a quadratic transformation ω^{-1} and a Cremona transformation ϕ whose order (i.e., the order of the curves of the net) is $n_1 < n$.

Operating on ϕ as has just been done for θ one can write θ as a product of two quadratic transformations and a Cremona transformation of order $< n_1$.

In this way, iterating the procedure, one obtains θ as a product of a finite number of quadratic transformations and a possible projectivity. □

Remark 9.3.11. The problem is much more complicated when there are base points with fixed tangents. The question was studied by Max Noether who believed that he had resolved it. But Noether's proof contained gaps that where pointed out by C. Segre [88] who observed that it is not obvious that the three points of maximal multiplicity of a homaloidal net can always be assumed to be base points of a net of conics not all of which are reducible, a circumstance which is certainly possible if the three points are distinct. The theorem does hold, however, in the general case as was later proved by Castelnuovo [24].

9.4 Cremona transformations between projective spaces of dimension 3

We start by considering the case of Cremona transformations between projective spaces S_r, S'_r of dimension $r > 2$. Consider an r-dimensional linear system Σ of hypersurfaces

$$\lambda_0 \varphi_0(x_0, \ldots, x_r) + \cdots + \lambda_r \varphi_r(x_0, \ldots, x_r) = 0, \qquad (9.24)$$

with base variety defined by the equations $\varphi_0 = \cdots = \varphi_r = 0$. If every choice of r independent hypersurfaces of Σ has intersection with only a single point in common outside the base variety, then Σ is said to be a *homaloidal system* of hypersurfaces of S_r. Such a homaloidal system Σ determines a Cremona transformation $\theta \colon S_r \to S'_r$. To Σ there is associated (as in the case of nets of conics) a homaloidal system Σ' in S'_r.

In general Σ and Σ' do not have the same order (that is, the hypersurfaces of Σ do not have the same order as those of Σ').

We will say that θ is a *Cremona transformation of type* (n, n') if n and n' are the orders of Σ and Σ'.

Assuming this terminology, let $\theta \colon S_r \to S'_r$ be a Cremona transformation of type (n, n'), and let Σ and Σ' be its two homaloidal systems. Let ℓ be a generic line of S_r.

Since Σ may always be supposed to be without fixed components, we can consider an open set U of S_r that contains the line ℓ and such that the restriction $\theta_{|U}: U \to \theta(U)$ is an isomorphism.

To the n intersections with ℓ of a generic hypersurface of Σ, with equation (9.24), there correspond bijectively the intersections of a generic hyperplane $\lambda_0 y_0 + \cdots + \lambda_r y_r = 0$ of S'_r with the curve $\mathcal{L}' = \theta_{|U}(\ell)$ of S'_r that corresponds to ℓ. Therefore n is the order of $\mathcal{L}'$. On the other hand, since ℓ is the intersection of $r - 1$ generic hyperplanes of S_r, the curve $\mathcal{L}'$ is the intersection of $r - 1$ generic hypersurfaces of Σ' (outside of the base variety). The curve $\mathcal{L}'$ is called the *proper transform* of ℓ. Thus:

- Given a Cremona transformation $\theta: S_r \to S'_r$ of type (n, n') let Σ, Σ' be its two homaloidal systems. Then the order of Σ is equal to the order of the intersection (outside the base variety) of $r - 1$ generic hypersurfaces of Σ', that is, of the proper transform of a generic line of S_r. The order of Σ' is defined similarly by interchanging S_r with S'_r and Σ with Σ'. In particular,

$$n' \leq n^{r-1}, \quad n \leq n'^{r-1}.$$

We now study the case $r = 3$ (and $n = 2$) more deeply. Let $\theta: S_3 \to S'_3$ be a Cremona transformation, Σ and Σ' its two homaloidal systems. Suppose that Σ is a linear system of quadrics ($n = 2$). We wish to determine the structure of Σ and Σ'.

We know that the order n' of the system Σ' is equal to the order of the variable part of the intersection of two quadrics of Σ. If $n' = 1$ then we must also have $n = 1$ (use, for example, the relation $n \leq n'^{r-1}$) and thus, since we suppose that θ is not a projectivity, n' can only be 2, 3, or 4.

In Σ we fix a generic quadric $\mathcal{Q}_0$. Bearing in mind that the quadrics of Σ have a single point in common outside of the base variety when taken three by three, the other quadrics of Σ meet $\mathcal{Q}_0$ in curves of order 4 whose variable parts form a homaloidal net σ (in $\mathcal{Q}_0$).

If σ is a homaloidal net of conics [which means that the variable part of the intersection of two quadrics of Σ is a conic and thus θ is of type $(2, 2)$], the quadrics of Σ all pass through a conic γ_0. Two conics of σ have two points in common (the intersections of $\mathcal{Q}_0$ with the line common to the two planes that contain the two conics). Since σ is a homaloidal net, one of these points must be a base point of σ.

If σ is a net of cubics C^3 (and hence θ is of type $(2, 3)$) Σ consists of quadrics passing through a line ℓ_0. Since the quartic C^4 which is the intersection of two quadrics of Σ is a curve of type $(2, 2)$ and $\ell_0 = (1, 0)$ on Q_0, the cubics C^3 are of the type $(1, 2)$. Thus every cubic of σ meets ℓ_0 in $(1, 2)(1, 0) = 2$ points. This means that ℓ_0 is a chord of every cubic of σ (cf. Section 7.3 for the curves on a quadric).

9.4. Cremona transformations between projective spaces of dimension 3

Moreover, two cubics on σ meet in $(1, 2)(1, 2) = 4$ points; since σ is a homaloidal net, three of these points must be base points for Σ.

If σ is a net of quartics (and thus θ is of type $(2, 4)$), σ must have a double base point O. Indeed, the curves of the net are rational (as was seen in Section 9.3) and a non-singular quartic that is the intersection of two quadrics is not rational (it has genus 1, cf. Corollary 7.2.5 and (7.9)).

The point O can not be double for all the quartics of Σ. Indeed, in that case the quadrics of Σ would be cones with vertex O and would intersect two by two in quartics split into lines; and then the curves of the net σ would all be reducible. It follows that the quadrics of Σ are tangent at O to a fixed plane.

Of the eight points common to two quartics of σ, four are absorbed by O and therefore three of the remaining four intersections must be base points of Σ.

So there are three possible types of homaloidal systems of quadrics and thus three possible types of Cremona transformations $\theta: S_3 \to S_3$ having as first homaloidal system a system of quadrics.

We now wish to determine the analytic structure of the Cremona transformation associated to each of these types.

9.4.1 Cremona transformations of type $(2, 2)$.

These transformations are called *quadratic transformations*; the two homaloidal systems Σ and Σ' are both linear systems of quadrics and the base variety consists of a conic γ and a further isolated point A.

We choose the reference system in such a way that the conic γ has equations $x_0 = \varphi(x_1, x_2, x_3) = 0$ and the point A is the point $A_0 = [1, 0, 0, 0]$. One then has

$$\lambda_0 \varphi(x_1, x_2, x_3) + x_0(\lambda_1 x_1 + \lambda_2 x_2 + \lambda_3 x_3) = 0$$

as the equation for Σ. The formulas for the transformation $\theta: S_3 \to S_3'$ are

$$\theta: \begin{cases} \rho y_0 = \varphi(x_1, x_2, x_3), \\ \rho y_1 = x_0 x_1, \\ \rho y_2 = x_0 x_2, \\ \rho y_3 = x_0 x_3, \end{cases} \quad \rho \in K^*, \tag{9.25}$$

and they are easily inverted. Indeed, it follows from these formulas that $x_i = \frac{\rho}{x_0} y_i$, $i = 1, 2, 3$, whence

$$\rho y_0 = \varphi\left(\frac{\rho}{x_0} y_1, \frac{\rho}{x_0} y_2, \frac{\rho}{x_0} y_3\right) = \frac{\rho^2}{x_0^2} \varphi(y_1, y_2, y_3).$$

Therefore

$$x_i = \frac{\rho}{x_0} y_i = \frac{\rho}{x_0 y_0} y_0 y_i, \quad i = 1, 2, 3; \quad x_0 = \frac{\rho}{x_0 y_0} \varphi(y_1, y_2, y_3).$$

Then, on setting $\frac{x_0 y_0}{\rho} = \rho'$, we obtain

$$\theta^{-1}: \begin{cases} \rho' x_0 = \varphi(y_1, y_2, y_3), \\ \rho' x_1 = y_0 y_1, \\ \rho' x_2 = y_0 y_2, \\ \rho' x_3 = y_0 y_3. \end{cases} \quad (9.26)$$

The system Σ' has as its base variety the conic γ' with equation $y_0 = \varphi(y_1, y_2, y_3) = 0$ and the isolated point $A'_0 = [1, 0, 0, 0]$.

Let $\pi : ax_1 + bx_2 + cx_3 = 0$, $a, b, c \in K$, be a plane in S_3 passing through A_0. The quadric of Σ' that corresponds to it has equation

$$y_0(ay_1 + by_2 + cy_3) = 0$$

and is composed of the plane $y_0 = 0$ that contains γ' and the plane π' with equation $ay_1 + by_2 + cy_3 = 0$. The planes π and π' correspond under a projectivity ω between the stars of planes having centers A_0 and A'_0.

The quadrics of Σ meet π in the conics of a homaloidal net of conics having A_0 and the two points Q and R at which π meets γ as base points. Thus θ induces a quadratic transformation ξ between the two planes with those three points as base points. The homaloidal net on π' cut out by Σ' has as its base points A'_0 and the intersections of γ' with π'. The exceptional lines of ξ in π are the traces on π of the plane of γ and the two generators $\langle A_0, Q \rangle$, $\langle A_0, R \rangle$ of the cone Γ projecting γ from A_0. Similarly, the exceptional lines in π' are the traces on π' of the plane of γ' and the two generators that are sections of π' with the cone Γ' projecting γ' from A'_0. In particular, the exceptional line corresponding to A_0 is the trace of π' in the plane of γ' and similarly the exceptional line corresponding to A'_0 is the trace of π in the plane of γ.

From this one deduces that to the base locus of Σ (the point A_0 and the conic γ) there correspond the plane of γ' (in which the first order neighborhood of A_0 is extended) and the cone Γ'. To each point P of γ there correspond the points of a generator g' of Γ', the image under the projectivity ω of the generator $\langle A_0, P \rangle$ of the cone Γ.

Analogous results may be obtained by interchanging S_3 and S'_3.

9.4.2 Cremona transformations of type (2, 3).
The first homaloidal system Σ is a system of quadrics passing through a line and through three further points, and the second homaloidal system Σ' is a system of cubic surfaces.

If we choose the reference system so that the base line of Σ contains the points $A_1 = [0, 1, 0, 0]$, $A_2 = [0, 0, 1, 0]$, and the three isolated base points become $A_0 = [1, 0, 0, 0]$, $A_3 = [0, 0, 0, 1]$ and $U = [1, 1, 1, 1]$, then the system Σ has the equation

$$\lambda_0 x_0(x_2 - x_3) + \lambda_1 x_3(x_1 - x_0) + \lambda_2 x_0(x_1 - x_3) + \lambda_3 x_3(x_2 - x_0) = 0.$$

9.4. Cremona transformations between projective spaces of dimension 3

The equations of the transformation $\theta: S_3 \to S_3'$ are

$$\theta: \begin{cases} y_0 = x_0(x_2 - x_3), \\ y_1 = x_3(x_1 - x_0), \\ y_2 = x_0(x_1 - x_3), \\ y_3 = x_3(x_2 - x_0). \end{cases}$$

One verifies that these formulas are inverted by the following:

$$\theta^{-1}: \begin{cases} x_0 = (y_2 - y_0)(y_2 y_3 - y_0 y_1), \\ x_1 = (y_2 - y_1)(y_3 - y_1)(y_2 - y_0), \\ x_2 = (y_1 - y_3)(y_2 - y_0)(y_3 - y_0), \\ x_3 = (y_1 - y_3)(y_2 y_3 - y_0 y_1). \end{cases}$$

Therefore, the surfaces in the homaloidal system Σ' having equation

$$\lambda_0 (y_2 - y_0)(y_2 y_3 - y_0 y_1) + \lambda_1 (y_2 - y_1)(y_3 - y_1)(y_2 - y_0) \\ + \lambda_2 (y_1 - y_3)(y_2 - y_0)(y_3 - y_0) + \lambda_3 (y_1 - y_3)(y_2 y_3 - y_0 y_1)$$

are the ruled cubics having as double line the line $y_1 - y_3 = y_2 - y_0 = 0$ and also having in common the three generators $y_0 = y_2 = 0$, $y_1 = y_2 = 0$, $y_1 - y_2 = y_0 - y_3 = 0$, which are supported by the double line in three distinct points (cf. 5.8.22).

9.4.3 Cremona transformations of type (2, 4).

The first homaloidal system Σ is a system of quadrics passing through four points and having a given fixed tangent plane at one of them, while the second homaloidal system Σ' is a system of surfaces of order 4. We choose the reference system so that the four base points of Σ are $A_0 = [1, 0, 0, 0]$, $A_1 = [0, 1, 0, 0]$, $A_2 = [0, 0, 0, 1]$ and $A_3 = [0, 0, 0, 1]$ and let $a_1 x_1 + a_2 x_2 + a_3 x_3 = 0$ be the plane to which the quadrics of Σ are tangent at A_0. Thus one has for Σ the equation

$$\lambda_0 x_0 (a_1 x_1 + a_2 x_2 + a_3 x_3) + \lambda_1 x_2 x_3 + \lambda_2 x_3 x_1 + \lambda_3 x_1 x_2 = 0.$$

The Cremona transformation $\theta: S_3 \to S_3'$ is given by the equations

$$\theta: \begin{cases} y_0 = x_0(a_1 x_1 + a_2 x_2 + a_3 x_3), \\ y_1 = x_2 x_3, \\ y_2 = x_3 x_1, \\ y_3 = x_1 x_2, \end{cases}$$

which are inverted by the following relations

$$\theta^{-1}: \begin{cases} x_0 = y_0 y_1 y_2 y_3, \\ x_1 = y_2 y_3 (a_1 y_2 y_3 + a_2 y_3 y_1 + a_3 y_1 y_2), \\ x_2 = y_1 y_3 (a_1 y_2 y_3 + a_2 y_3 y_1 + a_3 y_1 y_2), \\ x_3 = y_1 y_2 (a_1 y_2 y_3 + a_2 y_3 y_1 + a_3 y_1 y_2). \end{cases}$$

One sees immediately that all the surfaces of the second homaloidal system Σ' have the triple point $A'_0 = [1, 0, 0, 0]$ and the double lines $r_{A'_0 A'_1}, r_{A'_0 A'_2}$, and $r_{A'_0 A'_3}$. Hence these surfaces are Steiner surfaces, a particular type of rational surfaces that we will study in Exercise 10.5.6. The base variety of Σ' consists of the three double lines and the conic with equation $y_0 = a_1 y_2 y_3 + a_2 y_3 y_1 + a_3 y_1 y_2 = 0$.

9.5 Exercises

9.5.1. *In a Euclidean plane we choose a fixed point O and decree that two points P and P' are in correspondence when they are collinear with O and have distances from O whose product is a given constant. The transformation that one finds in this way (which is called an **inversion** or **transformation by reciprocal radius vectors**) is a quadratic transformation.*

Describe the two homaloidal nets and study the images of the conics under this transformation.

In orthogonal affine Cartesian coordinates, let $P = (x, y)$ and $P' = (x', y')$ be two corresponding points. Then

$$\frac{y'}{x'} = \frac{y}{x}; \quad \sqrt{x^2 + y^2} \sqrt{x'^2 + y'^2} = k, \quad k \in \mathbb{R}^*.$$

Put $y' = \rho y$, $x' = \rho x$. We then have $\rho \sqrt{x^2 + y^2} \sqrt{x^2 + y^2} = k$ and so $\rho = \frac{k}{x^2 + y^2}$. The equations of the transformation are then

$$x' = \frac{kx}{x^2 + y^2}; \quad y' = \frac{ky}{x^2 + y^2}.$$

The inverse formulas are

$$x = \frac{kx'}{x'^2 + y'^2}; \quad y = \frac{ky'}{x'^2 + y'^2}. \tag{9.27}$$

Using the relations (9.27) we see that the lines of each of the two planes (superimposed) correspond in the other plane to the circles passing through the origin O (possibly split into a line p passing through O and the line at infinity).

This can also be seen immediately by elementary means as follows. Let A be the orthogonal projection of the point O on the line r and let A' be the corresponding point of A. If P is another point of r and P' is its image we have $OA \cdot OA' = OP \cdot OP'$, or $\frac{OA}{OP} = \frac{OP'}{OA'}$. This means that the two triangles OAP and $OP'A'$ are similar. Therefore the two angles OAP and $OP'A'$ are equal and so both are right angles. The point P' thus belongs to the circle with diameter OA' (Figure 9.4).

The conics are mapped into circular quartics (that is, having two double points in the two circular points at infinity) with a double point at O. The two circular points at infinity are biflecnodes (that is, points at which there are two ordinary flexes with distinct inflectional tangents).

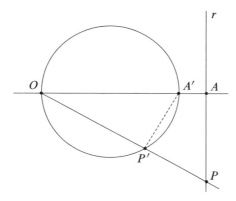

Figure 9.4

9.5.2. *The construction in 9.5.1 can be extended to any n-dimensional Euclidean space.*

9.5.3. *Let π and π' be two planes of S_3 and let a and b be two lines placed in general position with respect to π and π'. We say that two points $P \in \pi$ and $P' \in \pi'$ correspond to each other when the line $r_{PP'}$ is supported by a and b. In this way we define a quadratic transformation between π and π'. Find the base points and the exceptional lines.*

If P runs over the line r in π, the line passing through P and supported by a and b runs over the quadric having a, b, and r as directrices. Then P' runs over the conic r' which is the section of that quadric by π'. Thus one has a bijective correspondence (in general) that transforms lines in one plane into conics in the other.

Let A and B (respectively A' and B') be the intersections of the lines a and b with π (respectively π'); then let $C = r_{A'B'} \cap \pi$, and $C' = r_{AB} \cap \pi'$ (Figure 9.5).

324   Chapter 9. Cremona Transformations

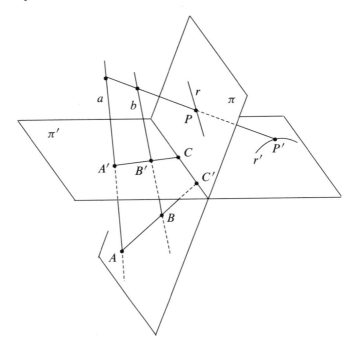

Figure 9.5

The fundamental points (to which there correspond all the points of an exceptional line in the other plane) are, for example in π, the points P for which one of the following two situations holds.

(1) The line issuing from P and meeting a and b lies on π'; there is only the point C in which the line $r_{A'B'}$ meets π and to C there corresponds the line $r_{A'B'}$.

(2) The line issuing from P and supported by a and b is indeterminate (i.e., there are infinitely many); the points with this property are A and B (for example to A there corresponds the intersection of π' with the plane joining A and b, that is the line $r_{B'C'}$).

9.5.4. *Let $\mathcal{Q}$ be a non-specialized quadric in S_3. We fix two of its points A and B. If π and π' are two planes, we say that two points $P \in \pi$ and $P' \in \pi'$ correspond to each other when the lines r_{AP}, $r_{BP'}$ meet $\mathcal{Q}$ (outside of A and B) in the same point.*

Describe the Cremona transformation between π and π' obtained in this way.

Given P in π we consider the line r_{AP} and its further intersection M with the quadric $\mathcal{Q}$. The point P' is the intersection with π' of the line r_{MB}. If P runs over a line r in the plane π, the line r_{AP} sweeps out a plane that cuts Q in a conic (traced by the point M): to r there corresponds the conic which is the projection of the former conic in the plane π' from the point B. Thus lines are transformed into conics (Figure 9.6).

The point P can be indeterminate in the following two cases.

(1) If the line r_{MB} is indeterminate, that is, if $M = B$ and thus P is the intersection N of the line r_{AB} with π. In this case r_{MB} is an arbitrary tangent to Q at B and so to N there corresponds the trace on π' of the tangent plane at B to Q.

(2) If the point M is indeterminate, which means that if the line r_{AP} lies on the quadric, and thus if $P = T_1$ or $P = T_2$ where T_1 and T_2 are the traces in π of the two generators g_1 and g_2 of Q passing through A. For example, the point T_1 joined with A gives the line g_1 on Q whose projection from B on π' is the line corresponding to T_1. This line passes through the point $N' = \pi' \cap r_{AB}$. Moreover, the line g_1 meets one of the two generators of Q passing through B and thus the plane determined by B and g_1 contains this generator b; and the projection from B onto π' passes through the trace T_2' of b in π'. Similarly, the projection from B of g_2 into π' gives the line that corresponds to T_2 (and which passes through the trace T_1' of the generator a of Q passing through B and meeting g_2).

The base points of the two homaloidal nets are N, T_1, T_2 and N', T_1', T_2'.

9.5.5. Let π and π' be two real Euclidean planes. We say that two points $P = (x, y)$ and $P' = (x', y')$ correspond to each other when, on setting $z = x + iy$, $z' = x' + iy'$, we have

$$z' = \frac{az + b}{cz + d}, \qquad (9.28)$$

with $a, b, c, d \in \mathbb{C}$, $ad - bc \neq 0$. Show that this transformation sends the system of circles of π onto the system of circles of π' (possibly degenerating into a line and the line at infinity).

Since

$$\frac{az + b}{cz + d} = \frac{a}{c} + \frac{\frac{bc - ad}{c^2}}{z + \frac{d}{c}},$$

the passage from z to z' can be decomposed into four steps

$$z \mapsto z + \frac{d}{c} \mapsto \frac{1}{z + \frac{d}{c}} \mapsto \frac{bc - ad}{c^2} \cdot \frac{1}{z + \frac{d}{c}} \mapsto \frac{a}{c} + \frac{bc - ad}{c^2} \cdot \frac{1}{z + \frac{d}{c}} = z'.$$

326   Chapter 9. Cremona Transformations

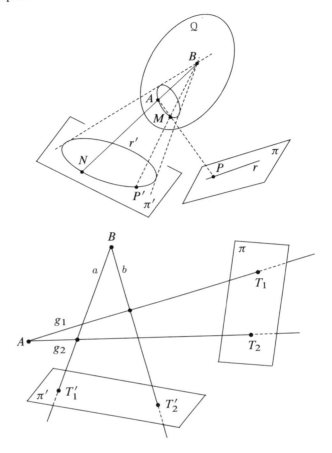

Figure 9.6

Thus one sees that the transformation (9.28) is the composition of transformations of the following types:

$$z' = z + \lambda, \quad z' = \mu z, \quad z' = \frac{1}{z}, \quad \lambda, \mu \in \mathbb{C}, \ \mu \neq 0. \tag{9.29}$$

On the other hand, if $\lambda = \lambda_1 + i\lambda_2$, the relation $z' = z + \lambda$ means that

$$x' + iy' = x + iy + \lambda_1 + i\lambda_2 = x + \lambda_1 + i(y + \lambda_2),$$

that is, $z' = z + \lambda$ is a translation defined by

$$\begin{cases} x' = x + \lambda_1, \\ y' = y + \lambda_2. \end{cases}$$

If $\mu = \mu_1 + i\mu_2$, the transformation $z' = \mu z$ may be rewritten
$$x' + iy' = (\mu_1 + i\mu_2)(x + iy) = \mu_1 x - \mu_2 y + i(\mu_2 x + \mu_1 y).$$
This means that $z' = \mu z$ is a particular direct similitude (which we will also call a *rotohomothety*, that is, a rotation around the origin followed by a homothety, cf. [13, Vol. I, Chapter 2]) defined by
$$\begin{cases} x' = \mu_1 x - \mu_2 y, \\ y' = \mu_2 x + \mu_1 y. \end{cases}$$
The structure of the transformation $z' = \mu z$ is best seen by writing $z = re^{i\theta}$, $z' = r'e^{i\theta'}$, $\mu = \rho e^{i\omega}$. Then $r'e^{i\theta'} = re^{i\theta}\rho e^{i\omega} = r\rho e^{i(\theta+\omega)}$, that is rotation through the angle ω and multiplication of the length by ρ.

Finally $z' = \frac{1}{z}$ may be rewritten
$$x' + iy' = \frac{1}{x+iy} = \frac{x-iy}{x^2+y^2},$$
or
$$\begin{cases} x' = \dfrac{x}{x^2+y^2}, \\ y' = -\dfrac{y}{x^2+y^2}, \end{cases}$$
and so here one has a transformation by reciprocal radii (cf. 9.5.1) followed by a symmetry with respect to the x-axis.

The three transformations (9.29) transform lines and circles into lines and circles, and preserve the angles (and their sense); and so (9.28) is in general a quadratic transformation (but can be a projectivity in special cases) that preserves angles and the sense of angles. The transformations (9.28) are called *Möbius transformations of the second kind*.

9.5.6. Let π and π' be two real Euclidean planes. We say that two points $P = (x, y)$ and $P = (x', y')$ correspond to each other when, on setting $z = x + iy$ and $z' = x' + iy'$, one has
$$z' = \frac{a\bar{z} + b}{c\bar{z} + d}, \tag{9.30}$$
where $a, b, c, d \in \mathbb{C}$, $ad - bc \neq 0$, and $\bar{z} = x - iy$. Show that this transformation changes the system of circles of π (possibly degenerating into a line and the line at infinity) into the system of circles of π'.

Reasoning as in Exercise 9.5.5 one sees that (9.30) are quadratic transformations (in general) which preserve angles (but not the sense of angles); they are obtained as a product of translations, rotohomotheties, transformations by reciprocal radii, and symmetries. One says that (9.30) gives the *Möbius transformations of the first kind*.

9.5.7. *Prove that the necessary and sufficient condition for the points z_1, z_2, z_3, z_4 in the Argand–Gauss plane to lie on a circle is that their cross ratio $\mathcal{R}(z_1, z_2, z_3, z_4)$ be real.*

Let z_1, z_2, z_3, z_4 be four points of a circle C. The Möbius transformation $z' = \frac{\alpha z + \beta}{\gamma z + \delta}$ that sends z_1, z_2, z_3 into three arbitrary points z'_1, z'_2, z'_3 of the real axis will also send z_4 into a point z'_4 of the real axis because it maps circles into circles or lines. We then have

$$\mathcal{R}(z_1, z_2, z_3, z_4) = \mathcal{R}(z'_1, z'_2, z'_3, z'_4) \in \mathbb{R}.$$

Conversely if $\mathcal{R}(z_1, z_2, z_3, z_4) \in \mathbb{R}$, a Möbius transformation that maps z_1, z_2, z_3 into three points z'_1, z'_2, z'_3 of the real axis will also map z_4 into a point z'_4 of the real axis (because $\mathcal{R}(z'_1, z'_2, z'_3, z'_4) \in \mathbb{R}$ and $z'_1, z'_2, z'_3 \in \mathbb{R}$). Then z_1, z_2, z_3, z_4, corresponding to four collinear points, belong to a circle.

9.5.8. *Examine the Cremona transformations θ_1, θ_2 between spaces S_3 which are represented in affine Cartesian coordinates by the equations*

$$\theta_1 : \begin{cases} x' = x, \\ y' = \dfrac{y}{x}, \\ z' = \dfrac{z}{x} \end{cases} \quad \text{and} \quad \theta_2 : \begin{cases} x' = x, \\ y' = y, \\ z' = \dfrac{z}{x}. \end{cases}$$

Study the compositions $\theta_1 \circ \theta_2$, $\theta_2 \circ \theta_1$ and for each of them the two homaloidal systems of surfaces.

We write the equations of θ_1 in homogeneous coordinates to obtain

$$\frac{x'_1}{x'_0} = \frac{x_1}{x_0} = \frac{x_1^2}{x_0 x_1}; \quad \frac{x'_2}{x'_0} = \frac{x_2}{x_1} = \frac{x_2 x_0}{x_1 x_0}; \quad \frac{x'_3}{x'_0} = \frac{x_3}{x_1} = \frac{x_3 x_0}{x_1 x_0},$$

whence

$$\theta_1 = \begin{cases} x'_0 = x_0 x_1, \\ x'_1 = x_1^2, \\ x'_2 = x_0 x_2, \\ x'_3 = x_0 x_3. \end{cases}$$

The first homaloidal system of quadrics has equation

$$\lambda_0 x_0 x_1 + \lambda_1 x_1^2 + \lambda_2 x_0 x_2 + \lambda_3 x_0 x_3 = 0,$$

that is,

$$x_0(\lambda_0 x_1 + \lambda_2 x_2 + \lambda_3 x_3) + \lambda_1 x_1^2 = 0.$$

These are quadrics that pass (simply) through $A_0 = [1, 0, 0, 0]$ and that are tangent to the plane $x_0 = 0$ along the line $x_0 = x_1 = 0$; thus they are cones. The base variety then consists of a point and a conic (given by two coinciding lines). The Cremona transformation θ_1 is a particular case of a transformation of the type (2, 2).

The equations of θ_1 may be inverted to obtain the following expression for the inverse transformation:

$$\theta_1^{-1} : \begin{cases} x = x', \\ y = x'y', \\ z = x'z'. \end{cases}$$

In homogeneous coordinates this becomes

$$\frac{x_1}{x_0} = \frac{x_1'}{x_0'} = \frac{x_1' x_0'}{x_0'^2}; \quad \frac{x_2}{x_0} = \frac{x_1' x_2'}{x_0'^2}; \quad \frac{x_3}{x_0} = \frac{x_1' x_3'}{x_0'^2},$$

and thus

$$\theta_1^{-1} : \begin{cases} x_0 = x_0'^2, \\ x_1 = x_0' x_1', \\ x_2 = x_1' x_2', \\ x_3 = x_1' x_3'. \end{cases}$$

The second homaloidal system is

$$\lambda_0 x_0'^2 + \lambda_1 x_0' x_1' + \lambda_2 x_1' x_2' + \lambda_3 x_1' x_3' = 0$$

and consists of quadric cones passing (simply) through $A_1' = [0, 1, 0, 0]$ and tangent to the plane $x_1' = 0$ along the line $x_1' = x_0' = 0$.

We write the equations of θ_2 in homogeneous coordinates to obtain

$$\frac{x_1'}{x_0'} = \frac{x_1}{x_0} = \frac{x_1^2}{x_0 x_1}; \quad \frac{x_2'}{x_0'} = \frac{x_2}{x_0} = \frac{x_2 x_1}{x_0 x_1}; \quad \frac{x_3'}{x_0'} = \frac{x_3}{x_1} = \frac{x_3 x_0}{x_1 x_0},$$

whence

$$\theta_2 : \begin{cases} x_0' = x_0 x_1, \\ x_1' = x_1^2, \\ x_2' = x_2 x_1, \\ x_3' = x_0 x_3. \end{cases}$$

The first homaloidal system of quadrics has equation

$$\lambda_0 x_0 x_1 + \lambda_1 x_1^2 + \lambda_2 x_2 x_1 + \lambda_3 x_0 x_3 = 0,$$

or also

$$x_1(\lambda_0 x_0 + \lambda_1 x_1 + \lambda_2 x_2) + \lambda_3 x_0 x_3 = 0.$$

The base variety consists of a pair of distinct lines (given by $x_1 = x_0 x_3 = 0$) having the point $A_2 = [0, 0, 1, 0]$ in common. (The other base point is the point $A_3 = [0, 0, 0, 1]$, contained in one of the two lines since the quadrics of the system all have the same tangent plane $x_0 = 0$ at A_3.) The transformation θ_2 is also of type $(2, 2)$.

The formulas for the inverse transformation are

$$\theta_2^{-1} = \begin{cases} x = x', \\ y = y', \\ z = x'z'. \end{cases}$$

The second homaloidal system is

$$\lambda_0 x_0'^2 + \lambda_1 x_0' x_1' + \lambda_2 x_0' x_2' + \lambda_3 x_1' x_3' = 0$$

and consists of quadrics passing through the pair of lines $x_0' = x_1' x_3' = 0$ and tangent at $A_3' = [0, 0, 0, 1]$ to the plane $x_1' = 0$.

For the description of the composed maps $\theta_1 \circ \theta_2$, $\theta_2 \circ \theta_1$ one proceeds in analogous fashion.

9.5.9. *In what cases can the composition of two quadratic transformations between planes be a Cremona transformation of order ≤ 2?*

Let $\theta_1 \colon \pi' \to \pi''$, and $\theta_2 \colon \pi'' \to \pi'''$ be the two quadratic transformations.

If Σ_1, Σ_2 are the two homaloidal nets of θ_1 and Σ_3, Σ_4 those of θ_2, in order that $\theta_2 \circ \theta_1 \colon \pi' \to \pi'''$ be a Cremona transformation of order ≤ 2 (i.e., either a projectivity or a quadratic transformation) it is necessary and sufficient that at least two of the three base points of Σ_3 are also base points of Σ_2. Indeed, in this way the conics of Σ_2 are transformed by θ_2 into lines (if the base points of Σ_3 are the same three base points as those of Σ_2), in which case $\theta_2 \circ \theta_1$ is a projectivity, or else into conics (if just two base points of Σ_3 are also base points of Σ_2); cf. Proposition–Definition 9.2.2.

9.5.10. *A quadratic transformation $\theta \colon S_2 \to S_2'$ between two planes is the intersection of two reciprocities in the sense that two points P and P' correspond if they are reciprocal in two given reciprocities (i.e., P' is the intersection of the lines that correspond to P under the two reciprocities).*

Therefore, if we represent algebraic correspondences between two planes as subvarieties of the Segre variety X_4^6 realizing the inclusion $\mathbb{P}^2 \times \mathbb{P}^2 \subset \mathbb{P}^8$ (see Chapter 11), the quadratic transformations appear as sections of X_4^6 by the spaces S_6 in $\mathbb{P}^8$.

In Section 9.1 we have defined a quadratic transformation by way of two bilinear equations; each of these represents a reciprocity inasmuch as each point of S_2 is made to correspond to a line of S_2' (see equations (9.3) and [52, Vol. 1, Chapter IX]), and two corresponding points under the quadratic transformation are reciprocal in both.

Suppose that the two bilinear equations are written in the form

$$\begin{cases} \sum_{i,j=0}^{2} a_{ij} x_i y_j = 0, \\ \sum_{i,j=0}^{2} b_{ij} x_i y_j = 0 \end{cases} \quad (9.31)$$

(where $A = (a_{ij})$ and $B = (b_{ij})$ are the matrices that define the two reciprocities). Let $P = [x_0, x_1, x_2]$ and $P' = [y_0, y_1, y_2]$ be two corresponding points under θ, that is, two points whose coordinates satisfy (9.31). To the pair $(P, P') \in \mathbb{P}^2 \times \mathbb{P}^2$ there corresponds in the Segre variety $X_4^6 = \mathbb{P}^2 \times \mathbb{P}^2 \subset \mathbb{P}^8$ the point Q with coordinates

$$Q = [x_0 y_0, x_0 y_1, x_0 y_2, x_1 y_0, x_1 y_1, x_1 y_2, x_2 y_0, x_2 y_1, x_2 y_2].$$

The fact that the coordinates of P and P' satisfy equations (9.31) means that the point Q belongs to the two hyperplanes with equations $\sum_{i,j=0}^{2} a_{ij} T_{ij} = 0$, $\sum_{i,j=0}^{2} b_{ij} T_{ij} = 0$, where the T_{ij} are the coordinates in $\mathbb{P}^8$. Therefore the quadratic transformations between two planes are the sections of X_4^6 by linear spaces S_6 in $\mathbb{P}^8$.

We note that the quadratic transformations between two planes are in number ∞^{14}. Indeed, the nets of conics in $\mathbb{P}^2$ are in number ∞^6, the same as the number of triples of points in the plane. Moreover, to each net of conics there correspond ∞^8 quadratic transformations, since there are ∞^8 projectivities between the net Σ ($\cong \mathbb{P}^2$) and the plane S_2'. Of course, the subspaces S_6 in $\mathbb{P}^8$ are also ∞^{14} in number, since 14 is the dimension $\dim(\mathbb{G}(6, 8)) = (8-6)(6+1)$ of the Grassmann variety $\mathbb{G}(6, 8)$ of the S_6 in $\mathbb{P}^8$; see Chapter 12 for further details.

9.5.11. *Let two non-degenerate conics γ and δ be given in the plane. We decree that two points P and P' correspond to each other if they are reciprocal with respect to both conics (this means that the point $P' = \gamma(P) \cap \delta(P)$ is the intersection of the polars of P with respect to the two conics). Study the (involutory) correspondence $\theta: \pi \to \pi'$ defined by $P \mapsto P'$.*

If P runs over the line r, the polar of P with respect to γ traces the pencil of lines (with center the pole of r with respect to γ), and similarly for the polar of P with respect to δ. The two pencils are projective (that is, in bijective algebraic correspondence) and so the locus of the points common to corresponding lines is a conic r' (cf. §7.4.5). This is the correspondent of r under θ. As r varies one obtains a net of conics.

Let A, B, C, and D be the points common to γ and δ and let LMN be the diagonal triangle of the quadrangle $ABCD$.

Let X be the intersection of r with one of the sides of the triangle LMN, for example LM. The polars of X with respect to γ and δ both pass through N which is the pole of the line r_{LM} with respect to both conics. Thus the conic $r' = \theta(r)$ passes through L, M, and N. The vertices of the diagonal triangle of the quadrangle $ABCD$ inscribed in both conics are thus base points of the quadratic transformation.

9.5.12. *Let $\mathcal{C}$ be a plane quartic having three cusps. Verify that the three cuspidal tangents belong to a pencil.*

The quadratic transformation of the plane into itself with fundamental points in the three cusps A, B, and C transforms $\mathcal{C}$ into a conic γ tangent to the lines $a = BC$, $b = AC$ and $c = AB$ in the points A', B', and C'. The lines AA', BB', and CC' (the images of the cuspidal tangents) have a common point by Brianchon's theorem applied to the hexilateral $aabbcc$ where $aa = A'$, $bb = B'$, $cc = C'$ (cf. [13, Vol. II, Chapter 16] and also [31, Chapter XIV]).

9.5.13. *Let J be the Jacobian curve of the homaloidal net $\Sigma = C^n(P_1^{s_1}, \ldots, P_h^{s_h})$. Prove that J consists of the fundamental curves of Σ.*

We know that J has order $3(n-1)$ and passes with multiplicity $3s_i - 1$ through the point P_i, $i = 1, \ldots, h$ (cf. Section 6.4). Bearing in mind the relations (9.12), the number of intersections of J with the generic C^n of Σ that are absorbed by the base points is at least

$$\sum_i s_i(3s_i - 1) = 3\sum_i s_i^2 - \sum_i s_i = 3(n^2 - 1) - 3n + 3 = 3n(n-1).$$

Therefore J and the generic curve of Σ meet only in the base points.

The curves of Σ that pass through a point P of J other than the base points thus have a common component which is also a component of J. Moreover, each of these components imposes a single condition on the curves of Σ which must contain it. Thus each of these components is a fundamental curve of Σ. Hence the Jacobian curve consists of all the fundamental curves of Σ' (which agrees with the fact that the order of J is $3n - 3 = \sum_i s_i$, with s_i the orders of the fundamental curves). Each point of a fundamental curve is thus a double point for some curve of the net (cf. 9.3.8). Therefore if P is a point of the fundamental curve α different from the base points of Σ, the curves of Σ passing through P split into the curve α and a further curve passing through P.

9.5.14. *Write the equation of the Jacobian curve J of the homaloidal net $\Sigma = C^n(A^{n-1}, B_1, \ldots, B_{2n-2})$, given by*

$$(x_2\psi_{n-2} + \psi_{n-1})(\lambda_0 x_0 + \lambda_1 x_1) + x_2\varphi_{n-1} + \varphi_n = 0,$$

of a De Jonquières transformation (cf. §9.3.5).

We know from Exercise 9.5.13 that the Jacobian curve is the union of the fundamental curves of the net. The fundamental curves are the curve Γ with equation

$x_2\psi_{n-2} + \psi_{n-1} = 0$ and the lines joining the $(n-1)$-fold point A with the other base points $B_1, \ldots, B_{2n-2}$, that is, the points (distinct from A) common to the two curves

$$x_2\psi_{n-2} + \psi_{n-1} = 0; \quad x_2\varphi_{n-1} + \varphi_n = 0.$$

The equation for the union of the lines $\langle A, B_i \rangle$ is obtained by eliminating x_2 from these two equations, and so is $\psi_{n-2}\varphi_n - \varphi_{n-1}\psi_{n-1} = 0$. Hence J is the curve with equation

$$(x_2\psi_{n-2} + \psi_{n-1})(\psi_{n-2}\varphi_n - \varphi_{n-1}\psi_{n-1}) = 0.$$

The degree of this equation is (as it should be) $n - 1 + 2n - 2 = 3n - 3$.

9.5.15. Let $\theta: S_3 \to S_3'$ be a Cremona transformation of type $(2, 2)$ and let γ be the base conic of the homaloidal system Σ in S_3. Study the images of the lines of S_3 that are supported by γ.

If γ has equations $x_0 = \gamma(x_1, x_2, x_3) = 0$, and $A_0 = [1, 0, 0, 0]$ is the further isolated point, the equations of θ are

$$\begin{cases} x_0' = \gamma(x_1, x_2, x_3), \\ x_1' = x_0 x_1, \\ x_2' = x_0 x_2, \\ x_3' = x_0 x_3. \end{cases} \tag{9.32}$$

Let $P = [0, a, b, c]$ be a point of γ and let

$$r: \begin{cases} x_0 = t, \\ x_1 = a + lt, \\ x_2 = b + mt, \\ x_3 = c + nt \end{cases}$$

be a line passing through P (but not lying on the cone Γ with equation $\gamma(x_1, x_2, x_3) = 0$). Substituting in (9.32) one has, for suitable constants A and B,

$$\begin{cases} x_0' = \gamma(a + lt, b + mt, c + nt) = \gamma(a, b, c) + At + Bt^2 = t(A + Bt), \\ x_1' = t(a + lt), \\ x_2' = t(b + mt), \\ x_3' = t(c + nt), \end{cases}$$

and so, suppressing the common factor t,

$$r': \begin{cases} x_0' = A + Bt, \\ x_1' = a + lt, \\ x_2' = b + mt, \\ x_3' = c + nt. \end{cases}$$

Hence the transform of the line r is the line r'.

The generators of the cone Γ are exceptional in this regard. If $P = [0, a, b, c]$ is a point of the conic, the generator of the cone that passes through P is the locus of the point $[t, a, b, c]$ to which there corresponds under θ the point P' with coordinates

$$\begin{cases} x'_0 = \gamma(a, b, c) = 0, \\ x'_1 = at, \\ x'_2 = bt, \\ x'_3 = ct, \end{cases}$$

that is, $P' = [0, a, b, c]$.

Note that to a generic line of the space there corresponds a conic; if the line r meets the conic γ, the corresponding conic $\theta(r)$ splits into a generator of the cone Γ (corresponding to the point at which r meets γ) and a second line r'.

9.5.16. *Examine the Cremona transformation* $\theta \colon S_3 \to S'_3$ *under which the two points* $P = [x_0, x_1, x_2, x_3]$ *and* $P' = [y_0, y_1, y_2, y_3]$ *correspond when*

$$\begin{cases} f(x_0, x_1, x_2, x_3; y_0, y_1, y_2, y_3) = 0, \\ g(x_0, x_1, x_2, x_3; y_0, y_1, y_2, y_3) = 0, \\ h(x_0, x_1, x_2, x_3; y_0, y_1, y_2, y_3) = 0, \end{cases}$$

where f, g, *and* h *are bilinear forms in the series of variables* x_i *and* y_j, $i, j = 0, 1, 2, 3$. *One finds a Cremona transformation of type* $(3, 3)$. *To the lines of one of the two spaces there correspond curves of order* 6 *in the other space.*

We rewrite the three equations ordering them, for example, with respect to the variables y_j:

$$A_t(x_0, x_1, x_2, x_3)y_0 + B_t(x_0, x_1, x_2, x_3)y_1 \\ + C_t(x_0, x_1, x_2, x_3)y_2 + D_t(x_0, x_1, x_2, x_3)y_3 = 0,$$

where A_t, B_t, C_t, and D_t are linear forms for $t = 1, 2, 3$. Solving the system with respect to the y_j one finds

$$y_0 : y_1 : y_2 : y_3$$

$$= \begin{vmatrix} B_1 & C_1 & D_1 \\ B_2 & C_2 & D_2 \\ B_3 & C_3 & D_3 \end{vmatrix} : - \begin{vmatrix} A_1 & C_1 & D_1 \\ A_2 & C_2 & D_2 \\ A_3 & C_3 & D_3 \end{vmatrix} : \begin{vmatrix} A_1 & B_1 & D_1 \\ A_2 & B_2 & D_2 \\ A_3 & B_3 & D_3 \end{vmatrix} : - \begin{vmatrix} A_1 & B_1 & C_1 \\ A_2 & B_2 & C_2 \\ A_3 & B_3 & C_3 \end{vmatrix}.$$

Thus θ is a Cremona transformation of type $(3, n')$, and thus of type $(3, 3)$ by the symmetry between the two series of variables.

The homaloidal system Σ of S_3 has as base variety the locus of points that endow the matrix

$$M = \begin{pmatrix} A_1 & B_1 & C_1 & D_1 \\ A_2 & B_2 & C_2 & D_2 \\ A_3 & B_3 & C_3 & D_3 \end{pmatrix}$$

with rank two. They form a curve C^6 of order 6; indeed, the system of equations

$$\begin{vmatrix} A_1 & B_1 & C_1 \\ A_2 & B_2 & C_2 \\ A_3 & B_3 & C_3 \end{vmatrix} = \begin{vmatrix} A_1 & B_1 & D_1 \\ A_2 & B_2 & D_2 \\ A_3 & B_3 & D_3 \end{vmatrix} = 0$$

defines a curve of order nine that splits into a curve C^6 and a cubic C^3 which is the locus of the points that annul the minors of order 2 of the matrix

$$\begin{pmatrix} A_1 & B_1 \\ A_2 & B_2 \\ A_3 & B_3 \end{pmatrix}.$$

Discarding this cubic there remains the curve C^6 mentioned above.

In particular, to each line in S_3 there corresponds the curve of third order which is the intersection (outside the base curve C^6) of the two cubic surfaces corresponding to the two planes whose intersection is the line.

9.5.17. *Among the Cremona transformations θ as in Exercise 9.5.16 there are in particular those of equations*

$$y_0 : y_1 : y_2 : y_3 = \frac{1}{x_0} : \frac{1}{x_1} : \frac{1}{x_2} : \frac{1}{x_3}.$$

Study them. How is an elliptic cubic cone passing through the four vertices of the reference tetrahedron transformed? (An elliptic cubic cone is a cone that projects an elliptic cubic.)

We rewrite the equations of θ in the form

$$y_0 : y_1 : y_2 : y_3 = x_1 x_2 x_3 : x_0 x_2 x_3 : x_0 x_1 x_3 : x_0 x_1 x_2.$$

A surface F of order n with equation $f(y_0, y_1, y_2, y_3) = 0$ is transformed into the surface G with equation

$$g(x_0, x_1, x_2, x_3) = f(x_1 x_2 x_3, x_0 x_2 x_3, x_0 x_1 x_3, x_0 x_1 x_2) = 0$$

of order $3n$. In the equation for G each of the variables appear to degree at most n; G has order $3n$ and has four $2n$-fold points. Moreover, consider one of the edges of the fundamental tetrahedron, for example the line $x_0 = x_1 = 0$. It is clear that $g \in (x_0, x_1)^n$ and so the edge is a line of multiplicity n. Hence the surface G passes with multiplicity $2n$ through every vertex A_i, $i = 0, 1, 2, 3$, and with multiplicity n through each edge of the tetrahedron.

Suppose now that F passes through A_0 with multiplicity s_0, so that its equation may be rewritten in the form:

$$f = y_0^{n-s_0}\alpha_{s_0}(y_1, y_2, y_3) + y_0^{n-s_0-1}\alpha_{s_0+1}(y_1, y_2, y_3) + \cdots + \alpha_n(y_1, y_2, y_3) = 0.$$

Then the surface $G = \theta(F)$ has equation

$$g(x_0, x_1, x_2, x_3) = (x_1 x_2 x_3)^{n-s_0} x_0^{s_0} \alpha_{s_0}(x_2 x_3, x_1 x_3, x_1 x_2) + \cdots$$
$$\cdots + x_0^n \alpha_n(x_2 x_3, x_1 x_3, x_1 x_2) = 0,$$

that is

$$g(x_0, x_1, x_2, x_3) = x_0^{s_0} g^*(x_0, x_1, x_2, x_3) = 0,$$

where

$$g^*(x_0, x_1, x_2, x_3) = (x_1 x_2 x_3)^{n-s_0} \alpha_{s_0}(x_2 x_3, x_1 x_3, x_1 x_2) + \cdots$$
$$\cdots + x_0^{n-s_0} \alpha_n(x_2 x_3, x_1 x_3, x_1 x_2).$$

Having discarded the plane $x_0 = 0$ which must be counted s_0 times (and which corresponds to the point A_0) there remains an equation $g^*(x_0, x_1, x_2, x_3) = 0$ of degree $3n - s_0$, in which the variable x_0 appears at most to degree $n - s_0$ and the variables x_1, x_2, and x_3 still to degree n. Thus the surface G^* of equation $g^*(x_0, x_1, x_2, x_3) = 0$ has in the points A_0, A_1, A_2, and A_3 the multiplicities $3n - s_0 - (n - s_0) = 2n$, $3n - s_0 - n = 2n - s_0$, $2n - s_0$, and $2n - s_0$ respectively. Furthermore, one sees that

$$g^* \in (x_1, x_2)^n \cap (x_1, x_3)^n \cap (x_2, x_3)^n \cap (x_0, x_1)^{n-s_0} \cap (x_0, x_2)^{n-s_0} \cap (x_0, x_3)^{n-s_0},$$

whence G^* has the three edges of the tetrahedron that issue from A_0 as n-fold lines, and the other three edges as $(n - s_0)$-fold lines.

The conclusion is that if f passes with multiplicity s_0, s_1, s_2, and s_3 through A_0, A_1, A_2, and A_3 respectively, then the proper transform G^* has order

$$3n - s_0 - s_1 - s_2 - s_3,$$

and multiplicities

$$2n - s_1 - s_2 - s_3, \quad 2n - s_0 - s_2 - s_3, \quad 2n - s_0 - s_1 - s_3, \quad 2n - s_0 - s_1 - s_2$$

in A_0, A_1, A_2, and A_3 respectively. Moreover, f passes through the lines $r_{A_0 A_1}$, $r_{A_0 A_2}$, $r_{A_0 A_3}$, $r_{A_1 A_2}$, $r_{A_1 A_3}$, and $r_{A_2 A_3}$ with respective multiplicities

$$n - s_2 - s_3, \quad n - s_1 - s_3, \quad n - s_1 - s_2, \quad n - s_0 - s_3, \quad n - s_0 - s_2, \quad n - s_0 - s_1.$$

Therefore, the cubic cone (the case $n = 3$, $s_0 = s_1 = s_2 = s_3 = 1$) is transformed into a surface G^* of order $3 \times 3 - 1 - 1 - 1 - 1 = 5$ which has five triple points, namely the points A_0, A_1, A_2, A_3 all having multiplicity $2 \times 3 - 1 - 1 - 1 = 3$, and also the transform of the vertex of the cone. The surface G^* also passes simply through the six lines $r_{A_i A_j}$ (in agreement with the fact that on each of these it has two triple points and so must contain that line).

9.5.18. *Study the Cremona transformations φ (of type (n, n)) between two spaces S_n realized as intersections of n reciprocities, in the sense that two points correspond if they are reciprocal with respect to n given reciprocities (cf. Theorem 5.4.6).*

If we represent the algebraic correspondences between two spaces S_n as subvarieties of the Segre variety X_{2n}^d giving $\mathbb{P}^n \times \mathbb{P}^n \subset \mathbb{P}^{n(n+2)}$ (so that X_{2n}^d has order $d = (2n)!/n!\,n!$), then the transformations φ appear as sections of X_{2n}^d by linear spaces $S_{n(n+1)}$.

This is an extension of Exercises 9.5.16 and 9.5.10. See Chapter 11 for general results on Segre varieties.

The equations of the transformation may be written in the form

$$A_{i0}(x_0, \ldots, x_n) y_0 + A_{i1}(x_0, \ldots, x_n) y_1 + \cdots + A_{in}(x_0, \ldots, x_n) y_n = 0,$$

where the $A_{ij}(x_0, \ldots, x_n)$ are linear forms, $i = 1, \ldots, n$, $j = 0, \ldots, n$. The variables y_j are proportional to the minors of maximal order of the $n \times (n+1)$ matrix

$$M = \begin{pmatrix} A_{10} & \cdots & A_{1n} \\ \vdots & & \vdots \\ A_{n0} & \cdots & A_{nn} \end{pmatrix}.$$

The first homaloidal system Σ of the transformation has as base variety B_{n-2} the locus of the points that annihilate the minors of order n of M. To find the equations of B_{n-2} it suffices to annul the two minors made up of the columns number $(0, 1, 2, \ldots, n-2, n-1)$ and $(0, 1, 2, \ldots, n-2, n)$. It is an $(n-2)$-dimensional variety which contains the variety (to be discarded, cf. 9.5.16) which is the locus of the zeros of the $n-1$ minors of maximal order in the $n \times (n-1)$ matrix consisting of columns number $(0, 1, 2, \ldots, n-2)$ of M.

If $\delta(n)$ is the order of B_{n-2} we have

$$\delta(2) = 3,$$
$$\delta(3) = 3^2 - \delta(2),$$
$$\delta(4) = 4^2 - \delta(3),$$
$$\vdots$$
$$\delta(n) = n^2 - \delta(n-1).$$

Thus, e.g., for $n = 4$, $\delta(2) = 3$, $\delta(3) = 9 - 3 = 6$, and $\delta(4) = 16 - 6 = 10$.

9.5.19. *Consider the matrix*

$$A = (a_{ij}) = \begin{pmatrix} x_0 & x_1 & x_2 \\ x_1 & x_3 & x_4 \\ x_2 & x_4 & x_5 \end{pmatrix}$$

and the matrix $A^* = (A_{ij})$ *having as its elements the cofactors of the elements a_{ij} of A. If B_{ij} is the cofactor of A_{ij} in A^*, we have*

$$B_{ij} = a_{ij} \det(A).$$

From this deduce that the quadrics of $\mathbb{P}^5$ passing through a Veronese surface form a homaloidal linear system (cf. Example 10.2.1).

If $(x_0, x_1, x_2, x_3, x_4, x_5)$ are coordinates in S_5 and $(y_0, y_1, y_2, y_3, y_4, y_5)$ are coordinates in a second space S'_5, consider the two transformations defined by the equations

$$\varphi: \begin{cases} y_0 = A_{11} = x_3 x_5 - x_4^2, \\ y_1 = A_{12} = x_1 x_5 - x_2 x_4, \\ y_2 = A_{13} = x_1 x_4 - x_2 x_3, \\ y_3 = A_{22} = x_0 x_5 - x_2^2, \\ y_4 = A_{23} = x_0 x_4 - x_1 x_2, \\ y_5 = A_{33} = x_0 x_3 - x_1^2 \end{cases} \quad \text{and} \quad \theta: \begin{cases} x_0 = y_3 y_5 - y_4^2, \\ x_1 = y_1 y_5 - y_2 y_4, \\ x_2 = y_1 y_4 - y_2 y_3, \\ x_3 = y_0 y_5 - y_2^2, \\ x_4 = y_0 y_4 - y_1 y_2, \\ x_5 = y_0 y_3 - y_1^2. \end{cases}$$

The equations of $\theta \circ \varphi$ are

$$x'_i = x_i \det(A), \quad i = 0, \ldots, 5,$$

which means that $\theta \circ \varphi = \mathrm{id}_{S_5}$. Analogously one has $\varphi \circ \theta = \mathrm{id}_{S'_5}$; thus φ and θ are mutually inverse, that is, they are birational isomorphisms. It follows that the linear system of quadrics with equation $\sum_{i,j=0}^{5} \lambda_{ij} A_{ij} = 0$ is a homaloidal system.

On the other hand, it is known that $\sum_{i,j=0}^{5} \lambda_{ij} A_{ij} = 0$ is the linear system of the quadrics in S_5 that pass through the Veronese surface (see Example 10.2.1 and Section 10.4 for further details).

9.5.20. *Consider the linear system Σ of hypersurfaces in S_r with equation*

$$\sum_{i=0}^{r} \lambda_i x_0 x_1 \ldots x_{i-1} x_{i+1} \ldots x_r = 0.$$

Is this a homaloidal system?

Consider the rational transformation $\theta: S_r \to S'_r$ with equations

$$y_i = x_0 \ldots x_{i-1} x_{i+1} \ldots x_r, \quad i = 0, \ldots, r. \tag{9.33}$$

On the open set where $x_0 x_1 \ldots x_i \ldots x_r \neq 0$, that is, outside the faces of the fundamental $(r+1)$-hedron, the equations (9.33) can be written in the form

$$y_i = \frac{1}{x_i}, \quad i = 0, \ldots, r,$$

and one can immediately verify that they are rationally inverted by

$$x_i = \frac{1}{y_i}, \quad i = 0, \ldots, r.$$

Thus θ is a birational isomorphism and so Σ is a homaloidal system.

Chapter 10
Rational Surfaces

Rational surfaces, that is, the images of a plane under a rational transformation, are of particular interest. In virtue of an important result due to Castelnuovo which extends Lüroth's theorem (cf. Theorems 6.6.2, 6.6.3 and 7.4.1) to surfaces, a rational surface is birationally isomorphic to a plane π. Thus the algebraic geometry of a rational surface $\mathcal{F}$ is reduced to that of the plane, so that all problems regarding curves traced on $\mathcal{F}$ can be translated into questions regarding algebraic plane curves. In particular to the linear system of the hyperplane sections of $\mathcal{F}$ there corresponds in π a linear system Σ of algebraic curves, defined up to a Cremona transformation. One says that Σ represents the surface, and that the latter is the projective image of Σ. The birational invariants of Σ are projective invariants of the surface: for example the dimension of Σ, its degree, and the genera of its curves are the dimension of the space in which $\mathcal{F}$ is embedded, the order of $\mathcal{F}$, and the genera of its hyperplane sections respectively. In general the algebraic correspondence between $\mathcal{F}$ and π is bijective only generically: the exceptions to bijectivity are related to the base points and fundamental curves of Σ.

The numerous examples drawn from the classical repertoire and the exercises have the goal not only of introducing the reader to some surfaces of particular interest (for example surfaces of minimal order, Veronese surfaces, Del Pezzo surfaces) but also and especially offer the occasion to enter more deeply into the theory which is expounded here only in its essential points. In particular, the results of Section 10.3 furnish a complete classification of the surfaces of minimal order (cf. Proposition 4.5.6).

10.1 Planar representation of rational surfaces

We return now, in the case of the plane, to the discussion carried out in Section 6.6. Let u_0, u_1, u_2 be projective coordinates in $\mathbb{P}^2 = \mathbb{P}^2(K)$ and $x_0, x_1, \ldots, x_r$ projective coordinates in $\mathbb{P}^r = \mathbb{P}^r(K)$.

Given $r+1$ homogeneous polynomials of the same degree $\varphi_j = \varphi_j(u_0, u_1, u_2)$ in $K[u_0, u_1, u_2]$, $j = 0, \ldots, r$, consider the rational map $\varphi \colon \mathbb{P}^2 \to \mathbb{P}^r$ defined by:

$$\begin{cases} x_0 = \varphi_0(u_0, u_1, u_2), \\ \quad \vdots \\ x_r = \varphi_r(u_0, u_1, u_2). \end{cases} \quad (10.1)$$

The image of φ is an irreducible algebraic variety V whose dimension is ≤ 2 and

10.1. Planar representation of rational surfaces

the condition that the dimension of V be exactly 2 is that the Jacobian matrix of the polynomials φ_j have rank $\varrho = 3$ at all points of an open set in $\mathbb{P}^2$; indeed, ϱ is the dimension of the affine cone associated to V, and thus $\dim(V) = 2$ if $\varrho = 3$ (cf. Section 3.2). This means that there do not exist two polynomials $P, Q \in K[u_0, u_1, u_2]$ such that $\varphi_j \in K[P, Q]$, $j = 0, \ldots, r$; or equivalently that the linear system Σ of algebraic curves with (system) equation

$$\lambda_0 \varphi_0 + \lambda_1 \varphi_1 + \cdots + \lambda_r \varphi_r = 0 \qquad (10.2)$$

is not composed with a pencil (cf. Section 6.5). The rational map φ is defined outside of the base points of the system Σ.

In agreement with the definitions given in Section 6.6, if $\dim(\mathrm{Im}\,\varphi) = 2$, we say that $\mathcal{F} := \mathrm{Im}\,\varphi$ is a *rational surface* in $\mathbb{P}^r$ and that (10.1) is one of its parametric representations; we also say that $\mathcal{F}$ is the *projective image* (or *projective model*) of the linear system Σ.

If Σ is not simple but rather composed with an involution of order > 1 it is possible, as Castelnuovo proved, to obtain the same surface $\mathcal{F}$ as the image of a simple linear system Σ' (cf. Theorem 6.6.3).

In the sequel we will suppose that Σ is simple and thus that there exists an open set $U \subset \mathbb{P}^2$ such that $\varphi_{|U} : U \to \varphi(U)$ is an isomorphism.

We note a few preliminary facts.

10.1.1. Substituting the basis $\{\varphi_0, \ldots, \varphi_r\}$ of Σ with another basis is equivalent to replacing the right-hand sides of (10.1) with suitable linear combinations of them, and so with replacing $\mathcal{F}$ by its image under a non-degenerate projectivity of $\mathbb{P}^r$.

This means that the projective model of Σ depends on Σ and not on the choices of a basis for Σ.

10.1.2. Carrying out a Cremona transformation in the plane $\mathbb{P}^2_{[u_0, u_1, u_2]}$ is equivalent to a change of parameters in the parametric representation of $\mathcal{F}$. Therefore, two linear systems such that each may be obtained from the other via a Cremona transformation have the same projective model.

Hence we will be interested only in the birational invariants of the system Σ, that is, those characteristics of the system which are invariant under Cremona transformations.

Thus, for example, the order of Σ, that is, the order of its curves, is of no interest. A problem that has been amply treated in the literature (but which we shall not discuss) is that of seeking a linear system of minimal order that represents a given surface.

However, the dimension of Σ will be of interest, as will its degree and its *genus*, the former being the number of variable points (that is, points not absorbed by the base points) common to two generic curves of Σ, and the latter being the genus of the generic curve of Σ.

10.1.3. If the curves with equations $\varphi_0(u_0, u_1, u_2) = 0, \ldots, \varphi_r(u_0, u_1, u_2) = 0$ are linearly dependent, then to each relation of linear dependence

$$a_0\varphi_0(u_0, u_1, u_2) + \cdots + a_r\varphi_r(u_0, u_1, u_2) = 0$$

among them, there corresponds a hyperplane

$$a_0 x_0 + a_1 x_1 + \cdots + a_r x_r = 0$$

which contains every point of $\mathcal{F}$. Thus the surface $\mathcal{F}$ is contained in a subspace of $\mathbb{P}^r$.

Suppose that the linear system (10.2) has dimension $h < r$. If, for example, the forms $\varphi_0, \varphi_1, \ldots, \varphi_h$ are linearly independent, then for suitable $\lambda_{j0}, \ldots, \lambda_{jh} \in K$ we have

$$\varphi_j(u_0, u_1, u_2) = \lambda_{j0}\varphi_0(u_0, u_1, u_2) + \cdots + \lambda_{jh}\varphi_h(u_0, u_1, u_2), \quad j = h+1, \ldots, r,$$

so that the surface $\mathcal{F}$ belongs to the intersection S_h of the $r - h$ hyperplanes

$$x_j = \lambda_{j0} x_0 + \cdots + \lambda_{jh} x_h, \quad j = h+1, \ldots, r.$$

In this linear space S_h we may suppose that $x_0, x_1, \ldots, x_h$ are projective coordinates and $\mathcal{F}$ is then represented by the equations

$$\begin{cases} x_0 = \varphi_0(u_0, u_1, u_2), \\ \quad \vdots \\ x_h = \varphi_h(u_0, u_1, u_2). \end{cases}$$

Thus, the surface appears as the projective image of the linear system, of dimension h,

$$\Sigma: \lambda_0\varphi_0 + \cdots + \lambda_h\varphi_h = 0.$$

In the sequel we will suppose that the linear system with equations (10.2) has dimension r; and then the projective model $\mathcal{F}$ will belong to S_r but not to any linear space of lower dimension. Usually this fact is expressed by saying that $\mathcal{F}$ is *embedded* in $\mathbb{P}^r$ (cf. 6.6.4).

10.1.4. We prove that if Σ, Σ_0 are two linear systems of curves in $\mathbb{P}^2_{[u_0, u_1, u_2]}$ and one has $\Sigma \subset \Sigma_0$, then the surface $\mathcal{F}$, projective image of Σ, is the projection of the image $\mathcal{F}_0$ of Σ_0.

If $\dim \Sigma = r$ and $\dim \Sigma_0 = r_0$, to define Σ_0 we can take $r_0 + 1$ linearly independent curves chosen such that the first $r + 1$ belong to Σ and thus so that the equation of Σ_0 is

$$\lambda_0\varphi_0 + \cdots + \lambda_r\varphi_r + \lambda_{r+1}\varphi_{r+1} + \cdots + \lambda_{r_0}\varphi_{r_0} = 0,$$

where $\varphi_0 = 0, \ldots, \varphi_r = 0$ are curves of Σ.

The surface $\mathcal{F}_0$ is represented in S_{r_0} by the equations

$$\begin{cases} x_i = \varphi_i(u_0, u_1, u_2), & i = 0, 1, \ldots, r, \\ x_j = \varphi_j(u_0, u_1, u_2), & j = r+1, \ldots, r_0, \end{cases}$$

and its projection from the space defined by $x_0 = x_1 = \cdots = x_r = 0$ into the linear space S_r with equations $x_{r+1} = \cdots = x_{r_0} = 0$ (in which $x_0, \ldots, x_r$ are homogeneous projective coordinates) is represented there by the equations

$$x_i = \varphi_i(u_0, u_1, u_2), \quad i = 0, 1, \ldots, r,$$

and hence is a projective image of Σ.

In the sequel we shall suppose that the linear system Σ of equation (10.2) has all base points ordinary.

If n is the order of Σ and $B_1, B_2, \ldots, B_q$ are its base points with multiplicities $s_1, s_2, \ldots, s_q$, the degree D of Σ is expressed by the relation, cf. Theorem 4.2.1,

$$D = n^2 - \sum_{i=1}^{q} s_i^2. \tag{10.3}$$

Bearing in mind that by Bertini's first theorem (Theorem 6.3.11) the generic curve of Σ is non-singular outside of the base points and recalling Proposition 7.2.7, the genus p of Σ is given by

$$p = \frac{1}{2}(n-1)(n-2) - \frac{1}{2}\sum_{i=1}^{q} s_i(s_i - 1). \tag{10.4}$$

To the points of $\mathbb{P}^2_{[u_0, u_1, u_2]}$ that satisfy the equation of the curve

$$a_0 \varphi_0 + \cdots + a_r \varphi_r = 0$$

of Σ there correspond the points of $\mathcal{F}$ that belong to the hyperplane

$$a_0 x_0 + \cdots + a_r x_r = 0$$

and one has a bijection between the curves of Σ and the hyperplane sections of $\mathcal{F}$. Moreover, a curve of Σ and the corresponding hyperplane section of $\mathcal{F}$ are birationally isomorphic. Therefore:

- *The genus of Σ coincides with the genus of the hyperplane sections of $\mathcal{F}$.*

Furthermore, as a particular case of §6.6.5, one has:

344   Chapter 10. Rational Surfaces

- *The order of $\mathcal{F}$ coincides with the degree of Σ.*

10.1.5 (Curves of a linear system having multiple points that are not base points). Let Σ be a linear system of plane curves of order n in $\mathbb{P}^2_{[u_0,u_1,u_2]}$, and let r be its dimension.

If $r \geq 3$, the Jacobian locus of Σ coincides with the entire space $\mathbb{P}^2$ (cf. Section 6.4). Bearing in mind that a double point imposes three linearly independent conditions on plane curves (cf. Section 6.2), every generic point P of the plane is double for ∞^{r-3} curves of Σ. To these curves there correspond ∞^{r-3} curves on the surface $\mathcal{F}$, the projective image of Σ, all passing through $\varphi(P)$ and having a double point there, that is, the sections of $\mathcal{F}$ by the tangent hyperplanes at $\varphi(P)$. The latter ∞^{r-3} hyperplanes are those of the star having as center the tangent plane to $\mathcal{F}$ at $\varphi(P)$. If $r = 3$, $\mathcal{F}$ is a surface of $\mathbb{P}^3$ and to the unique curve of Σ having P as double point there corresponds the section of $\mathcal{F}$ by the tangent plane at $\varphi(P)$.

Among the curves of Σ having P as double point there can be some which have P as (at least) a triple point. These exist, for any choice of P, if $r \geq 6$, because a triple point in a given position involves six independent linear conditions on the curves of the plane.

If $r < 6$ there can be particular points in the plane that are triple for the curves of Σ. For example, if $r = 5$ and $\Sigma : \sum_{i=0}^{r} \lambda_i \varphi_i = 0$, the system of six linear equations in the parameters $\lambda_0, \lambda_1, \ldots, \lambda_5$,

$$\frac{\partial^2 (\sum_{i=0}^{5} \lambda_i \varphi_i)}{\partial u_j \partial u_k} = 0, \quad \text{that is,} \quad \sum_{i=0}^{5} \lambda_i \frac{\partial^2 \varphi_i}{\partial u_j \partial u_k} = 0 \quad (j, k = 0, 1, 2),$$

has non-trivial solutions if

$$\begin{vmatrix} \frac{\partial^2 \varphi_0}{\partial u_0^2} & \frac{\partial^2 \varphi_0}{\partial u_0 \partial u_1} & \cdots & \frac{\partial^2 \varphi_0}{\partial u_2^2} \\ \frac{\partial^2 \varphi_1}{\partial u_0^2} & \frac{\partial^2 \varphi_1}{\partial u_0 \partial u_1} & \cdots & \frac{\partial^2 \varphi_1}{\partial u_2^2} \\ \cdots & \cdots & \cdots & \cdots \\ \frac{\partial^2 \varphi_5}{\partial u_0^2} & \frac{\partial^2 \varphi_5}{\partial u_0 \partial u_1} & \cdots & \frac{\partial^2 \varphi_5}{\partial u_2^2} \end{vmatrix} = 0.$$

Let $\mathcal{C}$ be the plane curve defined by the above equation and let $\mathcal{L}$ be the curve, of degree $\leq 6(n-2)$, consisting of the components of $\mathcal{C}$ which are not fundamental curves, cf. §10.1.6. Let $\mathcal{L}'$ be the corresponding curve on $\mathcal{F}$. Thus $\mathcal{L}'$ is the locus of points of $\mathcal{F}$ wherein the tangent plane meets $\mathcal{F}$ in a curve with a triple point.

If $r = 4$ instead of a curve there are isolated points (finite in number) that are triple for the curves of Σ.

If $r = 3$ in general there are no triple points for curves of Σ. But for particular choices of Σ there may be curves with triple points. For example, a (non-ruled) cubic surface in $\mathbb{P}^3$ can have points P such that the tangent plane at P has in common with $\mathcal{F}$ a cubic having P as triple point, that is, a triple of lines issuing from P.

A point P that offers this particularity is called an *Eckardt point* (cf. Exercises 10.5.22 and 13.1.36).

10.1.6 Fundamental curves of a linear system. Now assume that ω is a curve in $\mathbb{P}^2_{[u_0,u_1,u_2]}$ which does not meet the curves of Σ (of which it is not a component) outside of the base points of Σ. We will say that ω is a *fundamental curve* of Σ. The curves of Σ that pass through a point of ω different from the base points must therefore contain ω as component. The residual components of such curves vary in a linear system Ω that is said to be *partially contained* in Σ. Since Ω has dimension $r - 1$ (inasmuch as a curve of Σ must satisfy only one condition in order to contain ω), the surface $\mathcal{F}_0$, the projective image of Ω, is embedded in $\mathbb{P}^{r-1}$. Moreover, the image of the curve ω under (10.1) is a single point $O \in \mathcal{F}$, since if P is a point of ω and $\varphi(P) = O$, all the hyperplane sections that pass through O contain $\varphi(\omega)$.

If we adjoin the curve ω to the curves of Ω we obtain a linear system Ω' ($= \Omega + \omega$), which is also of dimension $r - 1$, and is contained in Σ; we will also say that ω is the *fixed component* of Ω'. Under (10.1) the curves of Ω' correspond to the curves arising as sections of $\mathcal{F}$ by the hyperplanes of the star with center O. Therefore, the projective image $\mathcal{F}_0$ of Ω is the surface which is the projection of $\mathcal{F}$ from O onto a hyperplane (note that $\mathcal{F}_0$ coincides with the projective image of the linear system Ω', cf. §6.6.6).

The order of $\mathcal{F}_0$, which means the degree D_0 of Ω, is the number of intersections of $\mathcal{F}_0$ with a generic S_{r-3} in $\mathbb{P}^{r-1}$ and so is the number of intersections (other than O) of $\mathcal{F}$ with a generic $S_{r-2} \subset \mathbb{P}^r$ passing through O. The multiplicity of $\mathcal{F}$ at O is thus (cf. Section 3.4)

$$\mu_O(\mathcal{F}) = \deg(\mathcal{F}) - \deg(\mathcal{F}_0) = D - D_0,$$

where D is the degree of Σ.

To a fundamental curve ω of Σ there thus corresponds under (10.1) a point O of the surface $\mathcal{F}$, which is called the *fundamental point* of $\mathcal{F}$ corresponding to ω, and to the points of ω there correspond "the points of $\mathcal{F}$ belonging to the neighborhood of O".

All this can be rediscovered very easily with analytic means if one takes $r + 1$ linearly independent curves to define Σ, of which r belong to Ω', and thus have equations of the form $(\varphi_j :=)\theta f_j = 0$, $j = 0, \ldots, r - 1$, where $\theta = 0$ is the equation of ω. We will then have, cf. (6.6),

$$\Sigma: \quad \theta(\lambda_0 f_0 + \cdots + \lambda_{r-1} f_{r-1}) + \lambda_r \varphi_r = 0,$$
$$\Omega': \quad \theta(\lambda_0 f_0 + \cdots + \lambda_{r-1} f_{r-1}) = 0.$$

The second equation represents the surface that is locus of the point $[f_0, \ldots, f_{r-1}]$, and this surface appears as the projection (from the point $[0, 0, \ldots, 0, 1]$ onto the hyperplane $x_0 = 0$) of the surface which is the locus of the point $[\theta f_0, \ldots, \theta f_{r-1}, \varphi_r]$, that is, the projective image of Σ (with equation $\lambda_0 \varphi_0 + \cdots + \lambda_{r-1} \varphi_{r-1} + \lambda_r \varphi_r = 0$).

One then sees immediately that to the curve ω there corresponds on the surface $\mathcal{F}$ (the image of Σ) the point $[0, 0, \ldots, 0, 1]$, and that the surface $\mathcal{F}_0$ represented by Ω is the projection of $\mathcal{F}$ from that point.

Note that the fundamental curves of Σ are finite in number (≥ 0). Otherwise, for each point P of $\mathbb{P}^2$ there would exist a fundamental curve ω_P containing P. Every curve C of Σ would then meet infinitely many fundamental curves ω_P, as P varies in C, away from the base points. All these curves would be components of C, but this is absurd.

10.1.7 Exceptional curves corresponding to base points. We again consider the rational map φ as in (10.1), that sends the point $P = [u_0, u_1, u_2] \in \mathbb{P}^2$ to the point

$$\varphi(P) = [\varphi_0(u_0, u_1, u_2), \ldots, \varphi_r(u_0, u_1, u_2)] \in \mathcal{F}.$$

It is defined away from the base points of the system Σ and, by the simplicity hypothesis on Σ, is bijective on an open set U of $\mathbb{P}^2$.

Suppose, for example, that B is an s-fold ordinary base point, and let t be a line of $\mathbb{P}^2$ passing through B. The curves of Σ for which t is one of the s tangents in B form a linear system $\Sigma_{(t)}$ of dimension $r - 1$ to which there corresponds a linear system of the same dimension consisting of hyperplane sections of $\mathcal{F}$ and thus a point T of $\mathcal{F}$. As t varies in the pencil of lines of $\mathbb{P}^2$ passing through B, the point T describes a rational curve β on $\mathcal{F}$. We say that β is the *exceptional curve* corresponding to the base point B.

It is easy to see that β is a rational normal curve of order s. The order of β is s because s is the number of points of β belonging to the generic hyperplane section. Every curve C of Σ has, in fact, s distinct tangent $t_1, \ldots, t_s$ at B and each of them corresponds, as described above, to a point of intersection of β with the hyperplane section of $\mathcal{F}$ corresponding to C (note that the curve C belongs to the s linear systems $\Sigma_{(t_i)}$ corresponding to the tangents t_i, $i = 1, \ldots, s$).

It follows that in order to impose on a hyperplane section $\mathcal{L}$ of $\mathcal{F}$ that it contain β as a component it suffices to impose on $\mathcal{L}$ that it contain $s + 1$ points of β, that is, to be the curve corresponding to a curve Γ of Σ such that Γ has B as $(s+1)$-fold point. Conversely, if a curve Γ of Σ has B as $(s+1)$-fold point, the corresponding hyperplane section $\mathcal{L}$ of $\mathcal{F}$ meets β in $s+1$ points and thus contains β.

Hence the hyperplanes of $\mathbb{P}^r$ that contain β are those and only those corresponding to curves of Σ having B as $(s+1)$-fold point, and which are thus subject to $s+1$ further independent linear conditions (cf. Section 6.2). These linear conditions mean that those hyperplanes must pass through $s+1$ linearly independent points, and so must contain the space S_s which is their join.

See Exercise 10.5.14 for the analysis of a remarkable example where the base point B is not ordinary.

10.1.8 Curves on a rational surface. The planar representation (10.1) of the surface $\mathcal{F}$ permits an easy analysis of the algebraic curves traced on $\mathcal{F}$.

In the plane π consider a linear system Σ of curves C^n of order n passing through the points $B_1, \ldots, B_q$ with the multiplicities $s_1, \ldots, s_q$. We will say that Σ is a system *complete with respect to the base group* $G = \{B_1^{s_1}, \ldots, B_q^{s_q}\}$ if it contains *all* curves C^n in π having multiplicities s_j at the points B_j. In this case we will write $\Sigma = C^n(B_1^{s_1}, \ldots, B_q^{s_q})$.

Let $\mathcal{F}$ be the projective image of a linear system Σ, not necessarily complete. On the surface $\mathcal{F}$, take an algebraic curve $\mathcal{L}$ of order m having, for $j = 1, \ldots, q$, a total of h_j points in common with the exceptional curves β_j corresponding to the base points B_j, and also having multiplicity μ in the fundamental point O of $\mathcal{F}$ corresponding to the fundamental curve ω of $\mathbb{P}^2_{[u_0, u_1, u_2]}$. The curve $\mathcal{L}$ is the image of a curve $\mathcal{L}_0$ in $\mathbb{P}^2_{[u_0, u_1, u_2]}$ which meets the curve ω, away from the base points, in μ points and passes through the points B_j with multiplicities h_j.

In general, the curves of the system Σ (by hypothesis non-composite) intersect $\mathcal{L}_0$ in a simple linear series of groups of points, and then the order m of $\mathcal{L}$, that is, the number of its intersections with the generic hyperplane section of $\mathcal{F}$, is the number of intersections of $\mathcal{L}_0$ with the generic curve of Σ which are not absorbed by the base points. Therefore, if m_0 is the order of $\mathcal{L}_0$, one has, by Bézout's theorem,

$$m = m_0 n - \sum_{j=1}^{q} h_j s_j. \tag{10.5}$$

- If $\mathcal{L}$ is the complete intersection of $\mathcal{F}$ with a hypersurface $\mathcal{G}$ of order d, the curve $\mathcal{L}_0$ (possibly completed by adjoining fundamental curves) belongs to the linear system $d\Sigma$. Assuming for convenience that Σ is complete, $d\Sigma$ is the linear system of curves of order dn that is complete with respect to the base group $\{B_1^{ds_1}, \ldots, B_q^{ds_q}\}$. For example, if ω is the fundamental curve of Σ that corresponds to the point O of $\mathcal{F}$, and $\mathcal{L}_0 + \lambda\omega$, $\lambda \in K$, is a curve of $d\Sigma$, the curve $\mathcal{L}$ will be the complete intersection of $\mathcal{F}$ with a hypersurface $\mathcal{G}$ of order d passing through O with multiplicity λ.

- It can happen that on a particular curve $\mathcal{L}_0$ of the plane π of the parameters the system Σ cuts out, away from the base points, a linear series of order m composed with an involution i_t of order t. This means that all the curves of Σ that pass through a generic point P of $\mathcal{L}_0$ are constrained to pass through other $t - 1$ points of the same curve. In this case each group of i_t corresponds to a single point of $\mathcal{L}$ and $\mathcal{L}$ is a multiple curve, more specifically, a curve of order $\frac{m}{t}$ counted t times.

In the same way, if the linear system Σ is composed with an involution of order t, which means that all the curves of Σ that pass through a generic point P of the plane are constrained to pass through other $t-1$ points, the surface which is the projective image of Σ has order $\frac{D}{t}$, where D is the degree of Σ (cf. §6.6.5 and 10.1.4).

10.1.9 The multiple points of a rational surface under its planar representation. Let $\mathcal{F}$ be the projective image of a linear system Σ with degree D and dimension r (so that $\mathcal{F}$ has order D and is embedded in $\mathbb{P}^r$) and let O be an s-fold point of $\mathcal{F}$. Then to the star ∞^{r-1} of hyperplanes passing through O there corresponds in π a linear system $\Sigma' \subset \Sigma$ of dimension $r-1$ and degree $D-s$ because two generic hyperplanes passing through O have a subspace S_{r-2} in common which meets $\mathcal{F}$ outside of O in $D-s$ other points. The system Σ' represents the surface $\mathcal{F}'$ which is the projection of $\mathcal{F}$ from O into an S_{r-1} not passing through O.

Conversely, if Σ contains a linear system Σ' of dimension $r-1$ and degree $D-s$, to this system there corresponds an s-fold point of $\mathcal{F}$ in $\mathbb{P}^r$, namely the point common to the hyperplanes corresponding to the curves of Σ'.

This situation which, as we have seen in §10.1.6, can occur when Σ has fundamental curves, can also happen if Σ does not have fundamental curves (in the following example its base points are not ordinary in the sense of Section 6.3). E.g., consider the linear system

$$\Sigma: \lambda_0 u_0^{n-s} \varphi_s(u_1, u_2) + \lambda_1 u_0^{n-s-1} \varphi_{s+1}(u_1, u_2) + \cdots = 0$$

whose curves, of order n, all have the same s tangents $\varphi_s(u_1, u_2) = 0$ at the s-fold base point $A_0 = [1, 0, 0]$.

Let $\Sigma' \subset \Sigma$ be the system consisting of the curves of Σ that are obtained for $\lambda_0 = 0$, and which thus have A_0 as $(s+1)$-fold point. If D and D' are the degrees of Σ and Σ' respectively, we have, cf. (4.9),

$$D = n^2 - (s^2 + s) - c,$$

and

$$D' = n^2 - (s+1)^2 - c,$$

where c is the contribution given by the base points distinct from A_0 in the calculation of the intersections of the curves of Σ and of Σ'. Thus

$$D' = D - [(s+1)^2 - (s^2 + s)] = D - (s+1). \tag{10.6}$$

In conclusion, the hyperplanes of $\mathbb{P}^r$ that correspond to the curves of Σ' have in common an $(s+1)$-fold point of $\mathcal{F}$.

Examples 10.1.10. We now give a few examples.

(1) The linear system Σ of conics tangent at a given point to a line has dimension 3 and degree 2; the projective image is a quadric cone in $\mathbb{P}^3$ (cf. Remark 7.3.3).

(2) The linear system Σ of cubics tangent at a given point to a given line has dimension 7 and degree 7; the projective image is a surface $\mathcal{F}$, of order 7, in $\mathbb{P}^7$ with a double point. (Use the relation (10.6) with $s = 1$ and $D = 9$.)

(3) Consider the linear system of cubics tangent to three given lines in three given points A, B, C. The projective image is a cubic surface $\mathcal{F}^3$ in $\mathbb{P}^3$ with (at least) three double points. The surface $\mathcal{F}^3$ has a fourth double point if the cubics of Σ are tangent to a conic in the three points A, B, and C. The conic is a fundamental curve (cf. §10.1.6).

The cubic $\mathcal{F}^3$ with four nodes may be also be represented by a linear system Σ' of plane cubics passing through the six vertices of a complete quadrilateral (cf. §1.1.6). This can be seen by transforming Σ under a quadratic transformation having A, B, and C as fundamental points (cf. Exercise 10.5.20).

Indeed, in the plane π consider a cubic Γ and a conic γ, both tangent to three lines a, b, and c in the same points A, B, and C respectively. Consider the quadratic transformation $\omega: \pi \to \pi'$ having A, B, C as fundamental points. Let A', B', and C' be the base points of the homaloidal net of conics Σ' in π' defined by ω (cf. Section 9.1). The cubic Γ is transformed into a cubic Γ' passing through the points A', B', and C'. To the conic γ there corresponds a line γ'.

Since Γ and γ are mutually tangent at A, the curves Γ' and γ' both pass through a point A_1 belonging to the line $a' = \langle B', C' \rangle$ whose points are in projective correspondence with the directions issuing from A.

Similarly Γ' and γ' have in common a point B_1 of the line $b' = \langle A', C' \rangle$ and a point of the line $c' = \langle A', B' \rangle$. Thus Γ' also passes through three vertices of the complete quadrilateral $a'b'c'\gamma'$ which belong to γ'.

Thus we have established that the system of dimension 3 of cubics tangent to a conic in three fixed points is equivalent under a Cremona transformation to the system of cubics that pass through the six vertices of a complete quadrilateral.

10.2 Linearly normal surfaces and their projections

In the plane π of parameters we consider a complete linear system $\Sigma = C^n(B_1^{s_1}, \ldots, B_q^{s_q})$ of curves C^n of order n passing through the points $B_1, \ldots, B_q$ with the multiplicities $s_1, \ldots, s_q$ (cf. §10.1.8). We will say that the projective image $\mathcal{F}$ of the complete linear system Σ is a *linearly normal* surface in $\mathbb{P}^r$, or also that $\mathcal{F}$ is *normal* in $\mathbb{P}^r$, if $r = \dim \Sigma$. If the conditions imposed on the curves C^n of the plane by the base group are linearly independent we have (cf. Example–Definition 6.2.2)

$$r = \frac{n(n+3)}{2} - \sum_{i=1}^{q} \frac{s_i(s_i+1)}{2},$$

and Σ is regular. If on the contrary Σ is superabundant we have

$$r > \frac{n(n+3)}{2} - \sum_{i=1}^{q} \frac{s_i(s_i+1)}{2}.$$

It is important to note that

- *every surface $\mathcal{F}$ which is not linearly normal is the projection of a linearly normal surface $\mathcal{F}_0$ of the same order.*

Indeed, a non-complete linear system Σ of curves of order n, having certain base points with assigned multiplicities, is contained in the linear system Σ_0 consisting of *all* the curves of the same order and with the same multiplicities in the base points. Moreover, the projective image $\mathcal{F}$ of Σ is the projection of the image $\mathcal{F}_0$ of Σ_0. Since Σ and Σ_0 have the same degree, the two surfaces $\mathcal{F}$ and $\mathcal{F}_0$ have the same order, and therefore, if $r_0 = \dim \Sigma_0$ and $r = \dim \Sigma$, then $\mathcal{F}$ is the projection of $\mathcal{F}_0$ from a linear space S_{r_0-r-1} that *does not* meet $\mathcal{F}_0$.

Therefore we focus our attention on linearly normal surfaces. By the foregoing remarks, a surface $\mathcal{F}$ that is normal in $\mathbb{P}^r$ may be defined as a surface that can *not* be obtained as the projection of a surface $\mathcal{F}_0$, of the same order, and embedded in a space of dimension $> r$.

Let $\Sigma_0 = C^n(B_1^{s_1}, \ldots, B_q^{s_q})$ be a linear system of degree D_0 and dimension r_0, complete with respect to the base group $\{B_1^{s_1}, \ldots, B_q^{s_q}\}$, and let $\mathcal{F}_0$ be the normal surface in S_{r_0} represented by Σ_0.

We impose a new base point P of multiplicity s on the curves of Σ_0, and consider the linear system $\Sigma = C^n(B_1^{s_1}, \ldots, B_q^{s_q}, P^s)$ which is also complete with respect to its base group.

The surface $\mathcal{F}$, the projective image of Σ, is also linearly normal and is the projection of $\mathcal{F}_0$ (since $\Sigma \subset \Sigma_0$). But $\deg \Sigma = \deg \Sigma_0 - s^2$. Therefore we have

$$\deg(\mathcal{F}) = \deg(\mathcal{F}_0) - s^2,$$

and $\mathcal{F}$ is the projection of $\mathcal{F}_0$ from a linear space that meets $\mathcal{F}_0$ (cf. Example 10.2.3).

We also note that if $\mathcal{F}$ is the image surface of a linear system Σ, of dimension r, then every surface $\mathcal{F}'$ that is obtained as a projection of $\mathcal{F}$ is the image of a linear system Σ' contained (perhaps only partially) in Σ, and Σ' may be obtained from Σ by imposing a certain number ρ of linear conditions on the curves of Σ, namely

10.2. Linearly normal surfaces and their projections

by requiring that the hyperplane sections of $\mathcal{F}$ should belong to the hyperplanes that contain ρ points of S_r. If the ρ conditions are linearly independent, $\mathcal{F}'$ is the projection of $\mathcal{F}$ from a space of dimension $\rho - 1$ onto a linear space $S_{r-\rho}$.

For example, if Σ' is obtained from Σ by imposing ρ new simple base points in general position (that is, such that the curves of Σ are subjected to ρ new independent linear conditions) then the surface $\mathcal{F}'$ is the projection of $\mathcal{F}$ onto an $S_{r-\rho}$ from a ρ-secant $S_{\rho-1}$. Thus we have

$$\deg(\mathcal{F}') = \deg(\mathcal{F}) - \rho = n^2 - \sum_{i=1}^{q} s_i^2 - \rho. \tag{10.7}$$

If $\mathcal{F}$ is normal in S_r, then $\mathcal{F}'$ is normal in $S_{r-\rho}$.

The situation is not so simple when the new base points are not generic, or not all non-singular, or also if at the original base points of Σ one imposes higher multiplicities than those of the curves of Σ.

Here are some examples.

Example 10.2.1 (The Veronese surface [110]). A particularly remarkable surface is the surface $\mathcal{F}$, called the *Veronese surface*, which is the projective image of the linear system

$$\Sigma\colon \lambda_0 u_0^2 + \lambda_1 u_1^2 + \lambda_2 u_2^2 + \lambda_3 u_0 u_1 + \lambda_4 u_0 u_2 + \lambda_5 u_1 u_2 = 0$$

of all the conics in the plane (see also Section 10.4). It is a complete linear system of degree 4 and dimension 5 whence $\mathcal{F}$ is a surface of order 4 embedded in $\mathbb{P}^5$ with parametric equations

$$\begin{cases} x_0 = u_0^2, \\ x_1 = u_1^2, \\ x_2 = u_2^2, \\ x_3 = u_0 u_1, \\ x_4 = u_0 u_2, \\ x_5 = u_1 u_2. \end{cases}$$

On it there are no curves of odd order (because if m is the order of a curve on $\mathcal{F}$ and m_0 that of the corresponding plane curve, one has $m = 2m_0$ by (10.5)). In particular the Veronese surface does not contain lines.

We impose two base points A and B on the conics of Σ, that is, we consider the linear system Σ' (it too complete with respect to its base group) of conics of Σ that pass through A and B. Since $\deg \Sigma' = 2$ and $\dim \Sigma' = 3$, Σ' represents a quadric $\mathcal{Q}$ in S_3: $\mathcal{Q}$ is the projection of the Veronese surface from one of its chords.

If $A \neq B$ there are on $\mathcal{Q}$ two exceptional lines a, b that correspond to the two base points A and B of Σ'. They meet in the point O that corresponds to the line

r_{AB}, the unique fundamental curve of Σ'. On the line r_{AB} the neighborhood of O is expanded.

If $A = B$, that is, if Σ' is the linear system of conics tangent at a given point to a given line a, then $\mathcal{Q}$ is a quadric cone and may be obtained as the projection of the Veronese surface from one of its tangent lines. In this case the given tangent a is again the unique fundamental curve of Σ'; to it there corresponds the fundamental point $O \in \mathcal{Q}$.

Note that the planar representation of a quadric $\mathcal{Q}$ by way of a linear system of conics passing through two points is immediately acquired by projecting $\mathcal{Q}$ onto a plane π from one of its simple points. This is precisely the stereographic projection of the quadric studied in detail in §7.3.1, and used for the classification of the algebraic curves traced on $\mathcal{Q}$.

Example 10.2.2. Let $\Sigma = C^n(B_1^{s_1}, \ldots, B_q^{s_q})$ be a complete linear system of dimension r and let $\mathcal{F}$ be its projective image. Let Σ' be the linear system of curves of Σ that pass through n collinear points $P_1, \ldots, P_n$ all on a generic line ℓ. (Since ℓ is generic it does not contain base points of Σ.) Let σ be the $(n-1)$-dimensional space in $\mathbb{P}^r$ that joins the images under (10.1) of these points. Suppose that $r \geq n + 3$.

If D is the degree of Σ, then, again by 10.7, the projective image $\mathcal{F}'$ of Σ' is a surface of order $\deg(\mathcal{F}') = D - n$ embedded in a subspace S_{r-n}. Indeed, if $\mathcal{F}$ is the projective image of Σ, then $\mathcal{F}'$ is the projection of $\mathcal{F}$ on our S_{r-n} from the space σ.

The surface $\mathcal{F}'$ possesses a multiple point O, the image of the line ℓ, which is fundamental for the system Σ'. To calculate the multiplicity of $\mathcal{F}'$ at O, we note first that the curves of Σ' that pass through an additional point of ℓ distinct from the points $P_1, \ldots, P_n$ must contain ℓ as a component. The components distinct from ℓ of those curves constitute a linear system $\Omega = C^{n-1}(B_1^{s_1}, \ldots, B_q^{s_q})$ of curves of order $n - 1$ having the same base group as the initial system Σ. By what we have seen in §10.1.6, the multiplicity of the point O for $\mathcal{F}'$ is $s = \deg(\mathcal{F}') - \deg(\mathcal{F}_0)$, where $\mathcal{F}_0$ is the projective image of the linear system Ω, the projection of $\mathcal{F}' \subset \mathbb{P}^{r-n}$ from the point O onto a hyperplane S_{r-n-1} of S_{r-n}. Thus

$$s = D - n - \left((n-1)^2 - \sum_{i=1}^{q} s_i^2 \right) = n - 1.$$

Since the new base points $P_1, \ldots, P_n$ (of the system Σ') belong to ℓ, their images belong to the line $\mathcal{L} = \varphi(\ell)$ of $\mathcal{F}$, the image of ℓ under the rational map φ defined by Σ.

The curve $\mathcal{L}$ has order n (by equation (10.5)) and is a rational normal curve. Indeed, it is contained in just as many hyperplane sections of S_r as there are curves of Σ containing the line ℓ; and the linear system of such curves has the same

10.2. Linearly normal surfaces and their projections 353

dimension as that of the linear system Ω. Hence $\mathcal{L}$ is contained in

$$\dim \Omega + 1 = \frac{(n-1)(n+2)}{2} - \sum_{i=1}^{q} \frac{s_i(s_i+1)}{2} + 1 = r - n$$

independent hyperplanes and thus in the linear space S_n which is their intersection.

Every generic S_n passing through σ, and which meets $\mathcal{F}$ in a further point Q, meets the S_{r-n} in a simple point of $\mathcal{F}'$. Indeed, the space $S_n = J(\sigma, Q)$ meets $\mathcal{F}$ (away from σ) only in the point Q in virtue of the hypothesis $n \leq r - 3$. The linear space S_n that contains $\mathcal{L}$ is exceptional; it meets S_{r-n} in the point O, and O is the projection of every point of $\mathcal{L}$.

Example 10.2.3. Let Σ be a linear system of dimension r and degree D, and let Σ' be the linear system of curves of Σ having a further multiple point P, of multiplicity s, which implies $\frac{s(s+1)}{2}$ independent linear conditions for the curves of Σ. Let $\mathcal{F}$, $\mathcal{F}'$ be the projective image surfaces of Σ and Σ' respectively.

The surface $\mathcal{F}'$ is the projection of $\mathcal{F}$ onto a space $S_{r'}$, where $r' = r - \frac{s(s+1)}{2}$, from a space of dimension $\frac{s(s+1)}{2} - 1$.

The surface $\mathcal{F}'$, which has order $D - s^2$, acquires a new exceptional curve that represents the neighborhood of P. This curve is a rational normal C^s in $\mathbb{P}^s$.

Example 10.2.4. Let Σ be a linear system of dimension r and degree D. Let B be one of the base points of Σ, s the corresponding multiplicity, and Σ' the linear system of the curves of Σ having multiplicity $s + t$ at B (the preceding example comes under this case if we allow t to assume the value zero). Let $\mathcal{F}$, $\mathcal{F}'$ be the projective image surfaces of Σ, Σ' respectively. If $D = \deg \Sigma$ and $D' = \deg \Sigma'$, we have, by equation (10.3), that

$$D - D' = (t+s)^2 - t^2 = s(2t+s).$$

The number of independent conditions that it is necessary to impose on the curves of Σ (which already have multiplicity s at B) in order that they pass through B with multiplicity $s + t$ is

$$\rho = (s+1) + \cdots + (s+t) = ts + \frac{t(t+1)}{2},$$

and so $\mathcal{F}'$ is the projection of $\mathcal{F}$ from a space $S_{\rho-1}$ onto a space $S_{r-\rho}$.

Example 10.2.5. Let Σ be a linear system of algebraic plane curves and let the dimension r of Σ be at least 6. Let $\mathcal{F}$ be its projective image and let A be a generic point of $\mathcal{F}$, and α the tangent plane to $\mathcal{F}$ at A. The hyperplane sections of the surface $\mathcal{F}'$ obtained by projecting $\mathcal{F}$ from α onto a space S_{r-3} are the projections on that S_{r-3} of the curves C arising as sections of $\mathcal{F}$ by hyperplanes through α, that is, sections by the hyperplanes tangent to $\mathcal{F}$ at A. The curves C are the hyperplane sections of $\mathcal{F}$ having A as double point.

The surface $\mathcal{F}'$ is therefore the projective image of the linear system Σ' of the curves of Σ having the point A' as a double point, where A' corresponds to the point A. Hence we have $\deg \Sigma' = \deg \Sigma - 4$.

10.3 Surfaces of minimal order

As a particular case of Proposition 4.5.6, one has that $\mathbb{P}^r$ does not admit algebraic surfaces (not contained in spaces of dimension $< r$) whose order is $< r - 1$. This may also be easily seen as follows. For $r - 1$ generic points of a surface $\mathcal{F}$ there passes a linear space S_{r-2} that has at least $r - 1$ points in common with $\mathcal{F}$; hence $\deg(\mathcal{F}) \geq r - 1$. On the other hand, there certainly exist surfaces of order $r - 1$ in $\mathbb{P}^r$: for example the cones that project a curve C^{r-1} in S_{r-1} from an external point.

Like the rational normal curves discussed in Section 7.4, so too the surfaces of minimal order $\mathcal{F}^{r-1}$ hold particular interest and enjoy remarkable properties. For instance, one sees immediately that

- a surface $\mathcal{F}^{r-1}$ of $\mathbb{P}^r$ is rational.

Indeed, taking $r - 2$ generic points on the surface and then the space S_{r-3} that they span, one has that every generic S_{r-2} passing through this S_{r-3} has in common with $\mathcal{F}^{r-1}$ only one further point, and so the surface is rational (cf. Corollary 2.6.6). In order to obtain a generically bijective representation of $\mathcal{F}^{r-1}$ over a plane it then suffices to project it from S_{r-3} onto a generic plane.

For what follows it will be useful to place some general facts in evidence.

Lemma 10.3.1. *An irreducible surface F that has a 2-dimensional system of conics is embedded in a space of dimension at most 5.*

Proof. Let γ be a generic conic of F. For each point of γ there pass ∞^1 conics of F and so the conics of F supported by γ are ∞^2, that is, they are all the conics of F. Thus one sees that any two conics of F have non-empty intersection.

Fix three conics γ_1, γ_2, and γ_3 on F and let P_{ij} be a point of $\gamma_i \cap \gamma_j$. If A_i is a further point of γ_i, every S_5 that contains the six points $P_{12}, P_{23}, P_{31}, A_1, A_2, A_3$ contains the three conics γ_i since any such space S_5 has three points in common with each of them, and so contains every conic supported by $\gamma_1, \gamma_2, \gamma_3$. Therefore every conic of F, and so the entire surface, is contained in any such space S_5. □

Lemma 10.3.2. *An irreducible surface $\mathcal{F}^4$ embedded in $\mathbb{P}^5$ and containing a 2-dimensional system of conics is the projective image of the linear system of all the conics of the plane (i.e., $\mathcal{F}^4$ is the Veronese surface).*

Proof. Indeed, let σ be the plane of a conic γ lying on $\mathcal{F}^4$. A generic S_3 passing through σ meets $\mathcal{F}^4$ in a single further point. In fact, the common points of S_3 and $\mathcal{F}^4$ are the points common to curves that are sections of $\mathcal{F}^4$ by two hyperplanes passing through S_3. Each of these hyperplanes meets $\mathcal{F}^4$ along a curve of order 4 containing the conic γ. Hence each of them contains another conic of $\mathcal{F}^4$ and these two conics (which are two generic conics of the surface) have a single point in common. In fact, if two generic conics had (at least) two points in common they would be contained in a space S_3 which would then contain all the conics of the surface, and hence the surface itself.

Therefore the surface $\mathcal{F}^4$ is projected bijectively from σ onto a plane π and one sees immediately that the hyperplane sections of $\mathcal{F}^4$ have as projections the conics of π. The section of $\mathcal{F}^4$ by a hyperplane H has in fact two points in common with the center σ of projection: they are the points in which the line $H \cap \sigma$ meets γ. □

Lemma 10.3.3. *A surface $\mathcal{F}^{r-1}$ of order $r - 1$ embedded in $\mathbb{P}^r$ that has no 2-dimensional systems of conics and which is projected from its generic point O onto a ruled surface $\mathcal{F}^{r-2}$ in S_{r-1} is itself a ruled surface.*

Proof. Let $\mathcal{F}'$ be the projection of $\mathcal{F}^{r-1}$ from one of its generic points O onto a hyperplane S_{r-1}.

If $r = 2k + 1$, we take k generic generators of the ruled surface $\mathcal{F}'$ and the hyperplane S_{2k} that joins these k lines with O. This S_{2k} meets $\mathcal{F}^{2k}$ along k curves, all of order d, lying in k planes passing through O. So we must have $dk \leq 2k$ and thus $d \leq 2$. On the other hand $\mathcal{F}^{2k}$ does not possess ∞^2 conics; hence $d = 1$ and $\mathcal{F}^{2k}$ is a ruled surface.

If $r = 2k$, we take $k - 1$ generic generators on $\mathcal{F}'$ and then a point $P \in \mathcal{F}^{2k-1}$. The hyperplane S_{2k-1} in $\mathbb{P}^r$ that joins these $k - 1$ lines with P and O meets $\mathcal{F}^{2k-1}$ in $k - 1$ curves all of order d and lying in planes passing through O as well as in another curve (whose projection contains P). Thus $d(k - 1) < 2k - 1$ and again $d \leq 2$. As above we conclude that $\mathcal{F}^{2k-1}$ is a ruled surface. □

We can now prove the structure theorem for surfaces of minimal order.

Theorem 10.3.4. *A surface $\mathcal{F}$ of order $r - 1$ embedded in $\mathbb{P}^r$ is either ruled or else is the Veronese surface.*

Proof. If $r = 3$, then $\mathcal{F}$ is a quadric whence it is ruled and has ∞^3 conics.

If $r = 4$ a tangent hyperplane to $\mathcal{F}$ at one of its generic points meets $\mathcal{F}$ in a space cubic endowed with a double point and hence split into a line and a further conic γ (cf. §5.2.4). Each S_3 passing through γ contains a line of $\mathcal{F}$ and so $\mathcal{F}$ is ruled. Moreover, it also has ∞^2 conics as residual intersections with the spaces S_3 passing through one of its lines ℓ.

If $r = 5$ the surface $\mathcal{F}'$ arising as the projection of $\mathcal{F}$ from one of its generic chords AB onto a subspace S_3 is a quadric $\mathcal{Q}$ in S_3. Moreover, the space S_4 joining

AB with a plane containing two generators of Q has two curves in common with $\mathcal{F}$, each contained in a space S_3 passing through AB. If one of these two curves is a line, the surface $\mathcal{F}$ is ruled. If neither of the two curves is a line they are conics (because the sum of their orders can not be greater than 4) passing through A and B. Since A and B are two generic points of $\mathcal{F}$, if $\mathcal{F}$ is not ruled it contains ∞^2 conics. Hence by Lemma 10.3.2 it is a Veronese surface.

If $r = 6$ the projection $\mathcal{F}'$ of $\mathcal{F}$ from one of its generic points O onto S_5 is a surface of order $r - 2 = 4$ containing a line (the trace on S_5 of the tangent plane to $\mathcal{F}$ at O) and so is not a Veronese surface. Thus, applying the preceding case $r = 5$ to $\mathcal{F}' \subset S_5$, we conclude that the surface $\mathcal{F}'$ is ruled. By Lemma 10.3.3 it then follows that the surface $\mathcal{F}$ (which does not contain ∞^2 conics in view of Lemma 10.3.1) is ruled.

If $r \geq 7$ we proceed by induction on r assuming the theorem to be true for the surface $\mathcal{F}^{r-2}$ in S_{r-1}. The projection of $\mathcal{F}$ from one of its generic points is then a ruled surface. Moreover, by Lemma 10.3.1, the surface $\mathcal{F}$ can not contain ∞^2 conics. Hence Lemma 10.3.3 shows that $\mathcal{F}$ is a ruled surface. □

Remark 10.3.5. We note explicitly that in $\mathbb{P}^5$ there do exist both ruled $\mathcal{F}^4$'s and non-ruled $\mathcal{F}^4$'s. If $r \neq 5$ then in $\mathbb{P}^r$ there are only ruled $\mathcal{F}^{r-1}$'s.

10.3.6 Planar representation of ruled surfaces $\mathcal{F}^{r-1} \subset \mathbb{P}^r$.

Let $\mathcal{F}$ be a ruled surface of order $r - 1$ embedded in $\mathbb{P}^r$. Take $r - 2$ generic points $P_1, \ldots, P_{r-2}$ on $\mathcal{F}$, and the space S_{r-3} that joins them. If g is a generic generator of $\mathcal{F}$, the space S_{r-1} which joins S_{r-3} and g meets $\mathcal{F}$ in a curve C^{r-1} split into g and a curve C' of order $r - 2$ which is therefore contained in a space σ of dimension $r - 2$. This curve must pass through the points $P_1, \ldots, P_{r-2}$.

For each of the points P_i there passes a generator p_i of $\mathcal{F}$, and a generic hyperplane H of $\mathbb{P}^r$ has a point Q_i in common with each of the lines p_i, $i = 1, \ldots, r-2$, and $r - 2$ points in common with C'. Let Γ be the curve which is the hyperplane section of $\mathcal{F}$ by H.

Consider the projection of $\mathbb{P}^r$ from the space S_{r-3} onto a plane π skew to it, and let $\psi: \Gamma \to \pi$ be the restriction to Γ. Since S_{r-3} does not meet Γ, the image $\psi(\Gamma)$ is a curve Γ' of order $r - 1$. Note that the projecting space S_{r-2} containing the point Q_i contains the line p_i as well, so that the line p_i is projected onto a point $A_i \in \Gamma'$, the trace in π of the space S_{r-2} joining S_{r-3} with p_i; and so $\psi^{-1}(A_i) = p_i \cap \Gamma = Q_i$. Thus A_i is a non-singular point for Γ', $i = 1, \ldots, r-2$. Moreover, the trace on π of the projecting space $\sigma = S_{r-2}$ containing the curve C' is an $(r-2)$-fold point O for Γ' since its inverse image $\psi^{-1}(O)$ consists of the $r - 2$ points that the hyperplane H has in common with C' on Γ.

It follows that the surface $\mathcal{F}$ is the projective image of a linear system Σ (which, *a posteriori*, turns out to be complete) of curves of order $r - 1$ with an $(r - 2)$-fold base point O and $r - 2$ non-singular base points $A_1, \ldots, A_{r-2}$. Note in this regard

10.3. Surfaces of minimal order

that the complete linear system $C^{r-1}(O^{r-2}, A_1, \ldots, A_{r-2})$ contains Σ and has dimension r and degree $r - 1$; that is, the same dimension and degree as Σ, whence $\Sigma = C^{r-1}(O^{r-2}, A_1, \ldots, A_{r-2})$.

Note further that this linear system is not of minimal order. Indeed, the same surface can be represented via the linear system Σ' of curves of order $< r - 1$ which is the image of Σ under a quadratic transformation (of type I, cf. §9.1.1 and Proposition–Definition 9.2.2) corresponding to the net of conics $\mathcal{R} = OA_i A_j$ having as base points O and two of the points $A_1, \ldots, A_{r-2}$.

Let $\varphi \colon \pi \to \mathcal{F}^{r-1}$ be the rational map associated to the linear system Σ. The image $\varphi(\ell)$ of a (generic) line ℓ of π is a directrix of the ruled surface. Indeed, the lines t of the pencil with center O (not passing through $A_1, \ldots, A_{r-2}$) correspond to lines g_t of $\mathcal{F}$ because they meet the curves of Σ only in one point away from O (cf. (10.5)). As t varies in the pencil, the lines g_t cover $\mathcal{F}$ and constitute its generators (the lines of the pencil with center O and passing through one of the points $A_1, \ldots, A_{r-2}$ are fundamental curves for Σ). Thus, since ℓ meets the lines of the pencil in a single point, the image curve $\varphi(\ell)$ has a single point of $\mathcal{F}$ in common with the generators and is therefore a directrix of $\mathcal{F}$.

More generally, in order to have a directrix $\mathcal{L}$ of $\mathcal{F}$ it suffices to take a curve C of order m with multiplicity $m - 1$ at the point O in the plane π. Since C meets the lines of the pencil with center O in only one point (away from O), the image curve $\mathcal{L} = \varphi(C)$ has only one point in common with the generators of $\mathcal{F}$ and is thus a directrix. Let $x \geq 0$ be the number of those among the points $A_1, \ldots, A_{r-2}$ that belong to C. Then by (10.5) the order of $\mathcal{L}$, given by the number of intersections of C with the curves of Σ away from the base points, will be

$$\deg(\mathcal{L}) = (r-1)m - (r-2)(m-1) - x = m + r - x - 2.$$

So we have $\deg(\mathcal{L}) \geq m$, with $\deg(\mathcal{L}) = m$ if C passes through all the points $A_1, \ldots, A_{r-2}$. Moreover, there do exist curves C of order m with O as an $(m-1)$-fold point and passing simply through the points $A_1, \ldots, A_{r-2}$ if, with $\Delta = C^m(O^{m-1}, A_1, \ldots, A_{r-2})$, we have

$$\dim \Delta \geq \frac{m(m+3)}{2} - \frac{(m-1)m}{2} - (r-2) = 2m - r + 2 \geq 0, \tag{10.8}$$

that is, $m \geq \frac{r-2}{2}$.

Note that the directrices $\mathcal{L}^m$ are embedded in linear spaces S_m since the curves C of the system $C^m(O^{m-1}, A_1, \ldots, A_{r-2})$ have m intersections with the curves of Σ away from the base points, and so a curve of Σ contains C as component as soon as it has $m + 1$ common points with C outside of the base points. Thus $\mathcal{L}^m = \varphi(C)$ is contained in

$$\dim \Sigma - (m+1) + 1 = r - m$$

independent hyperplanes and so too in the space S_m which is their intersection.

358   Chapter 10. Rational Surfaces

Suppose that $\mathcal{F}$ is of *general type*, that is, that the base points of the system Σ are in general position.

If $r = 2k$, then $\mathcal{F}^{2k-1}$ has a directrix $\mathcal{L}^{k-1}$ of order $k - 1$ and ∞^2 directrices $\mathcal{L}^k$ of order k. Indeed, if $r = 2k$, then by (10.8) one has dim $\Delta = 0$ for $m = k - 1$ and dim $\Delta = 2$ for $m = k$. The spaces in which $\mathcal{L}^{k-1}$ and an arbitrary $\mathcal{L}^k$ are embedded are spaces S_{k-1} and S_k that are mutually skew: otherwise their join $J(S_{k-1}, S_k)$ would have dimension $\leq 2k - 1 < r$ and would contain the ruled surface (because for each point of $\mathcal{F}$ there passes a generator that is supported by $\mathcal{L}^{k-1}$ and by $\mathcal{L}^k$ and which thus intersects the two spaces S_k and S_{k-1}, and so is contained in their join).

For each directrix $\mathcal{L}^k$, the correspondence obtained by associating to each point P of $\mathcal{L}^{k-1}$ the point P' of $\mathcal{L}^k$ at which the generator of $\mathcal{F}^{2k-1}$ passing through P is supported, is algebraic and bijective and so is a projectivity between the two rational normal curves $\mathcal{L}^{k-1}$ and $\mathcal{L}^k$ (cf. §1.1.3). In conclusion, the surface $\mathcal{F}^{2k-1}$ is the locus of the lines that join the pairs of corresponding points under a projectivity between two rational normal curves $\mathcal{L}^{k-1}$ and $\mathcal{L}^k$ belonging to independent spaces.

If $r = 2k + 1$, $\mathcal{F}^{2k}$ possesses ∞^1 directrices $\mathcal{L}^k$ (indeed, for $r = 2k + 1$ and $m = k$ equation (10.8) gives us dim $\Delta = 1$). The spaces S_k in which the directrices $\mathcal{L}^k$ are embedded are mutually skew, since otherwise their join would have dimension $\leq 2k < r$. The surface $\mathcal{F}^{2k}$ is the locus of the lines that join the pairs of points corresponding under a projectivity between the two rational normal curves of order k belonging to two independent spaces S_k.

We observe that for a ruled surface of general type $\mathcal{F}^{r-1}$, the reduction process of the system Σ to a system of lower order may be iterated until one finds linear systems with at least three base points. The process will stop when we have a linear system $\Sigma' = C^k(O^{k-1})$ if $r = 2k$ (and to the point O there corresponds the directrix $\mathcal{L}^{k-1}$, while the directrices $\mathcal{L}^k$ are images of the lines of π), or when $\Sigma' = C^{k+1}(O^k, A_1)$ if $r = 2k + 1$ (and the ∞^1 directrices $\mathcal{L}^k$ are the images of the lines of the pencil with center A_1).

If the points $A_1, \ldots, A_{r-2}$ are in general position, the minimal order h of the directrices is thus the integer part $\left[\frac{r-1}{2}\right]$ of $\frac{r-1}{2}$.

If the points $A_1, \ldots, A_{r-2}$ are not in general position one has ruled surfaces $\mathcal{F}^{r-1}$ for which $\left[\frac{r-1}{2}\right]$ is not the minimal order of the directrices, but there exists *one* directrix of order $h < \left[\frac{r-1}{2}\right]$. This happens if among the $r - 2$ simple base points A_i there are $r - 1 - h$ lying on a line ℓ. The image of that line is the unique directrix of order h. The other directrices of $\mathcal{F}^{r-1}$ will all have order $\geq r - h - 1$. Moreover, a directrix of order $r - h - 1$ is the image of a curve of order $r - h - 1$ in the plane π having O as an $(r - h - 2)$-fold point and passing through all the points $A_1, \ldots, A_{r-2}$.

Thus we have the following characterization of the ruled surfaces $\mathcal{F}^{r-1} \subset \mathbb{P}^r$.

10.3. Surfaces of minimal order

Theorem 10.3.7. *Let h be the minimal order of the directrices of a ruled surface $\mathcal{F}^{r-1} \subset \mathbb{P}^r$. Then the ruled surface is the locus of the lines joining pairs of corresponding points under a projectivity between two rational normal curves C^h and C^{r-h-1} contained in independent linear spaces. In particular, if $h = 0$, then $\mathcal{F}^{r-1}$ is the cone which projects C^{r-1} from the center V $(= C^0)$.*

10.3.8 Matricial representation of the ruled surfaces $\mathcal{F}^{r-1} \subset \mathbb{P}^r$.

The foregoing theorem allows us to give a simple analytic representation for the ruled surface $\mathcal{F}^{r-1} \subset \mathbb{P}^r$.

Let h be the minimal order of the directrices. If one chooses a reference system in $\mathbb{P}^r$ in such a way as to ensure that two corresponding points under the projectivity $\psi : C^h \to C^{r-h-1}$ are

$$P = [t^h, t^{h-1}, \ldots, t, 1, 0, 0, \ldots, 0, 0]$$

and

$$P' = [0, 0, \ldots, 0, 0, t^{r-h-1}, t^{r-h-2} \ldots, t, 1],$$

then the generic point of the ruled surface has coordinates

$$\begin{cases} x_i = t^{h-i} & (i = 0, \ldots, h), \\ x_j = \mu t^{r-j} & (j = h+1, \ldots, r), \end{cases} \quad (10.9)$$

for some $\mu \in K^*$. The surface $\mathcal{F}^{r-1}$ then appears as the locus of the points $[x_0, \ldots, x_r]$ of $\mathbb{P}^r$ that give rank one to the matrix

$$\begin{pmatrix} x_0 & x_1 & \cdots & x_{h-1} & x_{h+1} & x_{h+2} & \cdots & x_{r-1} \\ x_1 & x_2 & \cdots & x_h & x_{h+2} & x_{h+3} & \cdots & x_r \end{pmatrix}.$$

It is in fact known that the second order minors of this matrix provide a minimal system of generators for the homogeneous ideal of polynomials in $K[x_0, \ldots, x_r]$ that vanish on $\mathcal{F}^{r-1}$ (cf. [11]).

If t and μ are interpreted as non-homogeneous coordinates in $\mathbb{P}^2$, equations (10.9) show that $\mathcal{F}^{r-1}$ is the projective image of the linear system

$$\Sigma : P_h(t) + \mu Q_{r-h-1}(t) = 0,$$

where $P_h(t), Q_{r-h-1}(t) \in K[t]$ are polynomials of degrees h and $r-h-1$ respectively and with indeterminate coefficients.

Introducing homogeneous coordinates, that is, setting $t = \frac{u_1}{u_0}$, $\mu = \frac{u_2}{u_0}$, we have for Σ the equation

$$u_0^{r-2h} \theta_h(u_0, u_1) + u_2 \theta'_{r-h-1}(u_0, u_1) = 0, \quad (10.10)$$

where $\theta_h(u_0, u_1), \theta'_{r-h-1}(u_0, u_1) \in K[u_0, u_1]$ are binary forms of degrees h, $r - h - 1$. The curves of Σ have order $r - h$ and the multiple point $A_2 = [0, 0, 1]$ of multiplicity $r - h - 1$.

If $r = 2h + 1$, equation (10.10) becomes $u_0 \theta_h(u_0, u_1) + u_2 \theta'_h(u_0, u_1) = 0$ and so A_1 is a (simple) base point, A_2 is an h-fold base point, and one has the system $\Sigma = C^{h+1}(A_2^h, A_1)$. If $r = 2h$, equation (10.10) may be rewritten as $\theta_h(u_0, u_1) + u_2 \theta'_{h-1}(u_0, u_1) = 0$ and so in this case there is only the $(h - 1)$-fold base point A_2 and one obtains the system $\Sigma = C^h(A_2^{h-1})$. This is in agreement with §10.3.6.

The types of projectively distinct ruled surfaces $\mathcal{F}^{r-1}$ in $\mathbb{P}^r$ are obtained in correspondence with the order h of the "minimal" directrices. In particular, if $h = 0$ the minimal directrix is a point and $\mathcal{F}^{r-1}$ is the cone projecting from that point a curve C^{r-1} of S_{r-1}.

10.4 The conics of a plane as points of $\mathbb{P}^5$ and the Veronese surface

The Veronese surface is a very important example of a rational surface. Here we give a further description of it, and we study some of its remarkable properties.

Consider a plane π, with homogeneous coordinates x_0, x_1, x_2, and a space S_5 with homogeneous coordinates X_{00}, X_{01}, X_{11}, X_{02}, X_{12}, X_{22}. Let Σ_5 be the 5-dimensional linear system of conics in π. We associate to the conic γ with equation

$$a_{00} x_0^2 + 2a_{01} x_0 x_1 + a_{11} x_1^2 + 2a_{02} x_0 x_2 + 2a_{12} x_1 x_2 + a_{22} x_2^2 = 0$$

the point $P = [a_{00}, a_{01}, a_{11}, a_{02}, a_{12}, a_{22}] \in S_5$.

To the conics γ of a linear system of dimension h, $0 \leq h \leq 5$, there correspond the points of a subspace S_h of S_5. In particular, pencils and nets of conics "are" lines and planes of S_5.

The degenerate conics are the points of the cubic hypersurface M_4^3 with equation

$$\det(X_{ij}) = \det \begin{pmatrix} X_{00} & X_{01} & X_{02} \\ X_{10} & X_{11} & X_{12} \\ X_{20} & X_{21} & X_{22} \end{pmatrix} = 0 \quad (X_{ij} = X_{ji})$$

and to the ∞^2 doubly degenerate conics (i.e., double lines) there correspond the points of the surface $\mathcal{F}$ which is the locus of points that give rank one to the matrix (X_{ij}). One sees immediately that $\mathcal{F}$ is a Veronese surface. Indeed, a parametric representation of $\mathcal{F}$ is obtained by observing that the coefficients a_{ij} of the equation $(u_0 x_0 + u_1 x_1 + u_2 x_2)^2 = 0$ for a double line are $(u_0^2, u_0 u_1, u_1^2, u_0 u_2, u_1 u_2, u_2^2)$. Thus one rediscovers the parametric representation for $\mathcal{F}$ described in the example 10.2.1.

10.4. The conics of a plane as points of $\mathbb{P}^5$ and the Veronese surface

On M_4^3 there are two arrays of ∞^2 of planes, but no linear spaces of dimension > 2. Indeed, each point A of π is a double base point of a net $\Sigma_2(A)$ of conics, to which there corresponds a plane situated on M_4^3. Thus one has a first ∞^2 family of planes, which we will call *planes* $[A]$. If a is a line of π, the net $\Sigma_2(a)$ of the pairs of lines one of which is the fixed line a leads to a plane, which we will call the *plane* $[a]$. Each such plane $[a]$ lies on M_4^3 and, as a varies, sweeps out a second array of ∞^2 planes. In M_4^3 there are no linear spaces of dimension ≥ 3 because a linear space S_h belonging to M_4^3 must come from a linear system of conics all of which are degenerate. By Bertini's first theorem (Theorem 6.3.11), such a linear system consists of pairs of lines issuing from a fixed point, or of pairs of lines one of which is fixed. Therefore $h \leq 2$.

The following properties hold.

(1) Two planes belonging to the same system have a point in common; that point belongs to $\mathcal{F}$ if (and only if) the two planes are of type $[A]$.

   Indeed, the two nets having A and B as centers have in common only the line r_{AB} counted twice, and so $[A] \cap [B]$ is a point of the Veronese surface $\mathcal{F}$.

   In analogous fashion one sees that $[a] \cap [b]$ is the point that represents the pair of (distinct) lines a and b, and thus does not belong to $\mathcal{F}$.

(2) Two planes $[A]$ and $[a]$ not belonging to the same system are skew or otherwise have a line in common according to whether $A \notin a$ or $A \in a$.

   If $A \notin a$ no pair of lines passing through A can contain a. If however $A \in a$ the two nets have in common the pencil of conics split into the line a and a line of the pencil with center A. This pencil contains a doubly degenerate conic (the line a counted twice) and thus, if $A \in a$, the intersection $[A] \cap [a]$ is a line that has a point in common with the Veronese surface.

(3) Through a point of M_4^3 not belonging to $\mathcal{F}$ there pass a plane $[A]$ and two planes $[a]$.

(4) Every point of $\mathcal{F}$ belongs to ∞^1 planes $[A]$ and to only one plane $[a]$.

(5) The surface $\mathcal{F}$ does not contain lines.

   Indeed, by taking linear combinations of two doubly degenerate conics one obtains a pencil of conics which does not contain others that are doubly degenerate.

Proposition 10.4.1. *The cubic hypersurface M_4^3 is the locus of the chords of the Veronese surface $\mathcal{F}$.*

Proof. By taking linear combinations of two doubly degenerate conics α and β one obtains a pencil of conics all of which are degenerate. Thus the chord of $\mathcal{F}$ that joins the points which represent the two conics α, β, and whose points represent the conics of the pencil, is contained in M_4^3. Hence the chords of $\mathcal{F}$ lie in M_4^3.

To show that each point of M_4^3 belongs to some chord of $\mathcal{F}$ it suffices to show that every degenerate conic $L_1 L_2 = 0$ of the representative plane π (L_1 and L_2 linear polynomials in x_0, x_1, x_2) is a linear combination of two doubly degenerate conics. But this is true since

$$L_1 L_2 = \frac{1}{4}(L_1 + L_2)^2 - \frac{1}{4}(L_1 - L_2)^2. \qquad \square$$

With regard to Proposition 10.4.1, note that the Veronese surface is the unique surface embedded in S_r, $r \geq 5$, the locus of whose chords is a variety of dimension < 5 (cf. [14] and also [96]).

Proposition 10.4.2. *Let A and ℓ respectively be a point and a line of the plane π. Then:*

(1) *The plane $[A]$ contains a conic of the Veronese surface $\mathcal{F}$ (and thus $\mathcal{F}$ contains ∞^2 conics). Furthermore, through two points of $\mathcal{F}$ there passes one and only one of its conics, and two conics of $\mathcal{F}$ have in common only the point in which their planes meet.*

(2) *The plane $[\ell]$ is the tangent plane to $\mathcal{F}$ in the point that represents the line ℓ counted twice.*

Proof. Consider a plane $[A]$. The corresponding net $\Sigma_2(A)$ contains ∞^1 double lines and thus $[A]$ contains ∞^1 points of $\mathcal{F}$. The locus $\mathcal{L}_A$ of these points is a conic because every line of the plane $[A]$ represents a pencil of pairs of lines issuing from A, and in this pencil there are two (and only two) double lines. Thus every line of the plane $[A]$ contains two points of $\mathcal{L}_A$.

Therefore $\mathcal{F}$ contains ∞^2 conics $\mathcal{L}_A$ and M_4^3 is the locus of the planes of those conics. Furthermore, two conics of $\mathcal{F}$ have in common only the point in which their planes meet, and so have in common a single point of $\mathcal{F}$.

Now let ℓ be a line of π, γ the doubly degenerate conic consisting of the line ℓ counted twice, and P the point of $\mathcal{F}$ that represents γ in S_5. Through P there pass ∞^1 planes $[A]$ (one for every point $A \in \ell$) and thus ∞^1 conics of $\mathcal{F}$. The tangent at P to each of these conics belongs to the plane of that same conic, and so to M_4^3. Thus the tangent plane to $\mathcal{F}$ at P is contained in M_4^3; and is the plane $[\ell]$. $\qquad \square$

We remark that the following further properties (consequences of what has just been said) characterize the Veronese surface.

(1) Any two tangent planes to $\mathcal{F}$ are incident. [And it is known that the only surface (other than a cone) embedded in S_r, $r \geq 5$, for which any pair of tangent planes are incident is the Veronese surface.]

This follows immediately from the previous property (1) and Proposition 10.4.2 (2).

10.4. The conics of a plane as points of $\mathbb{P}^5$ and the Veronese surface

(2) $\mathcal{F}$ is the only surface embedded in S_r, $r \geq 5$, whose generic projections in S_4 are non-singular.

By Proposition 10.4.1 the generic S_{r-5} is skew to the variety of chords of $\mathcal{F}$ and so the various projecting spaces S_{r-4} (with such S_{r-5} as center of the projection) do not contain any chord.

(3) The Veronese surface is the locus of double points of M_4^3.

Indeed, the first partial derivatives of the polynomial $\det(X_{ij})$ are linear combinations of second order minors of the matrix (X_{ij}).

10.4.3 The Steiner surface. From a generic line p we project the Veronese surface $\mathcal{F}^4$ into a subspace S_3 which we call Π. Thus p has three distinct points in common with M_4^3. Let them be called L, M, and N. Through each of these points there passes a plane containing a conic of $\mathcal{F}^4$. Let λ, μ, and ν be the three conics and let M_0 be the common point of λ and ν. The line r_{LM_0} (which belongs to the plane of λ) contains another point N_0 of λ. In analogous fashion, the line r_{NM_0} (which belongs to the plane of ν) meets ν in a second point L_0.

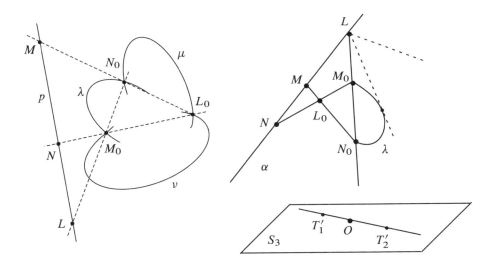

Figure 10.1.

The three points L_0, M_0, N_0 belong to the plane $J(M, L, L_0)$, that is, to a plane containing the two lines p and $r_{L_0N_0}$. Thus these two lines meet, and one sees immediately that their common point is M. Indeed, the three lines $r_{M_0N_0}$, $r_{L_0N_0}$,

$r_{L_0M_0}$, which are chords of $\mathcal{F}^4$, are contained in the variety M_4^3 that meets $\mathcal{F}^4$ only in L, M, and N.

The plane $\alpha = J(L_0, M_0, N_0)$ is a trisecant plane of $\mathcal{F}$ containing p. Each of the three conics λ, μ, ν is contained in a space S_3 passing through p (and containing α) because the planes of the three conics meet p. Thus the three conics project onto three lines of Π each of which passes through the point $O := \Pi \cap \alpha$. Thus one finds three lines l, m, and n in Π all passing through O and each of which is double for the surface $\mathcal{G}$ obtained by projection. For example two points of λ collinear with L belong to a plane passing through p and thus are projected into the same point of l: on l the surface $\mathcal{G}$ has two pinch-points, T_1', T_2', the traces in Π of the planes that join p with the tangents of λ passing through L.

The projection $\mathcal{G}$ of $\mathcal{F}^4$ onto Π is thus a surface of order 4 (since p does not meet $\mathcal{F}^4$) having three double lines that are concurrent in a triple point O (and having on each of the three double lines two pinch-points, cf. 5.8.22). This surface was discovered by Steiner in 1844 during a visit to Rome, and therefore has been called *Steiner's Roman surface*.

10.5 Complements and exercises

In this section we describe further properties of rational surfaces by means of exercises and the study of some illuminating examples.

10.5.1. *Show that the Veronese surface has class* 3.

The class of a surface in $\mathbb{P}^r$ is the number of tangent hyperplanes belonging to a pencil. Hence we must count the hyperplane sections belonging to a pencil and endowed with a double point. This is equivalent to seeking, in the representative plane, the conics of a pencil that have a double point, which means that are degenerate, and this number is precisely three.

10.5.2. *Show that every curve belonging to the Veronese surface $\mathcal{F}$ is a curve of contact of $\mathcal{F}$ with a hypersurface in $\mathbb{P}^5$.*

The Veronese surface is the projective image of the linear system of all the conics in the plane. On the other hand, the square of a homogeneous polynomial $f \in K[u_0, u_1, u_2]$ is a homogeneous polynomial $g \in K[u_0^2, u_1^2, u_2^2, u_0u_1, u_0u_2, u_1u_2]$ (of degree equal to that of f in the new variables). Hence the curve $\mathcal{L}$ on $\mathcal{F}$ which is the image of the plane curve with equation $f = 0$ is the set-theoretic intersection of $\mathcal{F}$ with the hypersurface $\mathcal{G}$ of $\mathbb{P}^5$ having equation $g = 0$; and in fact, $2\mathcal{L} = \mathcal{F} \cap \mathcal{G}$ is the algebraic complete intersection of $\mathcal{F}$ and $\mathcal{G}$. In this case the curve $\mathcal{L}$ is a *contact curve* of $\mathcal{F}$ with $\mathcal{G}$, in the sense that for every generic point P of $\mathcal{L}$, the tangent plane to $\mathcal{F}$ at P is contained in the tangent hyperplane to $\mathcal{G}$ at P.

In analogous fashion, if $V_{n,2}$ is the surface in $\mathbb{P}^{\binom{n+2}{2}-1}$ which is the projective image of the linear system of all the plane curves of order n, one sees that the curve

$\mathcal{L}$ of $V_{n,2}$ which is the image curve of the plane curve with equation $f = 0$ is the set-theoretic intersection of $V_{n,2}$ with a hypersurface $\mathcal{G}$ in $\mathbb{P}^{\binom{n+2}{2}-1}$. Moreover, we have $n\mathcal{L} = \mathcal{F} \cap \mathcal{G}$, that is, $n\mathcal{L}$ is the complete algebraic intersection of $V_{n,2}$ and $\mathcal{G}$. In this case $\mathcal{L}$ is the contact curve of order $n-1$ of $V_{n,2}$ with $\mathcal{G}$.

10.5.3. *Use the planar representation in Example 10.2.1 to prove that the Veronese surface belongs to six linearly independent quadrics.*

It suffices to observe that the quadrics in $\mathbb{P}^5$ constitute a linear system of dimension 20, while the curves corresponding to the sections of the Veronese surface by quadrics are the plane quartics and so constitute a linear system of dimension only 14. Thus there must be six linearly independent quadrics containing the Veronese surface.

10.5.4. *Study the surface $\mathcal{F}$ represented in a plane π by the linear system of conics that pass through a point A, and prove that it is the residual intersection of two quadrics of S_4 containing a common plane.*

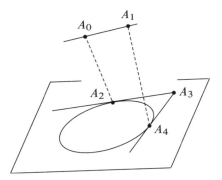

Figure 10.2

Bearing in mind (10.7), we have that $\mathcal{F}$ is the cubic in S_4 which is the projection of the Veronese surface from one of its points O. By what we have seen in Section 10.3, the surface $\mathcal{F}^3 \subset S_4$ is ruled and the generators are images of the lines of π that pass through A. The ruled cubic $\mathcal{F}^3 \subset S_4$ has a linear directrix (the exceptional line corresponding to the point O, cf. §10.1.7) and ∞^2 conic directrices, namely the images of the (generic) lines of the plane. Moreover, the surface $\mathcal{F}^3$ is the locus of the lines that join the pairs of points corresponding under a projectivity ω between the rectilinear directrix and any one of these conics.

The sections of the surface $\mathcal{F}^3$ with the quadric hypersurfaces of S_4 are represented by the plane quartics having A as double point. Since these make up a linear system of dimension $r = 11$, while the quadrics in S_4 form a space of dimension 14, the surface $\mathcal{F}^3$ belongs to three linearly independent quadrics. The intersection of two of these quadrics is the union of the surface $\mathcal{F}^3$ with a plane.

We can choose the projective reference system in S_4 in such a way that the rectilinear directrix is the line $\langle A_0, A_1 \rangle$, that $x_0 = x_1 = x_2 x_4 - x_3^2 = 0$ (Figure 10.2) is a conic directrix, and that the two points $[1, t, 0, 0, 0]$ and $[0, 0, 1, t, t^2]$ correspond under the projectivity ω.

Therefore, as seen in §10.3.8, the surface $\mathcal{F}^3$ appears as the locus of points that give rank one to the matrix
$$\begin{pmatrix} x_0 & x_2 & x_3 \\ x_1 & x_3 & x_4 \end{pmatrix}.$$

10.5.5. *Let $\mathcal{F}$ be the surface which is the image of a 4-dimensional linear system Σ of conics without base points. What special properties does $\mathcal{F}$ have if Σ is the linear system of conics that meet a line r in pairs of points belonging to a given involution $\omega : r \to r$?*

Since Σ has both dimension and degree 4, the surface $\mathcal{F}$ is of order 4 in $S_4 = \mathbb{P}^4$, and is obtained as the projection of the Veronese surface from a point P of $\mathbb{P}^5$. If the conics of Σ intersect the line r in pairs of points belonging to the involution ω, two points M and $M' = \omega(M)$ of r have the same image under the morphism φ associated to Σ because every conic of Σ that passes through one of them also contains the other.

The image $\varphi(r)$ is a line. Indeed, it meets every hyperplane of S_4 in the unique point that corresponds to the two intersections of a conic of Σ with r. In fact, $\varphi(r)$ is a double line for $\mathcal{F}$. Indeed, if A is a point of $\varphi(r)$ and π is a generic plane containing A, two of the four intersections of $\mathcal{F}$ with π are absorbed by A inasmuch as to two hyperplane sections passing through A there correspond two conics of Σ that cut out the same pair of points $\varphi^{-1}(A)$ on r, and which consequently have only two points in common away from r.

The fact that $\mathcal{F}$ has double points means that it is the projection of the Veronese surface $V_{2,2}$ from a point of the variety W locus of the chords of $V_{2,2}$. The double points of $\mathcal{F}$ are all the points of a line, which is in agreement with the fact that for each point P of W that does not belong to $V_{2,2}$ there passes the plane of a conic γ of $V_{2,2}$ and r is the trace of the plane of γ in S_4. [Recall that W is the locus of the planes of the conics of $V_{2,2}$, cf. Proposition 10.4.2.] The traces of the two tangents to γ that pass through P are the two pinch-points that $\mathcal{F}$ has on $\varphi(r)$ (cf. 5.8.22). They are the images of the fixed points of ω.

We note in passing that a surface of order 4 in $\mathbb{P}^4$ that is the projection of a Veronese surface is either non-singular or has a double line.

10.5.6 (The Steiner surface). *Study the surface $\mathcal{F}$ which is the projective image of the most general 3-dimensional linear system Σ of conics without base points.*

By what we have seen in Section 10.2 it follows that $\mathcal{F}$ is a surface of order 4 (not linearly normal) and is the projection of the Veronese surface $V_{2,2}$ from a line ℓ

of $\mathbb{P}^5$ to a subspace S_3, with ℓ chosen to meet the variety of the chords M_4^3 of $V_{2,2}$ in three distinct points. Hence ℓ does not meet $V_{2,2}$ since the latter is the locus of the double points for M_4^3, cf. Section 10.4.

Represent the conics of the plane π of parameters with the points of $\mathbb{P}^5$ by associating to the conic with equation

$$a_{00}u_0^2 + 2a_{01}u_0u_1 + 2a_{02}u_0u_2 + a_{11}u_1^2 + 2a_{12}u_1u_2 + a_{22}u_2^2 = 0$$

the point $[a_{00}, a_{01}, a_{02}, a_{11}, a_{12}, a_{22}]$. We choose as the line ℓ the line that intersects the hypersurface M_4^3 with equation

$$\begin{vmatrix} x_0 & x_1 & x_2 \\ x_1 & x_3 & x_4 \\ x_2 & x_4 & x_5 \end{vmatrix} = 0$$

in the three distinct points

$$L = [1, 0, 0, -1, 0, 0], \quad M = [1, 0, 0, 0, 0, -1], \quad N = M - L = [0, 0, 0, 1, 0, -1],$$

and as the space S_3 that having equations $x_3 = x_5 = 0$.

The plane that joins ℓ with the point $P = [u_0^2, u_0u_1, u_0u_2, u_1^2, u_1u_2, u_2^2]$ is the locus of the point $P + \lambda L + \mu M = [u_0^2 + \lambda + \mu, u_0u_1, u_0u_2, u_1^2 - \lambda, u_1u_2, u_2^2 - \mu]$ which belongs to our S_3 if $\lambda = u_1^2$, $\mu = u_2^2$. The projection of P from ℓ into our S_3 is thus the point

$$[u_0^2 + u_1^2 + u_2^2, u_0u_1, u_0u_2, 0, u_1u_2, 0].$$

The surface $\mathcal{F}$ is then the projective image of the linear system

$$\Sigma: \lambda_0(u_0^2 + u_1^2 + u_2^2) + \lambda_1 u_0u_1 + \lambda_2 u_0u_2 + \lambda_3 u_1u_2 = 0.$$

Moreover Σ will then contain the homaloidal net $\lambda_1 u_0 u_1 + \lambda_2 u_0 u_2 + \lambda_3 u_1 u_2 = 0$ having the points $A = [1, 0, 0]$, $B = [0, 1, 0]$, and $C = [0, 0, 1]$ as its base points.

To the lines r_{BC}, r_{AC}, and r_{AB} there correspond three lines a, b, and c which are double for $\mathcal{F}$. Indeed, the conics of Σ meet each of the sides of the triangle ABC in pairs of points belonging to an involution. If, for example, ω is the involution induced by Σ on the line r_{AB}, the conics of Σ that pass through a point P of r_{AB} also pass through the point $\omega(P)$, and the two points $P, \omega(P)$ represent the same point of $\mathcal{F}$ (belonging to the double line c that corresponds to r_{AB}). To the three points A, B, C there corresponds the same point O, which is triple for the surface, and through which the three double lines pass. Thus we find the *Steiner surface*. The two fixed points of ω are $[1, 1, 0]$ and $[1, -1, 0]$. They correspond to the two pinch-points that $\mathcal{F}$ has along c. Naturally we have the analogous facts for the lines a and b.

368   Chapter 10. Rational Surfaces

One sees immediately that the four conics with equations $2r_i : (u_0 \pm u_1 \pm u_2)^2 = 0$, $i = 1, 2, 3, 4$, all belong to Σ. Using the description of the Veronese surface as the locus of the points corresponding to the ∞^2 doubly degenerate conics of the plane (as we have seen in Section 10.4) one immediately sees that Σ contains no other doubly degenerate conics, because a generic S_3 of S_5 has four points in common with a Veronese surface. By equation (10.5) to the lines $r_i : u_0 \pm u_1 \pm u_2 = 0$ there correspond conics γ_i on $\mathcal{F}$. Thus to the four doubly degenerate conics $2r_i$ there correspond four curves $2\gamma_i$ on $\mathcal{F}$, which are the contact locus of $\mathcal{F}$ with four *double tangent hyperplanes* (that is tangent along the conics γ_i). The class of $\mathcal{F}$ is three (as is the class of the Veronese surface, cf. Exercise 10.5.1) because to the curves that are sections of $\mathcal{F}$ by tangent planes belonging to a pencil there correspond in the representative plane the three conics of a pencil endowed with a double point. Thus one sees that the Steiner surface is the dual surface of a cubic surface with four double points, inasmuch as it has the projective characteristic dual to those of the latter (the double tangent plane is dual to the double point, and the class is dual to the order).

10.5.7 (Del Pezzo surfaces). *Study the surfaces that are the projective images of complete linear systems of cubics of genus 1.*

The linear system Σ of all plane curves of third order has dimension 9 and order 9 and so the projective image of Σ is a non-singular $\mathcal{F}^9$ in $\mathbb{P}^9$, called a *Del Pezzo surface*. If we require the curves of Σ to pass through $d \leq 6$ points in general position (that is, d distinct points, no three of which are collinear, and no six of which lie on a conic) we find a non-singular surface $\mathcal{F}^{9-d}$ in $\mathbb{P}^{9-d}$. This $\mathcal{F}^{9-d}$ is the projection of $\mathcal{F}^9$ from d of its generic points, and is the projective image of the linear system Σ' consisting of the curves of Σ that pass through the d points. While $\mathcal{F}^9$ does not contain any line, the surface $\mathcal{F}^{9-d}$ possesses at least the exceptional lines that correspond to the base points imposed on Σ (cf. §10.1.7). Using (10.5) one sees that if $d > 1$ then $\mathcal{F}^{9-d}$ also contains the lines corresponding to the lines of the plane that contain two of these points; if $d \geq 5$ there are also the lines corresponding to the conics that contain five of them.

If three of the base points belong to a line ℓ, the line ℓ is a fundamental curve and hence to ℓ there corresponds a single point P (cf. §10.1.6) which is a double point for the surface $\mathcal{F}^{9-d}$. Indeed, two hyperplane sections that pass through P are represented in the plane π by two cubics split into the line ℓ and a conic that contains the other $d - 3$ base points. Therefore, besides their intersection at P, they meet in only $4 - (d - 3)$ other points. This means that the point P absorbs $(9 - d) - (7 - d) = 2$ of the intersections of $\mathcal{F}^{9-d}$ with a generic S_{7-d} passing through P.

Similarly one sees that if Σ' has six base points belonging to a conic, that conic is a fundamental curve and to it there corresponds a double point O on $\mathcal{F}^{9-d}$ (which is a cubic surface in S_3). Note that in this case the representation of $\mathcal{F}^3$ on a plane by way of the system of conics through six points is immediately obtained by projecting $\mathcal{F}^3$ from its double point O onto a plane not passing through O.

Among the Del Pezzo surfaces (projections of the surface $\mathcal{F}^9$ of $\mathbb{P}^9$) a particularly interesting one is the surface $\mathcal{F}^4$ called C. Segre's surface which arises when $d = 5$. It contains sixteen lines (the five exceptional lines, the ten lines that correspond to the lines that join in pairs the five base points, and the line that corresponds to the conic containing the five points) and is the complete intersection of two quadric hypersurfaces of S_4. To see this, observe that the quadrics of S_4 are in number ∞^{14}, while the sections of $\mathcal{F}^4$ (the projective image of the linear system $\Sigma' = C^3(B_1, \ldots, B_5)$) with quadric hypersurfaces are represented in the plane by the curves of order 6 having double points in the five base points (cf. §10.1.8), and these constitute a linear system of dimension $\frac{6(6+3)}{2} - 5 \times 3 = 12$. On the other hand, we know that a linear system L of hypersurfaces intersects $\mathcal{F}^4$ in a linear system L' whose dimension is $\dim L' = \dim L - t$, where t is the number of linearly independent hypersurfaces passing through $\mathcal{F}^4$. Thus, $t = 2$.

We note explicitly that, in agreement with what we have seen in Section 10.2, the surfaces that are projective images of non-complete linear systems of cubics of genus 1 are surfaces which are (not linearly normal) projections of the Del Pezzo hypersurfaces described above from points not belonging to them.

10.5.8. *Write the equation of a Steiner surface $\mathcal{F}$ having a triple point at $A_0 = [1, 0, 0, 0]$ and double lines $r_{A_0 A_1}$, $r_{A_0 A_2}$, $r_{A_0 A_3}$.*

The conic which splits into the lines $r_{A_0 A_2}$ and $r_{A_0 A_3}$ belongs to the plane $x_1 = 0$ and therein has the equation $x_2 x_3 = 0$. Therefore, if $f = 0$ is the equation of $\mathcal{F}$, on setting $x_1 = 0$ in f one must find $x_2^2 x_3^2 = 0$. In analogous fashion, for $x_2 = 0$ one must find $x_3^2 x_1^2 = 0$, and similarly $x_1^2 x_2^2 = 0$ for $x_3 = 0$. Thus

$$f = ax_2^2 x_3^2 + bx_3^2 x_1^2 + cx_1^2 x_2^2 + x_1 x_2 x_3 (lx_0 + mx_1 + nx_2 + px_3) = 0, \quad (10.11)$$

for suitable coefficients $a, b, c, l, m, n, p \in \mathbb{C}$. With the change of coordinates expressed by the relations

$$X_0 : X_1 : X_2 : X_3 = lx_0 + mx_1 + nx_2 + px_3 : x_1 : x_2 : x_3,$$

equation (10.11) becomes

$$aX_2^2 X_3^2 + bX_3^2 X_1^2 + cX_1^2 X_2^2 + X_0 X_1 X_2 X_3 = 0.$$

Finally, if we put $X_0 = Y_0 \sqrt{abc}$, $X_1 = Y_1 \sqrt{a}$, $X_2 = Y_2 \sqrt{b}$, $X_3 = Y_3 \sqrt{c}$ (where $\sqrt{\zeta}$ denotes either choice of the complex square root of ζ), we find the following simple equation for the Steiner surface:

$$Y_2^2 Y_3^2 + Y_3^2 Y_1^2 + Y_1^2 Y_2^2 + Y_0 Y_1 Y_2 Y_3 = 0.$$

10.5.9. *Consider the Del Pezzo surface $\mathcal{F}^9$ in $\mathbb{P}^9$, two of its points A and P, and the plane α tangent to $\mathcal{F}^9$ at A. What does one obtain by projecting $\mathcal{F}^9$ onto a linear space S_5 from the space S_3 that joins P with α?*

The projection of the Del Pezzo surface $\mathcal{F}^9$ from the space S_3 that joins α with P may be realized via two successive projections: the projection from α to S_6 and (in that S_6) the projection from the point P' that is the projection into S_6 of P from α. The first projection leads to a ruled surface $\mathcal{F}^5$ in S_6; the second projection gives a ruled surface $\mathcal{F}^4$ in S_5 with ∞^1 conic directrices (cf. Example 10.2.5 and Section 10.3).

10.5.10. *A generic cubic surface $\mathcal{F}$ in S_3 is the projection of the Del Pezzo surface $\mathcal{F}^9$ from the space S_5 that joins six of its generic points. Use its planar representation to verify that $\mathcal{F}$ is non-singular, has twenty-seven lines, and forty-five tritangent planes (cf. Exercise 5.8.13).*

If $\mathcal{F}$ had a double point O the linear system ∞^2 of the planes passing through O would cut out a linear system of (rational) cubics on $\mathcal{F}$, any two of which would have in common, besides O, also a single point. That point is the further point of intersection of $\mathcal{F}$ with the line common to the planes of the cubics. The curves (rational and of order ≤ 3) that correspond to these ∞^2 plane sections would then form a homaloidal net σ partially contained in the linear system $\Sigma = C^3(B_1, B_2, B_3, B_4, B_5, B_6)$ that represents the surface. Note that the curves of σ can not be irreducible cubics since otherwise they would all have a double point (which can not be fixed). The locus of this point would be, in virtue of Bertini's first theorem (Theorem 6.3.11), a line δ which when adjoined to all the curves of σ would furnish curves of Σ. Then δ would be a fundamental curve of Σ and would represent a double point.

A detailed analysis of all the possible cases shows that Σ can partially contain the homaloidal net σ only if three of its base points are collinear (and σ is the net of the conics passing through the other three base points) or if the six points belong to a conic (and σ is the net of the lines). This contradicts the hypothesis that the base points $B_1, \ldots, B_6$ be in general position.

Besides the six exceptional lines b_i that correspond to the six base points B_i ($i = 1, \ldots, 6$), the surface $\mathcal{F}$ also contains the fifteen lines ℓ_{ij} represented by the lines $r_{B_i B_j}$ that contain two of the base points, and the six lines r_i represented by the conics that contain the five points different from B_i.

Note that the surface $\mathcal{F}$ does not contain other lines. To see this, it suffices to observe that (besides the fifteen lines and the six conics) there do not exist other plane curves C^n of given order n that satisfy the condition $1 = 3n - \sum_{j=1}^{6} h_j$ expressed by equation (10.5), where $h_j \geq 0$ is the multiplicity of C^n at B_j, $j = 1, \ldots, 6$. [There is an analogous argument for the case of C. Segre's quartic $\mathcal{F}^4 \subset \mathbb{P}^4$ as in Problem 10.5.7.]

A tritangent plane contains three lines of $\mathcal{F}$ (those that make up the cubic with three double points which is the section of $\mathcal{F}$ by the tritangent plane). To seek the tritangent planes, that is, the planes that intersect $\mathcal{F}$ in three lines, is equivalent to seeking the cubics of Σ that are split into three lines. It is easy to see that every line of $\mathcal{F}$ belongs to five tritangent planes. We count, for example, the triples of lines on $\mathcal{F}$ corresponding to the cubics of Σ split into the line $r_{B_1 B_2}$ and two other lines. There are five of them, and more specifically they are:

$$\ell_{12} \cup \ell_{34} \cup \ell_{56}, \quad \ell_{12} \cup \ell_{35} \cup \ell_{46}, \quad \ell_{12} \cup \ell_{36} \cup \ell_{45},$$
$$\ell_{12} \cup r_1 \cup b_2, \quad \ell_{12} \cup r_2 \cup b_1.$$

Since every tritangent plane contains three lines, the number of tritangent planes is $\frac{27 \times 5}{3} = 45$.

10.5.11. *Study the surfaces $\mathcal{F}$ represented in the plane π by linear systems Σ of rational cubics with distinct base points.*

If all the cubics of an irreducible linear system Σ have a double point that point will be a base point O by Bertini's first theorem. It follows that all the lines passing through O meet the cubics C^3 of Σ only in one point away from the base points, and so $\mathcal{F}$ is covered by at least ∞^1 lines and is thus a ruled surface.

If there are no other base points, $\Sigma = C^3(O^2)$ and $\mathcal{F}$ is a ruled surface of order five in S_6. To the base point of Σ there corresponds a conic and this is the only conic belonging to $\mathcal{F}$ (cf. §10.1.7).

If $\Sigma = C^3(O^2, B_1, \ldots, B_d)$ also possesses $d \leq 3$ (simple) distinct base points, we have deg $\Sigma = 5 - d$, dim $\Sigma = 6 - d$ and so $\mathcal{F}$ is a ruled surface of order $5 - d$ belonging to a space S_{6-d} (note that this is a planar representation different from that described by the equations (10.1)). We note that for $d = 3$ the conics of the representative plane passing through the four base points meet the cubics of Σ away from the base points in a single point and so $\mathcal{F}$ has infinitely many rectilinear directrices; therefore $\mathcal{F}$ is a quadric of S_3. On the other hand, the quadratic transformation having as its fundamental points the double point and two of the three base points transforms Σ into the linear system of conics passing through two points (cf. Proposition–Definition 9.2.2), and we know that two linear systems which are transformed into each other by a Cremona transformation represent surfaces that are projectively identical.

If $d = 1$, one has a surface $\mathcal{F}^4$ in S_5. The lines containing the simple base point represent conic directrices and $\mathcal{F}^4$ is the locus of the lines that join pairs of corresponding points under a projectivity between any two of these conics (cf. Section 10.1).

We choose our reference system so that the two conic directrices γ_1 and γ_2 are defined respectively by the equations $x_3 = x_4 = x_5 = x_1^2 - x_0 x_2 = 0$ and $x_0 = x_1 = x_2 = x_3 x_5 - x_4^2 = 0$, and moreover so that the two points $[1, t, t^2, 0, 0, 0]$ and $[0, 0, 0, 1, t, t^2]$ are in correspondence with each other (Figure 10.3). Then one

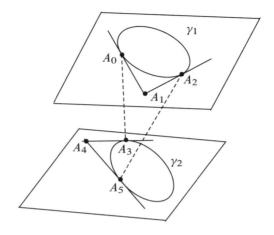

Figure 10.3

sees that $\mathcal{F}^4$ is the locus of points for which the matrix

$$\begin{pmatrix} x_0 & x_1 & x_3 & x_4 \\ x_1 & x_2 & x_4 & x_5 \end{pmatrix}$$

has rank one.

In paragraph 10.5.15 we will study another $\mathcal{F}^4$ in S_5.

Finally, if $d = 2$ one has a ruled cubic in S_4 which is the projection of the Veronese surface from one of its points. Indeed, the system Σ can be reduced via a quadratic transformation to the system Σ' (of dimension 4 and degree 3) of the conics that pass through a point (cf. Proposition–Definition 9.2.2).

10.5.12. *Study the surface $\mathcal{F}$ represented by the linear system Σ of the cubics with a node O and with a given tangent in a given point P.*

The system $\Sigma = C^3(O^2, P, P)$ can be transformed via a quadratic transformation (with only two fundamental points) into the system of conics passing through a point. The surface $\mathcal{F}$ is then the ruled cubic of S_4. If the cubics of Σ have as a double point the point $[0, 0, 1]$ and are tangent at the point $[0, 1, 0]$ to the line $u_0 = 0$, so that

$$\Sigma: \lambda_0 u_2^3 + \lambda_1 u_1^2 u_0 + \lambda_2 u_2^2 u_0 + \lambda_3 u_2^2 u_1 + \lambda_4 u_0 u_1 u_2 = 0,$$

the quadratic transformation that transforms Σ into a system of conics with a base point is that defined by $[u_0, u_1, u_2] \mapsto [y_2^2, y_0 y_1, y_0 y_2]$.

10.5.13. *Study the surface which is the projective image of the linear system of cubics having a double point O and a flex B with given tangent.*

10.5. Complements and exercises 373

If the system of coordinates in the representative plane π is chosen so that O is the point $[1, 0, 0]$, and $u_0 = 0$ is the equation of the tangent at the flex $B = [0, 1, 0]$, we then have for Σ the equation, cf. Section 5.7,

$$\lambda_0 u_2^3 + \lambda_1 u_1^2 u_0 + \lambda_2 u_2^2 u_0 + \lambda_3 u_0 u_1 u_2 = 0.$$

The quadratic transformation defined by $[u_0, u_1, u_2] \mapsto [y_2^2, y_0 y_1, y_0 y_2]$ changes Σ into the linear system Σ' of conics having a point in common as well as the tangent at that point. Thus Σ' represents a quadric cone in S_3 (cf. Remark 7.3.3).

10.5.14. *Describe the surface $\mathcal{F}$ represented in the plane π by a linear system Σ of cubics having a cusp at a given point P.*

The curves of Σ must all have the same tangent at P because by taking a linear combination of the equations of two curves with a cusp at P and different cuspidal tangents one finds a pencil of curves whose generic member has P as a non-cuspidal double point.

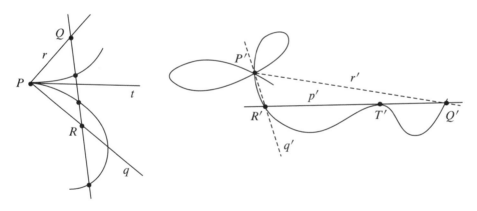

Figure 10.4

To the cuspidal tangent (which is a fundamental line of Σ) there is associated a point O that is a triple point for $\mathcal{F}$. This may be seen more clearly by transforming Σ via a quadratic transformation $\omega \colon \pi \to \pi'$ having one of its base points at P and the other two base points Q and R chosen arbitrarily. If p', q' and r' are the exceptional lines corresponding to P, Q and R respectively, the curves of Σ' are quartics having as triple point the point $P' = q' \cap r'$ and passing simply through $Q' = r' \cap p'$ and $R' = p' \cap q'$. Moreover they are all tangent to the line p' at the point T' which corresponds to the cuspidal tangent (cf. Proposition–Definition 9.2.2). The line p' is therefore fundamental for Σ' and the point O of $\mathcal{F}$ corresponding to it is triple for $\mathcal{F}$ because the curves of Σ' that contain it have as residual component a triple of lines issuing from P' (Figure 10.4).

The lines of π passing through P (and which ω changes into the lines of π' passing through P') represent lines of the surface passing through O. Thus $\mathcal{F}$ is a cubic cone or a quadric cone with vertex at O according to whether or not Σ has another base point distinct from P.

Moreover, one can arrive at the same conclusion immediately by analytic arguments. If $P = [1, 0, 0]$ and if the cuspidal tangent is $u_1 = 0$, one finds that Σ has the equation

$$\lambda_0 u_0 u_1^2 + \lambda_1 u_1^3 + \lambda_2 u_1^2 u_2 + \lambda_3 u_1 u_2^2 + \lambda_4 u_2^3 = 0.$$

Hence we get for $\mathcal{F}$ the parametric representation

$$x_0 : x_1 : x_2 : x_3 : x_4 = u_0 u_1^2 : u_1^3 : u_1^2 u_2 : u_1 u_2^2 : u_2^3,$$

or also, on setting $t = \frac{u_1}{u_2}$ and $v = \frac{u_0 u_1^2}{u_2^3}$,

$$x_0 : x_1 : x_2 : x_3 : x_4 = v : t^3 : t^2 : t : 1.$$

Then $\mathcal{F}$ appears as a cone that projects the cubic C from the point $[1, 0, 0, 0, 0]$ where C is the locus of the point $[0, t^3, t^2, t, 1]$.

If Σ has no other base points (and thus has dimension 4 and degree 3) an arbitrary line r not passing through P represents a rational normal cubic C belonging to the space S_3 associated to the cubic of Σ that splits into r and the cuspidal tangent counted twice. Moreover, $\mathcal{F}$ is the cubic cone in S_4 that projects C from O. Three arbitrary lines issuing from P give three coplanar generators since Σ contains the cubic split into the three lines.

Bearing in mind Bézout's theorem and the fact that two cubics of Σ have intersection multiplicity 6 at P (cf. Exercise 5.7.1), the system Σ can have, besides P, at most three other base points.

If Σ has a second base point P_2 (at which two generic curves of Σ have intersection multiplicity 1, and thus Σ has dimension 3 and degree 2) the surface $\mathcal{F}$ is a quadric cone in S_3.

If Σ has a third base point P_3 (of the same type as P_2), the system Σ is a net of degree 1, that is, a homaloidal net defining a Cremona transformation of degree 3 between two planes (the plane of Σ and $\mathcal{F}$).

A possible fourth base point would imply that the projective image of Σ was a curve.

10.5.15. *Examine the surface $\mathcal{F}$ represented in the plane π by the linear system Σ of cubics having a common double point O with one of the two tangents there being fixed.*

The system Σ has the same degree and dimension as the system $C^3(O^2, A)$ of cubics having O as a double point and passing through an additional point A, and so deg $\Sigma = 4$, dim $\Sigma = 5$. Therefore it represents a ruled surface of order 4 in S_5. A quadratic transformation that has a fundamental point at the double point and the other two in generic position transforms Σ into a linear system Σ' of quartics with a triple base point and three collinear simple base points (cf. Proposition–Definition 9.2.2). The degree and dimension of Σ' are respectively 4 and 5. Let $\mathcal{F}^4 \subset S_5$ be the projective image surface of Σ'. The triple base point of Σ' gives a cubic directrix $\mathcal{L}$ on $\mathcal{F}^4$. (Note that the three simple base points of Σ' are not in general position, cf. §10.3.6). Moreover, the line that contains the three simple base points gives a rectilinear directrix r skew to the S_3 containing $\mathcal{L}$, and $\mathcal{F}^4$ is the locus of the lines that join the pairs of points corresponding under a projectivity between $\mathcal{L}\ (\cong \mathbb{P}^1)$ and r.

One can choose the coordinate system in S_5 so that this ruled surface is the locus of the points that endow the matrix

$$\begin{pmatrix} x_0 & x_2 & x_3 & x_4 \\ x_1 & x_3 & x_4 & x_5 \end{pmatrix}$$

with rank one (cf. 10.3.8).

10.5.16. *Study the surface $\mathcal{F}$ represented in the plane π by the linear system Σ of conics passing through a given point P and meeting a given line r in pairs of points that correspond under an involution ω.*

The system $C^2(P)$ of conics through the point P has degree 3 and dimension 4. The hypothesis on the involution ω imposes a linear condition and thus the system (non-complete) Σ has dimension 3. Therefore $\mathcal{F}$ is a ruled cubic in S_3.

The exceptional curve $\mathcal{L} \subset \mathcal{F}$ corresponding to the base point P is a simple directrix. Indeed, the lines of the plane π passing through P meet the curves of Σ only in one point away from P and thus have as images on $\mathcal{F}$ lines that are supported by $\mathcal{L}$. The line r represents a double directrix; indeed, two points A and A' of r corresponding to each other under ω give the same point of $\mathcal{F}$ through which there pass the two generators that correspond to the lines r_{PA} and $r_{PA'}$. The double points of ω lead to the two pinch-points of $\mathcal{F}$.

If we take P to be the point $[1, 0, 0]$ in a projective reference system for π, and take the double points of ω to be the points $[0, 1, 0]$, $[0, 0, 1]$, one finds that Σ has equation

$$\lambda_0 u_0 u_1 + \lambda_1 u_0 u_2 + \lambda_2 u_1^2 + \lambda_3 u_2^2 = 0,$$

and $\mathcal{F}$ has the parametric representation

$$x_0 = u_0 u_1, \quad x_1 = u_0 u_2, \quad x_2 = u_1^2, \quad x_3 = u_2^2.$$

Eliminating the parameters one finds that our ruled cubic in S_3 has the equation

$$x_0^2 x_3 - x_1^2 x_2 = 0. \tag{10.12}$$

Conversely, let R be a ruled cubic in S_3, not a cone. If g is a generator and if the tangent plane to $\mathcal{F}$ along g is not fixed (that is, if g is a non-singular generator of $\mathcal{F}$, cf. Section 5.8) the correspondence between the points of g and the corresponding tangent planes is a projectivity by Chasles' theorem (see 5.8.18 (2)). On the other hand, every plane α passing through g intersects the ruled surface in an additional conic, and thus its section by R is a cubic with two double points. If R is of general type (i.e., if the base points of the linear system that represents it are in general position) and if α is a generic plane passing through g, these two points are distinct. Since only one of such points is a point of contact of R with a tangent plane passing through g (inasmuch as g is a non-singular generator), over every generator there is a double point of the ruled surface; this point (which can not be fixed) describes a line. Every ruled cubic thus possesses a double line. From each point of this line, there issue two generators s and s' of R and the plane σ that contains them meets the ruled surface in a third line d, skew to the double line. The intersection of another arbitrary generator of R with σ will belong to the line d which is therefore a simple directrix for the ruled surface. Each plane passing through d also intersects R in two generators (issuing from the point in which the plane meets the double line) which are supported by d in two points that correspond to each other via an involution ω. For each of the two double points of ω there issues a singular generator of R, namely a generator along which the tangent plane to R is fixed. Moreover, the point in which this line is supported by the double directrix is one of the pinch-points.

Now let O be a point of the double line a, and let b and c be the generators of R passing through O. Let π be a plane not passing through O, and finally let A, B, and C be the traces on π of a, b, and c. We project the section $\mathcal{L}$ of R by a generic plane β from O onto π. In π we obtain a cubic $\mathcal{L}'$ passing through A, B, and C and having A as double point with tangents the traces in π of the two tangent planes to R in the point $a \cap \beta$. The projections of the plane sections of R are thus the curves of the linear system Σ' of cubics passing doubly through A and simply through B and C, and such that the pairs of tangent lines at A belong to an involution (Figure 10.5). Since R is embedded in S_3, the system Σ' has dimension 3. The quadratic transformation having A, B and C as fundamental points changes Σ' into the linear system Σ of conics passing through a point P and cutting out an involution ω on a fixed line (see also Proposition–Definition 9.2.2).

To find the equation (10.12) for R it suffices to choose the projective reference system in S_3 in such a way that $[1, 0, 0, 0]$ and $[0, 1, 0, 0]$ are the double points of ω, $[0, 0, 1, 0]$ and $[0, 0, 0, 1]$ are the pinch-points and $U = [1, 1, 1, 1]$ belongs to R.

10.5.17. *In the plane π consider a point P and a line r containing P, and let ω be a projectivity between the line r and the pencil of lines with center P. Consider the conics of π passing through P and satisfying the property that under ω there correspond the tangent at P and the other intersection of the conic with r. Then such conics constitute a linear system Σ that represents a ruled cubic in S_3. Study this ruled surface (known as Cayley's ruled cubic, cf. §5.8.23).*

10.5. Complements and exercises

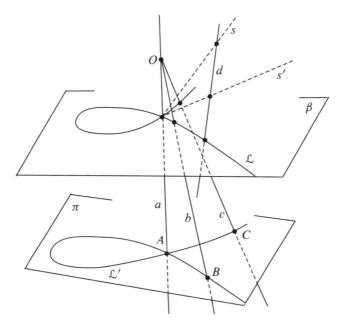

Figure 10.5

Let a system of projective coordinates be given in the plane with P as the point $[1, 0, 0]$ and r as the line $u_2 = 0$, and suppose that the conics of Σ that pass through the point $[1, a, 0]$ are tangent at P to the line $u_2 + au_1 = 0$ (that is, we suppose that to the generic point $[1, a, 0]$ of r there corresponds in ω the line $u_2 + au_1 = 0$ of the pencil with center P). One finds that Σ has the equation

$$\Sigma: \lambda_0(u_0u_2 + u_1^2) + \lambda_1 u_0 u_1 + \lambda_2 u_1 u_2 + \lambda_3 u_2^2 = 0,$$

and thus that the projective image $\mathcal{F}$ has the parametric representation

$$x_0 : x_1 : x_2 : x_3 = u_0 u_2 + u_1^2 : u_0 u_1 : u_1 u_2 : u_2^2$$

and the cartesian equation

$$x_2^3 + x_1 x_3^2 - x_0 x_2 x_3 = 0.$$

A generic plane that passes through the double line (which has equations $x_2 = x_3 = 0$) intersects this surface in three lines, two of which coincide with the double line; the only exception is the plane $x_3 = 0$ for which also the third line coincides with the double directrix (cf. Exercise 5.8.21).

10.5.18. *Let Σ be the linear system of curves of order 4 with two double base points. Show that its projective image is a surface $\mathcal{G}^8$ in S_8 projectively distinct from the surface $\mathcal{F}^8$ obtained in S_8 as the projection of the Del Pezzo $\mathcal{F}^9 \subset \mathbb{P}^9$ from one of its generic points.*

The two base points of Σ (which has degree and dimension 8) represent two exceptional conics on $\mathcal{G}^8$ and are the centers of two pencils of lines whose images are two pencils of conics lying on $\mathcal{G}^8$. By contrast, the projection of the Del Pezzo surface $\mathcal{F}^9 \subset S_9$ from one of its points O is a surface that contains only one pencil of conics that are the projections of the space cubics of $\mathcal{F}^9$ passing through O. Thus the two surfaces are projectively distinct. From this one has that the linear system Σ and the system of cubics with a base point are not mutually related by a Cremona transformation. In fact, Del Pezzo [73] proved that linearly normal surfaces of order n exist in S_n only for $n \leq 9$, and they are the surfaces $\mathcal{F}^9 \subset S_9$, their projections $\mathcal{F}^{9-d} \subset S_{9-d}$, and also the surface $\mathcal{G}^8$.

Note that projecting $\mathcal{G}^8$ from one its points onto a space S_7 one finds the same surface that is obtained by projecting $\mathcal{F}^9 \subset S_9$ from two of its points: indeed, the linear system of quartics with three base points, two double and one simple, is changed by the quadratic transformation having these three points as its fundamental points into the linear system of cubics with two base points (cf. Proposition–Definition 9.2.2).

If we use "$\longrightarrow$" to indicate the projection from a point of a surface onto a hyperplane, we have the following situation for linearly normal surfaces $\mathcal{F}^n$ in S_n:

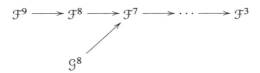

10.5.19. *Show that the surface $\mathcal{F}$ of order 4 in S_3 represented by a linear system Σ of cubics with five base points contains a double conic.*

The complete linear system Σ_0 of cubics passing through five points represents C. Segre's surface $\mathcal{G}^4 \subset S_4$. The system Σ is not complete, inasmuch as the surface $\mathcal{F}$ is embedded in S_3, and Σ is contained in Σ_0. The surface $\mathcal{F}$ is thus the projection of the surface $\mathcal{G}^4$ in S_4 from an external point O onto a hyperplane Π. As we have seen in Problem 10.5.7, the $\mathcal{G}^4$ is the complete intersection of two quadric hypersurfaces of S_4. In the pencil Φ of quadric hypersurfaces passing through $\mathcal{G}^4$ there is a quadric $\mathcal{Q}$ that passes through O. The space S_3 tangent at O to $\mathcal{Q}$ meets $\mathcal{Q}$ in a quadric cone whose generators are chords of $\mathcal{G}^4$ since they have two points in common with another quadric of Φ. The trace of this cone in Π is a double conic of the surface $\mathcal{F}$.

10.5.20. *Study the cubic surface $\mathcal{F}$ in S_3 represented on the plane π by the linear system of cubics that pass through the six vertices of a complete quadrilateral.*

10.5. Complements and exercises

Let r_1, r_2, r_3, r_4 be four lines in the plane π no three of which are concurrent in a point. These lines are fundamental lines for the linear system of cubics passing through the six points $P_{ij} := r_i \cap r_j$. To each of the lines there corresponds a point of $\mathcal{F}$, so that on $\mathcal{F}$ there are four fundamental points O_i (cf. §10.1.6). One sees immediately that these points are double for the surface. Indeed, two plane sections of $\mathcal{F}$ passing, for example, through the point O_1 corresponding to r_1 are represented on π by two cubics both of which contain the line r_1. The residual components are two conics that have in common a single point in addition to the three base points P_{23}, P_{34}, P_{42}. Hence, a generic line of S_3 passing through O_1 meets $\mathcal{F}$, away from O_1, in a single point; that is, O_1 is a double point for $\mathcal{F}$ (Figure 10.6).

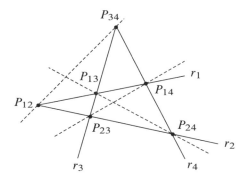

Figure 10.6

The diagonal trilateral of the quadrilateral $r_1 r_2 r_3 r_4$ represents the unique tritangent plane of $\mathcal{F}$. The lines composing the three sides of the diagonal trilateral (each having only one point in common with the curves of Σ away from the base points) are transformed into three lines of $\mathcal{F}$ contained in the tritangent plane; in addition to these lines, $\mathcal{F}$ also contains the six lines $r_{O_i O_j}$.

Taking as fundamental points of a projective reference system in S_3 the four double points, one finds for $\mathcal{F}$ the equation

$$\lambda_0 x_1 x_2 x_3 + \lambda_1 x_2 x_3 x_0 + \lambda_2 x_3 x_0 x_1 + \lambda_3 x_0 x_1 x_2 = 0 \quad (\lambda_0 \lambda_1 \lambda_2 \lambda_3 \neq 0),$$

which can be reduced to the form

$$x_1 x_2 x_3 + x_2 x_3 x_0 + x_3 x_0 x_1 + x_0 x_1 x_2 = 0.$$

Note that this $\mathcal{F}^3$ can also be represented by way of the system Σ' of conics tangent to three given lines in three given points. In fact we have seen in Example 10.1.10 (3) that Σ and Σ' are equivalent via a Cremona transformation.

10.5.21. *Describe the planar representation of a **monoid** (that is, a surface of order n with an $(n-1)$-fold point).*

We first observe that a monoid M of order n is embedded in S_3. If in fact M were embedded in a space S_r with $r \geq 4$, the space S_{r-2} joining the $(n-1)$-fold point with $r-2$ other generic points would have $n-1+r-2 \geq n+1$ intersections with M.

Let O be the $(n-1)$-fold point of the monoid M. A planar representation of M is obtained by projecting M from O onto a plane π. Since M contains the $n(n-1)$ lines that it has in common with the tangent cone at O, the linear system of plane sections of M is projected into a linear system Σ, of dimension 3 (since M is embedded in S_3), of curves of order n passing through the $n(n-1)$ points common to two curves of order $n-1$ and n. We may in fact suppose that $O = [1, 0, 0, 0]$ and thus that M has equation

$$x_0 \varphi_{n-1}(x_1, x_2, x_3) + \varphi_n(x_1, x_2, x_3) = 0,$$

where φ_{n-1}, φ_n are forms of order $n-1$ and n respectively. If π has equation $x_0 = 0$, then Σ is the system of curves of order n through the $n(n-1)$ points common to the two curves of orders $n-1$ and n with equations $\varphi_{n-1}(x_1, x_2, x_3) = x_0 = 0$ and $\varphi_n(x_1, x_2, x_3) = x_0 = 0$ respectively.

We note that the dimension of the (complete) linear system Φ of all curves of order n that pass through these points is 3 (and hence $\Phi = \Sigma$). Indeed, let $f = 0$ and $g = 0$ be the equations of the curves of orders n and $n-1$ whose intersections are the base points of Φ. Let $h = 0$ be the equation of a generic curve of Φ and let P be a point of the curve C with equation $g = 0$, distinct from the base points. The curve of the pencil $h - \mu f = 0$, $\mu \in K$, that passes through P splits into the curve C and a line r, and thus there is a value μ_0 of the parameter μ such that

$$h - \mu_0 f = g(\mu_0 u_0 + \mu_1 u_1 + \mu_2 u_2),$$

where $\mu_0 u_0 + \mu_1 u_1 + \mu_2 u_2 = 0$ is the equation of the line r. This means that every curve of Φ has equation of the form

$$h = \mu_0 f + g(\mu_0 u_0 + \mu_1 u_1 + \mu_2 u_2) = 0$$

and thus depends on three essential parameters μ_0, μ_1, μ_2.

Hence the system Σ is complete with respect to the base group and therefore the monoid is a linearly normal surface. Note that Σ is not regular if $n > 3$ (because $3 > \frac{n(n+3)}{2} - n(n-1)$ for $n > 3$).

10.5.22 (Clebsch's diagonal surface). *Let $\mathcal{F}$ be the cubic surface (known as **Clebsch's diagonal surface**) which is the projective image of the linear system Σ of curves of order 3 that pass through the vertices and the center of a regular pentagon. Recall that an Eckardt point on a cubic surface $\mathcal{F}$ is a point at which the tangent plane to*

$\mathcal{F}$ *has three lines of a pencil in common with the surface* (*cf.* §10.1.5). *Verify that* $\mathcal{F}$ *has ten Eckardt points.*

It suffices to note that if P is the point of the representative plane π to which there corresponds an Eckardt point, then through P there must pass three lines which together constitute a cubic of Σ. By observing the Figure 10.7 (where there is the pentagon $ABCDE$ with center the point indicated by "$\bigcirc$") one sees immediately that the ten points signed with the symbol "•" are Eckardt points.

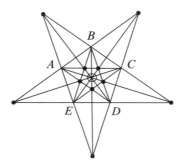

Figure 10.7

For another remarkable cubic surface endowed with Eckardt points see Exercise 13.1.36.

10.5.23. *Consider three planes passing through the three lines constituting the intersection of a general cubic* $\mathcal{F}$ *in* $\mathbb{P}^3$ *with a tritangent plane. Show that the resulting residual intersections* (*of the planes with* $\mathcal{F}$) *are three conics lying on a quadric.*

A tritangent plane intersects $\mathcal{F}$ in a cubic with three double points, and thus a cubic split into three lines. We represent $\mathcal{F}$ in a plane π via the complete linear system $\Sigma = C^3(P_1, P_2, P_3, P_4, P_5, P_6)$, and we consider the lines $r_{12} = P_1 P_2$, $r_{34} = P_3 P_4$, $r_{56} = P_5 P_6$ in π. Then to the cubic of Σ which splits into the lines r_{12}, r_{34}, r_{56} there corresponds on $\mathcal{F}$ the cubic (split into lines) which is a section of $\mathcal{F}$ by a tritangent plane (cf. §10.1.8).

Let γ_{12}, γ_{34}, and γ_{56} be three conics in π such that $r_{ij} \cup \gamma_{ij}$ corresponds to a plane section of $\mathcal{F}$ (that is, to the section of $\mathcal{F}$ by a plane through the line ℓ_{ij} which is the image of r_{ij}). Then the conic γ_{12} must pass through P_3, P_4, P_5, P_6, while γ_{34} must pass through P_1, P_2, P_5, P_6, and γ_{56} must pass through P_1, P_2, P_3, P_4. The sextic $\gamma_{12} \cup \gamma_{34} \cup \gamma_{56}$ passes doubly through the six points P_i and hence is the image of the section of $\mathcal{F}$ with a quadric.

In this way one finds all the quadric sections of $\mathcal{F}$ that are split into three lines.

10.5.24. *Prove that on a Veronese surface $\mathcal{F}$ there is a 9-dimensional linear system of non-singular elliptic curves.*

It suffices to observe that the non-singular elliptic curves on $\mathcal{F}$ are the images, under the plane representation of $\mathcal{F}$, of the non-singular elliptic curves of $\mathbb{P}^2$, namely the non-singular plane cubics. These form a linear system of dimension 9.

10.5.25 The Cremona birational transformations between projective spaces of dimension 3.

We now give a brief discussion of a method, first suggested by L. Cremona [29], [30], for determining (if they exist) all the homaloidal systems of surfaces in $\mathbb{P}^3$ that contain a given (rational) surface F_0 having a known representation over a plane π. Such systems furnish birational transformations $\theta \colon \mathbb{P}^3 \to \mathbb{P}^3$.

Let $F_0 \subset \mathbb{P}^3$ be a rational surface, with equation $F_0 = 0$, represented over a plane π (in which u_1, u_2, u_3 are homogeneous projective coordinates) by way of the linear system Σ (simple and without fixed part) consisting of curves of a given order n:

$$\lambda_0 f_0(u_1, u_2, u_3) + \lambda_1 f_1(u_1, u_2, u_3) + \lambda_2 f_2(u_1, u_2, u_3) + \lambda_3 f_3(u_1, u_2, u_3) = 0.$$

We know that the curves which are sections of F_0 by surfaces of order N are precisely those represented over π by the curves of the system $N\Sigma$, that is, by curves whose equation may be written in the form

$$F(f_0, f_1, f_2, f_3) = 0, \tag{10.13}$$

with $F(t_0, t_1, t_2, t_3)$ a homogeneous polynomial of degree N.

We assume that there exists a homaloidal system Ω of surfaces containing F_0, in such a way that F_0 is generic in Ω. The section curves of F_0 by the other surfaces of Ω form a net (of curves of F_0) to which there corresponds in π a net of curves of order n^2 each of which contains as a component a curve which comes from the base locus of Ω in π. The variable components (namely, the characteristic curves of F_0) form a homaloidal net of curves of order $\leq n^2$.

Given these premises, let there be given in π an arbitrary homaloidal net

$$\mathcal{R} \colon \mu_1 g_1(u_1, u_2, u_3) + \mu_2 g_2(u_1, u_2, u_3) + \mu_3 g_3(u_1, u_2, u_3) = 0 \tag{10.14}$$

of curves of order $\leq n^2$ as well as a curve h of equation $h(u_1, u_2, u_3) = 0$ such that the three polynomials $h(u_1, u_2, u_3) g_i(u_1, u_2, u_3)$ may be written in the form (10.13). The curves of F_0 represented in π by the curves

$$h(u_1, u_2, u_3) g_i(u_1, u_2, u_3) = F_i(f_0, f_1, f_2, f_3) = 0, \quad i = 1, 2, 3,$$

are the sections of F_0 by the three surfaces $F_i(x_0, x_1, x_2, x_3) = 0$, and

$$\mu_0 F_0 + \mu_1 F_1 + \mu_2 F_2 + \mu_3 F_3 = 0$$

is a homaloidal system containing the given surface F_0.

In the special case $n = 2$, let us see how to construct the homaloidal systems of non-singular quadrics of $\mathbb{P}^3$ described in Section 9.4.

We represent a quadric F_0 in the plane π by the 3-dimensional system $\Sigma = C^2(A, B)$ of conics with two base points A, B (infinitely near or distinct according to whether F_0 is or is not a cone, cf. Section 7.3), and let $\sigma : \pi \to F_0$ be the birational map defined by Σ. Note that the line $\ell = \langle A, B \rangle$ is the fundamental curve of Σ, to which there corresponds on F_0 the fundamental point O (see §10.1.6). The homaloidal nets $\mathcal{R}$, of curves of degree ≤ 4, which together with a fixed curve give quartics belonging to 2Σ (i.e., having A, B as two-fold points) are the following:

i) the net of lines;

ii) the net $C^2(A, B, P)$ of the conics through A, B, and through a third point P;

iii) the net $C^2(A, P, Q)$ of the conics passing through A and two points P, Q different from B;

iv) the net $C^2(P, Q, R)$ of the conics with three base points P, Q, R different from A and B;

v) the net $C^3(A^2, B, P, Q, R)$ of the cubics passing through A with multiplicity two and simply through the points B, P, Q, R;

vi) the net $C^4(A^2, B^2, H^2, P, Q, R)$ of the quartics with three two-fold base points A, B, H and three simple base points P, Q, R.

In view of Proposition–Definition 9.2.2, one sees that via a quadratic transformation the cases ii), v), vi) can be reduced to the cases i), iii), iv) respectively. Therefore we need only consider the cases i), iii), iv).

Let us consider case i). If to the net $\mathcal{R}$ of lines of π we add, as fixed components, the line $\ell = \langle A, B \rangle$ and a conic γ through A and B, we obtain a net of quartics belonging to 2Σ (with the notation as above, $h(u_1, u_2, u_3) = 0$ is the defining equation of the cubic $h = \ell \cup \gamma$). To the curves of $h\mathcal{R}$ there correspond the curves of a net cut out on F_0 by a net

$$\Omega_0 : \lambda_1 F_1 + \lambda_2 F_2 + \lambda_3 F_3 = 0$$

of quadrics. Then we get the homaloidal system $\Omega : \lambda_0 F_0 + \Omega_0$ having as base locus the fundamental point $O = \sigma(\ell)$ and the conic $\sigma(\gamma)$ (cf. (10.5)). Thus Ω defines a Cremona transformation of type $(2, 2)$.

In case iii), the net $C^2(A, P, Q)$, completed with the line $\ell = \langle A, B \rangle$ and a further line r through B, gives a net of quartics belonging to 2Σ. The procedure described above now leads to a homaloidal system Ω of quadrics whose base variety is constituted by the three points $\sigma(A)$, $\sigma(P)$, $\sigma(Q)$, and the line $\sigma(r)$. Thus Ω defines a Cremona transformation of type $(2, 3)$.

In case iv), the net $C^2(P, Q, R)$ completed with the line $\ell = \langle A, B \rangle$ counted twice, leads to a homaloidal system Ω of quadrics passing through the four points $\sigma(P)$, $\sigma(Q)$, $\sigma(R)$, O, and having at O the fixed tangent plane joining the lines a, b corresponding to the two base points A, B (cf. §10.1.7). Thus Ω defines a Cremona transformation of type $(2, 4)$.

Example 10.5.26. In order to give a better illustration of Cremona's procedure, we show how one can effectively obtain the equations of a birational transformation starting from a surface F_0 of the corresponding homaloidal system.

With the preceding notations we consider the quadric $F_0 = x_0 x_3 - x_1 x_2 = 0$, having parametric equations $x_i = f_i(u_1, u_2, u_3)$, $i = 1, 2, 3$, where

$$f_0 = u_1 u_2, \quad f_1 = u_1 u_3, \quad f_2 = u_2 u_3, \quad f_3 = u_3^2,$$

so that the quadric is represented over the plane $\pi = \mathbb{P}^2_{[u_1, u_2, u_3]}$ by the linear system

$$\Sigma \colon \lambda_0 u_1 u_2 + \lambda_1 u_1 u_3 + \lambda_2 u_2 u_3 + \lambda_3 u_3^2 = 0.$$

As our homaloidal net in π we take the net of lines

$$\mathcal{R} \colon \lambda_1 u_1 + \lambda_2 u_2 + \lambda_3 u_3 = 0.$$

If we wish to adjoin a common curve h to all the lines of $\mathcal{R}$ so as to obtain curves contained in 2Σ, that is, quartics passing doubly through A and B, we must choose h to be a cubic passing doubly through both A and B, and thus splitting into the line $\ell = \langle A, B \rangle$ (which has equation $u_3 = 0$) and a conic γ passing through A and B but otherwise arbitrary. The possible choices of γ lead to the possible projectively distinct types of linear systems Ω of quadrics passing through a point and containing a conic.

For example, suppose that γ is the line $\langle A, B \rangle$ counted twice, so that the curve $h = \ell \cup \gamma$ has equation $u_3^3 = 0$. One obtains the net

$$h\mathcal{R} \colon \lambda_1 u_1 u_3^3 + \lambda_2 u_2 u_3^3 + \lambda_3 u_3^4 = 0$$

(of quartics belonging to 2Σ) which may be written in the form

$$h\mathcal{R} \colon \lambda_1 f_1 f_3 + \lambda_2 f_2 f_3 + \lambda_3 f_3^2 = 0.$$

To it there corresponds the net cut out on F_0 by the net of quadrics

$$\Omega_0 \colon \lambda_1 x_1 x_3 + \lambda_2 x_2 x_3 + \lambda_3 x_3^2 = 0.$$

Thus we have the homaloidal system

$$\lambda_0 F_0 + \Omega_0 = \lambda_0 (x_0 x_3 - x_1 x_2) + \lambda_1 x_1 x_3 + \lambda_2 x_2 x_3 + \lambda_3 x_3^2 = 0, \quad (10.15)$$

which defines the Cremona transformation $\theta: S_3 \to S'_3$ of type (2,2) expressed analytically by

$$\begin{cases} y_0 = x_0 x_3 - x_1 x_2, \\ y_1 = x_1 x_3, \\ y_2 = x_2 x_3, \\ y_3 = x_3^2. \end{cases}$$

These formulas may be inverted as follows:

$$\begin{cases} x_0 = y_0 y_3 + y_1 y_2, \\ x_1 = y_1 y_3, \\ x_2 = y_2 y_3, \\ x_3 = y_3^2. \end{cases}$$

We note that the homaloidal system (10.15) obtained with this choice of the conic γ is that of the quadrics tangent at $A_0 = [1, 0, 0, 0]$ to the plane $x_3 = 0$ and having as common component the degenerate conic $x_3 = x_1 x_2 = 0$.

Chapter 11
Segre Varieties

If $S_p, S_q, \ldots$ are k linear spaces, it is useful to consider the k-tuples $(P_1, P_2, \ldots, P_k)$ of points taken respectively in $S_p, S_q, \ldots$ (that is, the elements of $S_p \times S_q \times \cdots$) as points of a variety W of dimension $p + q + \cdots$. In this chapter we shall study a simple projective model for W called the *Segre variety* to honor C. Segre, who introduced it in [85].

In Section 11.1 we study the case of two lines, which gives rise to W which is a quadric in $\mathbb{P}^3$. Then, in Sections 11.2 and 11.3 we consider the product of two projective spaces of arbitrary dimensions. The variety which arises is a special case of a determinantal variety. Indeed, its associated ideal is generated by second order minors of a suitable matrix. The extension to the case of several spaces, sketched in § 11.3.1, does not present new conceptual difficulties.

The chapter ends with some applications of the theory and, in Section 11.4, with some examples and illustrative exercises.

11.1 The product of two projective lines

Let r and s be two projective lines and on each of them let a homogeneous coordinate system be given: x_0, x_1 on r and y_0, y_1 on s. To the pair of points $A = [x_0, x_1] \in r$, $B = [y_0, y_1] \in s$ we associate the point P of $\mathbb{P}^3$ having coordinates

$$X_{00} = x_0 y_0, \quad X_{01} = x_0 y_1, \quad X_{10} = x_1 y_0, \quad X_{11} = x_1 y_1. \tag{11.1}$$

This point does not depend on the choice of coordinates of A and B: indeed, if x_0, x_1 are multiplied by a non-zero factor ρ and y_0, y_1 by a non-zero factor ρ', the coordinates X_{ij} are modified by the non-zero factor $\rho\rho'$. The equations

$$\frac{x_1}{x_0} = \frac{X_{10}}{X_{00}} = \frac{X_{11}}{X_{01}}, \quad \frac{y_1}{y_0} = \frac{X_{11}}{X_{10}} = \frac{X_{01}}{X_{00}},$$

which follow from equations (11.1), show that the point P determines the pair of points (A, B).

The locus described by P when A and B run over the line r and the line s respectively is the quadric $\mathcal{Q}$ with equation

$$X_{00} X_{11} - X_{01} X_{10} = 0,$$

the points of which correspond bijectively to the elements $(A, B) \in \mathbb{P}^1 \times \mathbb{P}^1$. Thus the quadric $\mathcal{Q}$ is a convenient projective model for the product of two projective lines. It has a parametric representation given by the equations (11.1).

Putting $x_0 = y_0 = z_0$ and $y_1 = z_1$, $x_1 = z_2$ in equations (11.1), one finds the representation

$$X_{00} = z_0^2, \quad X_{01} = z_0 z_1, \quad X_{10} = z_0 z_2, \quad X_{11} = z_1 z_2$$

of $\mathcal{Q}$ over a plane π (of equation $aX_{00} + bX_{01} + cX_{10} + dX_{11} = 0$) where z_0, z_1, z_2 are homogeneous projective coordinates (cf. Example 2.6.4). To the plane sections of $\mathcal{Q}$ there correspond on π the conics with equations

$$az_0^2 + bz_0 z_1 + cz_0 z_2 + dz_1 z_2 = 0, \quad a,b,c,d \in K,$$

that pass through the points $[0, 0, 1]$, and $[0, 1, 0]$.

To the pairs (A, B) with A fixed there correspond the points of the line r_x with equations

$$\frac{x_1}{x_0} = \frac{X_{10}}{X_{00}} = \frac{X_{11}}{X_{01}},$$

which describes one of the rulings of $\mathcal{Q}$ as A varies. We denote that ruling by $\{r_x\}$, and similarly to the pairs (A, B) with fixed B there correspond the points of the line with equations

$$\frac{y_1}{y_0} = \frac{X_{11}}{X_{10}} = \frac{X_{01}}{X_{00}},$$

which describes the other ruling $\{r_y\}$ as B varies. A line in $\{r_x\}$ is represented on $\mathcal{Q}$ by an equation of the form $ax_0 + bx_1 = 0$; a line of $\{r_y\}$ is represented on $\mathcal{Q}$ by an equation of the form $cy_0 + dy_1 = 0$, with $a, b, c, d \in K$.

More generally, a bihomogeneous equation of bidegree (β, α), that is, homogeneous of degree β with respect to x_0, x_1 and of degree α with respect to y_0, y_1,

$$G(x_0, x_1; y_0, y_1) = 0,$$

represents a curve $\mathcal{L}$ of type (α, β) on $\mathcal{Q}$. This curve has α points in common with a generic line r_x and β points in common with a generic line r_y.

Suppose that $\beta = \alpha + q$ ($q \geq 0$). To prove that $\mathcal{L}$ is an algebraic curve of $\mathbb{P}^3$ it is enough to find a homogeneal ideal $\mathfrak{a}$ in $K[X_{00}, X_{01}, X_{10}, X_{11}]$ such that $\mathcal{L}$ is the locus of zeros of $\mathfrak{a}$. To this end, consider the $q + 1$ polynomials

$$G_h(x_0, x_1; y_0, y_1) = y_0^h y_1^{q-h} G(x_0, x_1; y_0, y_1), \quad h = 0, 1, \ldots, q,$$

bihomogeneous of the same degree β with respect to both pairs of variables. They do not have common zeros other than zeros of G, and may be written as polynomials in the four variables $x_i y_j$. If we set

$$G_h(x_0, x_1; y_0, y_1) = F_h(x_0 y_0, x_0 y_1, x_1 y_0, x_1 y_1), \quad h = 0, 1, \ldots, q,$$

we see that $\mathcal{L}$ is the locus of the points that annul the $q + 2$ polynomials

$$X_{00} X_{11} - X_{01} X_{10}, \quad F_h(X_{00}, X_{01}, X_{10}, X_{11}), \quad h = 0, 1, \ldots, q.$$

which means that $\mathcal{L} = V(\mathfrak{a})$ where $\mathfrak{a} = (X_{00}X_{11} - X_{01}X_{10}, F_0, \ldots, F_q)$. A geometric argument shows that $\mathcal{L}$ is the set-theoretic complete intersection of the three surfaces with equations (cf., for the case $q = 2$, the last remark in Example 11.1.2)

$$X_{00}X_{11} - X_{01}X_{10} = 0, \quad F_0(X_{00}, X_{01}, X_{10}, X_{11}) = 0,$$
$$F_q(X_{00}, X_{01}, X_{10}, X_{11}) = 0,$$

that is, $\mathcal{L}$ is the locus of the points common to the three surfaces.

Note that if $q = 1$ these are all the generators of $\mathfrak{a}$; and we have that only when $\beta = \alpha + 1$ the ideal $\mathfrak{a}$ is generated by three polynomials (and not less).

If $q = 0$, i.e., $\alpha = \beta$, $\mathcal{L}$ is the complete intersection of the quadric with another surface and thus $\mathfrak{a}$ is generated by two polynomials.

Example 11.1.1. With the preceding notations, suppose that the equation of $\mathcal{L}$ on $\mathcal{Q}$ is $G = x_0^2 y_1 - x_1^2 y_0 = 0$. One then has $\alpha = q = 1$, $\beta = 2$ and

$$G_1 = y_0 G = y_0 x_0^2 y_1 - y_0 x_1^2 y_0 = (x_0 y_0)(x_0 y_1) - (x_1 y_0)^2,$$
$$G_0 = y_1 G = y_1 x_0^2 y_1 - y_1 x_1^2 y_0 = (x_0 y_1)^2 - (x_1 y_0)(x_1 y_1).$$

Therefore $\mathcal{L}$ is the curve of $\mathbb{P}^3$ which is the locus of the zeros of the ideal $\mathfrak{a}$ generated by the three polynomials

$$X_{00}X_{11} - X_{01}X_{10}, \quad X_{00}X_{01} - X_{10}^2, \quad X_{01}^2 - X_{10}X_{11},$$

that is, by the second order minors of the matrix

$$\begin{pmatrix} X_{00} & X_{10} & X_{01} \\ X_{10} & X_{01} & X_{11} \end{pmatrix}.$$

The ideal $\mathfrak{a}$ is the kernel of the homomorphism $f : K[X_{00}, X_{01}, X_{10}, X_{11}] \to K[t, u]$ defined by $X_{00} \mapsto t^3$, $X_{01} \mapsto tu^2$, $X_{10} \mapsto t^2 u$, $X_{11} \mapsto u^3$ (the inclusion $\mathfrak{a} \subset \ker(f)$ is obvious; the converse inclusion is seen via a standard argument of polynomial algebra which we have already used in Exercise 3.4.11 (2)). It follows that $\mathfrak{a} = I(\mathcal{L})$ is a prime ideal and the curve $\mathcal{L}$ is irreducible. The curve $\mathcal{L}$ is the twisted cubic, the locus of the points

$$\begin{cases} X_{00} = t^3, \\ X_{01} = tu^2, \\ X_{10} = t^2 u, \\ X_{11} = u^3. \end{cases}$$

Example 11.1.2. Let $\mathcal{L}$ be the curve represented on $\mathcal{Q}$ by the equation $G = x_0^3 y_1 - x_1^3 y_0 = 0$. We have $(\alpha, \beta) = (1, 3), q = 2$ and

$$G_0 = y_1^2(x_0^3 y_1 - x_1^3 y_0) = (x_0 y_1)^3 - (x_1 y_1)^2(x_1 y_0),$$
$$G_1 = y_0 y_1 (x_0^3 y_1 - x_1^3 y_0) = (x_0 y_0)(x_0 y_1)^2 - (x_1 y_1)(x_1 y_0)^2,$$
$$G_2 = y_0^2(x_0^3 y_1 - x_1^3 y_0) = (x_0 y_0)^2(x_0 y_1) - (x_1 y_0)^3.$$

Therefore, if

$$\mathfrak{a} := (X_{00} X_{11} - X_{01} X_{10}, X_{01}^3 - X_{11}^2 X_{10}, X_{00} X_{01}^2 - X_{11} X_{10}^2, X_{00}^2 X_{01} - X_{10}^3),$$

one has $\mathcal{L} = V(\mathfrak{a})$. But to obtain $\mathcal{L}$ as a complete intersection of three surfaces it suffices to take

$$f := X_{00} X_{11} - X_{01} X_{10}, \quad g := X_{01}^3 - X_{11}^2 X_{10}, \quad h := X_{00}^2 X_{01} - X_{10}^3,$$

since

$$gh + h X_{11}^2 X_{10} - g X_{10}^3 - 2 f X_{10}^2 X_{01}^2 = (X_{00} X_{01}^2 - X_{11} X_{10}^2)^2.$$

And we have $\mathcal{L} = V(\mathfrak{b})$ with $\mathfrak{b} = (f, g, h)$.

Note that the polynomial $F = X_{00} X_{01}^2 - X_{11} X_{10}^2$ although belonging to $I(\mathcal{L}) = \sqrt{\mathfrak{b}}$ does not belong to $\mathfrak{b}$.

Observe that the surface $X_{01}^3 - X_{11}^2 X_{10} = 0$ passes doubly through the line r with equations $X_{01} = X_{11} = 0$ whence it intersects $\mathcal{Q}$ in a curve that splits into $\mathcal{L}$ and the line r counted two times. Similarly, the surface $X_{00}^2 X_{01} - X_{10}^3 = 0$ intersects $\mathcal{Q}$ outside of $\mathcal{L}$ in the line r' with equations $X_{00} = X_{10} = 0$ counted twice. The two lines r and r' are generators of the same ruling of $\mathcal{Q}$ and thus are skew; therefore the zeros of $\mathfrak{b}$ are precisely the points of $\mathcal{L}$ (and only those). This provides a geometric version of the proof that $\mathcal{L} = V(\mathfrak{b})$.

Finally, a parametric representation of $\mathcal{L}$ is given by

$$X_{00} : X_{01} : X_{10} : X_{11} = t^4 : tu^3 : t^3 u : u^4.$$

11.2 Segre morphism and Segre varieties

What has been seen in Section 11.1 was generalized by C. Segre who introduced the varieties, called Segre varieties, that represent h-tuples of points taken from h projective spaces.

Let $\mathbb{P}^n$ and $\mathbb{P}^m$ be two projective spaces, with homogeneous coordinates $X_0, X_1, \ldots, X_n$ in $\mathbb{P}^n$ and $Y_0, Y_1, \ldots, Y_m$ in $\mathbb{P}^m$. Together with them consider the projective space $\mathbb{P}^N$ of dimension

$$N = (m+1)(n+1) - 1 = mn + m + n,$$

in which we fix a homogeneous coordinate system $Z_{ij}, i = 0,\ldots,n, j = 0,\ldots,m$. We will call the map

$$\varphi: \mathbb{P}^n \times \mathbb{P}^m \to \mathbb{P}^N$$

under which the pair of points $x = [x_0, x_1, \ldots, x_n]$, $y = [y_0, y_1, \ldots, y_m]$ is mapped to the point with coordinates

$$z_{ij} = x_i y_j \tag{11.2}$$

the *Segre morphism*. The morphism φ is well defined (and it is clearly injective). Indeed, on altering all the x_i by a common factor $\rho \neq 0$ and altering all the y_i by a common factor $\rho' \neq 0$, the z_{ij} are all changed by the common factor $\rho\rho' \neq 0$.

In $\mathbb{P}^N$ we consider the Zariski topology, and on $\mathbb{P}^n \times \mathbb{P}^m$ (instead of the product topology with respect to the Zariski topology of $\mathbb{P}^n$ and $\mathbb{P}^m$, cf. §2.1.3), we consider the coarsest topology that renders the Segre morphism $\varphi: \mathbb{P}^n \times \mathbb{P}^m \to \mathbb{P}^N$ continuous. We call this topology the *Zariski topology* on the product. The closed subsets of $\mathbb{P}^n \times \mathbb{P}^m$ are thus the inverse images of the closed sets in $\mathbb{P}^N$. They are defined by bihomogeneous equations, that is, equations homogeneous with respect to each series of variables.

The image of φ is the intersection of the quadrics with equations

$$Z_{ij} Z_{kl} - Z_{kj} Z_{il} = 0. \tag{11.3}$$

In fact it is clear that every point $z \in \varphi(\mathbb{P}^n \times \mathbb{P}^m)$ satisfies these equations. Conversely, if the point $z = [\ldots, z_{ij}, \ldots]$ satisfies equations (11.3) and if, for example, $z_{00} \neq 0$, from the relations $z_{ij} z_{00} - z_{0j} z_{i0} = 0$ it is evident that the points $x = [z_{00}, z_{10}, \ldots, z_{n0}] \in \mathbb{P}^n$ and $y = [z_{00}, z_{01}, \ldots, z_{0m}] \in \mathbb{P}^m$ are determined by z and are such that $\varphi(x, y) = z$ (indeed, $\varphi(x, y) = [\ldots, z_{i0} z_{0j}, \ldots] = [\ldots, z_{ij} z_{00}, \ldots] = z$).

From this it follows that $W := \varphi(\mathbb{P}^n \times \mathbb{P}^m)$ is an algebraic variety of $\mathbb{P}^N$; we say that W is the *Segre variety* of the product of the two projective spaces $\mathbb{P}^n \times \mathbb{P}^m$. The variety W is a projective model for the product space $\mathbb{P}^n \times \mathbb{P}^m$. It has a parametric representation

$$Z_{ij} = u_i v_j, \quad i = 0, \ldots, n, \; j = 0, \ldots, m. \tag{11.4}$$

Since the products $u_i v_j$ (with u_i, v_j indeterminates) are linearly independent over the base field K, the variety W does not belong to any hyperplane of $\mathbb{P}^N$, that is, it has $\mathbb{P}^N$ as its embedding space.

If in equations (11.4) we fix the u_i (or better, their ratios, which means a point $x = [u_0, u_1, \ldots, u_n]$ in $\mathbb{P}^n$) we find the parametric representation of a linear space $L_m = \varphi(x \times \mathbb{P}^m)$ of dimension m. This is the space that joins the points

$$P_0 = [u_0, u_1, \ldots, u_n, 0, 0, \ldots, 0, \ldots, 0, 0, \ldots, 0],$$
$$P_1 = [0, 0, \ldots, 0, u_0, u_1, \ldots, u_n, \ldots, 0, 0, \ldots, 0],$$
$$\vdots$$
$$P_m = [0, 0, \ldots, 0, 0, 0, \ldots, 0, \ldots, u_0, u_1, \ldots, u_n].$$

11.2. Segre morphism and Segre varieties

Therefore W possesses an m-dimensional system of n-dimensional linear projective spaces L_n, and similarly it also has an n-dimensional system of m-dimensional spaces L_m. For each point $\varphi(x \times y)$ of W there passes a unique space of each of the two systems: the space $L_m = \varphi(x \times \mathbb{P}^m)$ and the space $L_n = \varphi(\mathbb{P}^n \times y)$. Two spaces of the same system are skew; however, a space L_m and a space L_n have a point in common.

The morphisms $p_1 \colon W \to \mathbb{P}^n$ and $p_2 \colon W \to \mathbb{P}^m$ defined by

$$p_1(z) = [x_0, x_1, \ldots, x_n], \quad p_2(z) = [y_0, y_1, \ldots, y_m],$$

are called the *projections of W* respectively on the first and second factors of the product. The fibers of p_1 and p_2 are the spaces L_m and L_n respectively.

If

$$G(x; y) = G(x_0, x_1, \ldots, x_n; y_0, y_1, \ldots, y_m)$$

is a bihomogeneous polynomial of bidegree (β, α), which means homogeneous of degree β in the x_i and degree α in the y_j, then the equation $G(x; y) = 0$ defines a subvariety Γ of W which has codimension 1, and which intersects the generic linear projective space L_m in a hypersurface of order α, and the generic linear projective space L_n in a hypersurface of order β. Let $\beta = \alpha + q$ with $q \geq 0$. To obtain a system of equations for Γ as algebraic subvariety of $\mathbb{P}^N$, it suffices first to write the $m+1$ polynomials $y_h^q G(x; y)$, $h = 0, 1, \ldots, m$ (which are bihomogeneous of degree β with respect to both series of variables), as polynomials $F_h(\ldots, x_i y_j, \ldots)$ in the indeterminates $x_i y_j$ and then to adjoin the $m+1$ equations $F_h(\ldots, Z_{ij}, \ldots) = 0$ to the equations $Z_{ij} Z_{kl} - Z_{kj} Z_{il} = 0$ of W. The system of equations so obtained for the $m+n-1$ dimensional subvariety Γ is not necessarily minimal.

Consider two principal affine charts in $\mathbb{P}^n$ and $\mathbb{P}^m$, for example $U_0^n \subset \mathbb{P}^n$ defined by $X_0 \neq 0$ and $U_0^m \subset \mathbb{P}^m$ defined by $Y_0 \neq 0$; and also the chart U_{00} in $\mathbb{P}^N$ defined by $Z_{00} \neq 0$. We have

$$W_{00} := \varphi(U_0^n \times U_0^m) = U_{00} \cap \varphi(\mathbb{P}^n \times \mathbb{P}^m),$$

whence $\varphi(U_0^n \times U_0^m)$ is an open subset of W.

On introducing affine coordinates $T_i = \frac{X_i}{X_0}$ in U_0^n and $T_j' = \frac{Y_j}{Y_0}$ in U_0^m together with $Z_{ij}' = \frac{Z_{ij}}{Z_{00}}$ in U_{00}, we see that the morphism $\theta \colon \mathbb{A}^n \times \mathbb{A}^m \to W_{00}$ defined by setting

$$\theta(t_1, t_2, \ldots, t_n, t_1', t_2', \ldots, t_m')$$
$$:= (1, t_1', t_2', \ldots, t_m', t_1, t_1 t_1', t_1 t_2', \ldots, t_1 t_m', \ldots, t_n, t_n t_1', t_n t_2', \ldots, t_n t_m')$$

is an isomorphism $\mathbb{A}^{n+m} \cong \mathbb{A}^n \times \mathbb{A}^m \cong W_{00}$. Hence W is an irreducible variety of dimension $m+n$. It is not difficult to show that (cf. for example [48, 18.15])

$$\deg(W) = \frac{(m+n)!}{m! \, n!}.$$

Note that W can also be defined as the locus of points of $\mathbb{P}^N$ that give rank 1 to the matrix

$$(Z_{ij}) = \begin{pmatrix} Z_{00} & Z_{01} & Z_{02} & \cdots & Z_{0m} \\ Z_{10} & Z_{11} & Z_{12} & \cdots & Z_{1m} \\ \vdots & \vdots & \vdots & & \vdots \\ Z_{n0} & Z_{n1} & Z_{n2} & \cdots & Z_{nm} \end{pmatrix}$$

and one can show that the homogeneous ideal $I(W)$ is generated by the second order minors of this matrix (see, for instance, [48, Lecture 2]). The ideal $I(W)$ is prime, in agreement with the fact that W is irreducible.

11.2.1 Diagonal subvariety. In the particular case of the product $\mathbb{P}^n \times \mathbb{P}^n$ ($n = m$) we can consider the *diagonal subvariety* of the Segre variety $W \subset \mathbb{P}^N$ (where $N = n(n+2)$), namely the locus Δ of the points of the type $\varphi(x, x)$. The diagonal is the intersection of W with the subspace, of dimension $\binom{n+2}{2} - 1$, which is the intersection of the $\binom{n+1}{2}$ independent hyperplanes with equations

$$Z_{ij} - Z_{ji} = 0 \tag{11.5}$$

(note that $n(n+2) - \binom{n+1}{2} = \binom{n+2}{2} - 1$). Eliminating the coordinates that are located beneath the principal diagonal in the square matrix (Z_{ij}) from the equations for W and from (11.5), we obtain the projection Δ_0 of Δ onto a space S_r of dimension $r = \binom{n+2}{2} - 1$ from a space of dual dimension and skew to S_r. In S_r we can interpret Z_{ij} with $i \neq j$ as coordinates, and we see that Δ_0 is the Veronese variety which is the locus of the points of $\mathbb{P}^r$ which give rank 1 to the symmetric matrix (Z_{ij}) (see Section 6.7 and also Example 10.2.1 as well as Section 10.4 for definitions and further details on Veronese varieties). For example, if $n = 2$ so that $r = 5$, Δ_0 (and thus also Δ) is a Veronese surface.

11.3 Segre product of varieties

If $V \subset \mathbb{P}^n$ and $V' \subset \mathbb{P}^m$ are two projective varieties and $\varphi \colon \mathbb{P}^n \times \mathbb{P}^m \to \mathbb{P}^N$ is the Segre morphism, then $\varphi(V \times V')$ is a projective variety that is called the *Segre product* of the two varieties.

Let $I(V) = (f_1, f_2, \ldots)$ in $K[X_0, \ldots, X_n]$ and $I(V') = (g_1, g_2, \ldots)$ in $K[Y_0, \ldots, Y_m]$ be the homogeneous ideals of the polynomials vanishing respectively on V and V'.

A (not necessarily minimal) system of equations for $\varphi(V \times V')$ as a subvariety of $\mathbb{P}^N$ is obtained by taking the second order minors of the matrix (Z_{ij}) together with the homogeneous polynomials from which the polynomials

$$f_s^{(j)} = Y_j^{\deg(f_s)} f_s, \quad g_t^{(i)} = X_i^{\deg(g_t)} g_t, \quad i = 0, \ldots, n, \ j = 0, \ldots, m,$$

arise by way of replacement of the indeterminates Z_{ij} with the products $X_i Y_j$.

Note that these polynomials are bihomogeneous of the same degree with respect to both series of variables.

11.3.1 The Segre variety $\mathbb{P}^{n_1} \times \mathbb{P}^{n_2} \times \cdots \times \mathbb{P}^{n_r}$. The preceding considerations may be extended to the case of several projective spaces $\mathbb{P}^{n_1}, \mathbb{P}^{n_2}, \ldots, \mathbb{P}^{n_r}$.

Let $X_0^{(h)}, X_1^{(h)}, \ldots, X_{n_h}^{(h)}$ be homogeneous projective coordinates in $\mathbb{P}^{n_h}$, $h = 1, \ldots, r$. Set

$$N = (n_1 + 1)(n_2 + 1) \ldots (n_r + 1) - 1,$$

and consider the projective space $\mathbb{P}^N$ with coordinates $X_{i_1 i_2 \ldots i_r}$ ($i_h \in \{0, \ldots, n_h\}$, $h = 1, \ldots, r$) and the map

$$\varphi : \mathbb{P}^{n_1} \times \mathbb{P}^{n_2} \times \cdots \times \mathbb{P}^{n_r} \to \mathbb{P}^N$$

defined by

$$\varphi(x^{(1)}, x^{(2)}, \ldots, x^{(r)}) = [\ldots, x_{i_1 i_2 \ldots i_r}, \ldots] \quad \text{with } x_{i_1 i_2 \ldots i_r} = x_{i_1}^{(1)} x_{i_2}^{(2)} \ldots x_{i_r}^{(r)},$$

where $x^{(h)} = (x_0^{(h)}, x_1^{(h)}, \ldots, x_{n_h}^{(h)})$, $h = 1, \ldots, r$. An argument similar to that used in the case of the product of two projective spaces shows that $\varphi(\mathbb{P}^{n_1} \times \mathbb{P}^{n_2} \times \cdots \times \mathbb{P}^{n_r})$ is an algebraic variety in $\mathbb{P}^N$. It is called the *Segre variety of the product* $\mathbb{P}^{n_1} \times \mathbb{P}^{n_2} \times \cdots \times \mathbb{P}^{n_r}$, for which it is a convenient projective model.

As in the case $r = 2$, this variety is the locus of linear spaces; more precisely, if $d = n_1 + n_2 + \cdots + n_r$, it has r systems, of dimensions $d - n_h$, of linear spaces L_{n_h}, $h = 1, \ldots, r$. We have

$$d = \dim(X) = \sum_{h=1}^{r} n_h, \quad \deg(X) = \frac{(n_1 + n_2 + \cdots + n_r)!}{n_1! n_2! \ldots n_r!}.$$

Example 11.3.2. In the case of the product of two projective planes with coordinates respectively x_0, x_1, x_2 and y_0, y_1, y_2, the Segre morphism is given by

$$\varphi([x_0, x_1, x_2], [y_0, y_1, y_2])$$
$$= [x_0 y_0, x_0 y_1, x_0 y_2, x_1 y_0, x_1 y_1, x_1 y_2, x_2 y_0, x_2 y_1, x_2 y_2],$$

and $\varphi(\mathbb{P}^2 \times \mathbb{P}^2)$ is a variety of dimension 4 and order 6 embedded in $\mathbb{P}^8$.

In the case of the product $\mathbb{P}^2 \times \mathbb{P}^2 \times \mathbb{P}^1$, with respective coordinates x_0, x_1, x_2, y_0, y_1, y_2 and z_0, z_1, we have

$$\varphi([x_0, x_1, x_2], [y_0, y_1, y_2], [z_0, z_1])$$
$$= [x_0 y_0 z_0, x_0 y_1 z_0, \ldots, x_i y_j z_k, \ldots, x_2 y_1 z_1, x_2 y_2 z_1]$$

for $i, j = 0, 1, 2$, $k = 0, 1$, and $\varphi(\mathbb{P}^2 \times \mathbb{P}^2 \times \mathbb{P}^1)$ is a variety of dimension 5 and order 30 embedded in $\mathbb{P}^{17}$.

When $n_1 = n_2 = \cdots = n_r = n$, that is, if the Segre variety is the projective model of the product of r copies of $\mathbb{P}^n$, we can consider the diagonal Δ, the locus of the points of the type $\varphi(x, x, \ldots, x)$.

One sees easily that Δ is a Veronese variety of indices (n, r). It is embedded in a subspace of dimension $\binom{n+r}{r} - 1$ in $\mathbb{P}^{(n+1)^r - 1}$ (cf. Section 6.7).

11.3.3 Graph of a rational transformation. Let $\psi : \mathbb{P}^n \to \mathbb{P}^m$ be a rational transformation defined on an open set U of $\mathbb{P}^n$ by the equations

$$Y_0 : Y_1 : \cdots : Y_m = F_0 : F_1 : \cdots : F_m,$$

with $F_j = F_j(X_0, \ldots, X_n) \in K[X_0, \ldots, X_n]$ homogeneous polynomials of the same degree (cf. Section 2.6).

In $\mathbb{P}^n \times \mathbb{P}^m$ consider the set Γ_ψ of the pairs $(x, \psi(x))$, with $x \in U$. The projective closure of $\varphi(\Gamma_\psi)$ in $\mathbb{P}^{mn+m+n}$, where $\varphi : \mathbb{P}^n \times \mathbb{P}^m \to \mathbb{P}^{mn+m+n}$ is the Segre morphism, is called the *graph of* ψ. It has a parametric representation given by

$$Z_{ij} = u_i F_j(u_0, u_1, \ldots, u_m), \quad i = 0, \ldots, n, \ j = 0, \ldots, m,$$

where $[u_0, \ldots, u_n]$ varies in the open set U.

Similarly, one defines the graph of a rational transformation $\psi : V \to V'$, if $V \subset \mathbb{P}^n$ and $V' \subset \mathbb{P}^m$ are two arbitrary projective varieties.

11.3.4 (Blowing up). Let $V = V(\mathfrak{a}) \subset \mathbb{P}^n$ and let $\mathfrak{a} = (f_0, f_1, \ldots, f_r)$ be its homogeneous ideal in $\mathbb{C}[x_0, \ldots, x_n]$. We may assume that the homogeneous polynomials $f_0, f_1, \ldots, f_r$ are all of the same degree d. If such were not the case, after putting $d = \max \deg f_j$, $j = 0, \ldots, r$, it would suffice to take (for each index j) rather than the single polynomial f_j, all the polynomials of the form $x_0^{i_0} x_1^{i_1} \ldots x_n^{i_n} f_j$ with $\sum_{\alpha=0,\ldots,n} i_\alpha = d - \deg f_j$.

Assuming this, we consider the morphism $\mathbb{P}^n \setminus V \to \mathbb{P}^r$ with equations

$$X_j = f_j(x_0, x_1, \ldots, x_n), \quad j = 0, \ldots, r, \tag{11.6}$$

and its graph $\Gamma \subset \mathbb{P}^n \times \mathbb{P}^r$, which is the locus of the point

$$[x_0, \ldots, x_n; f_0(x), \ldots, f_r(x)],$$

$x = (x_0, \ldots, x_n)$. If $\mathbb{P}^n \times \mathbb{P}^r$ is represented by the Segre variety (with equations $X_{ij} = X_i Y_j$, $i = 0, \ldots, n$, $j = 0, \ldots, r$) one finds the following parametric representation for the closure W of Γ:

$$X_{ij} = x_i f_j(x_0, x_1, \ldots, x_n), \quad i = 0, \ldots, n, \ j = 0, \ldots, r. \tag{11.7}$$

The projection σ of Γ onto the first factor, that is, the map defined by

$$[x_0, \ldots, x_n; f_0(x), \ldots, f_r(x)] \mapsto [x_0, \ldots, x_n],$$

is clearly an isomorphism $\Gamma \cong \mathbb{P}^n \setminus V$. From equations (11.7) it is clear that W is the projective image of the linear system

$$\sum_{ij} \lambda_{ij} x_i f_j(x_0, x_1, \ldots, x_n) = 0$$

of hypersurfaces of order $d+1$ passing through V. Hence W is a blowing up of $\mathbb{P}^n$ with center V (cf. Section 6.8).

As we have seen in Section 6.8, if V is non-singular, then to each point P of V there corresponds a linear space L_P and $E := \bigcup_{P \in V} L_P$ is the exceptional divisor of the blowing up.

11.3.5 Algebraic correspondences between varieties. Let $V \subset \mathbb{P}^n$ and $V' \subset \mathbb{P}^m$ be two projective varieties. An *algebraic correspondence* between V and V' is an arbitrary closed subset Γ of the Segre product $\varphi(V \times V') \subset \varphi(\mathbb{P}^n \times \mathbb{P}^m)$. If $x^* \in V$ and $\varphi(x^*, y) \in \Gamma$ we will say that y is one of the (points) corresponding to x^*, or, briefly, one of the correspondents of x^*. The correspondents of $x^* \in V$ constitute the closed set $\Gamma \cap \varphi(x^* \times V')$, and similarly the correspondents of $y^* \in V'$ are the points of $\Gamma \cap \varphi(V \times y^*)$. Observe that $\varphi(x^* \times V')$ and $\varphi(V \times y^*)$ are contained in the two linear spaces $L_m = \varphi(x^* \times \mathbb{P}^m)$ and $L_n = \varphi(\mathbb{P}^n \times y^*)$.

A simple example of a correspondence between the space $\mathbb{P}^n$ and its dual $\mathbb{P}^{n*}$ is the incidence relation between point and hyperplane; it is a hyperplane section of the Segre variety which is a projective model for the product $\mathbb{P}^n \times \mathbb{P}^{n*}$.

11.4 Examples and exercises

For definitions and further details regarding Veronese varieties we refer the reader to Section 6.7.

11.4.1. *Study the projective correspondences between two copies of $\mathbb{P}^1$ as a curve on a non-specialized quadric $\mathcal{Q} = \varphi(\mathbb{P}^1 \times \mathbb{P}^1) \subset \mathbb{P}^3$, where φ is the Segre embedding.*

The points of $\mathcal{Q}$ that belong to the plane with equation

$$aZ_{00} + bZ_{01} + cZ_{10} + dZ_{11} = 0, \quad a, b, c, d \in K,$$

correspond to the pairs of points $A = [x_0, x_1]$, $B = [y_0, y_1]$ of $\mathbb{P}^1 \times \mathbb{P}^1$ which are related by the equation

$$ax_0 y_0 + bx_0 y_1 + cx_1 y_0 + dx_1 y_1 = 0,$$

namely, to the pairs of points that correspond under a projectivity between the two lines (cf. §1.1.3). If the plane is not tangent to $\mathcal{Q}$ that projectivity is non-degenerate.

Hence the plane sections of Ω "are" the projectivities between the two lines. In particular (if the two lines are superposed) the identity projectivity defined by the bilinear equation $x_0 y_1 - x_1 y_0 = 0$ corresponds to the plane π with equation $Z_{01} - Z_{10} = 0$, whose pole with respect to Ω is the center $[0, 1, 1, 0]$ of the star $aZ_{00} + b(Z_{01} - Z_{10}) + dZ_{11} = 0$ of planes that represent involutory projectivities between the two (coinciding) lines.

The non-degenerate parabolic projectivities between the two lines "are" the tangent planes to the conic which is a section of Ω by the plane π (that is, planes passing through a tangent line to the conic, but not tangent to Ω).

11.4.2. Let $\omega \colon \mathbb{P}^n \to \mathbb{P}^n$ be a projectivity given by the equations

$$Y_j = a_{j0} X_0 + a_{j1} X_1 + \cdots + a_{jn} X_n, \quad j = 0, \ldots, n,$$

and $\varphi(\mathbb{P}^n \times \mathbb{P}^n) \to \mathbb{P}^{n(n+2)}$ the Segre morphism. On $\varphi(\mathbb{P}^n \times \mathbb{P}^n)$ consider the graph $\mathcal{G}$ of ω, the locus of the point

$$Z_{ij} = u_i (a_{j0} u_0 + a_{j1} u_1 + \cdots + a_{jn} u_n), \quad i, j = 0, \ldots, n.$$

If ω is not degenerate, $\mathcal{G}$ is the image of the Veronese variety $V_{n,2} \subset \mathbb{P}^{\binom{n+2}{2}-1}$ under a projectivity $\theta \colon \mathbb{P}^{n(n+2)} \to \mathbb{P}^{n(n+2)}$ (subordinated by ω).

Examine the details in the case $n = 2$.

Consider the case $n = 2$. The graph $\mathcal{G}$ of ω is the subvariety of $\mathbb{P}^8$ with parametric equations

$$Z_{ij} = u_i (a_{j0} u_0 + a_{j1} u_1 + a_{j2} u_2), \quad i, j = 0, 1, 2.$$

If $A = (a_{ij})$ is the matrix associated to the projectivity ω, consider the projectivity $\theta \colon \mathbb{P}^8 \to \mathbb{P}^8$ defined by the matrix

$$\begin{pmatrix} A & 0 & 0 \\ 0 & A & 0 \\ 0 & 0 & A \end{pmatrix}.$$

Let F be the surface in $\mathbb{P}^8$ with parametric equations

$$Z_{ij} = u_i u_j, \quad i, j = 0, 1, 2.$$

One sees immediately that $\mathcal{G} = \theta(F)$ is contained in the linear subspace $S_5 \subset \mathbb{P}^8$ defined by the equations $Z_{01} - Z_{10} = Z_{02} - Z_{20} = Z_{12} - Z_{21} = 0$ (where we may interpret the Z_{ij} with $i \neq j$ as coordinates). In fact, $\mathcal{G}$ is the Veronese surface which is the image of the immersion

$$v \colon \mathbb{P}^2 \to S_5$$

defined by

$$[u_0, u_1, u_2] \mapsto [u_0^2, u_0 u_1, u_0 u_2, u_1^2, u_1 u_2, u_2^2].$$

11.4.3. *Consider a reciprocity ω between two superposed spaces S_n and S'_n. It may be regarded as a correspondence between two $\mathbb{P}^n$ where one says that a point of one of the two spaces corresponds to every point of the hyperplane that corresponds to it under ω (cf. Theorem 5.4.6). Such a correspondence ω is represented by a bilinear equation*

$$\sum_{i,j=0}^{n} a_{ij} x_i x_j = 0.$$

The reciprocity is non-degenerate if $\det(a_{ij}) \neq 0$.

After noting that the reciprocities between two $\mathbb{P}^n$ are the hyperplane sections of the Segre variety $\varphi(\mathbb{P}^n \times \mathbb{P}^n) \subset \mathbb{P}^{n(n+2)}$, study the remarkable special cases in which the matrix (a_{ij}) is symmetric (polarity with respect to a quadric) or antisymmetric (null polarity).

11.4.4. *Study the Segre variety of the product of a line by a plane.*

The Segre variety $\varphi(\mathbb{P}^1 \times \mathbb{P}^2) \subset \mathbb{P}^5$ has order and dimension 3 (it is known as C. Segre's X_3^3). It contains a 1-dimensional system of planes L and a 2-dimensional system of lines not contained in the planes L. It is easy to see that Segre's X_3^3 is the locus of the lines that join pairs of points corresponding under a non-degenerate projectivity ω between two mutually skew planes π and π'. Furthermore, it is the locus of the ∞^1 planes that are supported by three independent lines r_1, r_2, r_3 in triples of points P_1, P_2, P_3 such that there exist two projectivities

$$\varphi_{12}: r_1 \to r_2, \quad \varphi_{23}: r_2 \to r_3$$

with

$$\varphi_{12}(P_1) = P_2, \quad \varphi_{23}(P_2) = P_3.$$

One usually expresses this fact by saying that Segre's X_3^3 is the locus of the planes that join the triples of corresponding points of three projectively referred lines.

In the space $\mathbb{P}^5$ that joins π and π' we assume the reference system to be such that $\pi = J(A_0, A_1, A_2)$ and $\pi' = J(A_3, A_4, A_5)$ with $A_3 = \omega(A_0)$, $A_4 = \omega(A_1)$, and $A_5 = \omega(A_2)$. We consider a projectivity $\Omega: \mathbb{P}^5 \to \mathbb{P}^5$ whose restriction to π coincides with ω and that satisfies the further condition $\Omega([1,1,1,0,0,0]) = [0,0,0,1,1,1]$.

For each point $P = [x_0, x_1, x_2]$ of π we then have

$$\omega(P) = \Omega([x_0, x_1, x_2, 0, 0, 0]) = [0, 0, 0, x_0, x_1, x_2].$$

Moreover, the variable point on the line $\langle P, \omega(P) \rangle$ has coordinates

$$(y_0 x_0, y_0 x_1, y_0 x_2, y_1 x_0, y_1 x_1, y_1 x_2),$$

where y_0, y_1 are homogeneous parameters that can be taken as projective coordinates in $\mathbb{P}^1$. Thus the locus of the lines $\langle P, \omega(P) \rangle$ is the Segre variety $\varphi(\mathbb{P}^1 \times \mathbb{P}^2)$.

Chapter 11. Segre Varieties

The equations of $\varphi(\mathbb{P}^1 \times \mathbb{P}^2) \subset \mathbb{P}^5$ are obtained by annihilating the minors of the matrix
$$\begin{pmatrix} T_0 & T_1 & T_2 \\ T_3 & T_4 & T_5 \end{pmatrix};$$
and $\varphi(\mathbb{P}^1 \times \mathbb{P}^2)$ is the three dimensional cubic variety arising as the residual intersection of two quadric hypersurfaces in $\mathbb{P}^5$ ($T_0 T_4 - T_1 T_3 = T_0 T_5 - T_2 T_3 = 0$) which also have in common a linear space S_3 ($T_0 = T_3 = 0$).

It is a useful exercise to extend these results to the case $\mathbb{P}^1 \times \mathbb{P}^n$.

Chapter 12
Grassmann Varieties

In this chapter we introduce the Grassmann varieties (Grassmannians) that describe the geometry of the linear subspaces of a given projective space, and we study their basic properties.

Section 12.1 is dedicated to the simplest case of a Grassmann variety that is not a projective space: we introduce homogeneous Plücker line coordinates and study the Grassmann variety $\mathbb{G}(1, 3)$ parameterizing the lines in $\mathbb{P}^3$. It is a quadric in $\mathbb{P}^5$, called Klein's quadric.

In Sections 12.2, 12.3 and 12.4 we consider the complexes and the congruences of lines in $\mathbb{P}^3$ and ruled surfaces, with particular attention to the linear case; we study some of their geometric properties expressed in terms of the Klein quadric. We mention in particular the characterization of developable ruled surfaces given in Proposition 12.4.1.

In Section 12.5 we introduce the Grassmann coordinates of a linear space S_h in $\mathbb{P}^n$ (of which the homogeneous Plücker line coordinates are the particular case $n = 3, h = 1$). We then define Grassmann varieties and describe some of their noteworthy properties. In particular in Proposition–Definition 12.5.7 we prove that Grassmann varieties are always defined by quadratic forms. Further properties of Grassmannians are illustrated in Section 12.6, in the form of complete resolved exercises.

For further information we refer the reader to the texts [52, Vol. 2], [48, Lecture 6], [46, Chapter 1, §5] and to the memoir [98]. An excellent (albeit very concise) treatment of the properties of Grassmann coordinates and Grassmann varieties is found in [80].

12.1 The lines of $\mathbb{P}^3$ as points of a quadric in $\mathbb{P}^5$

In the projective space $\mathbb{P}^3 = \mathbb{P}^3(K)$, with a fixed homogeneous projective coordinate system t_0, t_1, t_2, t_3, we consider a line r and two distinct points $x = [x_0, x_1, x_2, x_3]$, $y = [y_0, y_1, y_2, y_3]$ on r. We use p_{ik} to denote the second order minors

$$p_{ik} = x_i y_k - x_k y_i$$

extracted from the matrix

$$\begin{pmatrix} x_0 & x_1 & x_2 & x_3 \\ y_0 & y_1 & y_2 & y_3 \end{pmatrix}$$

formed with the coordinates of these two points. Since

$$p_{ik} = -p_{ki}, \qquad (12.1)$$

we have essentially six distinct numbers which are not all zero.

One sees immediately that they (or better their ratios) depend only on the line r and not on the pair of points x, y taken on r. Indeed, if x and y are replaced by two other distinct points of r, say $x' = [x'_i]$ and $y' = [y'_i]$, where

$$x'_i = \lambda x_i + \mu y_i, \quad y'_i = \lambda' x_i + \mu' y_i, \quad i = 0, 1, 2, 3, \ \lambda\mu' - \lambda'\mu \neq 0,$$

rather than the numbers p_{ik} we find the numbers

$$p'_{ik} = \begin{vmatrix} \lambda x_i + \mu y_i & \lambda x_k + \mu y_k \\ \lambda' x_i + \mu' y_i & \lambda' x_k + \mu' y_k \end{vmatrix} = (\lambda\mu' - \lambda'\mu) p_{ik}$$

which are proportional to the preceding six. We then have

$$\begin{vmatrix} x_0 & x_1 & x_2 & x_3 \\ y_0 & y_1 & y_2 & y_3 \\ x_0 & x_1 & x_2 & x_3 \\ y_0 & y_1 & y_2 & y_3 \end{vmatrix} = 2(p_{01}p_{23} + p_{02}p_{31} + p_{03}p_{12}) = 0$$

and therefore, no matter how r is chosen, the p_{ik} satisfy the quadratic relation

$$\Phi(p_{ik}) = p_{01}p_{23} + p_{02}p_{31} + p_{03}p_{12} = 0. \qquad (12.2)$$

Thus to each line of $\mathbb{P}^3$ there are associated six numbers p_{ik} not all zero and defined up to a non-zero factor of proportionality and satisfying the equation of degree two (12.2).

We seek the points H_i at which the line r intersects the four coordinate planes $t_i = 0$. We find

$$H_0 = [\lambda x_0 + \mu y_0, \lambda x_1 + \mu y_1, \lambda x_2 + \mu y_2, \lambda x_3 + \mu y_3]$$

with $\lambda x_0 + \mu y_0 = 0$, that is, $-\lambda/\mu = y_0/x_0$. Therefore

$$H_0 = [0, p_{10}, p_{20}, p_{30}] \ (= [0, p_{01}, p_{02}, p_{03}]).$$

Similarly we find the points H_1, H_2, H_3. With the coordinates of these points we form the matrix

$$\begin{pmatrix} 0 & p_{01} & p_{02} & p_{03} \\ p_{10} & 0 & p_{12} & p_{13} \\ p_{20} & p_{21} & 0 & p_{23} \\ p_{30} & p_{31} & p_{32} & 0 \end{pmatrix}. \qquad (12.3)$$

12.1. The lines of $\mathbb{P}^3$ as points of a quadric in $\mathbb{P}^5$

It is antisymmetric and has rank two because the four points are collinear, and among them there certainly are at least two that are distinct.

To fix ideas, suppose that $p_{01} \neq 0$ and thus that $H_0 \neq H_1$. The second order minors p'_{ik} of the matrix

$$\begin{pmatrix} 0 & p_{01} & p_{02} & p_{03} \\ p_{10} & 0 & p_{12} & p_{13} \end{pmatrix}$$

formed with the coordinates of H_0 and H_1 are proportional to the same numbers p_{ik}. In fact, bearing in mind (12.1) and (12.2), we have

$$\begin{aligned}
p'_{01} &= -p_{01} p_{10} = p_{01}^2, \\
p'_{02} &= -p_{10} p_{02} = p_{01} p_{02}, \\
p'_{03} &= -p_{10} p_{03} = p_{01} p_{03}, \\
p'_{12} &= p_{01} p_{12}, \\
p'_{13} &= p_{01} p_{13}, \\
p'_{23} &= p_{02} p_{13} - p_{03} p_{12} = p_{02} p_{13} + p_{03} p_{21} = p_{01} p_{23}.
\end{aligned} \quad (12.4)$$

This shows that the line $r = r_{H_0 H_1}$ is determined by the numbers p_{ik}. Thus we have proved the following:

- There exists a bijective correspondence (without exceptions) between the lines of $\mathbb{P}^3$ and the homogeneous sextuples (p_{ik}), $i, k = 0, 1, 2, 3$, $i \neq k$, that satisfy equations (12.1), (12.2).

Therefore the numbers p_{ik} may be taken to be homogeneous coordinates for the lines in $\mathbb{P}^3$: one says that they are the *Plückerian homogeneous line coordinates*, or, more briefly, the *Plücker line coordinates*.

In a space $\mathbb{P}^5$ in which $p_{01}, p_{02}, p_{03}, p_{12}, p_{13}, p_{23}$ (in the order indicated) are homogeneous projective coordinates, the equation (12.2) represents a non-degenerate quadric $\mathfrak{Q}$ called the *Klein quadric*. Bearing in mind the relation

$$\begin{vmatrix} 0 & p_{01} & p_{02} & p_{03} \\ p_{10} & 0 & p_{12} & p_{13} \\ p_{20} & p_{21} & 0 & p_{23} \\ p_{30} & p_{31} & p_{32} & 0 \end{vmatrix} = (p_{01} p_{23} + p_{02} p_{31} + p_{03} p_{12})^2,$$

and the fact that the antisymmetric matrix (12.3) has even rank, one thus has that (cf. Exercise 12.6.4 for the extension to the case $\mathbb{P}^n$)

- the Klein quadric is the locus of the points of $\mathbb{P}^5$ that give rank 2 to the matrix (12.3).

In virtue of the bijection that we have thus established between the points of this quadric and the lines of $\mathbb{P}^3$, every proposition regarding geometric properties of the ordinary ruled space can be interpreted in $\mathbb{P}^5$ and becomes a proposition regarding the non-singular quadrics in $\mathbb{P}^5$; and conversely properties of a non-singular quadric in $\mathbb{P}^5$ can be read as properties of the lines in ordinary space.

In this regard, we have the following fundamental fact.

Proposition 12.1.1. *Let $\mathfrak{Q}$ be the Klein quadric. A necessary and sufficient condition for two lines r and r' in $\mathbb{P}^3$ to be incident is that the two points R and R' of $\mathfrak{Q}$ corresponding to them be reciprocals with respect to $\mathfrak{Q}$ (cf. Section 5.4).*

Proof. We consider two lines $r = r_{XY}$ and $r' = r_{ZW}$ where $X = [x_0, x_1, x_2, x_3]$, $Y = [y_0, y_1, y_2, y_3]$, $Z = [z_0, z_1, z_2, z_3]$, and $W = [w_0, w_1, w_2, w_3]$. They are incident if and only if X, Y, Z, W are four coplanar points, which means

$$\begin{vmatrix} x_0 & x_1 & x_2 & x_3 \\ y_0 & y_1 & y_2 & y_3 \\ z_0 & z_1 & z_2 & z_3 \\ w_0 & w_1 & w_2 & w_3 \end{vmatrix} = 0$$

that is, letting p_{ik} and p'_{ik} be the Plücker coordinates of r and r':

$$p_{01} p'_{23} + p_{02} p'_{31} + p_{03} p'_{12} + p_{12} p'_{03} + p_{13} p'_{20} + p_{23} p'_{01} = 0. \tag{12.5}$$

This is indeed the condition in order for the two points $R = [p_{ik}]$, $R' = [p'_{ik}]$ of $\mathbb{P}^5$ to be reciprocals with respect to $\mathfrak{Q}$. □

12.1.2 (Lines and planes of a quadric in $\mathbb{P}^5$). We will now see how one can deduce properties of the lines and planes of a quadric in $\mathbb{P}^5$ from simple properties of the lines and planes in $\mathbb{P}^3$ and how, conversely, one can deduce results concerning lines and planes in $\mathbb{P}^3$ from properties of the lines and planes of a quadric in $\mathbb{P}^5$.

If a and b are two incident lines of $\mathbb{P}^3$, to them there correspond in $\mathbb{P}^5$ two points A and B on $\mathfrak{Q}$ which are mutually reciprocal and thus joined by a line belonging to $\mathfrak{Q}$ (cf. §5.5.4, c)). Therefore, every point C of this line is the reciprocal of every point reciprocal to both A and B. Thus each such point C represents a line c incident with the ∞^2 lines of the star containing a and b, as well as with the ∞^2 lines of the ruled plane $\langle a, b \rangle$. To the points C collinear with A and B there thus correspond the lines c of the pencil that contains a and b.

Every pencil of lines in $\mathbb{P}^3$ gives a line of $\mathfrak{Q}$ (whose points "are" the lines of the pencil). Thus, the lines of $\mathfrak{Q}$ form a 5-dimensional family because the number of pencils of lines in $\mathbb{P}^3$ is ∞^5 (∞^2 in each of the ∞^3 planes).

We remark explicitly that the necessary and sufficient condition for the linear combination $\lambda p_{ik} + \mu p'_{ik}$ of the Plücker coordinates p_{ik} and p'_{ik} of two lines a and b to be the Plücker coordinates of a line r is that a and b be incident; and then r sweeps out, as λ and μ vary, the pencil defined by a and b.

A system ∞^2 of pairwise incident lines in $\mathbb{P}^3$ can only be a ruled plane or a star of lines. To such a system there corresponds in $\mathbb{P}^5$ an ∞^2 set of points on $\mathfrak{Q}$

that are pairwise reciprocal with respect to $\mathfrak{Q}$, namely the set of points of a plane contained in $\mathfrak{Q}$.

Thus we rediscover the two systems ∞^3 of planes on $\mathfrak{Q}$: the system $\{\alpha\}$ associated to the ruled planes in $\mathbb{P}^3$ and the system $\{\beta\}$ of planes that represent the stars of lines in $\mathbb{P}^3$ (cf. §5.5.9).

Two stars of lines, or two ruled planes, have in common a single line, and thus *two planes of $\mathfrak{Q}$ belonging to the same system have in common a single point.* However, a ruled plane and a star of lines do not have any line in common, or else they have in common a pencil of lines according to whether the center of the star does not belong or does belong to the plane; and then *two planes of $\mathfrak{Q}$ belonging to different systems are either skew or have a line in common.*

Since a pencil of lines of $\mathbb{P}^3$ belongs to a single ruled plane and to a single star of lines, *a line of $\mathfrak{Q}$ belongs to only one plane α and to only one plane β.*

For each point of $\mathfrak{Q}$ there pass ∞^1 planes of each system, because a line in $\mathbb{P}^3$ belongs to ∞^1 ruled planes and to ∞^1 stars of lines.

12.2 Complexes of lines in $\mathbb{P}^3$

With the notation of Section 12.1, consider a homogeneous projective coordinate system t_0, t_1, t_2, t_3 in $\mathbb{P}^3$ and the projective space $\mathbb{P}^5$ in which the Plücker line coordinates $p_{01}, p_{02}, p_{03}, p_{12}, p_{13}, p_{23}$ are (in the order indicated) homogeneous coordinates.

Consider then Klein's quadric $\mathfrak{Q}$ in $\mathbb{P}^5$, with equation (12.2), whose points are in bijective correspondence with the lines of $\mathbb{P}^3$.

The lines in $\mathbb{P}^3$ whose coordinates p_{ik} satisfy an equation

$$f(p_{ik}) = 0, \qquad (12.6)$$

with f a homogeneous polynomial of degree n, constitute an *algebraic complex* $\mathcal{K}$ (of lines in $\mathbb{P}^3$) of *degree n*. It is represented in $\mathbb{P}^5$ by the variety V_3^{2n} of order $2n$ and dimension 3, the section of $\mathfrak{Q}$ by the hypersurface of equation (12.6). Note the following characterization of the degree of an algebraic complex:

- The degree n of the algebraic complex $\mathcal{K}$ coincides with the number of lines of $\mathcal{K}$ that belong to a generic pencil.

The n intersections of the hypersurface having equation (12.6) with a line ℓ of $\mathfrak{Q}$ represent, in fact, the lines of the complex belonging to the pencil of lines represented by ℓ.

We observe that the same complex $\mathcal{K}$ may be defined via an arbitrary equation of the form

$$f(p_{ik}) + g(p_{ik})\Phi(p_{ik}) = 0,$$

where g is a form of degree $n-2$. Therefore an algebraic complex of degree n depends on

$$\binom{n+5}{5} - \binom{n+3}{5} - 1$$

constants and can be determined with the same number of generic lines.

An algebraic complex of degree $n=1$ is called a *linear complex*. A linear complex $\mathcal{K}$ is thus represented by a hyperplane section of $\mathfrak{Q}$. If the hyperplane $\mathcal{K}^*$ whose intersection with $\mathfrak{Q}$ represents $\mathcal{K}$ is not tangent to $\mathfrak{Q}$, we also say that $\mathcal{K}$ is a *general linear complex*. If on the contrary $\mathcal{K}^*$ is tangent to $\mathfrak{Q}$, we then say that $\mathcal{K}$ is a *special linear complex*.

Bearing in mind Proposition 12.1.1, we have that if $\mathcal{K}$ is special the point A of contact of $\mathcal{K}^*$ with $\mathfrak{Q}$ represents a line a belonging to $\mathcal{K}$ and incident with all the lines of $\mathcal{K}$; conversely, every line of $\mathbb{P}^3$ incident with a is represented by a reciprocal point of A which is therefore contained in the S_4 tangent to $\mathfrak{Q}$ at A. The lines of a special linear complex are thus all those and only those lines of $\mathbb{P}^3$ supported by a fixed line, called the *axis* of the complex.

In any case, a linear complex $\mathcal{K}$ consists of ∞^3 pencils of lines, represented by the lines of a quadric in $\mathbb{P}^4$, which are precisely ∞^3 (cf. §5.5.9).

In the sequel we will identify the linear complex $\mathcal{K}$, that is the section of $\mathfrak{Q}$ by the hyperplane $\mathcal{K}^*$ having equation

$$a_{23}p_{01} + a_{31}p_{02} + a_{12}p_{03} + a_{03}p_{12} + a_{20}p_{13} + a_{01}p_{23} = 0, \qquad (12.7)$$

where $a_{ik} = -a_{ki}$, with the point $A = [a_{01}, a_{02}, a_{03}, a_{12}, a_{13}, a_{23}]$ in $\mathbb{P}^5$. It is easily seen that $\mathcal{K}^*$ is the polar hyperplane of A with respect to the quadric $\mathfrak{Q}$, which means that A is the pole of $\mathcal{K}^*$ with respect to $\mathfrak{Q}$. We will also say that the point A *represents the complex* $\mathcal{K}$.

With these preliminaries in hand, the principal properties of a linear complex can be gathered into the following proposition.

Proposition 12.2.1. *Let $\mathcal{K}$ be a linear complex. Then:*

(1) *The lines of the complex $\mathcal{K}$ passing through a point x in $\mathbb{P}^3$ are contained in a plane and constitute a pencil (with center x).*

(2) *The lines of $\mathcal{K}$ that lie in a plane π pass through a point and constitute a pencil.*

(3) *The correspondence that associates to each point x in $\mathbb{P}^3$ the plane of the pencil of lines of the complex $\mathcal{K}$ passing through x is a null polarity $\varphi \colon \mathbb{P}^3 \to \mathbb{P}^{3*}$.*

(4) *The complex $\mathcal{K}$ is special if and only if the point A of $\mathbb{P}^5$ that represents it belongs to the quadric $\mathfrak{Q}$; and also if and only if φ is a degenerate projectivity. In this case each line of $\mathcal{K}$ meets the line represented by the point A.*

12.2. Complexes of lines in $\mathbb{P}^3$

Proof. To prove (1) and (2) it suffices to recall that the lines of $\mathbb{P}^3$ that pass through a point P (and similarly the lines of $\mathbb{P}^3$ that belong to a plane α) are represented by the points of a plane π lying in $\mathfrak{Q}$. To the points of the line in which π intersects the hyperplane $\mathcal{K}^*$ that represents the complex there correspond the lines of $\mathcal{K}$ passing through P (or the lines of $\mathcal{K}$ belonging to α). Thus these lines form a pencil.

Let r be a line of $\mathbb{P}^3$ passing through the point $x = [x_0, x_1, x_2, x_3]$. If $y = [y_0, y_1, y_2, y_3]$ is another point of r, the Plücker coordinates of r are $p_{ik} = x_i y_k - x_k y_i$. The necessary and sufficient condition for r to belong to $\mathcal{K}$ is that the numbers p_{ik} satisfy equations (12.7), namely that the point y satisfies the equation (linear in y_0, y_1, y_2, y_3):

$$a_{23}(x_0 y_1 - x_1 y_0) + a_{31}(x_0 y_2 - x_2 y_0) + a_{12}(x_0 y_3 - x_3 y_0) \\ + a_{01}(x_2 y_3 - x_3 y_2) + a_{20}(x_1 y_3 - x_3 y_1) + a_{03}(x_1 y_2 - x_2 y_1) = 0. \quad (12.8)$$

Then those and only those points y which when joined to x give lines of $\mathcal{K}$ are the points of the plane π_x with equation (12.8). Thus one again finds that the lines of $\mathcal{K}$ that pass through a point belong to a plane.

Ordering the summands of (12.8) with respect to y_0, y_1, y_2, y_3 we obtain

$$(-a_{23}x_1 - a_{31}x_2 - a_{12}x_3)y_0 + (a_{23}x_0 - a_{03}x_2 - a_{20}x_3)y_1 \\ + (a_{31}x_0 + a_{03}x_1 - a_{01}x_3)y_2 + (a_{12}x_0 + a_{20}x_1 + a_{01}x_2)y_3 = 0. \quad (12.9)$$

Hence the coordinates of π_x, that is, the coefficients of (12.9), are the numbers

$$\begin{cases} u_0 = -a_{23}x_1 - a_{31}x_2 - a_{12}x_3, \\ u_1 = a_{23}x_0 - a_{03}x_2 - a_{20}x_3, \\ u_2 = a_{31}x_0 + a_{03}x_1 - a_{01}x_3, \\ u_3 = a_{12}x_0 + a_{20}x_1 + a_{01}x_2. \end{cases} \quad (12.10)$$

From this it is evident that π_x is the plane that corresponds to x under a null polarity φ. The matrix of coefficients of (12.10) is in fact antisymmetric (cf. 1.1.13). Since the determinant of this matrix is

$$\begin{vmatrix} 0 & -a_{23} & -a_{31} & -a_{12} \\ a_{23} & 0 & -a_{03} & -a_{20} \\ a_{31} & a_{03} & 0 & -a_{01} \\ a_{12} & a_{20} & a_{01} & 0 \end{vmatrix} = (a_{01}a_{23} + a_{02}a_{31} + a_{03}a_{12})^2,$$

the polarity φ is degenerate if and only if the point $A = [a_{01}, a_{02}, a_{03}, a_{12}, a_{13}, a_{23}]$ belongs to the quadric $\mathfrak{Q}$. Since A is the pole of the hyperplane $\mathcal{K}^*$, the polarity degenerates if and only if the hyperplane $\mathcal{K}^*$ with equation (12.7) is tangent at A to the quadric $\mathfrak{Q}$. In that case, the complex $\mathcal{K}$ is special and all the points of the hyperplane $\mathcal{K}^* (= \mathfrak{Q}_1(A))$ are reciprocals of A (with respect to $\mathfrak{Q}$) and thus the lines of $\mathcal{K}$ are all supported by the line represented by the point A (cf. Proposition 12.1.1). □

Exercise 12.2.2. *Prove that the lines of an algebraic complex of order n that pass through a point $x \in \mathbb{P}^3$ form an algebraic cone of order n having vertex at x.*

The proof is analogous to that of Proposition 12.2.1 and is left to the reader.

12.2.3 Pencils of linear complexes. A *pencil of linear complexes* of lines of $\mathbb{P}^3$ is a line in $\mathbb{P}^5$ and so the pencils of linear complexes can be classified in relation to the various positions that a line r can assume in $\mathbb{P}^5$ with respect to $\mathfrak{Q}$: generic line, that is, line having two distinct points in common with $\mathfrak{Q}$, tangent line to $\mathfrak{Q}$, line situated on $\mathfrak{Q}$.

If the line r has two distinct points A and B in common with $\mathfrak{Q}$, the pencil of linear complexes contains two special complexes corresponding to the points A and B of $\mathfrak{Q}$ and having skew axes. Indeed, by what we have seen in §12.1.2, the points A and B are not reciprocal with respect to $\mathfrak{Q}$ (inasmuch as they are joined by a line not contained in $\mathfrak{Q}$) and thus the two lines of $\mathbb{P}^3$ which correspond to A and B (the axes of the two complexes) are skew.

If r is tangent to $\mathfrak{Q}$ at a point A, the pencil of linear complexes contains only one special complex, corresponding to the point A, and with axis represented by the point A.

If r lies on $\mathfrak{Q}$ we have a pencil of linear complexes all of which are special (and whose axes are the lines of a pencil).

12.2.4 (The reduced equation of a linear complex). One can immediately write the equation of the special complex of lines that has as its axis one of the edges of the fundamental tetrahedron.

For example, the linear complex of the lines supported by the line $r_{A_0 A_1}$ has equation $p_{23} = 0$. Indeed, the line $r_{A_0 A_1}$ has Plücker coordinates $[1, 0, 0, 0, 0, 0]$ since $p_{01} = \left|\begin{smallmatrix} 1 & 0 \\ 0 & 1 \end{smallmatrix}\right|$ is the only non-zero second order minor of the matrix

$$\begin{pmatrix} 1 & 0 & 0 & 0 \\ 0 & 1 & 0 & 0 \end{pmatrix}.$$

If $p'_{01}, \ldots, p'_{23}$ are the Plücker coordinates of a line supported by $r_{A_0 A_1}$, equations (12.5) yield $p_{01} p'_{23} = 0$ and so $p'_{23} = 0$.

One shows in analogous fashion that the special linear complex of lines with axis $r_{A_3 A_4}$ has equation $p_{01} = 0$, and so on.

It is then easily seen that given a linear complex $\mathcal{K}$ of general type one can choose the reference system in $\mathbb{P}^3$ in such a way that $\mathcal{K}$ has equation of the type

$$\lambda p_{01} + \mu p_{23} = 0.$$

Indeed, if A is the point of $\mathbb{P}^5$ that represents $\mathcal{K}$, then a generic line through A represents a pencil Σ of linear complexes (containing $\mathcal{K}$) and we can write $\mathcal{K}$ as a linear combination of the two special complexes of Σ. It then suffices to take the axes of these two complexes as the edges $A_3 A_4$ and $A_0 A_1$ of the fundamental tetrahedron in $\mathbb{P}^3$.

12.3 Congruences of lines in $\mathbb{P}^3$

An algebraic surface $\mathcal{S}$ contained in the Klein quadric $\mathfrak{Q}$ represents a totality, say Γ, of ∞^2 lines of $\mathbb{P}^3$ that is called an *algebraic congruence* of lines in $\mathbb{P}^3$.

In particular, the collection Γ of all lines in $\mathbb{P}^3$ that belong to two algebraic complexes without common components is an algebraic congruence of lines: the surface $\mathcal{S} \subset \mathfrak{Q}$ that represents the congruence is the intersection of $\mathfrak{Q}$ with the two hypersurfaces that represent the two complexes.

If the surface $\mathcal{S}$ that represents the congruence is a section of $\mathfrak{Q}$ by a linear space $\Pi = S_3$, the congruence is called *linear*. If α and β are two hyperplanes whose intersection is Π, the linear congruence Γ is the collection of the lines in $\mathbb{P}^3$ that belong to the two linear complexes represented by the hyperplanes α and β. We will also say, for simplicity, that Γ is represented by the space $\Pi = \alpha \cap \beta$, and thus that Γ belongs to all the linear complexes of the pencil Σ that they define.

Moreover, we will have three types of linear congruences of lines according to whether the polar line p of Π with respect to $\mathfrak{Q}$ (cf. §5.5.4) is skew to Π (*general linear congruence*), or meets Π in a point (*special linear congruence*), or is contained in Π (*degenerate linear congruence*).

12.3.1 (General linear congruence). If Γ is a general linear congruence, the line p polar to Π meets $\mathfrak{Q}$ in two distinct points A and B. The hyperplanes tangent to $\mathfrak{Q}$ at A and B represent the two special complexes of Σ (which can be taken to be the two linear complexes that define Σ, cf. §12.2.4). In this case Γ consists of all the lines in $\mathbb{P}^3$ which meet the two skew lines of $\mathbb{P}^3$ that correspond to the two points A and B. Two such skew lines are the axes of the two special complexes that define Σ, and are called *directrices* (or *axes*). The two arrays of lines of the quadric $\mathcal{Q} = \mathfrak{Q} \cap \Pi$ represent two arrays of pencils of lines of Γ, that is, the pencils of lines having centers on one of the two directrices and contained in planes passing through the other.

12.3.2 (Special linear congruence). We now suppose that p and Π have a single point A in common. This point will belong to $\mathfrak{Q}$, cf. §5.5.4. The space Π and the polar line p are contained in the space S_4 tangent at A to $\mathfrak{Q}$ (because Π is the intersection of the polar hyperplanes of the points of p, and p passes through A). Therefore p and Π are both tangent at A to $\mathfrak{Q}$. In this case we may take as the linear complexes that define the pencil Σ the special linear complex $\mathcal{K}$ cut out on $\mathfrak{Q}$ by the tangent hyperplane $\alpha = \mathfrak{Q}_1(A)$ (which corresponds to the unique intersection A of p with $\mathfrak{Q}$) and a linear complex $\mathcal{H}$ (not special) obtained as the intersection of $\mathfrak{Q}$ with an arbitrary hyperplane β passing through the space S_3 polar to p (and distinct from $\mathfrak{Q}_1(A)$).

The quadric $\mathcal{Q} = \mathfrak{Q} \cap \Pi$ in this case is a quadric cone of $\mathbb{P}^3$. Note that the cone $\mathcal{Q}$ is irreducible. Indeed, if $\mathcal{Q}$ split into two distinct planes, then they would be planes of different systems having a line in common: that is, a plane α that represents a ruled plane of $\mathbb{P}^3$ and a plane β that represents a star of lines in $\mathbb{P}^3$ (cf. §12.1.2). The ruled plane and the star of lines have in common a pencil of lines (corresponding to the line of intersection of the two planes α and β) and therefore the ruled plane would contain the center of the star. The points of the polar line p would then be represented by the lines of this pencil and so p would be contained in $\mathfrak{Q}$, in contradiction to the present hypothesis. Moreover, $\mathfrak{Q}$ can not be a double plane because $\mathfrak{Q}$ is not specialized (cf. Exercise 5.5.10).

The pencil Σ contains the only special linear complex $\mathcal{K}$, while the only directrix of Γ is the axis r of $\mathcal{K}$. The congruence Γ consists of the lines that are supported by r and belong to the linear complex $\mathcal{H}$, which is not special, and which contains r because r is a line of Γ (inasmuch as r is the line represented in $\mathbb{P}^5$ by the vertex of the cone $\mathcal{Q}$, section of $\mathfrak{Q}$).

If x is a point of r, the lines of Γ passing through x are those of the pencil of lines that has center x and which is contained in the plane π corresponding to x under the null polarity ω associated to $\mathcal{H}$ (cf. Proposition 12.2.1). Indeed, the plane π passes through r because r is a line of Γ passing through x. Associating to x the plane π thus yields a correspondence $\omega' : r \to \mathcal{F}$ (defined by $x \mapsto \pi = \omega(x)$) between the (pointed) line r traced by x and the pencil $\mathcal{F}$ of planes with axis r. The correspondence ω' is a projectivity because it is induced by ω.

Therefore the congruence Γ of the type under examination (i.e., with polar line p tangent to $\mathfrak{Q}$) is the set of pencils of lines whose centers and whose planes are corresponding elements under a projectivity between a projective line r and a pencil of planes with axis r (that is, the line and the pencil have the same support r).

An example of a special linear congruence is the collection of lines that are tangent to a non-degenerate quadric $\mathcal{Q}$ in $\mathbb{P}^3$ at a point of a given line of $\mathcal{Q}$. A further example is given by the collection of lines that are tangent to a ruled surface (non-developable) at the points of one of its non-singular generators.

12.3.3 (Degenerate linear congruence). If the polar line p is contained in the quadric $\mathfrak{Q}$, the space Π polar to p passes through p and meets $\mathfrak{Q}$ in a doubly specialized quadric $\mathcal{Q}$, that is, in a pair of distinct planes passing through p. The two planes do not belong to the same system of planes of $\mathfrak{Q}$ since two planes of the same system have only a point in common (cf. §12.1.2). The congruence thus degenerates into a ruled plane and a star of lines having in common the pencil of lines represented by p (and thus the center of the star belongs to the plane).

12.4 Ruled surfaces in $\mathbb{P}^3$

An algebraic curve $\mathcal{C}$ contained in the Klein quadric $\mathfrak{Q}$ represents an algebraic totality ∞^1 of lines of $\mathbb{P}^3$, which is called an *algebraic ruled surface*. This totality of lines is a ruled surface $\mathcal{R}$ in $\mathbb{P}^3$. One immediately sees that

12.4. Ruled surfaces in $\mathbb{P}^3$ 409

- the order of the algebraic ruled surface $\mathcal{R}$ coincides with the order of the algebraic curve $\mathcal{C}$ that represents it.

Indeed, the order of the ruled surface $\mathcal{R}$ coincides with the number of generators that are supported by a generic line, that is, with the number d of the lines that belong to a generic special linear complex $\mathcal{K}$. The hyperplane $\mathcal{K}^*$ that represents $\mathcal{K}$ is thus a hyperplane which is tangent to the quadric $\mathcal{Q}$ and intersects the curve $\mathcal{C}$ in d points. By the genericity of $\mathcal{K}$, the hyperplane $\mathcal{K}^*$ is a generic hyperplane tangent to $\mathcal{Q}$; this suffices to conclude that d is the order of $\mathcal{C}$.

Furthermore, the following general fact holds, for which we give an elementary proof based on the criterion for developability of a (not necessarily algebraic) ruled surface, cf. §5.8.14 (see also [79]).

Proposition 12.4.1. *A necessary and sufficient condition for an algebraic ruled surface $\mathcal{R}$ to be developable is that the tangents of the curve $\mathcal{C}$ that represents it in $\mathbb{P}^5$ should all be contained in the Klein quadric $\mathcal{Q}$.*

Proof. In an affine coordinate system in $\mathbb{P}^3$ we represent the ruled surface $\mathcal{R}$ in the form
$$\begin{cases} x = \alpha(u) + tl(u), \\ y = \beta(u) + tm(u), \\ z = t, \end{cases} \tag{12.11}$$

where $\alpha(u)$, $l(u)$, $\beta(u)$, $m(u)$ are algebraic (or, more generally, differentiable) functions and we suppose that the generator $g(u)$ that corresponds to the value u of the parameter is singular, which means that the tangent plane at $\mathcal{R}$ along $g(u)$ is fixed (cf. §5.8.14).

The Plücker coordinates of the generic generator of the ruled surface $\mathcal{R}$ with equation (12.11) are the minors extracted from the matrix

$$\begin{pmatrix} \alpha(u) & \beta(u) & 0 & 1 \\ l(u) & m(u) & 1 & 0 \end{pmatrix}.$$

The curve $\mathcal{C}$ that represents $\mathcal{R}$ is thus the locus of the point (for convenience, the Plücker coordinates of $\mathbb{P}^5$ are here taken to be in the order $p_{01}, p_{23}, p_{02}, p_{31}, p_{03}, p_{12}$)

$$p_{01} : p_{23} : p_{02} : p_{31} : p_{03} : p_{12}$$
$$= \alpha(u)m(u) - \beta(u)l(u) : -1 : \alpha(u) : m(u) : -l(u) : \beta(u).$$

Since $\mathcal{C}$ is contained in $\mathcal{Q}$, the tangent of $\mathcal{C}$ at one of its generic points is tangent there to the quadric $\mathcal{Q}$; it is contained in $\mathcal{Q}$ if and only if another one of its points belongs to Ω. Another point of that tangent line is

$$p_{01} : p_{23} : p_{02} : p_{31} : p_{03} : p_{12}$$
$$= \alpha'(u)m(u) + \alpha(u)m'(u) - \beta'(u)l(u) - \beta(u)l'(u) : 0$$
$$: \alpha'(u) : m'(u) : -l'(u) : \beta'(u),$$

where $\alpha'(u), \beta'(u), l'(u), m'(u)$ are the derivatives with respect to u. The necessary and sufficient condition in order that this point belong to Ω (namely, that it satisfies (12.2)) is

$$\alpha'(u)m'(u) - \beta'(u)l'(u) = 0.$$

This is the same condition that is necessary and sufficient in order for the ruled surface to be developable (cf. §5.8.14). □

12.4.2 Linear ruled surfaces. A *linear ruled surface* is a ruled surface $\mathcal{R}$ represented by a section $\mathcal{C}$ of Ω with a plane π not lying in Ω, that is, by a conic $\mathcal{C}$ not contained in a plane of Ω. The order of a linear ruled surface is thus $d = 2$. [The fact that the curve representing $\mathcal{R}$ in $\mathbb{P}^5$ is a plane section of Ω justifies the name "linear ruled surface". Note also that to a line of Ω there corresponds a pencil of lines of $\mathbb{P}^3$ and the ruled surface $\mathcal{R}$ is the plane of that pencil.]

The plane π is the intersection of three hyperplanes in $\mathbb{P}^5$, whose sections with the quadric Ω represent three linear complexes. Thus a linear ruled surface $\mathcal{R}$ is the set of lines common to three linear complexes, that is, the base variety of a net Σ of linear complexes. Observe that in the net of hyperplanes of $\mathbb{P}^5$ that pass through the plane π there are ∞^1 hyperplanes tangent to Ω and thus the net Σ contains ∞^1 special linear complexes, whose axes constitute a linear ruled surface (represented by the section of Ω with the polar plane of π).

Remark 12.4.3. A quadric cone or a plane envelope of the second class (the set of tangents to a conic, cf. Section 5.3) are not linear ruled surfaces but do always have as image in $\mathbb{P}^5$ a conic belonging to the quadric Ω, and contained in a plane of Ω.

Indeed, if x is the vertex of the cone, consider the star of lines with center x and the plane β of Ω represented by the star. To the generators of the cone there correspond the points of a curve C contained in the plane β. Since every generic pencil contained in the star contains two generators of the cone, every line of β has two points in common with C which is therefore a conic. However, C is not obtained as a plane section of Ω (inasmuch as it belongs to the plane β contained in Ω) and thus the quadric cone is not a linear ruled surface.

More generally, an algebraic cone of order n is represented by an algebraic curve of order n contained in a plane β.

To the linear ruled surface $\mathcal{R}$ there is naturally associated another linear ruled surface $\mathcal{R}'$ represented by the polar plane π' of π with respect to Ω. We must

12.4. Ruled surfaces in $\mathbb{P}^3$

distinguish three cases according to whether π and π' are generic (and so mutually skew), or have a single point in common, or intersect in a line. We denote by γ and γ' the conics which are the sections of $\mathfrak{Q}$ by the planes π and π' respectively.

We start with the most interesting case, namely that in which π and π' are in generic position, hence mutually skew. In this case the two conic sections γ and γ' are non-degenerate. Each point of γ (respectively, of γ') is the reciprocal with respect to $\mathfrak{Q}$ of every point of γ' (respectively, of γ) whence, by Proposition 12.1.1, the ∞^1 lines of each of the two ruled surfaces are supported by the ∞^1 lines of the other. The lines of $\mathcal{R}'$ are thus the axes of the ∞^1 special linear complexes contained in Σ. Every line of the surface S', which is the locus of the lines of $\mathcal{R}'$, thus has infinitely many points in common with the surface S which is the locus of the lines of $\mathcal{R}$, and therefore the surface S' is contained in S. Analogous reasoning shows that S' contains S, whence the surfaces S and S' coincide (but are to be considered as two different *ruled* surfaces). Thus we have a doubly ruled surface in $\mathbb{P}^3$, that is, a quadric, and $\mathcal{R}$ and $\mathcal{R}'$ are its two arrays of lines. *A linear ruled surface of general type (i.e., such that the plane π that represents it is in general position with respect to its polar plane π') is thus an array of lines of a quadric in $\mathbb{P}^3$.*

Now suppose that π and π' have only one point in common, so that their join is an S_4. This S_4 will be tangent to $\mathfrak{Q}$ at the point $A = \pi \cap \pi'$ (because that point is the reciprocal with respect to $\mathfrak{Q}$ of every point of S_4). The point A belongs to $\mathfrak{Q}$ and thus represents a line a common to the two rulings $\mathcal{R}$ and $\mathcal{R}'$. The two conics γ and γ', the sections of $\mathfrak{Q}$ by the two planes π and π' respectively (which in this case are tangent to $\mathfrak{Q}$ at A), both split into two lines issuing from A, and thus both the ruled surfaces consist of a pair of pencils of lines that are supported by a. Therefore each of these pencils has its center on a, and lie in planes passing through a.

Now let m and n be the two lines that compose γ. They represent the pencils of lines in $\mathbb{P}^3$ which are generators of $\mathcal{R}$. Let M and N be the centers of these pencils and μ and ν their planes. Every line of $\mathfrak{Q}$ belongs to a plane of the first array and also to a plane of the second array. Thus m belongs to the planes that represent the ruled plane μ and the star of lines with center M, while n belongs to the planes that represent the ruled plane ν and the star of lines with center N. The two planes μ and ν meet in a; the two stars of lines have in common the line a that supports M and N.

The two lines m' and n' that compose γ' represent two pencils of lines that must be incident with all the lines of the two pencils represented by m and n. Thus the pencils represented by m' and n' are the two pencils with centers M and N and lying in ν and μ respectively.

The two rulings appear as in the Figure 12.1, where the lines of one of them are solid, while those of the other are dotted.

If π and π' have a line ℓ in common, it is the locus of self-reciprocal points and thus is contained in $\mathfrak{Q}$. The line ℓ represents a pencil of lines in $\mathbb{P}^3$. The two conics

γ and γ' are doubly degenerate and consist of the line ℓ counted twice. Each of the two rulings $\mathcal{R}$ and $\mathcal{R}'$ is the pencil of lines represented by ℓ counted twice.

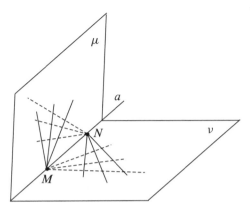

Figure 12.1

12.4.4 Ruled cubics in $\mathbb{P}^3$. Let $\mathcal{R}^3$ be a ruled cubic in $\mathbb{P}^3$, represented on $\mathfrak{Q}$ by an algebraic curve $\mathcal{C}$ of order 3.

The cubic $\mathcal{C}$ can be planar; the plane π that contains it then lies on $\mathfrak{Q}$ and so the generators of $\mathcal{R}^3$ (which do not all lie in one plane) pass through a point and $\mathcal{R}^3$ is a cubic cone.

If $\mathcal{C}$ is not planar, it belongs to an S_3 in $\mathbb{P}^5$. If this S_3 is not tangent to $\mathfrak{Q}$, its section with $\mathfrak{Q}$ is a non-degenerate quadric $\mathcal{Q}$ and on $\mathcal{Q}$ the curve $\mathcal{C}$ is of type $(2, 1)$. The polar line s of S_3 with respect to $\mathfrak{Q}$ meets $\mathfrak{Q}$ in two distinct points A and B which are not reciprocals with respect to $\mathfrak{Q}$. Thus they are images of two distinct lines a and b which are mutually skew and incident with all the generators of $\mathcal{R}^3$ (because A and B are reciprocals of every point of $\mathcal{C}$). Thus $\mathcal{R}^3$ has two rectilinear directrices a and b. In other words, $\mathcal{R}^3$ belongs to the non-special linear congruence represented by the quadric section $\mathcal{Q}$. One of the two directrices is certainly simple, because otherwise all the lines of the congruence would belong to $\mathcal{R}$.

Let a be the simple generator of $\mathcal{R}^3$ (cf. Section 5.8). A generic plane σ passing through a meets $\mathcal{R}^3$ (outside of a) in a conic γ. For each point of γ there passes a generator g of $\mathcal{R}^3$ supported by a (and thus lying on σ) and also supported by b (in the point S where b meets σ). Thus γ consists of a pair of lines issuing from a point of b (Figure 12.2), and b is a double generator for $\mathcal{R}^3$ (cf. Exercise 5.8.21).

Let $\{r\}$, $\{s\}$ be the two systems of lines on $\mathcal{Q}$, and let, for example, $\{s\}$ be the ruling of chords of $\mathcal{C}$.

Every line $s \in \{s\}$ represents a pencil of lines of S_3 containing two generators g_1, g_2 of $\mathcal{R}^3$. These generators are generally distinct and have a double point of $\mathcal{R}^3$

in common. The locus of these double points (which vary with s, since $\mathcal{R}^3$ is not a cone) is the double directrix b of the ruled surface.

Since the straight lines g_1, g_2 meet a (generally in distinct points), the planes $\langle g_1, g_2 \rangle$ all pass through a. In $\{s\}$ there are two tangent lines of $\mathcal{C}$. The points of contact, which are double points of the linear series g_2^1 cut out on $\mathcal{C}$ by $\{s\}$, represent two singular generators of $\mathcal{R}^3$ (that is, straight lines along which the tangent plane to $\mathcal{R}^3$ is fixed); they meet b in cuspidal points of the ruled surface.

Every line $r \in \{r\}$ represents a pencil of straight lines in S_3 containing only one generator of $\mathcal{R}^3$. The plane of this pencil describes the pencil with axis b.

The ruled surface $\mathcal{R}^3$ in this case is a general ruled cubic (cf. 5.8.22).

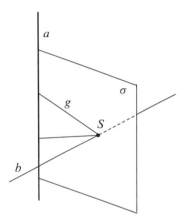

Figure 12.2

Now suppose that the space S_3 containing $\mathcal{C}$ is tangent to $\mathfrak{Q}$ at a point A. The polar line of S_3 with respect to $\mathfrak{Q}$ is now tangent to $\mathfrak{Q}$ at A, and the quadric Q is a cone with vertex A. Here $\mathcal{R}^3$ belongs to a special congruence with axis a.

The cubic $\mathcal{C}$ will pass through the point A, which is thus the image of a line a that is a rectilinear directrix of $\mathcal{R}^3$, but also a generator. Let g be a simple generator of $\mathcal{R}^3$ and G the corresponding point in $\mathbb{P}^5$. Since g and a are incident, the line r_{AG} is contained in $\mathfrak{Q}$ and through it there pass two planes of $\mathfrak{Q}$ (one for each system, cf. Section 12.1) which represent the ruled plane $\langle a, g \rangle$ and the star of lines with center $a \cap g$ (Figure 12.3).

The space S_3 joining the two planes is contained in the space S_4 tangent at A to $\mathfrak{Q}$ (such an S_4 necessarily contains $\mathcal{C}$). Thus it is tangent to $\mathcal{C}$ at A. Therefore this S_3 does not meet the cubic $\mathcal{C}$ outside of A and G. Hence the ruled plane $\langle a, g \rangle$ does not contain other lines of $\mathcal{R}^3$. Similarly, the star of lines with center $a \cap g$

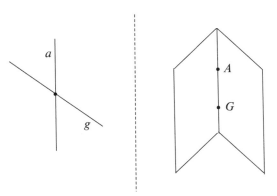

Figure 12.3

contain no lines of $\mathcal{R}^3$ except a and g. On the other hand, the section of $\mathcal{R}^3$ by the plane $\langle a, g \rangle$, containing a and g, must contain another line; this other line of $\mathcal{R}^3$ must therefore coincide with a (not with g because g is a generic line of $\mathcal{R}^3$). Thus every plane passing through a meets $\mathcal{R}^3$ in three lines, two of which coincide with a, whence the line a is double for $\mathcal{R}^3$. So each plane passing through a contains a generator of $\mathcal{R}^3$ (besides a) and in particular there is a plane for which this generator is the line a, and which therefore osculates $\mathcal{R}^3$ along a. In this case the surface $\mathcal{R}^3$ is a Cayley ruled cubic surface (cf. Exercise 5.8.23).

In conclusion, the ruled cubics in $\mathbb{P}^3$ are the following: cubic cones (elliptic or rational), ruled surfaces with two directrices, one simple and one double (general ruled cubics, whose equation may be written in the form $x_0 x_1^2 + x_2 x_3^2 = 0$), and Cayley ruled cubics (whose equation is usually written in the form $x_2^3 - 2x_1 x_2 x_3 + x_0 x_3^2 = 0$).

12.5 Grassmann coordinates and Grassmann varieties

In analogy with what we have done in Section 12.1 for the lines in $\mathbb{P}^3$, it is also possible to define coordinates, called Grassmann coordinates, for the subspaces S_h of a given dimension $h > 1$ in a projective space $\mathbb{P}^n$. The Plückerian coordinates for the lines in $\mathbb{P}^3$ are the simplest special case. Let $t_0, t_1, \ldots, t_n$ be homogeneous projective coordinates in $\mathbb{P}^n$.

We consider a matrix P of type $(h+1, n+1)$ with entries from the field K. We suppose that $h \leq n$ and denote the minor of order $h+1$ formed by the columns in positions $i_0, i_1, \ldots, i_h$ ($i_0 < i_1 < \cdots < i_h$) by $(i_0 i_1 \ldots i_h)_P$.

If P has rank $h + 1$, one obtains in this way $\binom{n+1}{h+1}$ elements of K, and not all of them are zero.

12.5. Grassmann coordinates and Grassmann varieties

If Δ is a square matrix of order $h+1$, the matrix $Q = \Delta P$ is again of type $(h+1, n+1)$ and we have

$$(i_0 i_1 \ldots i_h)_Q = \det(\Delta)(i_0 i_1 \ldots i_h)_P.$$

Assuming this, let $P_j = [z_0^{(j)}, z_1^{(j)}, \ldots, z_n^{(j)}]$, for $j = 0, \ldots, h$, be $h+1$ linearly independent points of $\mathbb{P}^n$ and suppose that P is the matrix (of type $(h+1, n+1)$) whose rows consist of the coordinates of the points $P_0, \ldots, P_h$.

In the space S_h defined by the points $P_0, \ldots, P_h$ consider other $h+1$ linearly independent points

$$Q_i = \left[\sum_{j=0}^{h} \lambda_{ij} z_0^{(j)}, \sum_{j=0}^{h} \lambda_{ij} z_1^{(j)}, \ldots, \sum_{j=0}^{h} \lambda_{ij} z_n^{(j)} \right], \quad i = 0, \ldots, h.$$

The matrix $\Delta = (\lambda_{ij}) \in M_{h+1}(K)$ has non-zero determinant and if we let Q be the matrix that has as its rows the coordinates of the points Q_i, we have $Q = \Delta P$. Therefore, the minors of order $h+1$ of the two matrices P and Q are proportional.

One is then naturally led to take as the homogeneous coordinates of $S_h = J(P_0, \ldots, P_h)$ the numbers $(i_0 i_1 \ldots i_h)_P$: indeed, they are not all zero, and their mutual ratios are determined by S_h and not by the particular choice of $h+1$ independent points that define it, nor by the choice of the coordinates of these points. We will say that $(i_0 i_1 \ldots i_h)_P$ are the *Grassmann coordinates* of S_h. They can be interpreted as homogeneous projective coordinates $X_{i_0 i_1 \ldots i_h}$ in a projective space $\mathbb{P}^N$, $N = \binom{n+1}{h+1} - 1$ (in the case $h = 1, n = 3$ we rediscover in this fashion the Plücker line coordinates, cf. Section 12.1).

Remark 12.5.1. Note that, in agreement with equations (12.1), the Grassmann coordinate $(i_0 i_1 \ldots i_h)_P$ of $S_h = J(P_0, \ldots, P_h)$ is obtained by substituting the coordinates of the points $P_0, \ldots, P_h$ into the minor $p_{i_0 i_1 \ldots i_h}$ consisting of the columns in positions $i_0, i_1, \ldots, i_h$ of the matrix

$$\begin{pmatrix} t_0^{(0)} & t_1^{(0)} & \cdots & t_n^{(0)} \\ t_0^{(1)} & t_1^{(1)} & \cdots & t_n^{(1)} \\ \vdots & \vdots & & \vdots \\ t_0^{(h)} & t_1^{(h)} & \cdots & t_n^{(h)} \end{pmatrix},$$

where $t_0^{(j)}, t_1^{(j)}, \ldots, t_n^{(j)}$ are to be considered as $h+1$ distinct copies of the series of variables $t_0, t_1, \ldots, t_n$ consisting of the homogeneous coordinates of the initial space $\mathbb{P}^n$.

Example 12.5.2. Let us consider the points $P_0 = [1, 0, 1, 0, 0]$, $P_1 = [2, 1, 0, -1, 1]$ and $P_2 = [0, 2, 1, 3, 0,]$ in $\mathbb{P}^4$. The coordinates of the space $S_2 = J(P_0, P_1, P_2)$ in $\mathbb{P}^9$ are given by the third order minors $(012)_P = \begin{vmatrix} 1 & 0 & 1 \\ 2 & 1 & 0 \\ 0 & 2 & 1 \end{vmatrix}$, $(013)_P$, $(014)_P$, $(023)_P$, $(024)_P$, $(034)_P$, $(123)_P$, $(124)_P$, $(134)_P$, $(234)_P$ extracted from the matrix

$$P = \begin{pmatrix} 1 & 0 & 1 & 0 & 0 \\ 2 & 1 & 0 & -1 & 1 \\ 0 & 2 & 1 & 3 & 0 \end{pmatrix}.$$

Note too that the plane S_2 that joins the three fundamental points A_i, A_j, A_k of $\mathbb{P}^n$ has $(ijk)_P$ as its only non-zero coordinate, where P is the matrix, of type $(3, n + 1)$, whose rows consist of the coordinates of the points A_i, A_j, A_k.

Here are some properties of Grassmann coordinates.

12.5.3 (Incidence condition for dual dimensional subspaces). We consider two subspaces $S_h = J(P_0, P_1, \ldots, P_h)$ and $S_{n-h-1} = J(Q_0, Q_1, \ldots, Q_{n-h-1})$ of $\mathbb{P}^n$ having dual dimensions. If P and Q are the matrices whose rows are formed with the coordinates of the points P_j and Q_i respectively, the matrix $\binom{P}{Q}$ is a square matrix of order $n + 1$. The necessary and sufficient condition for the two spaces S_h and S_{n-h-1} to be incident is obtained by writing the condition for $P_0, P_1, \ldots, P_h$, $Q_0, Q_1, \ldots, Q_{n-h-1}$ not to be $n+1$ linearly independent points, namely $\det \binom{P}{Q} = 0$. Using Laplace's rule to develop this determinant via minors extracted from the matrix P, one finds a bilinear relation between the Grassmann coordinates of S_h and those of S_{n-h-1}. More precisely we have

$$\sum (i_0 i_1 \ldots i_h)_P (i_0 i_1 \ldots i_h)_Q^* = 0, \tag{12.12}$$

where $(i_0 i_1 \ldots i_h)_Q^*$ denotes the cofactor in $\binom{P}{Q}$ of the minor $(i_0 i_1 \ldots i_h)_P$.

12.5.4. Let $J(P_0, \ldots, P_h) = S_h$ be the space defined by the $h + 1$ independent points $P_j = [z_0^{(j)}, z_1^{(j)}, \ldots, z_n^{(j)}]$, $j = 0, \ldots, h$, and let $Y = [y_0, y_1, \ldots, y_n]$ be the point at which S_h meets the S_{n-h} with equations

$$t_{i_0} = t_{i_1} = \cdots = t_{i_{h-1}} = 0.$$

Further suppose that

$$(y_0, y_1, \ldots, y_n) = (\lambda_0, \lambda_1, \ldots, \lambda_h) P,$$

that is,

$$y_i = \lambda_0 z_i^{(0)} + \lambda_1 z_i^{(1)} + \cdots + \lambda_h z_i^{(h)}, \quad i = 0, \ldots, n.$$

12.5. Grassmann coordinates and Grassmann varieties

The fact that $y_{i_0} = y_{i_1} = \cdots = y_{i_{h-1}} = 0$ means that $(\lambda_0, \lambda_1, \ldots, \lambda_h)$ is a non-trivial solution of the system of h linear homogeneous equations

$$\begin{cases} \lambda_0 z_{i_0}^{(0)} + \lambda_1 z_{i_0}^{(1)} + \cdots + \lambda_h z_{i_0}^{(h)} = 0, \\ \lambda_0 z_{i_1}^{(0)} + \lambda_1 z_{i_1}^{(1)} + \cdots + \lambda_h z_{i_1}^{(h)} = 0, \\ \quad \vdots \\ \lambda_0 z_{i_{h-1}}^{(0)} + \lambda_1 z_{i_{h-1}}^{(1)} + \cdots + \lambda_h z_{i_{h-1}}^{(h)} = 0, \end{cases}$$

and thus one can suppose that $\lambda_0, \lambda_1, \ldots, \lambda_h$ are the minors of order h extracted from the matrix of coefficients of that system (taken with alternating signs). So the coordinates of the point Y are

$$y_i = (i_0 i_1 \ldots i_h i)_P, \quad i = 0, \ldots, n.$$

The non-zero coordinates are those whose index i is any one of the numbers $i_{h+1}, i_{h+2}, \ldots, i_n$.

12.5.5. The space S_h that joins the fundamental points $A_{i_0}, A_{i_1}, \ldots, A_{i_h}$ has only one non-zero coordinate. It is the coordinate $(i_0 i_1 \ldots i_h)_P$ (cf. Example 12.5.2). It follows that there is no (non-zero) linear homogeneous polynomial in the indeterminates $X_{i_0 i_1 \ldots i_h}$ that vanishes when the same indeterminates are replaced by the coordinates of an arbitrary space S_h. In fact, if $\sum_{i_0 i_1 \ldots i_h} \lambda_{i_0 i_1 \ldots i_h} X_{i_0 i_1 \ldots i_h}$ is a linear homogeneous polynomial in the indeterminates $X_{i_0 i_1 \ldots i_h}$ that vanishes when the indeterminates are replaced by the coordinates of S_h then the coefficient $\lambda_{i_0 i_1 \ldots i_h}$ must be zero.

12.5.6. One can also arrive at the Grassmann coordinates of a space S_h via a dual procedure.

Namely, we define S_h as the intersection of $n - h$ linearly independent hyperplanes:

$$\begin{cases} a_{00} t_0 + a_{01} t_1 + \cdots + a_{0n} t_n = 0, \\ \quad \vdots \\ a_{n-h-1\,0} t_0 + a_{n-h-1\,1} t_1 + \cdots + a_{n-h-1\,n} t_n = 0. \end{cases}$$

Let $A = (a_{ij})$ be the matrix (of type $(n - h, n + 1)$) of coefficients. If one uses other hyperplanes to define S_h, one obtains a second system whose matrix B of coefficients is $B = \Delta A$, where Δ is a non-degenerate square matrix of order $n - h$. One can then define new homogeneous coordinates for S_h by taking the minors of order $n - h$ of A. Not all such minors are zero, and they have mutual ratios which are determined by our S_h. Note that the number of minors of order $n - h$ extracted from A is

$$\binom{n+1}{n-h} = \binom{n+1}{h+1}.$$

One can prove (cf. [80, pp. 132–134]) that these new coordinates (taken in suitable order and with suitable signs) are proportional to those derived above.

Proposition–Definition 12.5.7. *The Grassmann coordinates $(i_0 i_1 \ldots i_h)_P$ of a space S_h in $\mathbb{P}^n$ annul a system of quadratic forms Φ_α in the $\binom{n+1}{h+1}$ indeterminates $X_{i_0 i_1 \ldots i_h}$. Moreover, the homogeneous ideal $\mathfrak{p}$ generated by those quadratic forms is prime. The ideal $\mathfrak{p}$ is called the **ideal of the Plücker relations**.* (In the case $n = 3$, $h = 1$, that is, in the case of the lines in ordinary space, one finds only the quadratic form (12.2).)

Proof. Let $t_0, \ldots, t_n$ be homogeneous coordinates in the projective space $\mathbb{P}^n$. We consider $h + 1$ linearly independent points $P_0, \ldots, P_h$ in $\mathbb{P}^n$ such that $S_h = J(P_0, \ldots, P_h)$ and let P be the matrix whose rows consist of the coordinates of the points $P_0, \ldots, P_h$.

If a point $A = [a_0, \ldots, a_n]$ belongs to S_h, the matrix, of type $(h + 2, n + 1)$,

$$P' = \begin{pmatrix} a_0 & a_1 & \cdots & a_{n-1} & a_n \\ & & P & & \end{pmatrix},$$

has rank $h + 1$ and thus all of its minors of rank $h + 2$ are zero. Therefore, developing these minors with respect to the first row, one sees that in order for A to belong to S_h it is necessary and sufficient that suitable linear combinations of the Grassmann coordinates $(i_0 i_1 \ldots i_h)_P$ of S_h should be zero. Moreover, the coefficients of these linear combinations are $h + 2$ elements of K suitably chosen among $a_0, \ldots, a_n$. Note that these linear combinations are bilinear forms

$$\Phi_\alpha(t_0, \ldots, t_n, X_{i_0 i_1 \ldots i_h}) = 0$$

(linear in the $n + 1$ indeterminates $t_0, \ldots, t_n$ and linear in the $\binom{n+1}{h+1}$ indeterminates $X_{i_0 i_1 \ldots i_h}$) evaluated at the point A and at the space $S_h = J(P_0, \ldots, P_h)$ respectively.

We have seen in 12.5.5 that the points where a space S_h meets the $(n - h)$-dimensional faces of the fundamental $(n + 1)$-hedron $A_0 A_1 \ldots A_n$ (namely the spaces S_{n-h} that one obtains by joining $n - h + 1$ of the points A_j) have as non-zero coordinates suitable Grassmann coordinates of S_h. Therefore, if in the preceding discussion we take as the point A each of the points in which S_h meets the $(n - h)$-dimensional faces of the fundamental $(n + 1)$-hedron, we obtain the quadratic equations

$$\Phi_\alpha(X_{i_0 i_1 \ldots i_h}) = 0$$

in only the $\binom{n+1}{h+1}$ indeterminates $X_{i_0 i_1 \ldots i_h}$, which are satisfied by the Grassmann coordinates $(i_0 i_1 \ldots i_h)_P$ of S_h.

Consider the homogeneous ideal $\mathfrak{p} \subset K[\ldots, X_{i_0 i_1 \ldots i_h}, \ldots]$ generated by those quadratic forms. To prove that $\mathfrak{p}$ is prime consider the morphism

$$\varphi: K[\ldots, X_{i_0 i_1 \ldots i_h}, \ldots] \to K[\ldots, t_0^{(j)}, \ldots, t_n^{(j)}, \ldots]$$

defined by $X_{i_0 i_1 \ldots i_h} \mapsto p_{i_0 i_1 \ldots i_h}$, where $p_{i_0 i_1 \ldots i_h}$ is the minor that, evaluated in the points $P_0, \ldots P_h$, furnishes the Grassmann coordinates $(i_0 i_1 \ldots i_h)_P$ of S_h (cf. Remark 12.5.1). We have

$$K[\ldots, X_{i_0 i_1 \ldots i_h}, \ldots]/\ker(\varphi) \cong K[\ldots, p_{i_0 i_1 \ldots i_h}, \ldots] \subset K[\ldots, t_0^{(j)}, \ldots, t_n^{(j)}, \ldots].$$

Since the ring of polynomials $K[\ldots, t_0^{(j)}, \ldots, t_n^{(j)}, \ldots]$ is an integral domain, it follows that the kernel $\ker(\varphi)$ is a prime ideal.

We conclude by noting that $\mathfrak{p} = \ker(\varphi)$. The inclusion $\mathfrak{p} \subset \ker(\varphi)$ is obvious; the converse is easily obtained via a standard argument from the theory of polynomials (already discussed in Exercise 3.4.11 (2)).

We also observe that techniques of computational algebra permit the explicit calculation of a set of generators of the ideal $\mathfrak{p}$ (see [76, Tutorial 35]). □

12.5.8 Grassmann varieties. With the preceding notations, we put $N = \binom{n+1}{h+1} - 1$ and consider the projective space $\mathbb{P}^N$ where $X_{i_0 i_1 \ldots i_h}$ are homogeneous projective coordinates.

In $\mathbb{P}^N$ consider the irreducible variety $\mathbb{G}(h, n)$, associated to the homogeneous prime ideal generated by the quadratic forms Φ_α defined in Proposition–Definition 12.5.7. With an argument similar to that used in the case $n = 3, h = 1$, it is not difficult to prove the following assertion:

- There exists a bijective correspondence without exceptions between the subspaces S_h of $\mathbb{P}^N$ and the homogeneous $(N + 1)$-uples $X_{i_0 i_1 \ldots i_h}$ that annihilate the quadratic forms Φ_α; that is, between the $S_h \subset \mathbb{P}^N$ and the points of $\mathbb{G}(h, n)$.

Observation 12.5.5 ensures, moreover, that $\mathbb{G}(h, n)$ has $\mathbb{P}^N$ as its embedding space.

Thus we have constructed an irreducible projective variety, $\mathbb{G}(h, n)$, whose points are in bijective correspondence without exceptions with the subspaces S_h in $\mathbb{P}^n$: this variety is called the *Grassmann variety* (or *Grassmannian*) with *indices* n and h.

The dimension of $\mathbb{G}(h, n)$ coincides with the number $d(n, h)$ of parameters on which the determination of a space S_h in $\mathbb{P}^n$ depends. We have

$$\dim(\mathbb{G}(h, n)) = (n - h)(h + 1). \tag{12.13}$$

Indeed, since a linear space S_h in $\mathbb{P}^n$ is determined by the choices of $h + 1$ linearly independent points of $\mathbb{P}^n$, the number $d(n, h)$ of parameters on which the determination of a space S_h in $\mathbb{P}^n$ depends can not be more than $\binom{n+1}{h+1} - 1$. To calculate $d(n, h)$ we observe that the system $\infty^{d(n,h)}$ of S_h in $\mathbb{P}^n$ intersects a hyperplane $\mathbb{P}^{n-1}$ in the system of S_{h-1} of $\mathbb{P}^{n-1}$. On the other hand, every S_h in $\mathbb{P}^n$ passing through a given S_{h-1} of $\mathbb{P}^{n-1}$ may be obtained by joining S_{h-1} with the points of a space S_{n-h} skew to S_{h-1}. Thus one has the relations

$$d(n, h) = d(n-1, h-1) + n - h,$$
$$d(n-1, h-1) = d(n-2, h-2) + n - h,$$
$$\vdots$$
$$d(n-h, 0) = n - h,$$

from which one deduces that

$$d(n, h) = (n-h)(h+1),$$

that is, equation (12.13).

In $\mathbb{P}^n$ we consider a non-degenerate projectivity ω defined by

$$\begin{cases} t'_0 = a_{00}t_0 + a_{01}t_1 + \cdots + a_{0n}t_n, \\ t'_1 = a_{10}t_0 + a_{11}t_1 + \cdots + a_{1n}t_n, \\ \quad\vdots \\ t'_n = a_{n0}t_0 + a_{n1}t_1 + \cdots + a_{nn}t_n, \end{cases}$$

with $A = (a_{ij})$ a non-degenerate matrix. Consider the points $P_0, P_1, \ldots, P_h$ and their images $Q_0, Q_1, \ldots, Q_h$ under ω. If we set $P_j = [x_{j0}, x_{j1}, \ldots, x_{jn}]$, then we have

$Q_j = \omega(P_j)$
$= [a_{00}x_{j0} + \cdots + a_{0n}x_{jn}, a_{10}x_{j0} + \cdots + a_{1n}x_{jn}, \ldots, a_{n0}x_{j0} + \cdots + a_{nn}x_{jn}],$

and one sees easily that the numbers $(i_0 i_1 \ldots i_h)_Q$ are linear combinations of the numbers $(i_0 i_1 \ldots i_h)_P$. Therefore the projectivity ω induces a projectivity Ω in $\mathbb{P}^N$ which one shows to be non-degenerate. To ω there thus corresponds a projectivity Ω of $\mathbb{P}^N$ that maps the Grassmann variety $\mathbb{G}(h, n)$ into itself. In fact, since there are certainly projectivities ω of $\mathbb{P}^n$ that transform any given S_h of $\mathbb{P}^n$ into any other S_h, we have that given two points of $\mathbb{G}(h, n)$ there exist projectivities Ω of $\mathbb{P}^N$ which send one into the other. This means that the restrictions of the projectivities Ω to $\mathbb{G}(h, n)$ form a transitive group G; equivalently, the action of the group G

$$G \times \mathbb{G}(h, n) \to \mathbb{G}(h, n)$$

defined by $(\Omega, x) \mapsto \Omega(x)$ is transitive. In conclusion, the variety $\mathbb{G}(h, n)$ is a variety on which there acts a transitive group of projectivities. It then follows that $\mathbb{G}(h, n)$ is non-singular (this conclusion is almost obvious since G in our case is a group of projectivities; for a general statement see for example [104, Theorem (4.3.7)]).

12.5. Grassmann coordinates and Grassmann varieties

The "dual" Grassmann coordinates (cf. 12.5.6) are essentially obtained by considering, instead of the space S_h of $\mathbb{P}^n$, its dual space Σ_{n-h-1} (of dimension $n - h - 1$). Indeed, S_h is the center of a star ∞^{n-h-1} of hyperplanes of $\mathbb{P}^n$. Thus we see that, interchanging the roles of S_h and Σ_{n-h-1}, there is a natural identification

$$\mathbb{G}(h, n) = \mathbb{G}(n - h - 1, n)$$

between Grassmann varieties.

Remark 12.5.9. Note that the variety $\mathbb{G}(1, 3)$ studied in Section 12.1 is the simplest example of a Grassmann variety which is different from a projective space. The variety $\mathbb{G}(1, 2)$ is the dual projective plane $\mathbb{P}^{2*}$, and, in general, $\mathbb{G}(n-1, n) = \mathbb{P}^{n*}$. Moreover, obviously $\mathbb{G}(0, n) = \mathbb{P}^n$.

We limit ourselves here to stating the following properties; for the proofs, further developments, and complementary material we refer the reader to [48], [52, Vol. 2, pp. 309–387] and to [98].

(1) The Grassmann variety $\mathbb{G}(h, n)$ is a rational variety. This means that there exists a birational isomorphism $\mathbb{G}(h, n) \to \mathbb{P}^d$, where $d = d(n, h) = (n - h)(h + 1)$. Note, however, that if one excludes the extreme cases $h = 0$ and $h = n - 1$, for which $d(n, h) = n$ and $\mathbb{G}(h, n) = \mathbb{P}^n$, it is not possible to represent the variety $\mathbb{G}(h, n)$ birationally and without exceptions onto a linear space $\mathbb{P}^d$, that is, there is no surjective birational *morphism* $\mathbb{G}(h, n) \to \mathbb{P}^d$.

(2) If one excludes the cases $\mathbb{G}(0, n) = \mathbb{P}^n$ and $\mathbb{G}(n - 1, n) = \mathbb{P}^{n*}$ and the case $n = 3, h = 1$ (in which $\mathbb{G}(1, 3)$ is a non-specialized quadric in $\mathbb{P}^5$), the Grassmann variety $\mathbb{G}(h, n)$ is not a complete intersection, that is, it can not be realized as the intersection (not even in the set-theoretic sense) of $N - d$ hypersurfaces of $\mathbb{P}^N$ ($d = d(n, h)$).

(3) The Grassmann variety $\mathbb{G}(h, n)$ is a *factorial variety*. This means that each of its irreducible subvarieties V of codimension 1 is the complete intersection of $\mathbb{G}(h, n)$ with a hypersurface F of $\mathbb{P}^N$. In other words, if $\mathfrak{p}$ is the homogeneous prime ideal associated to $\mathbb{G}(h, n)$, generated by the quadratic forms Φ_α (cf. Proposition–Definition 12.5.7), the homogeneous ideal of V is of the type $I(V) = \mathfrak{p} + (f)$, with f a homogeneous polynomial; and $f = 0$ is the equation of the hypersurface F.

(4) The hyperplane sections of $\mathbb{G}(h, n)$ are also rational and factorial.

The hyperplane section of a Grassmann variety is a "Fano variety" (for example the quadric Q in $\mathbb{P}^4$ is a hyperplane section of $\mathbb{G}(1, 3)$ in $\mathbb{P}^5$). Observe also that (except in the trivial cases $\mathbb{G}(0, n) = \mathbb{G}(n - 1, n) = \mathbb{P}^n$) a Grassmann variety $\mathbb{G}(h, n)$ can not be obtained as a hyperplane section of a given non-singular variety X, with the unique exception of the variety $\mathbb{G}(1, 3)$, in which case X is a quadric in $\mathbb{P}^6$ (for this see [40]). We refer the reader to [41] and [12] for definitions, references, and a complete development of the "theory of hyperplane sections".

(5) On $\mathbb{G}(h,n)$ there are two systems of linear spaces:

   i) a first system, parameterized by $\mathbb{G}(h+1,n)$, consists of the spaces S_{h+1} each of which is the Grassmann variety of the subspaces S_h of a given S_{h+1} in $\mathbb{P}^n$;

   ii) a second system, parameterized by $\mathbb{G}(h-1,n)$, consists of the spaces S_{n-h} each of which is the Grassmannian variety of the spaces S_h that contain a given S_{h-1} in $\mathbb{P}^n$.

We note that in the case $n=3$, $h=1$, studied in Section 12.1, the first system, parameterized by $\mathbb{G}(2,3) = \mathbb{P}^3$, consists of ∞^3 ruled planes; while the second system, parameterized by $\mathbb{G}(0,3) = \mathbb{P}^3$, consists of ∞^3 planes that represent the stars of lines of $\mathbb{P}^3$ (cf. §12.1.2).

(6) The order of $\mathbb{G}(h,n)$ coincides with the number of S_h in $\mathbb{P}^n$ that are supported by $d(n,h)$ generic linear spaces S_{n-h-1}. Indeed, the order of $\mathbb{G}(h,n)$ is the number of points that $\mathbb{G}(h,n)$ has in common with a generic linear subspace of dimension $N - d(n,h)$ in $\mathbb{P}^N$, $N = \binom{n+1}{h+1} - 1$. Such a linear subspace is given by the intersection of $d(n,h)$ generic hyperplanes of $\mathbb{P}^N$. The incidence relation (12.12) between a space S_h and a space S_{n-h-1} in $\mathbb{P}^n$ may be interpreted as the equation of a hyperplane in $\mathbb{P}^N$. We then have (cf. [78])

$$\deg(\mathbb{G}(h,n)) = \frac{1!\,2!\ldots h!\,[(h+1)(n-h)]!}{(n-h)!\,(n-h+1)!\ldots n!}.$$

12.6 Further properties of $\mathbb{G}(1,n)$ and applications

Here we offer a few exercises which furnish interesting additions to the theory developed in this chapter. The notations are those of Section 12.5.

12.6.1. *Show that four lines of $\mathbb{P}^3$ have two common transversals (that is, two lines that are supported by the four given lines).*

A line r that meets four lines ℓ_i of $\mathbb{P}^3$ is represented in $\mathbb{P}^5$ by a point R belonging to the Klein quadric $\mathfrak{Q}$, which is reciprocal of the four points $L_i \in \mathfrak{Q}$ that represent the lines ℓ_i, $i = 1, 2, 3, 4$. The points L_i define a subspace S_3 in $\mathbb{P}^5$ whose polar line p meets $\mathfrak{Q}$ in two points, A and B. The lines a and b of $\mathbb{P}^3$ that correspond to the points A and B are the required common transversals. It is hardly necessary to note that what has been said refers to the case in which the lines ℓ_i are generic, that is, each of them does not belong to the quadric of the lines that are supported by the other three.

12.6. Further properties of $\mathbb{G}(1,n)$ and applications

12.6.2. *Let $\mathcal{K}$ be a linear complex of lines in $\mathbb{P}^3$. Prove that the lines of $\mathcal{K}$ that are supported by a line r_1 all meet a second line r_2.*

It suffices to observe that the lines of $\mathcal{K}$ that are supported by r_1 form a special linear complex $\mathcal{K}_1$, and in the pencil of complexes determined by $\mathcal{K}$ and $\mathcal{K}_1$ there is another special complex $\mathcal{K}_2$ (cf. §12.2.4). The lines of $\mathcal{K}$ that are supported by r_1 are thus contained in the two special complexes $\mathcal{K}_1$ and $\mathcal{K}_2$ and therefore are supported by the two axes r_1 and r_2 of $\mathcal{K}_1$ and $\mathcal{K}_2$.

12.6.3. (Curves in $\mathbb{P}^3$ whose tangents belong to a linear complex $\mathcal{K}$). The curves in $\mathbb{P}^3$ whose tangents belong to a linear complex $\mathcal{K}$ are called *curves of the linear complex*.

An example is the twisted cubic. We know that a twisted cubic C determines a null polarity of $\mathbb{P}^3$ under which a point of C and its osculating plane are homologous. This null polarity defines a linear complex to which the tangents of C belong (cf. Theorem–Definition 7.4.9 and Proposition 12.2.1 (2)).

Returning to the general case, we write the equation of $\mathcal{K}$ in the reduced form

$$p_{01} + k p_{23} = 0 \quad (k = 0 \text{ if and only if } \mathcal{K} \text{ is special}), \tag{12.14}$$

and suppose that all the tangents of the curve

$$\mathcal{L}: x = \varphi(z), \quad y = \theta(z)$$

belong to $\mathcal{K}$. The tangent line to $\mathcal{L}$ at the point P with homogeneous coordinates $(\varphi(z), \theta(z), z, 1)$ contains the point $[\varphi'(z), \theta'(z), 1, 0]$, where the primed symbols denote derivatives with respect to z. Therefore, its Plücker coordinates p_{01} and p_{23} are $p_{01} = \varphi(z)\theta'(z) - \varphi'(z)\theta(z)$ and $p_{23} = -1$, so that $\mathcal{L}$ belongs to $\mathcal{K}$ if and only if, for some $k \in \mathbb{C}$,

$$\varphi(z)\theta'(z) - \varphi'(z)\theta(z) = k. \tag{12.15}$$

If $k = 0$ (and if $\mathcal{L}$ does not belong to the plane $x = 0$) equation (12.15) gives $\frac{d}{dz}\left(\frac{\theta(z)}{\varphi(z)}\right) = 0$ and so $\theta(z) = \lambda\varphi(z)$ for some non-zero constant λ, and $\mathcal{L}$ is contained in the plane $y = \lambda x$. Hence a curve $\mathcal{L}$ of a special linear complex is either planar or it decomposes into plane curves (contained in planes passing through the axis of the complex). Thus we see that a non-planar curve can not belong to two linear complexes because in the pencil of two linear complexes there is at least one special complex.

If $k \neq 0$ equation (12.15) can be written in the form

$$\frac{d}{dz}\left(\frac{\theta(z)}{\varphi(z)}\right) = \frac{k}{\varphi^2}.$$

If H is a primitive of $\frac{k}{\varphi^2}$, we have

$$\left(\frac{\theta(z)}{\varphi(z)}\right)' = H'\left(= \frac{k}{\varphi^2}\right)$$

and therefore we obtain (where $\sqrt{\zeta}$ denotes either choice of the complex square root of ζ)

$$\frac{\theta(z)}{\varphi(z)} = H, \quad \varphi(z) = \sqrt{\frac{k}{H'}}.$$

Hence the space curves $\mathcal{L}$ of $\mathcal{K}$ are precisely the following:

$$\begin{cases} x = \sqrt{\dfrac{k}{H'}}, \\ y = H\sqrt{\dfrac{k}{H'}}, \end{cases}$$

with H an arbitrary differentiable function of z.

For example, if $H = z^3$ we have $x = \frac{a}{z}$, $y = az^2$ ($a = \sqrt{k/3}$) and $\mathcal{L}$ is a twisted cubic.

Another remarkable example is obtained by taking $H = \tan z$. Then $x = \sqrt{k}\cos z$, $y = \sqrt{k}\sin z$; hence among the curves of a linear complex there are the circular helices.

12.6.4. *The Grassmann variety* $\mathbb{G}(1, n)$ *of lines in* $\mathbb{P}^n$ *is a variety having dimension* $2n - 2$ *and order* $\frac{(2n-2)!}{(n-1)!n!}$ *embedded in* $\mathbb{P}^N$, $N = \binom{n+1}{2} - 1$.

Show that it is the locus of the points in $\mathbb{P}^N$ *whose coordinates* X_{ij} *give rank 2 to the antisymmetric matrix* $M = (X_{ij})$ *of order* $n + 1$ (*where* $X_{ij} = -X_{ji}$, $i, j = 0, \ldots, n$, $X_{ij} = 0$ *if* $i = j$).

The case $n = 3$ was discussed in Section 12.1.

Let $A = [a_0, a_1, \ldots, a_n]$ and $B = [b_0, b_1, \ldots, b_n]$ be two points of $\mathbb{P}^n$ and consider the matrix

$$P = \begin{pmatrix} a_0 & a_1 & \cdots & a_n \\ b_0 & b_1 & \cdots & b_n \end{pmatrix}.$$

Recall that the line r of $\mathbb{P}^n$ that joins A and B meets the hyperplane with equation $t_j = 0$ in the point Y_j with coordinates (cf. observation 12.5.4)

$$(0j)_P = \begin{vmatrix} a_0 & a_j \\ b_0 & b_j \end{vmatrix}, \quad (1j)_P = \begin{vmatrix} a_1 & a_j \\ b_1 & b_j \end{vmatrix}, \quad \ldots, \quad (nj)_P = \begin{vmatrix} a_n & a_j \\ b_n & b_j \end{vmatrix},$$

where $(ij)_P = -(ji)_P$, $i, j = 0, \ldots, n$. Thus the matrix $(x_{ij}) = ((ij)_P)$ has rank 2 because it has in each row the coordinates of a point of the line r.

We now show that, conversely, every point $[x_{ij}]$ of $\mathbb{P}^N$ such that the matrix (x_{ij}) has rank 2 belongs to $\mathbb{G}(1, n)$. If (x_{ij}) has rank two, then for every choice of indices $\alpha, \beta, \gamma, \delta$ ($0 \leq \alpha < \beta < \gamma < \delta \leq n$) the determinant of the antisymmetric matrix of order 4 formed with the rows and columns of positions $\alpha, \beta, \gamma, \delta$ of (x_{ij}),

$$\det(x_{ij}) = \begin{vmatrix} 0 & x_{\alpha\beta} & x_{\alpha\gamma} & x_{\alpha\delta} \\ -x_{\alpha\beta} & 0 & x_{\beta\gamma} & x_{\beta\delta} \\ -x_{\alpha\gamma} & -x_{\beta\gamma} & 0 & x_{\gamma\delta} \\ -x_{\alpha\delta} & -x_{\beta\delta} & -x_{\gamma\delta} & 0 \end{vmatrix} = (x_{\alpha\beta}x_{\gamma\delta} - x_{\alpha\gamma}x_{\beta\delta} + x_{\alpha\delta}x_{\beta\gamma})^2,$$

12.6. Further properties of $\mathbb{G}(1,n)$ and applications

is zero. Therefore we have the following relations:

$$x_{\alpha\beta}x_{\gamma\delta} + x_{\alpha\gamma}x_{\delta\beta} + x_{\alpha\delta}x_{\beta\gamma} = 0. \tag{12.16}$$

To reach the conclusion, it now suffices to calculate the minors of the matrix P of type $(2, n+1)$ formed with the two rows of index α, β in (x_{ij}),

$$P = \begin{pmatrix} x_{\alpha 0} & \cdots & x_{\alpha\alpha} & \cdots & x_{\alpha\beta} & \cdots & x_{\alpha i} & \cdots & x_{\alpha\gamma} & \cdots & x_{\alpha\delta} & \cdots & x_{\alpha n} \\ x_{\beta 0} & \cdots & x_{\beta\alpha} & \cdots & x_{\beta\beta} & \cdots & x_{\beta i} & \cdots & x_{\beta\gamma} & \cdots & x_{\beta\delta} & \cdots & x_{\beta n} \end{pmatrix}.$$

Keeping in mind (12.16) we have

$$(\alpha j)_P = \begin{vmatrix} 0 & x_{\alpha j} \\ x_{\beta\alpha} & x_{\beta j} \end{vmatrix} = x_{\alpha j} x_{\alpha\beta}$$

$$(\gamma\delta)_P = \begin{vmatrix} x_{\alpha\gamma} & x_{\alpha\delta} \\ x_{\beta\gamma} & x_{\beta\delta} \end{vmatrix} = x_{\alpha\gamma}x_{\beta\delta} - x_{\alpha\delta}x_{\beta\gamma} = x_{\alpha\beta}x_{\gamma\delta}.$$

It follows that in $\mathbb{P}^N$ the points $[(ij)_P]$ and $[x_{ij}]$ coincide; and therefore the points $[x_{ij}]$, at which the antisymmetric matrix $M = (X_{ij})$ has rank 2, belong to $\mathbb{G}(1,n)$.

Thus we have proved that $\mathbb{G}(1,n)$ is indeed the locus of points in $\mathbb{P}^N$ that give rank two to the matrix $M = (X_{ij})$.

The foregoing discussion shows that the cases n even and n odd turn out to be very different. In fact we know that the rank of an antisymmetric matrix is always even. Hence if n is even the matrix M (of order $n+1$) is degenerate. By contrast, when n is odd $\det(M)$ is the square of a homogeneous polynomial of degree $\frac{1}{2}(n+1)$ in the X_{ij}.

Note. Let $x = [x_0, \ldots, x_n]$, $y = [y_0, \ldots, y_n] \in \mathbb{P}^n$. One can also obtain the relations (12.16) by writing the Plücker coordinates of the projection of the line r_{xy} onto $S_3 = J(A_\alpha, A_\beta, A_\gamma, A_\delta)$ (from the opposite face of the fundamental $(n+1)$-hedron). One finds that the minors of the matrix

$$\begin{pmatrix} x_\alpha & x_\beta & x_\gamma & x_\delta \\ y_\alpha & y_\beta & y_\gamma & y_\delta \end{pmatrix}$$

satisfy the equation of the Klein quadric that represents the lines of S_3, namely equation (12.16).

12.6.5. *Prove that, for odd n, the variety $\mathbb{G}(1,n)$ is the locus of $\frac{1}{2}(n-1)$-fold points of a hypersurface F of order $\frac{1}{2}(n+1)$ in $\mathbb{P}^{\binom{n+1}{2}-1}$. (In particular the variety $\mathbb{G}(1,3)$ is a quadric in $\mathbb{P}^5$.)*

By Exercise 12.6.4, we know that $\mathbb{G}(1,n)$ is the locus of the points that give rank two to the antisymmetric matrix $M = (X_{ij})$ of order $n+1$, and thus $\mathbb{G}(1,n)$ is the locus of the points that annihilate all the minors of M of order 3. Furthermore, $\det(M) = f^2$ is the square of a form f of degree $\frac{1}{2}(n+1)$ in the variables X_{ij}.

In $\mathbb{P}^{\binom{n+1}{2}-1}$ we consider the hypersurface F with equation $f = 0$, of order $\frac{1}{2}(n+1)$. Since the $(n-2)^{\text{nd}}$ derivatives of $\det(M)$ belong to the ideal generated by the third order minors of M, we have that the $(n-2)^{\text{nd}}$ derivatives of $\det(M)$ vanish in every point of $\mathbb{G}(1, n)$. Hence the points of $\mathbb{G}(1,n)$ are all $(n-1)$-fold for the hypersurface with equation $\det(M) = f^2 = 0$ and so $\frac{1}{2}(n-1)$-fold for F.

To prove that, conversely, every $\frac{1}{2}(n-1)$-fold point for F belongs to $\mathbb{G}(1,n)$ one can proceed by induction on n, bearing in mind that for $n = 3$ the hypothesis holds.

First we examine the case $n = 5$. Together with the matrix P whose rows consist of the coordinates of two distinct points $x = [x_0, \ldots, x_n]$, $y = [y_0, \ldots, y_n]$ in $\mathbb{P}^5$, we also consider the matrix

$$\begin{pmatrix} P \\ P \end{pmatrix} = \begin{pmatrix} x_0 & x_1 & x_2 & x_3 & x_4 & x_5 \\ y_0 & y_1 & y_2 & y_3 & y_4 & y_5 \\ x_0 & x_1 & x_2 & x_3 & x_4 & x_5 \\ y_0 & y_1 & y_2 & y_3 & y_4 & y_5 \end{pmatrix}.$$

It is clearly of rank two, and putting $X_{ij} = x_i y_j - x_j y_i$, one sees that the minor of order 4 obtained by cancelling the columns in positions i_0, i_1 may be written in the form

$$Q_{i_0 i_1} = \sum X_{i_2 i_3} X_{i_4 i_5},$$

where $(i_0, i_1, i_2, i_3, i_4, i_5)$ is an even permutation of the numbers 0, 1, 2, 3, 4, 5. It is not difficult to prove that the fifteen quadratic forms Q_{ij} are generators of the ideal of $\mathbb{G}(1, 5)$ (see also [76, Tutorial 35]). Since

$$f = \sum X_{i_0 i_1} X_{i_2 i_3} X_{i_4 i_5},$$

we have $\frac{\partial f}{\partial X_{i_0 i_1}} = Q_{i_0 i_1}$, and that means that every double point of the hypersurface F belongs to $\mathbb{G}(1, 5)$.

Now let $n \geq 7$ and proceed by induction on n, assuming the result known for every $n' \leq n - 2$. That is, we suppose that if $\Delta = \det(M')$ is an antisymmetric determinant of order $(n-2)+1 = n-1$, every $(n-3)$-fold point of the hypersurface with equation $\Delta = 0$ is a point of $\mathbb{G}(1, n-2)$ and thus it is a point at which the matrix M' has rank two.

We note first that every second derivative of $\det(M)$ is, except for a numerical factor ± 2, a principal minor (and thus antisymmetric) of order $n - 1$ of M. More precisely,

$$\frac{\partial^2 \det(M)}{\partial X_{i_0} \partial X_{i_1}} \tag{12.17}$$

is the determinant of the matrix $M_{i_0 i_1}$ that one obtains from M by suppressing the rows and columns in the places i_0 and i_1.

Assuming this, let x be an $(n-1)$-fold point of the hypersurface $\det(M) = 0$. At the point x, which annihilates all the $(n-2)^{\text{nd}}$ order derivatives of $\det(M)$, all the $(n-4)^{\text{th}}$ order derivatives of every second order derivative of $\det(M)$ are zero. Therefore x is an $(n-3)$-fold point of the hypersurface with equation $\det(M_{i_0 i_1}) = 0$ (and so, putting $\det(M_{i_0 i_1}) := f_{i_0 i_1}^2$, x is an $\frac{1}{2}(n-3)$-fold point of the hypersurface $f_{i_0 i_1} = 0$). Then, by the inductive hypothesis, x gives rank two to the matrix $M_{i_0 i_1}$.

Thus, at x all the third order minors contained in some principal minor of M (of order $n-1$) are zero. But if $n \geq 7$ every minor of order 3 is contained in some principal minor of M of order $n-1$ that is, a minor of the type (12.17). Hence at x all minors of third order of M are zero (i.e., M has rank two at x). Recalling Exercise 12.6.4, we may thus conclude that $x \in \mathbb{G}(1, n)$.

Other interesting results may be obtained by considering together with the matrix

$$P = \begin{pmatrix} x_0 & x_1 & \cdots & x_n \\ y_0 & y_1 & \cdots & y_n \end{pmatrix}$$

formed by the coordinates of two distinct points $x = [x_0, \ldots, x_n]$, $y = [y_0, \ldots, y_n]$ in $\mathbb{P}^n$, the $\frac{1}{2}(n+1)$ matrices

$$P, \begin{pmatrix} P \\ P \end{pmatrix}, \begin{pmatrix} P \\ P \\ P \end{pmatrix}, \ldots, \begin{pmatrix} P \\ P \\ \vdots \\ P \\ P \end{pmatrix}, \tag{12.18}$$

the last of which is a square matrix of order $n+1$ and all of which are of rank two. The minors of maximal order of the first matrix are the coordinates x_{ij} of the line r_{xy}. A minor of maximal order of the sth of these matrices (which is a matrix of type $(2s, n+1)$) may be written in the form $\sum x_{i_0 i_1} x_{i_2 i_3} \cdots x_{i_{2s-2} i_{2s-1}}$ and is zero. Hence it provides a hypersurface of order s in $\mathbb{P}^N$ having equation

$$\sum X_{i_0 i_1} X_{i_2 i_3} \cdots X_{i_{2s-2} i_{2s-1}} = 0,$$

which has multiplicity $s-1$ at every point of $\mathbb{G}(1, n)$. In particular, the determinant of the last matrix in (12.18) is the square of a form of degree $\frac{1}{2}(n+1)$ in the variables X_{ij} of which $\mathbb{G}(1, n)$ is the locus of $\frac{1}{2}(n-1)$-fold points.

12.6.6 (Linear complexes of lines in $\mathbb{P}^4$). A *linear complex* $\mathcal{K}$ *of lines in* $\mathbb{P}^4$ is the 5-dimensional family of lines of $\mathbb{P}^4$ whose Grassmann coordinates p_{ik} satisfy a linear equation

$$a_{34} p_{01} + \cdots + a_{01} p_{34} = 0. \tag{12.19}$$

It is represented in $\mathbb{P}^9$ by a hyperplane section of $V_6^5 = \mathbb{G}(1, 4)$. The lines of $\mathcal{K}$ that pass through a *generic* point x of $\mathbb{P}^4$ are contained in the S_3 corresponding to x under the null polarity $\varphi \colon \mathbb{P}^4 \to \mathbb{P}^{4*}$ given by the antisymmetric matrix $M = (a_{ij})$. The singular points of φ are exceptional: all the lines of $\mathbb{P}^4$ that pass through a singular point of φ belong to $\mathcal{K}$ (cf. [52, Vol. 1, Chapter IX]).

Since the rank $\varrho(M)$ of M is even, in $\mathbb{P}^4$ there are two types of linear complexes of lines. If $\varrho(M) = 4$, $\mathcal{K}$ is a general linear complex and there is only one singular point (the center of the complex, cf. Problem 12.6.8). If $\varrho(M) = 2$, $\mathcal{K}$ is a special linear complex and all the points of a plane σ (called the *center-plane*) are singular; in this case $\mathcal{K}$ consists of all the lines in $\mathbb{P}^4$ supported by σ.

12.6.7. *Let $\mathcal{K}$ be a linear complex of lines in $\mathbb{P}^4$. Prove that $\mathcal{K}$ is special if and only if the coefficients a_{ij} of its equation, taken in suitable order, are the coordinates of a line (cf. (12.19)).*

If $u = [u_0, \ldots, u_4]$, $v = [v_0, \ldots, v_4]$, and $w = [w_0, \ldots, w_4]$ are three independent points of the center-plane σ of $\mathcal{K}$, in order for two points $[x_0, \ldots, x_4]$ and $[y_0, \ldots, y_4]$ to be joined by a line of $\mathcal{K}$ it is necessary and sufficient that

$$\begin{vmatrix} u_0 & u_1 & \cdots & u_4 \\ v_0 & v_1 & \cdots & v_4 \\ w_0 & w_1 & \cdots & w_4 \\ x_0 & x_1 & \cdots & x_4 \\ y_0 & y_1 & \cdots & y_4 \end{vmatrix} = 0.$$

Using Laplace's rule to develop the minors extracted from the first three rows we have the equation $a_{34} p_{01} + \cdots + a_{01} p_{34} = 0$ for the complex. Moreover, the numbers a_{ij} (given by the third order determinant extracted from the first three rows by eliminating the columns in positions i and j) are the Grassmann coordinates of σ and thus also of a line r in $\mathbb{P}^4$ since $\mathbb{G}(1, 4) = \mathbb{G}(2, 4)$.

12.6.8 (Linear complexes of lines in $\mathbb{P}^n$). *A **linear complex of lines** in $\mathbb{P}^n$ is the set $\mathcal{K}$ of the lines of $\mathbb{P}^n$ whose Grassmann coordinates in $\mathbb{P}^N$, $N = \binom{n+1}{2} - 1$, satisfy a linear equation, $\sum_{i,j=0}^{N} a_{ij} X_{ij} = 0$. In other words, $\mathcal{K}$ is a hyperplane section of the variety $\mathbb{G}(1, n)$ (where $X_{ij} = -X_{ji}$, $i, j = 0, \ldots, n$, $X_{ij} = 0$ if $i = j$).*

Let $\mathcal{K}$ be a linear complex of lines in $\mathbb{P}^n$. After having proved that the lines of $\mathcal{K}$ that pass through a point x of $\mathbb{P}^n$ belong to a hyperplane, study the correspondence $\mathbb{P}^n \to \mathbb{P}^{n*}$ that associates to each point x of $\mathbb{P}^n$ the hyperplane of $\mathbb{P}^n$ that contains the lines of $\mathcal{K}$ passing through x.

The case $n = 3$ is treated in Proposition 12.2.1.

Suppose then that $n \geq 4$ and consider all the lines of $\mathcal{K}$ that contain the point $x = [x_0, \ldots, x_n]$. If r is one of these lines and $y = [y_0, \ldots, y_n]$ is another point of r, the Grassmann coordinates of r are $(ij)_P = x_i y_j - x_j y_i$. Hence the points y that when joined with x give lines of $\mathcal{K}$ belong to the hyperplane in $\mathbb{P}^n$ with equation

$$\sum_{i,j=0}^{n} a_{ij}(t_i x_j - t_j x_i) = 0. \tag{12.20}$$

12.6. Further properties of $\mathbb{G}(1,n)$ and applications

We put $a_{ji} = -a_{ij}$. Then equation (12.20) can be written in the form

$$\sum a_{ij} x_j t_i = 0, \qquad (12.21)$$

with the sum extended over the pairs (i, j) of the numbers $0, 1, 2, \ldots, n$ such that $i \neq j$.

Ordering the summands of (12.21) with respect to the t_i one sees that the locus of the points y that when joined with x give lines of $\mathcal{K}$ is the hyperplane u with equation

$$u : \sum_{i=0}^{n} u_i t_i = 0, \quad \text{where } u_i = \sum_{j=0}^{n} a_{ij} x_j.$$

Thus one finds a projectivity $\varphi \colon \mathbb{P}^n \to \mathbb{P}^{n*}$, defined by the matrix (a_{ij}), under which a point $x \in \mathbb{P}^n$ corresponds to the hyperplane $u \in \mathbb{P}^{n*}$ that contains the line of $\mathcal{K}$ passing through x.

The matrix (a_{ij}) of coefficients is antisymmetric and thus, when n is odd, the projectivity φ is in general non-degenerate and is therefore a non-degenerate null polarity. [It is interesting to examine the particular cases in which the determinant of (a_{ij}) is zero. To study them it is convenient to regard (a_{ij}) as a point of $\mathbb{P}^N$. Examine the various positions that it assumes with respect to $\mathbb{G}(1,n)$, bearing in mind that $\mathbb{G}(h,n) = \mathbb{G}(n-h-1,n)$.]

If n is even, we have $\det(a_{ij}) = 0$. Thus, for each index j, the hyperplanes $\sum_{i=0}^{n} a_{ij} t_i = 0$ of $\mathbb{P}^n$, and hence all the hyperplanes with equation (12.21) which are linear combinations of these (indeed $\sum_{i,j} a_{ij} x_j t_i = \sum_j x_j \sum_i a_{ij} t_i$) pass through a fixed space O, *center* of the complex $\mathcal{K}$.

If, for example, the rank of (a_{ij}) is n, that is the maximum possible rank, then the space O consists of a single point. Under the hypothesis that $O = A_n = [0, 0, \ldots, 0, 1]$ we have $a_{in} = 0, i = 0, \ldots, n-1$, and the equation of the complex $\mathcal{K}$ thus has the form, cf. (12.19),

$$\sum_{i,j=0}^{n-1} a_{ij} X_{ij} = 0. \qquad (12.22)$$

If n is even, the lines of $\mathbb{P}^n$ that belong to a general linear complex $\mathcal{K}$ are thus all those and only those that are contained in the planes that project the lines of a linear complex of lines of $\mathbb{P}^{n-1}$ with equation (12.22) from the center O. Hence one deduces the properties of linear complexes of lines in even-dimensional spaces from properties of linear complexes of lines in odd-dimensional spaces.

12.6.9. *Study the surface $\mathcal{F}$ that represents the congruence of the chords of a space cubic C in $\mathbb{P}^3$ on the Klein quadric $\mathfrak{Q}$.*

We note first that the chords of C can not all be contained in the same linear complex. Indeed, if A is a point of C, the generators of the quadric cone that

430   Chapter 12. Grassmann Varieties

projects C from A are chords of C that issue from a point but do not belong to a plane. Thus $\mathcal{F}$ is embedded in $\mathbb{P}^5$. Furthermore:

(1) The lines of a pencil can not all be chords of C, so that $\mathcal{F}$ does not possess lines.

(2) C belongs to ∞^2 quadrics and the lines of one of the two rulings of each of these quadrics are chords of C; therefore $\mathcal{F}$ possesses ∞^2 conics.

Thus $\mathcal{F}$, being a non-ruled surface that possesses ∞^2 conics, is the Veronese surface (cf. Lemma 10.3.2).

This fact may be seen analytically as follows. Let $U = [u^3, u^2, u, 1]$, $V = [v^3, v^2, v, 1]$ be two points of the curve C (cf. Exercise 7.5.3). The minors p_{ik} of the matrix

$$\begin{pmatrix} u^3 & u^2 & u & 1 \\ v^3 & v^2 & v & 1 \end{pmatrix}$$

furnish the Plücker coordinates of the line r_{UV} that joins the two points. More precisely, $p_{01} = u^2 v^2 (u - v)$, $p_{02} = uv(u + v)(u - v)$, $p_{03} = (u - v)(u^2 + uv + v^2)$, $p_{12} = uv(u - v)$, $p_{13} = (u + v)(u - v)$, $p_{23} = u - v$. Thus r_{UV} is represented in $\mathbb{P}^5$ by the point

$$R = [u^2 v^2, uv(u + v), u^2 + uv + v^2, uv, u + v, 1].$$

Putting $uv = \lambda$, $u + v = \mu$, the point R may be rewritten in the form $R = [\lambda^2, \lambda\mu, \mu^2 - \lambda, \lambda, \mu, 1]$ and then, setting $\lambda = \frac{x_1}{x_0}$, $\mu = \frac{x_2}{x_0}$,

$$R = [x_1^2, x_1 x_2, x_2^2 - x_1 x_0, x_1 x_0, x_2 x_0, x_0^2].$$

Hence, applying the projectivity of $\mathbb{P}^5$ defined by

$$[p_{01}, p_{02}, p_{03}, p_{12}, p_{13}, p_{23}] \mapsto [p_{01}, p_{02}, p_{03} + p_{12}, p_{12}, p_{13}, p_{23}],$$

we see that the point R describes a Veronese surface (cf. Example 10.2.1).

From properties of the Veronese surface $\mathcal{F}$ one can deduce properties of the space cubic C, and conversely.

For example, the fact that the order of C is three implies that $\mathcal{F}$ is met in three points by the planes of one of the two arrays of planes of $\mathfrak{Q}$. More precisely, the three chords of C that connect in pairs the three points where C meets a (ruled) plane π give rise to the three points common to $\mathcal{F}$ and the plane α of $\mathfrak{Q}$ associated to π (cf. §12.1.2).

A generic plane β of the other array has a single point in common with $\mathcal{F}$. Indeed, the plane β is associated to a star of lines of $\mathbb{P}^3$ and for a generic point of $\mathbb{P}^3$ there passes one and only one chord of C, for otherwise the cubic C would have four intersections with the plane defined by two concurrent chords. This chord represents the point of intersection of $\mathcal{F}$ and β. But there are ∞^1 planes in this second array $\{\beta\}$ that contain a conic of $\mathcal{F}$. To see this it suffices to consider a star of lines having center in a point $x \in C$ and the plane β_x contained in the quadric $\mathcal{Q}$ which is represented by the star. The chords of C issuing from x represent the points of a conic contained in β_x (the cone projecting C from one of its points is a quadratic cone).

Again, the fact that the Veronese surface has order 4 implies that there are four chords of C that are supported by two generic lines, that is, belonging to a linear congruence Γ. The chords of C that belong to Γ are in fact the lines represented by the four intersections of $\mathcal{F}$ with the space S_3 that intersects the Klein quadric in Γ.

12.6.10. *Let C be a rational normal curve in $\mathbb{P}^n$, that is, the image of the Veronese morphism $\varphi_{1,n}: \mathbb{P}^1 \to \mathbb{P}^n$. The chords of C are in number ∞^2 and thus they are represented by the points of a surface $\mathcal{F}$ lying on the Grassmann variety $\mathbb{G}(1,n)$. Study the surface $\mathcal{F}$.*

The same analytic procedure used in Exercise 12.6.9 to prove that the surface $\mathcal{F}$ is the Veronese surface when $n = 3$ may be applied whenever $n \geq 2$ to show that $\mathcal{F}$ is the projective image of the linear system of all curves of order $n - 1$ in a plane. (Remember that a symmetric polynomial $f(u, v) \in \mathbb{C}[u, v]$ may be written as a polynomial of the ring $\mathbb{C}[u + v, uv]$.)

12.6.11 (*Algebraic complexes of S_k in $\mathbb{P}^n$*). An *algebraic complex of subspaces S_k in $\mathbb{P}^n$* is a family of linear subspaces S_k in $\mathbb{P}^n$ whose image under the embedding of $\mathbb{G}(k,n)$ in $\mathbb{P}^N$, $N = \binom{n+1}{k+1} - 1$, is an algebraic subvariety of codimension 1 in $\mathbb{G}(k,n)$, that is, the intersection of $\mathbb{G}(k,n)$ with an algebraic hypersurface F of $\mathbb{P}^N$ (cf. §12.5.8, Property (3)). We define the *degree* of the complex to be that of the hypersurface F.

A *linear complex of subspaces S_k in $\mathbb{P}^n$* is a hyperplane section of $\mathbb{G}(k,n)$. For example, the collection of all subspaces S_k meeting a given subspace S_{n-k-1} is a linear complex. Indeed, it is a special linear complex in the sense that it has an equation of type (12.12).

12.6.12. *Fix a group G of $2n - 2$ distinct points in $\mathbb{P}^1$. How many linear series g_n^1 have G as Jacobian group?*

We replace the line with a rational normal curve C^n of $\mathbb{P}^n$, the projective image of the series g_n^n on $\mathbb{P}^1$ (cf. Section 8.4).

A group of n points is a hyperplane section of C^n, and a series g_n^1 is cut out on C^n by a pencil of hyperplanes, that is, by the hyperplanes passing through some fixed subspace S_{n-2}. A double point P of this g_n^1 is a point of contact of C^n with a hyperplane of the pencil, that is, with a hyperplane containing the given subspace S_{n-2} and also the tangent line p to C^n at P.

Since the line p and the subspace S_{n-2} belong to a hyperplane they must meet. Conversely, if p and the subspace S_{n-2} meet, the hyperplane joining them passes through the S_{n-2} and is tangent to C^n in P.

Thus the number required is merely the number of subspaces S_{n-2} of $\mathbb{P}^n$ which meet $2n-2$ pairwise skew lines. [If two tangents of C^n were incident, the hyperplane defined by these two lines and $n-3$ arbitrary further points of the curve would have at least $n+1$ intersections with the curve C^n.]

Consider the Grassmannian variety $\mathbb{G}(n-2,n)[=\mathbb{G}(1,n)]$ whose points are the subspaces S_{n-2} of $\mathbb{P}^n$. Recall that $\mathbb{G}(n-2,n)$ belongs to a space $\mathbb{P}^N$ of dimension $N=\binom{n+1}{2}-1$. The subspaces S_{n-2} meeting a line form a linear complex, which means that they are the points common to the Grassmannian variety and a hyperplane of $\mathbb{P}^N$. Thus we need only count the points of $\mathbb{G}(n-2,n)$ which belong to $2n-2$ hyperplanes of $\mathbb{P}^N$. Note that $2n-2$ hyperplanes meet in a space of dimension

$$\binom{n+1}{2} - 1 - (2n-2) = \mathrm{cod}_{\mathbb{P}^N}\,\mathbb{G}(n-2,n),$$

since $\dim \mathbb{G}(n-2,n) = \dim \mathbb{G}(1,n) = 2n-2$. Thus the requested number of linear series g_n^1 is, see §12.5.8, Property (6),

$$\deg \mathbb{G}(1,n) = \frac{(2n-2)!}{n!\,(n-1)!}.$$

Chapter 13
Supplementary Exercises

As an adjunct to the theory developed earlier in this text, and with particular reference to the subjects discussed in Chapters 5, 6, 7, 8, 9, 10, we here propose some exercises to summarize and review that material.

The exercises in Section 13.1 present greater difficulty and are completely solved; those proposed in Section 13.2 are simpler and their solutions are left to the reader. The argument sketched in §13.1.42 could inspire a research project or senior thesis.

The exercises in Section 13.3 have been added for the English edition of the text. Most of them constitute results on linear series on curves which, although known, are not easy to find in the literature. They are placed in this chapter rather than in Chapter 8 since their solution makes use of the planar representation of rational surfaces as described in Chapter 10.

13.1 Miscellaneous exercises

13.1.1. *Prove that the rational transformation* $\mathbb{P}^2 \to \mathbb{P}^2$ *with equations* $y_i = x_i^2$, $i = 0, 1, 2$, *induces a birational isomorphism between the line with equation* $a_0 x_0 + a_1 x_1 + a_2 x_2 = 0$ *and its image.*

Consider two projective planes S_2 and S_2' and let x_0, x_1, x_2 be homogeneous coordinates in S_2 and y_0, y_1, y_2 homogeneous coordinates in S_2'. Let $X \subset S_2$ be the line with equation $a_0 x_0 + a_1 x_1 + a_2 x_2 = 0$ with $a_0 a_1 a_2 \neq 0$. Let $[\varphi] \colon X \to S_2'$ be the rational transformation for which a representative $\varphi \colon S_2 \to S_2'$ is given by the equations (cf. §2.6.7)
$$y_i = x_i^2, \quad i = 0, 1, 2.$$
Let E_φ be the exceptional locus of φ. Since $I(E_\varphi) = (y_0, y_1, y_2)$, we have $E_\varphi = \emptyset$ and so $[\varphi]$ is a morphism $S_2 \to S_2'$, that is, it is defined on all of $\mathbb{P}^2$. But clearly φ is not a birational isomorphism because almost all the fibers are quadruples of points.

We show however that $[\varphi] \colon X \to Y := \overline{\varphi(X)}$ is a birational isomorphism. To that end, consider the system of equations
$$\begin{cases} y_0 = x_0^2, \\ y_1 = x_1^2, \\ y_2 = x_2^2, \\ a_0 x_0 + a_1 x_1 + a_2 x_2 = 0, \end{cases}$$

Chapter 13. Supplementary Exercises

from which one deduces the formulas

$$\begin{cases} y_0 = -\dfrac{x_0}{a_0}(a_1 x_1 + a_2 x_2), \\ y_1 = -\dfrac{x_1}{a_1}(a_2 x_2 + a_0 x_0), \\ y_2 = -\dfrac{x_2}{a_2}(a_0 x_0 + a_1 x_1) \end{cases} \qquad (13.1)$$

that give another rational map $\theta: X \to Y$ with $\theta \in [\varphi]$. Since θ is a birational isomorphism between the two planes (that is, a Cremona transformation), $[\varphi]$ is a birational isomorphism since it is the restriction of that Cremona transformation. It is easy to verify the following facts.

(1) θ is an isomorphism between the open set of S_2 which is the locus of the points $[x_0, x_1, x_2]$ such that $x_0 x_1 x_2 \neq 0$ and the open set in S_2' defined by

$$(a_0^2 y_0 - a_1^2 y_1 - a_2^2 y_2)(-a_0^2 y_0 + a_1^2 y_1 - a_2^2 y_2)(-a_0^2 y_0 - a_1^2 y_1 + a_2^2 y_2) = 0.$$

(2) The inverse formulas for (13.1) are the following:

$$\begin{cases} x_0 = a_1 a_2 [a_0^4 y_0^2 - (a_1^2 y_1 - a_2^2 y_2)^2], \\ x_1 = a_2 a_0 [a_1^4 y_1^2 - (a_2^2 y_2 - a_0^2 y_0)^2], \\ x_2 = a_0 a_1 [a_2^4 y_2^2 - (a_0^2 y_0 - a_1^2 y_1)^2]. \end{cases} \qquad (13.2)$$

To derive them, one can resolve the system (13.1); however, it may be more convenient to consider the Cremona transformation $\omega: S_2 \to S_2''$ (where S_2'' is another plane, with coordinates z_0, z_1, z_2) given by the formulas

$$\begin{cases} z_0 = x_1 x_2, \\ z_1 = x_2 x_0, \\ z_2 = x_0 x_1 \end{cases} \quad \text{and} \quad \begin{cases} x_0 = z_1 z_2, \\ x_1 = z_2 z_0, \\ x_2 = z_0 z_1. \end{cases}$$

Observe, moreover, that θ is the product of ω with the non-degenerate projectivity $\xi: S_2'' \to S_2'$, given by the equations

$$\xi = \begin{cases} y_0 = -\dfrac{a_2}{a_0} z_1 - \dfrac{a_1}{a_0} z_2, \\ y_1 = -\dfrac{a_2}{a_1} z_0 - \dfrac{a_0}{a_1} z_2, \\ y_2 = -\dfrac{a_1}{a_2} z_0 - \dfrac{a_0}{a_2} z_1; \end{cases} \qquad \xi^{-1} = \begin{cases} z_0 = a_0(a_0^2 y_0 - a_1^2 y_1 - a_2^2 y_2), \\ z_1 = a_1(-a_0^2 y_0 + a_1^2 y_1 - a_2^2 y_2), \\ z_2 = a_2(-a_0^2 y_0 - a_1^2 y_1 + a_2^2 y_2). \end{cases}$$

13.1. Miscellaneous exercises   435

(3) The image Y of $[\varphi]$ coincides with the image of X under θ; and it is the conic Y defined by the circumscribed trilateral $y_0 y_1 y_2 = 0$ and its Brianchon point $B = [\frac{1}{a_0^2}, \frac{1}{a_1^2}, \frac{1}{a_2^2}]$, cf. [13, Vol. II, Chapter 16].

Replacing the (13.2) in the equation $a_0 x_0 + a_1 x_1 + a_2 x_2 = 0$ (and dividing by $a_0 a_1 a_2$) one obtains the equation of the conic

$$a_0^4 y_0^2 + a_1^4 y_1^2 + a_2^4 y_2^2 - 2a_0^2 a_1^2 y_0 y_1 - 2a_1^2 a_2^2 y_1 y_2 - 2a_2^2 a_0^2 y_2 y_0 = 0,$$

tangent to the lines $y_0 = 0$, $y_1 = 0$, $y_2 = 0$ at the points $[0, a_2^2, a_1^2]$, $[a_2^2, 0, a_0^2]$, $[a_1^2, a_2^2, 0]$.

13.1.2. *If a conic γ and a cubic $\mathcal{C}$ have the same tangents in three points A, B, and C, the three tangents have a further intersection with $\mathcal{C}$ in three collinear points.*

Indeed, six of the nine base points of the pencil of cubics defined by $\mathcal{C}$ and the trilateral of the three tangents are A, A, B, B, C, C and they belong to the conic γ. Therefore, the other three base points are collinear, cf. Exercise 5.7.15.

Note that, conversely, if through three collinear points of $\mathcal{C}$ one draws three tangents of $\mathcal{C}$, the three points of contact are points of contact of $\mathcal{C}$ with a tritangent conic. All the tritangent conics are obtained in this way.

13.1.3. *Show that an irreducible quartic surface $\mathcal{F}$ in $\mathbb{P}^3$ that passes doubly through a curve C of order 3 is a Steiner surface or a rational ruled surface.*

It follows from Bézout's theorem that, since $\mathcal{F}$ is irreducible, the curve C is not a plane cubic (every line of the plane of C would have six intersections with $\mathcal{F}$ and so would be contained it $\mathcal{F}$), is not split into a conic and a line r without common points (for each point P belonging to the plane π of the conic, the line $\ell = \pi \cap J(P, r)$ would have six intersections with $\mathcal{F}$) nor is it a triple of lines that are skew in pairs (every line that is supported by the three lines would still have six intersections with $\mathcal{F}$).

Hence C can only be either a triple of lines issuing from a point P (and then $\mathcal{F}$ is a Steiner surface, cf. Exercise 10.5.6), or a space cubic (possibly split into an irreducible conic γ and a line r that meets γ in a point A but is not contained in the plane of γ).

The existence of quartic surfaces $\mathcal{F}$ passing doubly through a space cubic C is proved by writing their equation. Indeed, C is the base curve of a net of quadrics with equation $\lambda_0 \varphi_0 + \lambda_1 \varphi_1 + \lambda_2 \varphi_2 = 0$ (cf. Exercise 7.5.3), and each surface of order 4 with equation

$$\sum_{i,j=0}^{2} a_{ij} \varphi_i \varphi_j = 0, \quad a_{ij} \in K,$$

passes doubly through C. One sees immediately that if $\mathcal{F}$ is not a Steiner surface, so that C is a space cubic, then $\mathcal{F}$ is ruled. Indeed, through any one of its points

P there passes a chord of C (cf. Exercise 7.5.2). Such a chord, having at least five intersections with $\mathcal{F}$ (one at P and two in each of the points at which C is supported) must be contained in $\mathcal{F}$ (if $C = \gamma \cup r$ is split into the conic γ and the line r, then the chord of C passing through P is the line that joins P with the point $B \neq A$ common to γ and the plane $J(P, r)$).

A generally bijective correspondence between a quartic surface $\mathcal{F}$ containing a double space cubic C and a plane σ is obtained immediately by associating to each point P of $\mathcal{F}$ the intersection of σ with the chord of C issuing from P.

13.1.4. *Determine the order of the ruled surface $\mathcal{R}$ that is the locus of the tangents of a space cubic C in $\mathbb{P}^3$.*

The order of $\mathcal{R}$ is the number of tangents of C that are supported by a generic line r, that is, the rank of C, which is exactly 4 (cf. Exercise 7.5.2).

Alternatively, one could also proceed as follows. If one projects the cubic C from a point P of r onto a plane π one obtains therein a cubic C' with a double point (the trace of the chord of C issuing from P). The tangents of C that meet r have as their projections the tangents of C' issuing from the trace of r in π. Since the class of C' is 4, the surface $\mathcal{R}$ is a ruled surface of order 4.

13.1.5. *Let x, y, and z be affine coordinates in $\mathbb{A}^3$. Write the equation of the ruled surface $\mathcal{R}$ of the tangents to the cubic C which is the locus of the point $P(t) = (t^3, t^2, t)$.*

The equation of $\mathcal{R}$ is obtained by eliminating t from the equations of the tangent to C at the point $P(t)$. One can also arrive at the same equation by recalling that the ruled surface of the tangents to a space cubic C is the envelope of the osculating planes of C (cf. Exercise 7.5.2).

One finds that $\mathcal{R}$ has the equation $x^2 + 4y^3 - 3y^2z^2 + 4xz^3 - 6xyz = 0$ (cf. (7.24)).

13.1.6. *Let Σ be the linear system of quadrics in $\mathbb{P}^3$ that pass both through a conic γ and a point P (not belonging to the plane of γ). Show that the Jacobian surface of Σ decomposes into the cone that projects γ from P and the plane of γ counted twice.*

We may suppose that γ has equation $x_0 = \varphi(x_1, x_2, x_3) = 0$ and that P is the point $[1, 0, 0, 0]$. Then Σ has equation

$$\lambda_0 \varphi(x_1, x_2, x_3) + x_0(\lambda_1 x_1 + \lambda_2 x_2 + \lambda_3 x_3) = 0.$$

Applying the theory expounded in Section 6.4 one obtains the desired result.

13.1.7. *Let Σ be the linear system of conics in $\mathbb{P}^2$ with respect to which a given point P and a given line p are pole and polar. Describe the projective image of Σ.*

The system Σ, which does not have base points and has degree 4, is composed with the harmonic homology ω having P and p as center and axis (this means that

Σ is composed with the involution of the pairs of points corresponding under ω). Thus the projective image of Σ is a quadric Q (cf. §10.1.8).

The only lines of this quadric are the images of the lines of the pencil with center P. Therefore Q is a quadric cone.

This may be seen immediately in analytic fashion by choosing P as the origin O of an affine coordinate system u, v and p as the improper line (at infinity). Then Σ is the system of the conics having O as center: $\lambda_0 u^2 + \lambda_1 uv + \lambda_2 v^2 + \lambda_3 = 0$. Furthermore, the projective image of Σ is the locus of the point with Cartesian coordinates (u^2, uv, v^2), namely the surface with equation $y^2 - xz = 0$.

13.1.8. *Let Q be a non-specialized quadric in $\mathbb{P}^3$, π a plane, and P a point belonging to neither Q nor π. Projecting Q from P onto π one obtains a morphism $\varphi : Q \to \pi$.*

Study the algebraic system of the conics in π that correspond to the plane sections of Q, and that of the lines in π that correspond to the lines of Q.

The fibers of φ are pairs of points that are, in general, distinct. The *branch curve*, that is the curve in π which is the locus of the points A such that the fiber $\varphi^{-1}(A)$ is a pair of coinciding points is the trace Γ of the cone that has its vertex at P and is circumscribed to Q. In other words, the branch curve is the conic which is the projection of the curve γ obtained as the section of Q by the polar plane of P with respect to Q.

Every curve traced on Q and having a point M in common with γ is projected into π from P onto a curve tangent to Γ at the point M', the projection of M. Therefore, the plane sections of Q (which have two points of intersection with γ) have projections that are the conics bitangent to the conic Γ (the condition of tangency is quadratic, and thus one obtains a non-linear algebraic system). To the lines of Q there correspond the tangents of Γ.

13.1.9. *Prove the existence of algebraic plane curves of order 6 with nine cusps, and examine the configuration of their cuspidal tangents.*

A sextic $\mathcal{L}$ with nine cusps has class 3 (cf. (5.34)) and is the dual curve of a non-singular plane cubic C, which has class 6 and possesses nine inflectional tangents (cf. Remark 5.3.2 and Example 5.3.3). Under duality, to these nine lines there correspond the nine cusps of $\mathcal{L}$ and to the points of contact (that is, to the flexes) there correspond the nine cuspidal tangents of $\mathcal{L}$.

The configuration of the cuspidal tangents has properties dual to those of the flexes of a cubic: for example, for the point of intersection of two of the nine cuspidal tangents there passes a third cuspidal tangent.

13.1.10. *The algebraic plane curves of order n passing through $\frac{n(n+3)}{2} - 1$ points P in general position, all pass through $\frac{(n-1)(n-2)}{2}$ further points.*

The points P define a pencil of curves of order n that has

$$n^2 - \frac{n(n+3)}{2} + 1 = \frac{n(n-3)}{2} + 1 = \frac{(n-1)(n-2)}{2}$$

further base points.

13.1.11. *Determine the double points of the surface $\mathcal{F}$ in $\mathbb{P}^3$ with equation*

$$L_1 L_2 \ldots L_{2n} + M^2 = 0,$$

where M is a generic homogeneous polynomial of degree n and $L_1, \ldots, L_{2n}$ are linear forms.

The $n\binom{2n}{2}$ points P_{ij} given by $L_i = L_j = M = 0$ are double points for $\mathcal{F}$, and for a generic choice of the polynomials $L_1, \ldots, L_{2n}, M$ there are no other double points.

13.1.12. *Let P, Q be two arbitrary points of an irreducible space cubic C in $\mathbb{P}^3$. Choose the reference system for the projective coordinates x_0, x_1, x_2, x_3 in such a way as to have*

(1) $P = A_0 = [1, 0, 0, 0]$, $Q = A_3 = [0, 0, 0, 1]$;

(2) $J(A_0, A_1, A_2)$ *and* $J(A_1, A_2, A_3)$ *are the osculating planes of C at P and Q;*

(3) $r_{A_0 A_1}$ *and* $r_{A_2 A_3}$ *are the tangent lines at P and Q;*

(4) *the unit point* $U = [1, 1, 1, 1]$ *belongs to C.*

How can C be represented analytically?

Let u, v be projective homogeneous coordinates in $\mathbb{P}^1$ and let

$$[u, v] \mapsto [x_0, x_1, x_2, x_3]$$

be the isomorphism $\varphi \colon \mathbb{P}^1 \to C$ defined as follows (cf. (7.14)):

$$x_i = a_{i0} u^3 + a_{i1} u^2 v + a_{i2} u v^2 + a_{i3} v^3, \quad i = 0, 1, 2, 3.$$

We may suppose that the system of coordinates u, v in $\mathbb{P}^1$ is given in such manner that $\varphi^{-1}(P) = [1, 0]$, $\varphi^{-1}(Q) = [0, 1]$, $\varphi^{-1}(U) = [1, 1]$ (Figure 13.1).

The fact that C passes through P and Q implies that $a_{10} = a_{20} = a_{30} = a_{03} = a_{13} = a_{23} = 0$. Since $x_3 = 0$ is the osculating plane at A_0 we must have $a_{31} = a_{32} = 0$ (because the equation giving the intersections of C with the plane $x_3 = 0$ must be $v^3 = 0$, cf. §7.1.1); similarly we find $a_{01} = a_{02} = 0$. Thus C has parametric equations

$$\begin{cases} x_0 = a_{00} u^3, \\ x_1 = a_{11} u^2 v + a_{12} u v^2, \\ x_2 = a_{21} u^2 v + a_{22} u v^2, \\ x_3 = a_{33} v^3. \end{cases}$$

13.1. Miscellaneous exercises

Since the tangent at P is the line $r_{A_0 A_1} : x_2 = x_3 = 0$ we must have $a_{21} = 0$ (so that the equation giving the intersections of C with the line $x_2 = x_3 = 0$ is $v^2 = 0$); similarly $a_{12} = 0$.

Finally, the hypothesis that U belongs to C implies that $a_{00} = a_{11} = a_{22} = a_{33}$.

In conclusion C is the locus of the point $[u^3, u^2v, uv^2, v^3]$.

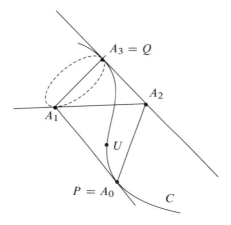

Figure 13.1.

13.1.13. *Prove that if two quadric cones F and G have a space cubic C in common (and thus also a chord of C) then there are no other cones in the pencil that they define.*

We may suppose that C is the locus of the point $[u^3, u^2v, uv^2, v^3]$ and that the two cones have vertices $[1, 0, 0, 0]$ and $[0, 0, 0, 1]$ (see Exercise 13.1.12). The two cones then have respectively the equations $x_1x_3 - x_2^2 = 0$ and $x_0x_2 - x_1^2 = 0$.

One sees immediately that the singular quadrics of the pencil $\lambda(x_1x_3 - x_2^2) + \mu(x_0x_2 - x_1^2) = 0$ are obtained for values of the ratio $\lambda : \mu$ that resolve the equation $\lambda^2\mu^2 = 0$.

13.1.14. *Determine the possible types of pencils of quadric cones in $\mathbb{P}^3$.*

A first type of pencil Σ of quadric cones is obtained by projecting the conics of a pencil contained in a plane from a point P not belonging to the plane. In this case, the cones of the pencil all have the same vertex. If however Σ is a pencil of quadric cones with variable vertex, the locus of the vertices is a curve C that is a base curve for Σ by Bertini's first theorem (Theorem 6.3.11). In particular, it follows that C has order ≤ 4.

The curve C can not be a space cubic because, as seen in Exercise 13.1.13, taking linear combinations of (the equations for) two quadric cones passing through a space cubic one obtains a pencil of quadrics which does not contain other cones. The curve C can not be a space quartic either since in that case the cones of the pencil (which are obtained by projecting C from one of its points) would all be cubics. It is then obvious that C can not be a plane curve of order ≥ 2. Thus C is a common generator of all the cones of Σ.

Let A and B be two arbitrary points of C. The pencil Σ can be obtained by taking linear combinations of the two cones α and β with vertices A and B respectively. Since A is a simple point for β, the tangent plane at A to the generic cone of Σ coincides with the tangent plane at A to β (cf. Remark 6.3.9); and therefore the cones of Σ are tangent along the line C to a fixed plane π. Intersecting Σ with a plane σ not passing through C we obtain a pencil Φ of conics tangent to the line $\sigma \cap \pi$ at the point where it meets C. Moreover, the correspondence which to a point P of C associates the trace in σ of the cone of Σ that has vertex P is a projectivity $C \to \Phi$ (because it is algebraic and bijective).

Conversely, let F and G be two quadric cones with distinct vertices A and B, and suppose that they have a line ℓ in common and the same tangent plane π along this line. At an arbitrary point P of ℓ the tangent planes of F and G coincide, and so, again by Remark 6.3.9, P is a double point for a quadric of the pencil determined by F and G. This pencil then has ∞^1 cones, and is therefore a pencil of quadric cones.

The conclusion is that in $\mathbb{P}^3$ there are only two types of pencils of quadric cones, namely

a) the pencils Σ obtained by projecting a pencil of conics of a plane σ from a point P and having base curve a quadruple of lines issuing from P;

b) the pencils Σ of cones tangent to a plane π along a line ℓ, and having base curve composed of the line ℓ counted twice together with a conic γ meeting ℓ at a point where γ is tangent to π.

This last statement is easily verified analytically. Suppose that ℓ has equations $x_1 = x_2 = 0$ and that one of the cones of the pencil is $x_2^2 - x_1 x_3 = 0$ (so that its vertex is $[1, 0, 0, 0]$) and suppose further that π is the plane $x_1 = 0$. A second cone of the pencil that has as vertex the point $[0, 0, 0, 1]$ will have equation of the form $x_1(ax_0 + bx_1 + cx_2) + x_2^2 = 0$ since it must be tangent to the plane π along ℓ. Then the cone of the pencil

$$x_1(ax_0 + bx_1 + cx_2) + x_2^2 + \lambda(x_2^2 - x_1 x_3) = 0$$

that is obtained when $\lambda = -1$ decomposes into the plane $x_1 = 0$ and the plane $ax_0 + bx_1 + cx_2 + x_3 = 0$. The latter plane intersects the generic cone of Σ along the conic with equation $ax_0 + bx_1 + cx_2 + x_3 = x_2^2 - x_1 x_3 = 0$ which is tangent to π at the point $[1, 0, 0, -a]$.

13.1.15. *Prove that the spaces S_3 that project the points of a Veronese surface from a plane tangent to it at an arbitrary point onto a plane do not exhaust all of $\mathbb{P}^5$, but rather only a quadric cone.*

Let $\mathcal{F}$ be the locus of the point $[x_0, x_1, x_2, x_3, x_4, x_5]$ where, cf. Example 10.2.1,

$$x_0 : x_1 : x_2 : x_3 : x_4 : x_5 = u^2 : 2uv : v^2 : 2uw : 2vw : w^2$$

and $P = [0, 0, 0, 0, 0, 1]$ (that is, P is given by $u = v = 0$). Then the space S_3 that joins the generic point of $\mathcal{F}$ with the three points $[0, 0, 0, 1, 0, 0]$, $[0, 0, 0, 0, 1, 0]$, $[0, 0, 0, 0, 0, 1]$ that define the tangent plane at P (cf. §3.2.1) is the locus of the point $[u^2, 2uv, v^2, \lambda, \mu, \nu]$, $\lambda, \mu, \nu \in K$, and is therefore contained in the quadric cone with equation $x_1^2 - 4x_0 x_2 = 0$.

13.1.16. *Let X be a variety in $\mathbb{P}^n$ and let $F_0, \ldots, F_s$ be linearly independent forms in the coordinate ring $K[x_0, \ldots, x_n]$ all having the same degree d and with no common zeros on X. Then the map $\varphi \colon X \to \varphi(X) \subset \mathbb{P}^s$, defined by $x \mapsto [F_0(x), \ldots, F_s(x)]$, is a finite morphism.*

Consider the Veronese immersion of degree d, $v_d \colon X \to \mathbb{P}^N$, $N = \binom{n+d}{d} - 1$, and let $L_0, \ldots, L_s$ be the linear forms that correspond to $F_0, \ldots, F_s$ (cf. Section 6.7). Let $p \colon \mathbb{P}^N \to \mathbb{P}^s$ be the projection of $\mathbb{P}^N$ from the linear space with equations $L_0 = \cdots = L_s = 0$. The morphism φ is the composition $p \circ v_d$. Since the restriction of p to $v_d(X)$ is a finite morphism (cf. Exercise 2.7.37), so too is φ.

13.1.17. *Consider a cubic surface $\mathcal{F}$ in $\mathbb{P}^3$ having four nodes. Show that every point P of $\mathcal{F}$ is the vertex of two quadric cones each of which is tangent to $\mathcal{F}$ along a cubic.*

By the Reciprocity Theorem 5.4.6 (see also Proposition 5.4.9), we know that if A is a simple point of $\mathcal{F}$ belonging to the quadric $\mathcal{Q}$ polar of P with respect to $\mathcal{F}$, the tangent plane to $\mathcal{F}$ at A passes through P. Conversely, the points of contact of $\mathcal{F}$ with tangent planes passing through P belong to the curve $\mathcal{L}$, the section of $\mathcal{F}$ by the quadric $\mathcal{Q}$. Therefore the cone Γ that projects $\mathcal{L}$ from P is tangent to $\mathcal{F}$ at every point of $\mathcal{L}$.

We now prove that the curve $\mathcal{L}$, which is of type $(3, 3)$, splits into two cubics. By Lemma 5.6.1, the polar quadric of P with respect to $\mathcal{F}$ passes through P and has the same tangent plane as $\mathcal{F}$ there. Then P is a double point for the curve $\mathcal{L}$, the section of $\mathcal{F}$ by the quadric $\mathcal{Q}$ and which has order 6; cf. 5.8.3. Hence Γ is a cone of order 4.

Moreover, the curve $\mathcal{L}$ has as double points the four nodes of $\mathcal{F}$, through which $\mathcal{Q}$ passes simply (cf. Lemma 5.6.2).

On the other hand, the cubics of type $(2, 1)$ on a quadric are in number ∞^5 and thus for the five double points of $\mathcal{L}$ there passes a cubic $C = (2, 1)$ contained in $\mathcal{Q}$. This cubic has (at least) ten intersections with $\mathcal{L}$, which contradicts (7.7) unless C is contained in $\mathcal{L}$. In that case the curve $\mathcal{L}$ of type $(3, 3)$ is a pair of cubics: one

(1, 2) and one (2, 1) whose five intersections are the double points of $\mathcal{L}$. It follows that the cone Γ (of order 4) decomposes into two quadric cones each of which is tangent to $\mathcal{F}$ along a cubic.

13.1.18. *Show that if a quadric $\mathcal{Q}$ in $\mathbb{P}^3$ is tangent to a cubic surface $\mathcal{F}$ along a cubic C, then $\mathcal{Q}$ is a cone with vertex on C and $\mathcal{F}$ has four double points (in general distinct) on C.*

On a non-specialized quadric a curve of type $(3, 3)$ can not be a cubic counted twice (since $2(2, 1) = (4, 2)$ and $2(1, 2) = (2, 4)$). Therefore, if $\mathcal{Q}$ and $\mathcal{F}$ are tangent along a cubic C, then $\mathcal{Q}$ is a cone and C passes through its vertex and meets every generator in a further point.

Consider a point P in $\mathbb{P}^3$. We seek the points of C for which $\mathcal{F}$ has a tangent plane passing through P. Since $\mathcal{F}$ and the quadric cone $\mathcal{Q}$ have the same tangent plane at every point of C, this is equivalent to seeking the tangent planes of $\mathcal{Q}$ that pass through P. (Those planes are tangent to $\mathcal{Q}$ along a generator, and so also at the point of C that lies on that generator.) The number of these planes is two since through a generic point of $\mathbb{P}^3$ there pass two tangent planes to a quadric cone.

On the other hand, the points of C that are simple for $\mathcal{F}$ and at which $\mathcal{F}$ has tangent plane passing through P lie (by the reciprocity theorem) on the first polar $\mathcal{F}_1(P)$ of P with respect to $\mathcal{F}$. Hence there are two simple points of $\mathcal{F}$ common to the cubic C and the quadric $\mathcal{F}_1(P)$.

But C and $\mathcal{F}_1(P)$ have six points in common; therefore, the four missing intersections are absorbed by the double points of $\mathcal{F}$. The quadric $\mathcal{F}_1(P)$ passes simply through each of the double points of $\mathcal{F}$ with tangent plane distinct from the tangent plane to $\mathcal{F}$, cf. Lemma 5.6.2 and Exercise 5.8.2. Therefore every double point corresponds to a single intersection. Thus in general $\mathcal{F}$ has four double points on C.

13.1.19. *Let C be a space curve of order n and genus p in $\mathbb{P}^3$. Represent the lines of $\mathbb{P}^3$ as points of a quadric hypersurface $\mathcal{Q} \subset \mathbb{P}^5$ (cf. Section 12.1). Determine the order of the surface $\mathcal{F} \subset \mathcal{Q}$ whose points "are" the chords of C.*

Each plane in $\mathbb{P}^3$ contains $\binom{n}{2}$ chords of C. Moreover, every star of lines contains $\binom{n-1}{2} - p$ chords of C, since on projecting C from the center of the star one obtains a plane curve, of genus p, having $\binom{n-1}{2} - p$ nodes that correspond to the chords of C, cf. 7.2.9.

It follows that the surface $\mathcal{F}$ meets every plane π_1 of one of the two systems of planes of $\mathcal{Q}$ in $\binom{n}{2}$ points, while meeting every plane π_2 of the other system of $\mathcal{Q}$ in $\binom{n-1}{2} - p$ points (cf. §5.5.9 and §12.1.2). In particular, if α is a generic plane of $\mathbb{P}^3$ and A is a generic point of α, to them there correspond on $\mathcal{Q}$ a plane π_1 and a plane π_2 having a common line (the image of the pencil of lines of α passing through A), and so belonging to a space S_3. The order of $\mathcal{F}$ is thus

$$\binom{n}{2} + \binom{n-1}{2} - p.$$

13.1.20. *Show that the surface $\mathcal{F}$, the projective image of the linear system of all algebraic curves of order n in the plane π, does not have trisecants.*

The surface $\mathcal{F}$ is embedded in a projective space $\mathbb{P}^{\frac{n(n+3)}{2}}$. If A, B, and C were three collinear points of $\mathcal{F}$, all the hyperplanes of $\mathbb{P}^{\frac{n(n+3)}{2}}$ passing through A and B would also pass through C. Consequently, in the representing plane π, all the algebraic curves of order n passing through two points A_0 and B_0 corresponding to A and B would also pass through the point C_0 whose image is C. This, however, is absurd (if $n > 1$).

13.1.21. *A reduced, irreducible algebraic curve has non-singular models in $\mathbb{P}^3$ and plane models which are either non-singular or whose only singularities are nodes (cf. §8.1.2).*

Let $C \subset \mathbb{P}^r$ be a reduced and irreducible algebraic curve. If it is not planar we can replace it with the plane curve birationally isomorphic to C (which we will again call C) obtained via projection on a plane π from a subspace S_{r-3} chosen so that the projection is simple (cf. §7.1.1).

If d is the order of C its Veronese immersion $C^* = \varphi_n(C)$ taken via the curves of order n in π has order nd and belongs to the surface $\mathcal{F}^{n^2} = \varphi_n(\pi) \subset \mathbb{P}^N$, where $N = \binom{n+2}{2} - 1$ (cf. Section 6.7).

Since in the plane we have $\binom{n+2-d}{2}$ linearly independent curves of order n that contain C as component, the curve C^* belongs to $\binom{n+2-d}{2}$ linearly independent hyperplanes of $\mathbb{P}^N$ and so is immersed in a space $\mathbb{P}^{N'}$ with

$$N' = \binom{n+2}{2} - 1 - \binom{n+2-d}{2} = nd - \frac{1}{2}(d-1)(d-2).$$

Since the Veronese immersion is an isomorphism, C^* has singularities if and only if C does too. Let P^* be an s-fold point of C^* ($s > 1$). Projecting C^* from P^* onto a hyperplane one finds there a curve Γ of order $\leq nd - 2$ in a space of dimension $N' - 1$.

By Exercise 13.1.20 we know that $\mathcal{F}^{n^2}$ does not have trisecants and so the projection is simple (and the multiple points of Γ come only from the multiple points of C^*).

Hence every projection from a multiple point produces a reduction by 1 of the dimension of the ambient space and by at least 2 in the order of the curve. With a finite number of projections one arrives at a non-singular curve Γ', because the order of a curve is never inferior to the dimension of the minimal space that contains it (cf. Proposition 4.5.6).

If Γ' is embedded in $\mathbb{P}^h$, its projection on S_3 from a space S_{h-4} that does not meet the V_3 of its chords (for the study of the variety of chords of a curve, see for example [48, (11.24), (11.25)]) is a non-singular curve C' of S_3 that is birationally isomorphic to C. Thus a plane model with only nodes is obtained by projecting C' from a generic O in S_3 (O should not belong to the ruled surface of the tangents, the ruled surface of trisecants, and the ruled surface of the lines that join pairs of contact points of bitangent planes).

13.1.22. *How many copies of corresponding points can one assign arbitrarily in order to determine a quadratic transformation between two planes π and π' ?*

Represent the pairs of points (P, P') with $P \in \pi$ and $P' \in \pi'$ by the points of the Segre variety $V_4^6 \subset \mathbb{P}^8$ which is the image of $\mathbb{P}^2 \times \mathbb{P}^2$ (cf. Section 11.2). The pairs of corresponding points under a reciprocity between two planes are the points of a hyperplane section of V_4^6. Since the pairs of corresponding points under a quadratic transformation are corresponding points for two distinct reciprocities, they are the points in which V_4^6 is intersected by two hyperplanes, and thus by the space S_6 which is their intersection (cf. 9.5.10).

Thus one may arbitrarily assign seven pairs of corresponding points (since this is the number of points needed to generate S_6) in order to define a quadratic transformation.

13.1.23. *How many common pairs of corresponding points are there for two quadratic transformations between two planes π and π'?*

By Exercise 13.1.22 we know that to the pairs of points P and P' that correspond under two quadratic transformations there correspond on the Segre variety $V_4^6 \subset \mathbb{P}^8$ (the Segre product of the two planes π and π') the points that belong to two spaces S_6, and thus to the space S_4 in which they intersect.

Hence the number of these pairs of points is the order of V_4^6, namely 6.

13.1.24. *Describe the hyperplane representation of a **monoid** (that is, of a variety V_{r-1}^n in $\mathbb{P}^r$ having an $(n-1)$-fold point, cf. 10.5.21).*

If $x_0, \ldots, x_r$ are projective coordinates in $\mathbb{P}^r$, the equation of a monoid F of order n having as an $(n-1)$-fold point the point $A_0 = [1, 0, \ldots, 0]$ is

$$x_0 \varphi_{n-1}(x_1, \ldots, x_r) - \varphi_n(x_1, \ldots, x_r) = 0,$$

where φ_{n-1} and φ_n are homogeneous polynomials of degrees $n-1$ and n respectively in the variables $x_1, \ldots, x_r$. Therefore the monoid has a parametric representation of the form

$$\begin{cases} x_0 = \dfrac{\varphi_n(u_1, \ldots, u_r)}{\varphi_{n-1}(u_1, \ldots, u_r)}, \\ x_i = u_i, \quad i = 1, \ldots, r, \end{cases}$$

or also, for $i = 1, \ldots, r$,

$$\begin{cases} x_0 = \varphi_n(u_1, \ldots, u_r), \\ x_i = u_i \varphi_{n-1}(u_1, \ldots, u_r). \end{cases}$$

Thus the monoid appears as the projective image of the linear system

$$\Sigma: \lambda_0 \varphi_n(u_1, \ldots, u_r) + \varphi_{n-1}(u_1, \ldots, u_r)(\lambda_1 u_1 + \lambda_2 u_2 + \cdots + \lambda_r u_r) = 0$$

of hypersurfaces in S_{r-1} with order n and passing through the complete intersection of two hypersurfaces with equations $\varphi_{n-1} = 0$ and $\varphi_n = 0$.

13.1.25. *Let $\mathcal{F}$ be a cubic surface with four nodes in $\mathbb{P}^3$. Represent it in a plane π by projection from one of the four nodes. Study in particular the space cubics belonging to $\mathcal{F}$ and passing through the four nodes, and show that each of them is a contact curve of $\mathcal{F}$ with a quadric cone.*

Suppose that the projective reference system is taken so that $\mathcal{F}$ has equation

$$x_0(x_1 x_2 + x_2 x_3 + x_1 x_3) = x_1 x_2 x_3,$$

whence its four nodes are the fundamental points of the reference tetrahedron. We project $\mathcal{F}$ from the point $A_0 = [1, 0, 0, 0]$ onto the plane $\pi = J(A_1, A_2, A_3)$ with equation $x_0 = 0$.

The projection on π of the section of $\mathcal{F}$ by a generic plane $\lambda_0 x_0 + \lambda_1 x_1 + \lambda_2 x_2 + \lambda_3 x_3 = 0$ is the curve with equations

$$\begin{cases} x_0 = 0, \\ \lambda_0 x_1 x_2 x_3 + (x_1 x_2 + x_2 x_3 + x_1 x_3)(\lambda_1 x_1 + \lambda_2 x_2 + \lambda_3 x_3) = 0, \end{cases}$$

which describes in π the linear system Σ of which $\mathcal{F}$ is the projective image.

The system Σ consists of the cubics passing through A_1 with tangent b_1 ($x_0 = x_2 + x_3 = 0$), through A_2 with tangent b_2 ($x_0 = x_1 + x_3 = 0$), and through A_3 with tangent b_3 ($x_0 = x_1 + x_2 = 0$).

The three lines b_1, b_2, and b_3 represent three lines a_1, a_2, a_3 in $\mathcal{F}$. Each of them is in fact tangent to the curves of Σ at one of the base points and so meets the curves of Σ in only a single point away from the base points.

Since the trilateral $b_1 b_2 b_3$ is a curve of Σ, the trilateral $a_1 a_2 a_3$ is a plane section of $\mathcal{F}$. The plane σ that contains a_1, a_2, a_3 is tritangent to $\mathcal{F}$. One checks easily that

$$a_1: \begin{cases} x_0 + x_1 = 0, \\ x_2 + x_3 = 0, \end{cases} \quad a_2: \begin{cases} x_0 + x_2 = 0, \\ x_1 + x_3 = 0, \end{cases} \quad a_3: \begin{cases} x_0 + x_3 = 0, \\ x_1 + x_2 = 0, \end{cases}$$

and

$$\sigma: x_0 + x_1 + x_2 + x_3 = 0.$$

The points B_1, B_2 and B_3 where a_1, a_2 and a_3 meet the plane $x_0 = 0$ (and which coincide with their projections on π) thus all lie on the line $r = \sigma \cap \pi$ (with equations $x_0 = x_1 + x_2 + x_3 = 0$).

They are points of $\mathcal{F}$ and belong to the section of $\mathcal{F}$ by π, that is, to the trilateral $x_1 x_2 x_3 = 0$, whose vertices A_1, A_2, and A_3 are three of the four nodes of $\mathcal{F}$. The fourth node is A_0 (B_1 is the point defined by $x_0 = x_1 = x_2 + x_3 = 0$, so that $B_1 \in b_1$; analogously $B_2 \in b_2$ and $B_3 \in b_3$. Moreover, for $i = 1, 2, 3$, the point B_i is contained in the plane $x_i = 0$).

Thus we have shown that the three fixed tangents of the cubics of Σ cut the sides of the triangle $A_1 A_2 A_3$ in three collinear points along the line r. Note that r is the Pascal line of the simple hexagon $A_1 A_1 A_2 A_2 A_3 A_3$ inscribed in the conic γ which is the section of π with the tangent cone to $\mathcal{F}$ at A_0 (cf. [13, Vol. II, Chapter 16] and Figure 13.2). The equations of γ are $x_0 = x_1 x_2 + x_2 x_3 + x_1 x_3 = 0$ and γ is tangent to b_i at A_i, $i = 1, 2, 3$.

The contact points of the plane σ with $\mathcal{F}$ are the diagonal points of the quadrilateral determined by the lines $A_1 A_2$, $A_2 A_3$, $A_1 A_3$, r (Figure 13.2).

The projection from A_0 of A_i is the same point A_i: to A_i there corresponds on $\mathcal{F}$ the line $A_0 A_i$, $i = 1, 2, 3$ (cf. §10.1.7).

The net of plane sections passing through A_0 and thus having A_0 as double point, projects into the net of cubics in π that split into γ and another line c. In particular, to a line c there corresponds on $\mathcal{F}$ a cubic C passing through A_0 (cf. expression (10.5)).

The lines $A_1 A_2$, $A_2 A_3$, $A_1 A_3$, which are contained in $\mathcal{F}$, each coincides with its own projection (for example, $A_1 A_2$ is the image of the cubic split into the three lines $A_0 A_1$, $A_0 A_2$, $A_1 A_2$ and similarly for the lines $A_1 A_3$ and $A_2 A_3$). The point A_1 is the projection from A_0 of all the points of the line $A_0 A_1$ and thus is an exceptional point to which there corresponds the line $A_0 A_1$; the analogous facts hold for A_2 and A_3.

The conic γ is fundamental for Σ. It is the trace on π of the tangent cone to $\mathcal{F}$ at A_0 and hence is the image of the double point A_0. (On γ the first order neighborhood of A_0 on $\mathcal{F}$ is opened up.)

The space cubics $\mathcal{L}$ passing through the four nodes of the surface are represented by the plane quartics split into the conic γ (which, being a fundamental curve, corresponds to a point of $\mathcal{F}$) and a conic $\mathcal{L}'$ which is the projection of $\mathcal{L}$ from A_0 onto the plane $J(A_1, A_2, A_3)$. The conic $\mathcal{L}'$ passes through the base points A_1, A_2, and A_3.

The sections of $\mathcal{F}$ with the quadrics are the images of the curves of the linear system Φ of curves of order 6 having A_1, A_2, and A_3 as tacnodes and the lines a_1, a_2, and a_3 as tacnodal tangents.

If $\mathcal{L}$ is a cubic lying on $\mathcal{F}$ and passing through the four nodes and $\mathcal{L}'$ is the conic of π that (taken together with γ) represents it, the system Φ contains the sextic which is split into the conic $\mathcal{L}'$ counted twice and γ. Thus there is a quadric tangent to $\mathcal{F}$ along $\mathcal{L}$.

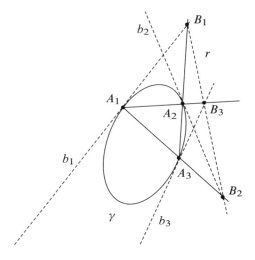

Figure 13.2

13.1.26. *After having confirmed that the plane quartic C with equation $x_0^2 x_2^2 = x_0^2 x_1^2 + x_1^2 x_2^2$ is rational, determine a rational parametric representation for it using a suitable quadratic transformation.*

The curve C is rational since it has three double points $[1, 0, 0]$, $[0, 1, 0]$, and $[0, 0, 1]$. The quadratic transformation

$$\begin{cases} y_0 = x_1 x_2, \\ y_1 = x_0 x_2, \\ y_2 = x_0 x_1 \end{cases}$$

transforms it into the conic with equation $y_1^2 = y_0^2 + y_2^2$, the locus of the point $[2t, 1 + t^2, 1 - t^2]$. Bearing in mind that $[x_0, x_1, x_2] = [y_1 y_2, y_0 y_2, y_0 y_1]$, we then have the following parametric representation for C:

$$\begin{cases} x_0 = 1 - t^4, \\ x_1 = 2t(1 - t^2), \\ x_2 = 2t(1 + t^2). \end{cases}$$

13.1.27. *Repeat the preceding exercise for the curves C_1 and C_2 in the affine plane having equations respectively $y^2 - x^4 + x^5 = 0$ and $(y - x^2)^2 = x^5$.*

Making use of the quadratic transformation $x = x'$, $y = x'y'$, we find rational parametric representations

$$C_1: \begin{cases} x = 1 - t^2, \\ y = t(1-t^2)^2 \end{cases} \quad \text{and} \quad C_2: \begin{cases} x = (t-1)^2, \\ y = t(t-1)^4. \end{cases}$$

13.1.28 (G. Salmon's theorem). *For $i = 1, 2, 3$, let $\mathcal{L}_i$ be three algebraic curves in $\mathbb{P}^3$ of orders ℓ_i with $\mathcal{L}_1 \cap \mathcal{L}_2 \cap \mathcal{L}_3 = \emptyset$. Suppose that $\mathcal{L}_i$ and $\mathcal{L}_j$ have r_{ij} points P_{ij} in common. Show that the algebraic ruled surface R of the lines in $\mathbb{P}^3$ supported by $\mathcal{L}_1$, $\mathcal{L}_2$, $\mathcal{L}_3$ has order (excluding the cones that project the curves $\mathcal{L}_k$ from P_{ij})*

$$2\ell_1\ell_2\ell_3 - \ell_1 r_{23} - \ell_2 r_{13} - \ell_3 r_{12}.$$

The multiplicity of $\mathcal{L}_1$ for R is $\ell_2\ell_3 - r_{23}$, that of $\mathcal{L}_2$ is $\ell_1\ell_3 - r_{13}$, and that of $\mathcal{L}_3$ is $\ell_1\ell_2 - r_{12}$.

We first assume that $r_{12} = r_{23} = r_{31} = 0$. The order of R is the number of points that R has in common with a generic line r, that is, the number of lines incident with $\mathcal{L}_1$, $\mathcal{L}_2$, $\mathcal{L}_3$, and r, or, again, the number of points that $\mathcal{L}_1$ has in common with the ruled surface of the lines supported by $\mathcal{L}_2$, $\mathcal{L}_3$, and r. Let $[\mathcal{A}, \mathcal{B}, \mathcal{C}]$ denote the order of the ruled surface that is the locus of the lines supported by three algebraic curves $\mathcal{A}$, $\mathcal{B}$, and $\mathcal{C}$. One thus has

$$[\mathcal{L}_1, \mathcal{L}_2, \mathcal{L}_3] = \ell_1[\mathcal{L}_2, \mathcal{L}_3, r] = \ell_1\ell_2[\mathcal{L}_3, r, r'] = \ell_1\ell_2\ell_3[r, r', r''],$$

where r, r', and r'' are three generic lines in $\mathbb{P}^3$; that is, $[\mathcal{L}_1, \mathcal{L}_2, \mathcal{L}_3] = 2\ell_1\ell_2\ell_3$.

Now suppose that the numbers r_{ij} are not all zero. If P is a point common to $\mathcal{L}_1$ and $\mathcal{L}_2$, the ruled surface $\mathcal{R}$ of the lines supported by $\mathcal{L}_1$, $\mathcal{L}_2$, $\mathcal{L}_3$ contains as component the cone of order ℓ_3 which projects $\mathcal{L}_3$ from P. We conclude in analogous fashion if there exist common points to $\mathcal{L}_1$ and $\mathcal{L}_3$ or to $\mathcal{L}_2$ and $\mathcal{L}_3$.

The residual component of the set of these cones in $\mathcal{R}$ is thus a ruled surface R of order $2\ell_1\ell_2\ell_3 - \ell_1 r_{23} - \ell_2 r_{13} - \ell_3 r_{12}$.

The multiplicity of $\mathcal{L}_1$ for R is the number of generators issuing from a generic point P of $\mathcal{L}_1$. The two cones with vertex P that project $\mathcal{L}_2$ and $\mathcal{L}_3$ have in common $\ell_2\ell_3$ lines; but one must discard the lines that project the r_{23} points of $\mathcal{L}_2 \cap \mathcal{L}_3$ from P. There remain $\ell_2\ell_3 - r_{23}$ lines. Similarly, one calculates the multiplicities of $\mathcal{L}_2$ and $\mathcal{L}_3$ for R.

13.1.29. *Use Salmon's theorem, 13.1.28, to prove that a general cubic surface $\mathcal{F}$ contains twenty-seven lines (cf. Exercises 5.8.13 and 10.5.10).*

The ruled surface R of the lines supported by three plane sections $\mathcal{L}_i$ of the cubic $\mathcal{F}$ has order $2 \times 3 \times 3 \times 3 - 3 \times 3 \times 3 = 27$ and contains three plane sections $\mathcal{L}_i$ as 6-fold curves ($\ell_1 = \ell_2 = \ell_3 = r_{ij} = 3$ in the notation of Exercise 13.1.28). Thus the surfaces R and $\mathcal{F}$ intersect in a curve of order 81 which has as components the three curves $\mathcal{L}_i$ each counted six times. There remains a curve $\mathcal{L}$ of order $81 - 3 \times 3 \times 6 = 27$.

If P is a generic point of $\mathcal{L}$ the generator of R passing through P has at least four points in common with $\mathcal{F}$ (the point P and the three points at which it meets the sections $\mathcal{L}_i$). Hence it is contained in $\mathcal{F}$ and is therefore a component of $\mathcal{L}$. The curve $\mathcal{L}$ is thus split into twenty-seven lines.

13.1.30. Let $\mathcal{F}$ be a general cubic surface in $\mathbb{P}^3$. We know that it can be represented over the plane π by way of the linear system Σ of cubics that pass through six generic points $B_1, \ldots, B_6$ (cf. Problem 10.5.7). If a, b are two skew lines of $\mathcal{F}$, there are five lines on $\mathcal{F}$ which are supported by a and b.

For example, we verify this fact for the two exceptional lines corresponding to two of the six base points of Σ, say B_1 and B_2. The lines incident with them are the image line of the line r that joins B_1 and B_2 and the four lines that are images of the four conics each of which contains B_1, B_2, together with three of the other base points. The line r and the four conics in question meet the curves of Σ in a single point away from the base points, and thus are transformed into lines.

13.1.31. Let a and b be two lines on a general cubic surface $\mathcal{F}^3$ in $\mathbb{P}^3$, and let a_0, a_1, a_2, a_3, and a_4 be the five lines of $\mathcal{F}^3$ supported by a and b (cf. 13.1.30). Let π be a plane passing through a_0.

A point $P \in \mathcal{F}^3$ and a point $P' \in \pi$ are said to correspond if the line $r_{PP'}$ is supported by a and b. In this way we obtain a planar representation of $\mathcal{F}^3$. Examine the linear system Σ of the curves in π that correspond to the plane sections of $\mathcal{F}^3$.

Let $\mathcal{L}$ be a plane section of $\mathcal{F}^3$. The line a has in common with $\mathcal{L}$ the point at which it intersects the plane of $\mathcal{L}$. Similarly for the line b. The lines a and b are skew. By 13.1.28, the ruled surface of the lines supported by a, b, and $\mathcal{L}$ then has order $(2 \times 3 \times 1 \times 1) - (3 \times 0) - (1 \times 1) - (1 \times 1) = 4$. It meets the plane π in a curve of order 4, one of whose components is the line a_0. The residual cubic passes through the traces on π of the lines a, b, a_1, a_2, a_3, and a_4. Hence, Σ is the linear system of the cubics passing through six points.

13.1.32 (Surfaces that represent Laplace equations). Consider the map $\omega \colon \mathbb{P}^2 \to \mathbb{P}^n$ defined by
$$x_i = \omega_i(u, v), \quad i = 0, \ldots, n,$$
with $\omega_i(u, v)$ continuous functions having first and second derivatives continuous in an open set U of $\mathbb{P}^2$, at each point of which the Jacobian matrix of the $\omega_i(u, v)$ with respect to u and v has rank 2. Let $\mathcal{F}$ be the image surface (since the function ω_i is not necessarily algebraic, we consider here a much wider class of surfaces than those studied up to now).

If $P_0 = [u_0, v_0] \in U$ and $P = \omega(P_0)$, we define the *h-osculating space* of $\mathcal{F}$ at P_0 to be the minimal space that contains together with the point P_0 all the "derived points of P'' up to order h inclusive, that is, the points $P_u, P_v, P_{uv}, \ldots$ defined as follows:

$$P_u = \left[\ldots, \frac{\partial \omega_i}{\partial u}, \ldots\right], \ P_v = \left[\ldots, \frac{\partial \omega_i}{\partial v}, \ldots\right], \ P_{uu} = \left[\ldots, \frac{\partial^2 \omega_i}{\partial u \partial u}, \ldots\right], \ldots$$

We know that the 1-osculating space (i.e., the tangent plane of $\mathcal{F}$ at P) is the locus of the tangent lines at P to the curves lying on $\mathcal{F}$, passing through P and having a tangent line there.

If $h = 2$ (and $n \geq 5$) one finds, in general, a space S_5 (the 2-osculating space of $\mathcal{F}$ at P). This S_5 contains the osculating planes at P to the curves of $\mathcal{F}$ passing through P (and having an osculating plane there). The points of these planes do not exhaust S_5 but fill out a quadric cone (called the *Del Pezzo cone*).

An important class of surfaces of $\mathbb{P}^n$ for $n \geq 5$, studied by C. Segre in his memoir [90], consists of the surfaces for which it happens that at every point the 2-osculating space is S_4. For such a surface $\mathcal{F}$ the six points $P, P_u, P_v, P_{uu}, P_{uv}$, and P_{vv} are not independent, but rather the six functions $\omega_i(u, v)$ are six integrals of a partial differential equation

$$A(u, v)\frac{\partial^2 f}{\partial u^2} + B(u, v)\frac{\partial^2 f}{\partial u \partial v} + C(u, v)\frac{\partial^2 f}{\partial v^2} \\ + D(u, v)\frac{\partial f}{\partial u} + E(u, v)\frac{\partial f}{\partial v} + F(u, v) = 0. \tag{13.3}$$

We say that $\mathcal{F}$ represents the Laplace equation (13.3). The study of surfaces of this type encounters notable difficulty.

All ruled surfaces are of this type. Moreover, Togliatti [106] determined all the non-ruled algebraic surfaces in $\mathbb{P}^n$, $n \geq 5$, and of order ≤ 6, that represent Laplace equations. There are four of them, all embedded in $\mathbb{P}^5$ and all of order 6. Three are rational surfaces with sectional curves of genus 1, and Togliatti found their planar representations. They are the projective images of linear systems of cubics, whence they are projections of a Del Pezzo surface $\mathcal{G}^6$, cf. Section 10.2 and Problem 10.5.7. The fourth surface, with sectional curves of genus 2, is the intersection of the cone V_3^3 that projects a twisted cubic of $\mathbb{P}^3$ from a line r with a quadric not passing through r.

Among these there is the particularly remarkable surface $\mathcal{F}^6$, called the *Togliatti surface*, which is the locus of the point $[u, v, u^2, v^2, u^2v, uv^2]$ that represents the Laplace equation

$$u^2 \frac{\partial^2 f}{\partial u^2} + v^2 \frac{\partial^2 f}{\partial v^2} + uv \frac{\partial^2 f}{\partial u \partial v} - 2u \frac{\partial f}{\partial u} - 2v \frac{\partial f}{\partial v} + f = 0.$$

It is the projection from the point $O = [0, 0, 0, 0, 0, 0, 1]$ of the Del Pezzo surface $\mathcal{G}^6$ which is the locus of the point $[u, v, u^2, v^2, u^2v, uv^2, uv]$. The fact that the 2-osculating spaces of $\mathcal{F}^6$ are S_4's depends on the very remarkable fact (discovered by Togliatti) that the 2-osculating spaces of $\mathcal{G}^6$ all pass through O (see also [63, §4]).

Note. The surface $\mathcal{F}^6$ which is the locus of $[u, v, u^2, v^2, u^2v, uv^2]$ should not be confused with the more famous "Togliatti surface", $\mathcal{F}^5$ in $\mathbb{P}^3$ with thirty-one double points, that is, having the maximum number of isolated double points (cf. Section 5.8).

13.1.33. *A cubic hypersurface $\mathcal{F}$ in $\mathbb{P}^4$ with a double plane π is the locus of ∞^1 planes that intersect π in the tangents to a conic γ. The tangent cone to $\mathcal{F}$ at each point of γ is a double hyperplane passing through π.*

If π is the plane $x_0 = x_1 = 0$ we may suppose that $\mathcal{F}$ has equation

$$x_0^2 x_2 - 2x_0 x_1 x_3 + x_1^2 x_4 = 0,$$

and thus $\mathcal{F}$ is the locus of the ∞^1 planes that meet π in the tangent lines of the conic $\gamma : x_0 = x_1 = x_3^2 - x_2 x_4 = 0$. Indeed, the tangent to γ at its generic point $[0, 0, \lambda^2, \lambda\mu, \mu^2]$ is the line with equations $x_0 = x_1 = \mu^2 x_2 - 2\lambda\mu x_3 + \lambda^2 x_4 = 0$.

The tangent cone to $\mathcal{F}$ at the generic point $[0, 0, \lambda^2, \lambda\mu, \mu^2]$ of γ is, cf. §5.8.6,

$$\lambda^2 x_0^2 - 2x_0 x_1 \lambda\mu + \mu^2 x_1^2 = (\lambda x_0 - \mu x_1)^2 = 0,$$

that is, a pair of two coinciding hyperplanes.

13.1.34. In $\mathbb{P}^4$ consider the cubic hypersurface $\mathcal{F}$ with equation

$$x_0 x_2^2 + x_1 x_3^2 + \varphi_3(x_2, x_3, x_4) = 0.$$

Examine the singularities of $\mathcal{F}$ along the line r that joins the points $A_0 = [1, 0, 0, 0, 0]$ and $A_1 = [0, 1, 0, 0, 0]$ (cf. §5.8.7).

The line $r : x_2 = x_3 = x_4 = 0$ is double for $\mathcal{F}$. At the generic point $[\lambda, \mu, 0, 0, 0]$ of r the tangent cone is $\lambda x_2^2 + \mu x_3^2 = 0$ (cf. §5.8.6).

One can also argue as follows. The equation $x_0 x_2^2 + x_1 x_3^2 = 0$ is that of the cone Γ projecting a ruled cubic surface $\mathcal{R}$ (of general type) contained in the S_3 defined by $x_4 = 0$ from the point $A_4 = [0, 0, 0, 0, 1]$. The cone Γ passes doubly through the plane $x_2 = x_3 = 0$ and thus also through r.

The equation $\varphi_3(x_2, x_3, x_4) = 0$ defines a cubic cone with vertex r and so passing triply through r. It follows that the tangent cone to $\mathcal{F}$ at a point P of r is the tangent cone at P to Γ (cf. Remark 6.3.9). Hence it is composed of the two hyperplanes that project the two tangent planes to $\mathcal{R}$ at P from A_4.

13.1.35. *After having observed that for a generic point P of a cubic hypersurface $\mathcal{F}$ in $\mathbb{P}^4$ there pass six lines belonging to $\mathcal{F}$, prove that if $\mathcal{F}$ has a double point $Q(\neq P)$ belonging to the hyperplane tangent at P then two of the six lines coincide with the line r_{PQ}.*

If $P = A_0 = [1, 0, 0, 0, 0]$, then the equation of $\mathcal{F}$ is of the type

$$x_0^2 \varphi_1 + 2x_0 \varphi_2 + \varphi_3 = 0,$$

with $\varphi_j \in \mathbb{C}[x_1, x_2, x_3, x_4]$ forms of degree j and $\varphi_j(1, 0, 0, 0) = 0$. Moreover, the six lines $\varphi_1 = \varphi_2 = \varphi_3 = 0$ pass through P.

If $Q = A_1 = [0, 1, 0, 0, 0]$ is the double point of $\mathcal{F}$, the forms φ_2, φ_3 are of degree one with respect to x_1 and thus the three surfaces with equations $\varphi_j = 0$, $j = 1, 2, 3$, contain the line $r_{PQ} : x_2 = x_3 = x_4 = 0$ with respective multiplicities 1, 1, and 2. Thus r_{PQ} counts doubly in the group of the six lines.

13.1.36. *Prove that the cubic surface $\mathcal{F} \subset \mathbb{P}^3$ with equation $\sum_{i=0}^{3} x_i^3 = 0$ has eighteen Eckardt points.*

Let $L_i = 0$ be three planes of a pencil, and $M_i = 0$ three planes of another pencil. Let the axes of the two pencils be skew. One sees immediately that on each of the two axes there are three Eckardt points of the (non-singular) surface $\mathcal{F}$ with equation

$$L_1 L_2 L_3 + M_1 M_2 M_3 = 0 \tag{13.4}$$

(for example the plane $L_1 = 0$ intersects $\mathcal{F}$ in the three lines $L_1 = M_i = 0$ that belong to a pencil).

It then suffices to note that $x_0^3 + x_1^3 + x_2^3 + x_3^3 = 0$ can be written in three ways in the form (13.4) $[(x_0^3 + x_1^3) + (x_2^3 + x_3^3) = 0, \ldots]$. Each of the edges of the fundamental tetrahedron contains three Eckardt points (cf. Exercise 10.5.22).

13.1.37. *Determine all the homaloidal nets of plane curves of given order n and not having base points of multiplicity > 2.*

If h_1 and h_2 are respectively the number of double base points and the number of simple base points, the two Cremona equations (9.12) give

$$4h_1 + h_2 = n^2 - 1; \quad 2h_1 + h_2 = 3n - 3.$$

Hence $2h_1 = (n-1)(n-2)$ and $h_2 = 3n - 3 - (n-1)(n-2) = -(n-1)(n-5)$. Therefore $n \leq 5$, and one then concludes the exercise easily. Note, however, that the relation $2h_1 = (n-1)(n-2)$ also follows immediately from the fact that the curves C^n of a homaloidal net are rational.

13.1.38. *In $\mathbb{P}^{2n+1}$ consider three pairwise disjoint subspaces α, β, and γ of dimension n. Determine the dimension and the order of the variety V swept out by the lines supported by the subspaces.*

For each point P in α there passes one (and only one) line r supported also by β and γ. It is the intersection of the two spaces S_{n+1} that join P with β and γ. Therefore $\dim(V) = \dim(\alpha) + 1 = n + 1$.

Let δ be a fourth space S_n, in general position with respect to α, β, γ. For each point P of α consider the two $(n+1)$-dimensional spaces $J(P, \beta)$ and $J(P, \gamma)$ and the two points Q and Q' that they have in common with δ. If $Q = Q'$, then Q belongs to V.

13.1. Miscellaneous exercises   453

The common points of δ and V are thus the fixed points of the correspondence $\omega: \delta \to \delta$ by which two points like Q and Q' correspond. We see immediately that ω is a projectivity (of general type by the genericity of the spaces S_n, α, β, and γ, cf. 1.1.14). Therefore $\deg(V) = n + 1$.

13.1.39. *Let α, β, γ be three n-dimensional varieties in $\mathbb{P}^{2n+1}$, of orders a, b, and c respectively. If these varieties have pairwise only a finite number of points in common, then the locus of the lines supported by α, β, and γ is a variety V_{n+1}^N, where $N = (n + 1)abc - ap - bq - cr$, and p, q, and r are respectively the number of points of $\beta \cap \gamma$, of $\gamma \cap \alpha$, and of $\alpha \cap \beta$.*

Bear in mind Exercise 13.1.38 and imitate the reasoning of Exercise 13.1.28.

13.1.40. *Study the rational surface that is the projective image of the linear system of curves of type (α, β) of a quadric $\mathfrak{Q} \subset \mathbb{P}^3$, which is not a cone.*

Use the representation of the curves on a quadric described in Section 11.1.

Equivalently, using the stereographic projection described in Section 7.3, study the rational surface that is the projective image of the linear system of plane curves of order $\alpha + \beta$ having both an α-fold point and a β-fold point.

13.1.41. *Consider two skew S_n's, say α and β, in $\mathbb{P}^{2n+1}$, and a projectivity $\omega: \alpha \to \beta$. Prove that the locus of the lines that join pairs of corresponding points is a variety V_{n+1}^{n+1} which is the locus of ∞^1 linear spaces S_n. In fact, V_{n+1}^{n+1} may be defined as the locus of the lines incident with three of these S_n (and thus is the same variety as that described in 13.1.38).*

Try assuming that the two S_n are the joins

$$J(A_0, A_1, \ldots, A_n) \quad \text{and} \quad J(A_{n+1}, A_{n+2}, \ldots, A_{2n+1})$$

respectively, where, $A_0, \ldots, A_{2n+1}$ are the fundamental points. If, as is permissible, one supposes that two corresponding points are

$$P = [\lambda_0, \ldots, \lambda_n, 0, \ldots, 0] \quad \text{and} \quad P' = [0, \ldots, 0, \lambda_0, \ldots, \lambda_n],$$

one sees that the locus of the lines that join the points P and P' is the locus of the points that give rank one to the matrix

$$\begin{pmatrix} x_0 & x_1 & \cdots & x_n \\ x_{n+1} & x_{n+2} & \cdots & x_{2n+1} \end{pmatrix}.$$

For any choice of $k \in \mathbb{C} \cup \{\infty\}$, the subspace S_n with equations

$$\begin{cases} x_0 = kx_{n+1}, \\ x_1 = kx_{n+2}, \\ \vdots \\ x_n = kx_{2n+1} \end{cases}$$

belongs to V_{n+1}^{n+1}. Thus one obtains the ∞^1 spaces S_n lying on V_{n+1}^{n+1}. In particular, for $k = \infty$ and $k = 0$ one recovers the spaces α and β.

13.1.42 Birational transformations between quadrics. We conclude this section by suggesting a possible topic for further study.

Let Q and Q' be two non-specialized quadrics in $\mathbb{P}^3$, and let S_1 and S_2 be the two systems of lines of Q while S_1' and S_2' are those of Q'.

On Q consider two pencils Σ_1 and Σ_2 of rational curves of type (α_1, β_1) and (α_2, β_2). Suppose that two curves of Σ_1 (or else two curves of Σ_2) do not meet outside the base points (which, without loss of generality, we may assume to be the same for the two pencils, admitting the possibility that the multiplicity of the base points can also be zero). Suppose further that a curve of Σ_1 and a curve of Σ_2 have only one point in common away from the base points. The numbers α_i, β_i, $i = 1, 2$, and the multiplicities of the base points of Σ_1 and Σ_2 must satisfy suitable relations, analogous to the Cremona equations for a homaloidal net of plane curves (cf. Section 9.3).

A birational transformation between Q and Q' can be obtained by setting up two projectivities $\omega_1 : S_1' \to \Sigma_1$, $\omega_2 : S_2' \to \Sigma_2$; more precisely, to the common point of the two lines $r \in S_1'$ and $s \in S_2'$ we associate the common point (away from the base points) of the two rational curves $\omega_1(r)$ and $\omega_2(s)$.

We propose the question of extending the Cremona theory discussed in Chapter 9, to the case of Cremona transformations between quadrics. [On Q' there will be two pencils Σ_1' and Σ_2' consisting of rational curves and analogous to Σ_1 and Σ_2 seen before; there will also be exceptional curves on the two quadrics, and so on.]

13.2 Further problems

13.2.1. *Study the singularities at $O = (0, 0)$ of the following affine plane curves:*

$$x + 2x^2 + y^3 = 0; \quad 2x + 2x^2 + y^4 = 0; \quad x - y = y^3;$$
$$3xy = x^3 + y^3; \quad xy - y^2 = x^3; \quad xy = x^4 + xy^3 + y^3;$$
$$2xy + x^4 + y^4 = 0; \quad 8xy = (x^2 + y^2)^2; \quad x^2 - y^3 = 0;$$
$$x^2 - yx^2 + y^4 = 0; \quad xy^2 = x^4 + y^4.$$

13.2.2. *Find the intersections of the two curves in the affine plane having equations*

$$y^2 - x^2 = x^3 \quad \text{and} \quad y^2 - 2x^2 = x^3.$$

13.2.3. *Determine the Jacobian curve of the net of conics with equation $\lambda x_0 x_1 + \mu x_1 x_2 + \nu x_2 x_0 = 0$ in $\mathbb{P}^2$.*

13.2. Further problems   455

13.2.4. Study the singularities at the point $O = (0,0,0)$ of the following algebraic surfaces in $\mathbb{A}^3$:

$z = xy + x^3;$ $\qquad xy - z^2 + x^3 - y^3 = 0;$ $\qquad xy + z^3 = 0;$
$y^2 + x^3 + z^2 = 0;$ $\qquad xyz + x^3 + y^3 = 0;$ $\qquad xyz = x^2y^2 + y^2z^2 + z^2x^2;$
$x^3 + xy^2 + z^3 = 0;$ $\qquad x^2 + xy^2 + z^3 = 0;$ $\qquad z(x^2 + y^2) = xy;$
$z^2 = x(x^2 + y^2);$ $\qquad x + y^2 + z^3 = 0;$ $\qquad xyz = (x + y + z)^3;$
$xyz = 1.$

13.2.5. Write the equation of a plane quartic with three nodes.

13.2.6. Write the equation of a plane quartic with three cusps. What can be said about the three cuspidal tangents?

13.2.7. Determine the multiple points of the algebraic plane curve with equation $y^2 = p(x)$, where $p(x)$ is a polynomial.

13.2.8. Determine the cubics in $\mathbb{P}^2$ tangent to three lines in three collinear points.

13.2.9. Let $\mathcal{F}$ be a cubic surface in $\mathbb{P}^3$ with two double points A and B. Determine the tangent plane at a point P of the line r_{AB}.

13.2.10. Study the singularity at $A_0 = [1,0,0,0]$ of the surface in $\mathbb{P}^3$ with equation $x_0x_1^2 + x_2^3 + x_1x_2^2 = 0$.

13.2.11. Prove that if a plane intersects a cubic surface in $\mathbb{P}^3$ in a line counted twice, then the plane is a component of the Hessian surface.

13.2.12. Use quadratic transformations to desingularize a curve C^5 in $\mathbb{P}^2$ having six or five double points on a conic or a cubic respectively.

13.2.13. Consider the pencil Σ of cubics with equation $x_0^3 + x_1^3 + x_2^3 + 6\lambda x_0 x_1 x_2 = 0$. The Hessian curve of the generic curve of the pencil Σ again belongs to Σ. Therefore, the curves of Σ all have the same flexes (since the Hessian belongs to the pencil it has in common with the curves of Σ the nine base points; on the other hand, the flexes of the generic curve C of Σ are the non-singular points of C through which the Hessian passes).

Write the coordinates of these nine flexes and verify the properties of their configurations (see the exercises on plane cubics in Section 5.7).

13.2.14. Let $\mathcal{F}^n$ be a surface of order n in $\mathbb{P}^3$ that is tangent to a plane along a curve C^m of order m. How many double points does $\mathcal{F}^n$ have on C^m?

13.2.15. Write the equation of a surface of order 4 in $\mathbb{P}^3$ with a double conic. Determine the pinch-points (cf. §5.8.4).

13.2.16. *Determine the possible types of linear systems of plane quartics all of whose members split.*

13.2.17. *Prove that the linear system of plane curves of third order passing through seven points in general position is composed with an involution.*

13.2.18. *Prove that in $\mathbb{P}^3$ there exists only one quadric passing through an irreducible conic γ, having a given point P as pole of the plane of γ, and passing through a given point A.*

13.2.19. *Let ω be a Cremona transformation of order $n \leq 4$ between planes. Determine its factorization as a product of quadratic transformations and projectivities.*

13.2.20. *Prove that the section of a Steiner surface by a quadric is a curve of genus 3.*

13.2.21. *Determine the hypersurface of order n in $\mathbb{P}^n$ having as s-fold points the points of the space $S_k = J(A_0, A_1, \ldots, A_k)$.*

13.2.22. *Let C be an algebraic plane curve and let σ be a quadratic transformation having one of its fundamental points at a flex of C. What can be said about the image of the curve C under the quadratic transformation?*

13.2.23. *Let Σ be a net of algebraic plane curves tangent at a given fixed point P to a given line p. Show that the Jacobian curve of Σ has P as a triple point.*

13.2.24. *Find the singular points and the flexes of the plane cubic with parametric representation*

$$x_0 : x_1 : x_2 = 1 : t^3 - a^2 t : t^2 + a, \quad a \in \mathbb{C}, \ t \in \mathbb{C} \cup \infty.$$

[Solution. A double point, $[1, 0, a + a^2]$, and three flexes: $[0, 1, 0]$, and the points P_t with $t = \pm i a/\sqrt{3}$]

13.2.25. *Find the singular points and the flexes of the plane cubic with parametric representation*

$$x_0 : x_1 : x_2 = t : t^3 + a^2 t : t^2 + at, \quad a \in \mathbb{C}^*, \ t \in \mathbb{C} \cup \infty.$$

[Solution. A double point, $[0, 1, 0]$, and three flexes corresponding to the cubic roots of $t_1^3 + a^2 t_0^3 = 0$, where $t = t_0/t_1$]

13.2.26. *How many quadric cones contain an elliptic quartic curve $C \subset \mathbb{P}^3$?*

13.2.27. *Prove that a developable surface $\mathcal{F} \subset \mathbb{P}^3$, which is not a cone, is of degree ≥ 4.*

13.2.28. *Let C be an algebraic curve belonging to a quadric cone $F \subset \mathbb{P}^3$. Prove that there exist surfaces G in $\mathbb{P}^3$ such that $F \cap G = 2C$.*

13.2.29. *Write the equation of the algebraic complex of the tangent lines of the quadric in $\mathbb{P}^3$ of equation $x_0^2 + x_1^2 + x_2^2 + x_3^2 = 0$.*

13.2.30. *Consider in $\mathbb{P}^3$ the quadric $\mathcal{Q}$: $x_0 x_3 - x_1 x_2 = 0$ and the line r: $x_1 = x_3 = 0$. Write the equation of the linear congruence of the lines tangent to $\mathcal{Q}$ at points of r.*

13.2.31. *Find the special linear complexes belonging to the pencil*

$$3p_{01} + 2p_{03} + \lambda(2p_{01} + 3p_{23}) = 0,$$

where the numbers p_{ik}'s are the Plücker line coordinates (cf. Section 12.1).

13.2.32. *Prove that the Jacobian curve J of the net of conics*

$$\lambda(x_0^2 + 2x_1 x_2) + \mu(x_1^2 + x_0 x_2) + \nu(x_2^2 + x_0 x_1) = 0$$

is a trilateral, that is, J consists of three lines.
[Hint: consider the Hessian curve H of J ...]

13.2.33. *Let F and G be two ruled surfaces and let g be a common non-singular generator. At how many points of g are F and G tangent?*

13.2.34. *Prove that the algebraic system of the conics $\gamma \subset \mathbb{P}^2$ tangent to a given line r is represented in $\mathbb{P}^5$ by the quadric cone projecting the Veronese surface $\mathcal{F}^4$ from a plane α. Notice that α is tangent to $\mathcal{F}^4$ and represents the net of conics having r as component.*

13.3 Exercises on linear series on curves

The following result expresses the index of speciality of a linear series on a plane algebraic curve in terms of the superabundance of adjoint linear systems.

13.3.1. Let $\mathcal{C}$ be an irreducible algebraic plane curve of order d, having as singular points only $\mathfrak{d}$ ordinary double points, and let A be a divisor of $\mathcal{C}$ consisting of k non-singular points. Let $\mathcal{L}$ be an adjoint curve of sufficiently large order N that contains the divisor A and that cuts $\mathcal{C}$ not only in the double points but also in a further divisor H. Let us suppose that in the double points of $\mathcal{C}$ the curves $\mathcal{L}$ and $\mathcal{C}$ have intersection multiplicity exactly 2. Then the divisor H consists of $h = Nd - 2\mathfrak{d} - k$ points. Thus prove that *the index of speciality, $i(A)$, of A coincides with the superabundance σ of the linear system $\mathcal{A}_N$ of adjoints of order N passing through H.*

In fact, one has (cf. Example–Definition 6.2.2),

$$\sigma = \dim \mathcal{A}_N - \frac{N(N+3)}{2} + \mathfrak{d} + h,$$

whence
$$\dim \mathcal{A}_N = \frac{N(N+3)}{2} - Nd + \mathfrak{d} + k + \sigma.$$

We know that $\mathcal{A}_N$ cuts out on $\mathcal{C}$, away from the nodes and the divisor H, the complete series $|A|$ (see the argument as in 8.6.5 and Remark 8.6.7). On the other hand, the dimension of the series $|A| = g_k^r$ coincides with the dimension of the linear series $\mathcal{D}$ of the divisors cut out on $\mathcal{C}$ by the curves of $\mathcal{A}_N$, and is (see §6.3.7)
$$\dim \mathcal{D} = \dim \mathcal{A}_N - \dim \Sigma_0 - 1,$$
where Σ_0 is the linear system of the curves of $\mathcal{A}_N$ containing $\mathcal{C}$ as component. Since $\dim \Sigma_0 = \frac{(N-d)(N-d+3)}{2}$, a simple calculation shows that

$$\dim |A| = \frac{N(N+3)}{2} - Nd + \mathfrak{d} + k + \sigma - \frac{(N-d)(N-d+3)}{2} - 1$$
$$= k - \left(\frac{(d-1)(d-2)}{2} - \mathfrak{d}\right) + \sigma = k - p + \sigma.$$

But by the Riemann–Roch theorem we have $\dim |A| = k - p + i(A)$, and so $i(A) = \sigma$.

13.3.2. *Study the linear series g_d^r on an irreducible planar quintic $\mathcal{C}$ with a node O.*

The genus of $\mathcal{C}$ is $p = 5$. The canonical series is cut out by the conics passing through O, so that $|K_\mathcal{C}| = g_8^4$. Thus a divisor D belonging to a special series has degree ≤ 8, the special series being contained in the canonical series.

We note first that $\mathcal{C}$ is linearly normal; indeed, there are no curves of order 5 and genus > 2 in $\mathbb{P}^3$ (see §8.10.17). On the other hand, the series $\mathcal{L}$ cut out by the straight lines has degree 5 and index of speciality 2, because five collinear points belong to a pencil of canonical adjoints (split into the straight line of the five points and another line of the pencil with center O). This means that $\dim |K_\mathcal{C} - \mathcal{L}| = 1$, and hence $i(\mathcal{L}) = 2$ by the Riemann–Roch theorem. The dimension of the complete series defined by a rectilinear section is thus $2 (= 5 - p + 2)$, as it is the dimension of the net of lines in the plane.

One sees immediately that the series cut out on $\mathcal{C}$ by the linear system Σ_n of curves of order $n \geq 2$ is non-complete (but rather has defect of completeness equal to 1). Indeed, if $n \geq 2$, the $5n$ intersections of $\mathcal{C}$ with a generic curve of order n form a divisor of degree $> 2p - 2$ which is therefore non-special. On the other hand, cf. §6.3.7,

$$r = \dim \Sigma_n - \dim \Sigma_0 - 1 = \frac{n(n+3)}{2} - \frac{(n-5)(n-2)}{2} - 1 = 5n - 6,$$

where Σ_0 is the system of curves of order n that contain the curve $\mathcal{C}$, while the complete series that contains it has dimension $5n - 5$ and may be constructed by use of sufficiently high order adjoints that contain the $5n$ intersections of $\mathcal{C}$ with a generic curve of order n (cf. Remark 8.6.7).

By definition of the genus (see Definition 8.5.1), a divisor D consisting of $d \leq 5$ *generic* points of $\mathcal{C}$ is isolated. On the other hand, the complete series $|K_{\mathcal{C}} - D|$, cut out on $\mathcal{C}$ by the adjoint conics passing through O and containing D, has dimension $4 - d$. Therefore, by the Riemann–Roch theorem, D has index of speciality $i(D) = 5 - d$, so that the complete linear series defined by D has dimension $d - p + i(D) = 0$.

The complete series defined by a divisor of degree $d \geq 9$ has index of speciality $i = 0$ inasmuch as it has degree $d > 2p - 2 = 8$, and thus has dimension $r = d - p = d - 5$ (see Theorem 8.5.4). The same holds for a *generic* (and so non-special) divisor of degree $d = 5, 6, 7, 8$. For a special divisor, which, by the above remarks we know must be of degree $d = 5, 6, 7, 8$, one has rather $r = d - p + 1 = d - 4$ because the index of speciality can only be $i = 1$ (for a divisor D of degree $d \geq 5$ either $|K_{\mathcal{C}} - D| = \emptyset$ or else $\dim |K_{\mathcal{C}} - D| = 0$).

On the curve $\mathcal{C}$ there are no infinite series g_d^r of degree $d < 3$. Indeed, there are no g_2^2's (and thus neither are there g_1^1's) inasmuch as $\mathcal{C}$ is not rational. Nor are there g_2^1's. Indeed, otherwise, since the adjoint canonical conics are the conics passing through O, a divisor $D \in g_2^1$ would have index of speciality $i(D) = \dim |K_{\mathcal{C}} - g_2^1| + 1 = 3$. This, however, would lead to the contradiction $\dim |D| = d - p + i(D) = 2 - 5 + 3 = 0$. In particular the canonical series is not composite and the curve $\mathcal{C}$ is not hyperelliptic (cf. Section 8.7).

There thus remain to examine only the series g_d^r of degree $d = 3$ and $d = 4$.

Suppose that $d = 3$. We note immediately that there are no g_3^3's or g_3^2's, inasmuch as the residual divisors of a single point would constitute a (complete) series g_2^2 or g_2^1 (cf. Proposition 8.3.3).

If D is a triple of points not belonging to a line issuing from O, then D is an isolated divisor. Indeed, in this case D belongs to a pencil of canonical adjoints (the conics that contain it and which also pass through O). The same holds if D is a triple of collinear points, in which case the canonical adjoints that contain D are split into the line containing D and a line passing through O. By the Riemann–Roch theorem it follows that $i(D) = 2$ and so $\dim |D| = d - p + i(D) = 3 - 5 + 2 = 0$.

If however D is a triple of points collinear with O, then D belongs to ∞^2 canonical adjoints, composed of the line $\langle O, D \rangle$ and an arbitrary line. In this case it follows that $\dim |D| = 3 - 5 + 3 = 1$, and so $|D| = g_3^1$ is the linear series cut out on $\mathcal{C}$ by the lines passing through O.

From these observations it follows that the quintic $\mathcal{C}$ is a *trigonal* curve (see Example 8.7.5).

Finally, we consider the case $d = 4$. If D is a quadruple of points in general position, D belongs to one and only one conic passing through O and therefore its index of speciality is $i(D) = 1$ and $\dim |D| = 4 - 5 + 1 = 0$. Hence D is an isolated divisor. The non-isolated divisors of degree $d = 4$ are the quadruples of points three of which are collinear with O and the quadruples of points belonging to a line (not passing through O).

If three, say A_1, A_2, A_3, of the points of D are collinear with O, the divisor D belongs to ∞^1 canonical adjoints (split into the line a joining O with the three points A_i and a line passing through the fourth point P of D). Therefore $i(D) = 2$ and $\dim |D| = 4 - 5 + 2 = 1$.

For this series $g_4^1 = |D|$ the point P is fixed. Indeed, the divisor $G = A_1 + A_2 + A_3$ is clearly contained in some canonical divisor, but not every canonical divisor that contains G also contains the point P. To see this, it suffices to take a canonical divisor cut out on $\mathcal{C}$ by any conic through O which splits into the line a and a line not belonging to the pencil of center P. Then by Noether's reduction theorem, as reformulated after Proposition 8.6.10, we conclude that P is a fixed point of the series $|G + P| = |D|$.

On the other hand, to construct the series g_4^1 we take an adjoint of the second order passing through D; it is composed of the line $\langle O, D \rangle$ and of a line passing through P, and cuts the quintic in four other points which are collinear with P. Thus all the conical adjoints passing through these four points pass through P as well. The residual series of P with respect to the series $g_4^1 = |D|$ is the series g_3^1 cut out on $\mathcal{C}$, away from O, by the pencil of lines having center O.

If the four points belong to a line ℓ not passing through O, the divisor D belongs to ∞^1 canonical adjoints, since the conics split into the line ℓ and a line through O. Then the series $|D|$ has dimension $\dim |D| = d - p + i(D) = 4 - 5 + 2 = 1$ and is cut out on $\mathcal{C}$ by the pencil of lines having as center the fifth point of intersection of $\mathcal{C}$ with the line ℓ.

13.3.3. *Let $\mathcal{C}$ be an elliptic irreducible planar quintic. Prove that it is the projection from a line of a linearly normal quintic $\Gamma \subset \mathbb{P}^4$. The curve Γ is non-singular, and has neither trisecant lines nor quadrisecant planes. The locus of the ∞^2 chords of Γ is a hypersurface of order 5. A generic line r of $\mathbb{P}^4$ does not belong to any trisecant plane; rather, through it there pass five bisecant planes. The quintic $\mathcal{C}$ possesses a triple point and two double points, or else five double points according to whether or not r belongs to a trisecant plane (a line can not belong to two trisecant planes). Study the projections of Γ into a subspace S_3.*

The lines of the plane cut out a non-special linear series g_5^2 on $\mathcal{C}$ which is contained in a complete series of dimension $r = 5 - 1 = 4$. Hence $\mathcal{C}$ is the projection of an elliptic quintic Γ of $\mathbb{P}^4$ on which the hyperplanes cut out a complete series g_5^4. Thus, Γ is linearly normal.

The curve Γ does not have any quadrisecant planes (and so has neither trisecant lines nor singular points) because the pencil of hyperplanes passing through a quadrisecant plane would cut out a g_1^1 on Γ.

The locus of the chords of Γ is a hypersurface of order 5. Indeed, if $\mathfrak{d}$ is the number of chords supported by a generic line ℓ, then $\mathfrak{d}$ is also the number of planes passing through ℓ and containing two further points of Γ, so that $\mathfrak{d}$ is the number of double points of the projection of Γ from ℓ onto a plane π. Thus $\mathfrak{d} = 5$ because 5 is the number of double points of a planar elliptic quintic without any triple points.

For special positions of the straight line ℓ three (but not more than three) of the above mentioned five planes can coincide in a trisecant plane. From such a straight line ℓ, that is, from a line ℓ lying over a trisecant plane of Γ, the curve Γ is projected into the plane as a quintic curve with one triple point and two double points. It is clear that a straight line can not belong to two trisecant planes.

The quadrics of $\mathbb{P}^4$ cut out on Γ a linear series g_{10}^r which is non-special ($10 > 2p - 2$), and of dimension $r = 14 - h$, where $h - 1$ is the dimension of the linear system Σ of quadrics containing Γ (cf. §6.3.7). Thus, $14 - h \leq 10 - p = 9$, so that $h \geq 5$. Therefore there exist ∞^3 quadric cones belonging to Σ.

Let Γ' be the projection of Γ from a point O not belonging to Γ into a subspace S_3 of $\mathbb{P}^4$.

If O is not the vertex of a cone of Σ, then Γ' does not belong to any quadric surface and is the residual intersection of two cubics surfaces passing through a rational normal quartic $\mathcal{C}$. Let us represent one of the two cubic surfaces with the linear system $C^3(B_1, \ldots, B_6)$. Thus the projection Γ' is represented in the plane by a cubic passing through four of the six base points, while the quartic $\mathcal{C}$ is represented in the plane by a curve of order 6 passing doubly through those four base points B_j and with multiplicity 3 at the remaining two.

If O is the vertex of a quadric cone $\mathcal{Q}$ of Σ, the planes of one of the two systems of planes lying over $\mathcal{Q}$ are trisecant planes of Γ. Moreover, Γ' is a curve of type $(2, 3)$ with a double point (since $p = 1$) on the intersection of $\mathcal{Q}$ with this hyperplane.

The projection of Γ from any one of its points into a hyperplane S_3 is the base curve of a pencil of quadrics.

The following exercises make use of the planar representation of rational surfaces developed in Chapter 10.

13.3.4. *Let $\mathcal{C}$ be an irreducible planar quintic of genus $p = 2$. It is the projection of a non-singular quintic $\Gamma \subset \mathbb{P}^3$. The curve Γ belongs to a quadric $\mathcal{Q}$ and is of type $(2, 3)$. If $\mathcal{C}$ has a triple point, the center O of projection is a point of $\mathcal{Q}$. If instead $\mathcal{C}$ has no triple point, the center of projection is a double point common to the cubic surfaces of a pencil whose base curve consists of Γ and four chords of Γ issuing from O.*

For a planar quintic $\mathcal{C}$ of genus $p = 2$ two cases are possible:

(1) $\mathcal{C}$ has a triple point O and an ordinary double point A;

(2) $\mathcal{C}$ has four not necessarily distinct ordinary double points A, B, C, D (which are three by three non-collinear).

The canonical series $|K_{\mathcal{C}}| = g_2^1$ is cut out in the first case by the lines through O (since the canonical adjoints are the conics split into the line $\langle O, A \rangle$ and another line through O). In the second case it is cut out by the conics passing through A, B, C, D.

The linear series $\mathcal{H}_{\mathcal{C}}$ cut out by the lines has dimension 2 and is non-special (cf. Theorem 8.5.4). The dimension of the complete series that contains the series of rectilinear sections is thus $r = 5 - p = 3$. Therefore the series $\mathcal{H}_{\mathcal{C}}$ is not complete, that is, $\mathcal{C}$ is not linearly normal, and so there exists a quintic curve Γ in S_3 which has $\mathcal{C}$ as its projection (see Section 8.4). The curve Γ is linearly normal. In fact, one should note that an irreducible curve of order 5 belonging to S_4 has genus 1 (see 8.10.13), and irreducible quintics in S_5 are rational. Moreover, if $n \geq 6$, irreducible quintics in S_n do not exist.

A quintic Γ in S_3 having genus 2 is non-singular and belongs to one (and only one) quadric $\mathcal{Q}$. Indeed, 2 is the maximum genus for a space curve in $\mathbb{P}^3$ and the curves of maximum genus belong to quadrics (see §8.10.17).

Moreover, on Γ the quadrics of $\mathbb{P}^3$ cut out a linear series g_{10}^r which is non-special (since $10 > 2p - 2$) and whose dimension is $r = 9 - h$ if $h - 1$ is the dimension of the linear system of quadrics containing Γ (cf. §6.3.7). Thus $9 - h \leq 10 - p = 8$, and so $h \geq 1$; in fact, $h = 1$ because a quintic in S_3 can not belong to two quadrics.

From §8.10.17 we know that, if $\mathcal{Q}$ is non-singular, Γ is a curve of type $(2, 3)$; if $\mathcal{Q}$ is a cone, Γ passes through the vertex and meets each generator in two other points.

If $\mathcal{C}$ has a triple point, the center, say P, of projection belongs to $\mathcal{Q}$ and conversely. Indeed, if $P \in \mathcal{Q}$, the curve Γ meets the line a of type $(1, 0)$ passing through P in three points, and thus a intersects the plane in the triple point O of $\mathcal{C}$. The converse is clear.

Suppose now that $\mathcal{C}$ has no triple point, so that it is the planar projection of the curve Γ from a point P not belonging to $\mathcal{Q}$. The curve Γ certainly belongs to cubic surfaces (because the cubic surfaces are ∞^{19} in number and the cubic surfaces that pass through sixteen points of Γ must contain Γ). The linear series g_{15}^r cut out on Γ by the cubic surfaces has dimension $r = 19 - h$, where $h - 1$ is the dimension of the linear system Σ of cubic surfaces that contain Γ. Since the series is non-special, it follows that $19 - h \leq 15 - p = 13$ and so $h \geq 6$. Hence $\dim \Sigma \geq 5$.

The system Σ contains at least a pencil Φ of surfaces having P as a double point (inasmuch as imposing on the cubics of $\mathbb{P}^3$ to pass doubly through the point P constitutes a linear condition of dimension $\binom{3+1}{3} = 4$, cf. Section 6.2).

The four chords of Γ issuing from P, whose traces in the plane projection are the four nodes of $\mathcal{C}$, belong to all the surfaces of the pencil. Hence the base curve of the pencil consists of Γ and these four chords (which constitute the base curve of the pencil of cones tangent at P to the surfaces of Φ).

Conversely, we take a pencil Φ of cubic surfaces having a common double point P such that the four base lines b_i of the pencil Ψ of cones tangent at P to the surfaces of Φ belong to all the surfaces of Φ. The base curve of Φ consists of the four lines b_i and a quintic Γ of genus two whose projection from P into the plane π is a quintic $\mathcal{C}$ having nodes at the four points where the lines b_i meet π. Indeed, the ten intersections of Γ with an arbitrary cone of Ψ are distributed on the six lines of intersection of that cone with the corresponding cubic surface: eight of these points lie in pairs on the lines b_i; the other two belong to the remaining two lines (one on each of them).

Over the plane π let us represent an arbitrary cubic surface $\mathcal{F}$ of the pencil Φ via projection from the point P. The system of plane sections of $\mathcal{F}$ has as its projection on π the linear system Σ' of cubics passing through six points of a conic γ. Indeed, γ is cut out on π by the tangent cone to $\mathcal{F}$ at P and the six points are the traces of the six lines of $\mathcal{F}$ passing through P. Four of these six lines are the chords b_i of Γ issuing from P, while the other two meet Γ in a (single) point (away from P). In the plane π one has the quintic $\mathcal{C}$ whose four nodes A, B, C, D are four of the six base points of Σ'; the other two base points M, N are a pair of the linear series g_2^1 cut out on $\mathcal{C}$ (away from the points A, B, C, D) by the conics through A, B, C, D. To g_2^1 there corresponds on the curve Γ the linear series σ_2^1 cut out (away from the eight intersections of Γ with the chords b_i) by the pencil of cones tangent at P to the surfaces of Φ.

13.3.5. *Describe all the irreducible curves of order 6 and genus $p = 3$ having ordinary singularities.*

In the first place we observe that a curve of genus 3 and order 6 is either planar or else it lies in $\mathbb{P}^3$. Indeed, let X be a curve of order 6 contained in $\mathbb{P}^n$ and belonging to ρ independent hyperplanes. On X the hyperplanes of $\mathbb{P}^n$ cut out a non-special linear series $g_6^{n-\rho}$ (because $6 > 2p - 2 = 4$). Hence $n - \rho \leq 6 - p = 3$, which gives $\rho \geq n - 3$, and therefore X belongs to the intersection of at least $n - 3$ hyperplanes, a projective space of dimension ≤ 3.

If the curve X is planar, then one has only four possibilities for its singular points:

(1) seven double points;

(2) a triple point and four double points;

(3) two triple points and a double point;

(4) a quadruple point and a double point.

The (planar) curve X is not linearly normal. Indeed, the linear series cut out on it by the lines is a non-special g_6^2 ($6 > 2p - 2 = 4$) and is not complete inasmuch as it is contained in a series g_6^r with $r = 6 - p = 3$.

Let us then consider a space curve $X \subset \mathbb{P}^3$ of order 6 and genus 3. We note first that, by the preceding remarks, X is linearly normal.

If $h-1$ is the dimension of the linear system of quadrics containing X, the series g_{12}^r cut out on X by all quadrics has dimension $r = 9-h$, and is non-special ($12 > 2p-2$). Then $9-h \leq 12-p = 9$ and so $h \geq 0$. Since X can not belong to two quadrics, one has only two cases: $h = 0$ (if $g_{12}^r = g_{12}^9$ is complete) and $h = 1$ (if $g_{12}^r = g_{12}^8$ is not complete).

If $h = 1$ and the quadric $\mathcal{Q}$ that contains X is non-singular, then on $\mathcal{Q}$ the curve X is of type $(4,2)$ and non-singular, or else it is a curve of type $(3,3)$ with a double point. Its projection from a point of $\mathcal{Q}$ into a plane π possesses, respectively, a quadruple point and a node, or two triple points and a node. Since for a point not belonging to $\mathcal{Q}$ there do not pass any trisecants of X, the projection of X from a generic point of the space is a sextic with seven nodes.

On a quadric cone $\mathcal{Q}$ with vertex a given point O a curve of order 6 and of genus 3 is the complete intersection of the cone with a cubic surface $\mathcal{F}$. There are two possible cases: either the cubic $\mathcal{F}$ passes through O, or $\mathcal{F}$ does not contain O and is tangent to the cone at a point of X. A projection with ordinary singularities is obtained only by projecting X from a point not belonging to the cone, and it is a sextic with seven nodes.

Now let us consider the case $h = 0$. Let X be a sextic of genus 3 not belonging to a quadric. If $t-1$ is the dimension of the linear system Σ of cubic surfaces passing through X, the series g_{18}^{19-t} cut out on it by the linear system, of dimension 19, of the cubic surfaces of $\mathbb{P}^3$ is non-special ($18 > 2p-2$) and so $19-t \leq 18-p = 15$, giving $t \geq 4$. Hence $\dim \Sigma \geq 3$ and X is a base curve of Σ. Moreover, two surfaces of Σ also meet along a further (variable) space cubic.

Let $\mathcal{F}$ be a non-singular surface of Σ represented over a plane via the linear system $C^3(B_1, \ldots, B_6)$ of cubics passing through the six points B_i in general position (and so not belonging to a conic). One sextic X of genus 3 belonging to $\mathcal{F}$ is, for example, that having as image in the plane a non-singular quartic $\mathcal{C}^4$ passing through the points B_i, $i = 1, \ldots, 6$.

Such a sextic X does not belong to a quadric, because the sections of $\mathcal{F}$ by quadrics are represented in the plane by the curves of order 6 passing doubly through the points B_i. Thus, to obtain such a sextic, it would be necessary to adjoin to $\mathcal{C}^4$ a conic passing through the six points B_i.

The cubic curves of $\mathbb{P}^3$ which are residual intersections (away from X) of the sections of $\mathcal{F}$ with the other cubic surfaces containing X have as image the quintics for which the points B_i are double points. Indeed, adjoining such a quintic to $\mathcal{C}^4$ one obtains a curve of order 9 having the B_i as triple points. On $\mathcal{F}$ this new curve corresponds to the intersection curve with another surface of the system Σ.

We note that the plane curves of order 5 with the six points B_i as double points are rational curves and thus form a regular linear system of dimension $\frac{5(5+3)}{2} - 6 \times 3 = 2$ (cf. Example–Definition 6.2.2 and Lemma 7.2.14). This system represents the linear system cut out on the cubic surface $\mathcal{F}$ by the surfaces of Σ. On the other hand, the only surface of Σ passing through the sextic X is $\mathcal{F}$ itself. From this it follows in particular that dim $\Sigma = 3$ (cf. §6.3.7).

In conclusion, the sextic X is the residual intersection of two cubic surfaces passing through a cubic space curve. The projection of X into a plane from a point P is either a sextic with a triple point and four double points or a sextic with seven double points, depending on whether or not P belongs to the ruled surface formed by the trisecants of X.

Bibliography

[1] E. Arbarello, M. Cornalba, P. A. Griffiths and J. Harris, *Geometry of Algebraic Curves*, Vol. I, Grundlehren Math. Wiss. 267, Springer-Verlag, New York, 1984.

[2] L. Bădescu, *Algebraic Surfaces*, Universitext, Springer-Verlag, New York, 2001.

[3] L. Bădescu, *Notes on the Course: Istituzioni di Geometria Superiore 2*, 2007/08; English version available at http://www.dima.unige.it/~badescu.

[4] W. Barth, C. Peters and A. Van de Ven, *Compact Complex Surfaces*, Ergeb. Math. Grenzgeb. (3) 4, Springer-Verlag, Berlin, 1984.

[5] W. Barth, Two projective surfaces with many nodes, admitting the symmetries of the icosahedron, *J. Algebraic Geom.* **5** (1996), 173–186.

[6] A. Beauville, *Surfaces Algébriques Complexes*, Astérisque **54**, Soc. Math. France, 1978.

[7] A. Beauville, Sur le nombre maximum de points doubles d'une surface dans $\mathbb{P}^3$ ($\mu(5) = 31$), in *Algebraic Geometry, Angers 1979*, ed. by A. Beauville, Sijthoff and Noodhoff, Alphen aan den Rijn, Rockville, Md., 207–215.

[8] A. Bellatalla, Sulle varietà razionali normali composte di ∞^1 spazi lineari, *Atti Reale Accad. Sci. Torino* **XXXVI** (1900/01), 481–511.

[9] M. C. Beltrametti, E. Carletti, D. Gallarati and G. Monti Bragadin, *Letture su Curve, Superficie e Varietà Proiettive Speciali – Un'introduzione alla Geometria Algebrica*, Nuova Didattica Scienze, Bollati Boringhieri, Torino, 2002.

[10] M. C. Beltrametti, E. Carletti, D. Gallarati and G. Monti Bragadin, *Lezioni di Geometria Analitica e Proiettiva*, Nuova Didattica Scienze, Bollati Boringhieri, Torino, 2002.

[11] M. C. Beltrametti, Una osservazione sulle varietà razionali normali, *Boll. Un. Mat. Ital.* (4) **8** (1973), 465–467.

[12] M. C. Beltrametti and A. J. Sommese, *The Adjunction Theory of Complex Projective Varieties*, de Gruyter Exp. Math. 16, Walter de Gruyter, Berlin, 1995.

[13] M. Berger, *Geometry I, II*, Universitext, Springer-Verlag, Berlin, 1987.

[14] E. Bertini, *Introduzione alla Geometria Proiettiva degli Iperspazi*, Principato, Messina, 1923.

[15] E. Bertini, *Complementi di Geometria Proiettiva*, Zanichelli, Bologna, 1927.

[16] M. T. Bonardi, Sulle ipersuperficie algebriche aventi una varietà algebrica assegnata come luogo di punti singolari, *Le Matematiche* **27** (1967), 10–18.

[17] M. T. Bonardi, Sopra le ipersuperficie che passano con molteplicità per una data varietà algebrica, *Atti Sem. Mat. Fis. Modena* **XVII** (1968), 19–22.

[18] N. Bourbaki, *General Topology*, Chapters 1–4, Elem. Math. (Berlin), Springer-Verlag, Berlin, 1980.

[19] G. Canonero, Osservazioni su certe curve algebriche appartenenti ad un cono quadrico, *Rend. Mat.* (VI) **12** (1979), 49–53.

[20] G. Canonero, *La superficie di Veronese*, Collana di Monografie VI, Accad. Ligure di Scienze e Lettere, Genova, 1991.

[21] G. Castelnuovo, Ricerche di geometria sulle curve algebriche, *Atti Reale Accad. Sci. Torino* **24** (1889), 196–223.

[22] G. Castelnuovo, Sui multipli di una serie lineare di gruppi di punti appartenenti ad una curva algebrica, *Rend. Circ. Mat. Palermo* **VII** (1893), 95–113.

[23] G. Castelnuovo, Sulla razionalità delle involuzioni piane, *Math. Ann.* **XLIV** (1894), 125–155; also in *Memorie Scelte*, n. XX.

[24] G. Castelnuovo, Le trasformazioni generatrici del gruppo Cremoniano nel piano, *Atti Reale Accad. Sci. Torino* **XXXVI** (1900/01), 539–552.

[25] G. Castelnuovo, *Memorie Scelte*, Zanichelli, Bologna, 1937.

[26] H. Clemens and P. A. Griffiths, The intermediate Jacobian of the cubic threefold, *Ann. of Math.* **95** (1972), 281–356.

[27] A. Comessatti, *Lezioni di Geometria Analitica e Proiettiva*, Vol. I and II, Cedam, Padova, 1930/31.

[28] D. Cox, J. Little and D. O'Shea, *Ideals, Varieties, and Algorithms – An Introduction to Computational Algebraic Geometry and Commutative Algebra*, Undergrad. Texts Math., Springer-Verlag, New York, 1992.

[29] L. Cremona, Sulle trasformazioni razionali dello spazio, *Rend. Ist. Lombardo* **IV** (1871), 269–279.

[30] L. Cremona, Sulle trasformazioni razionali dello spazio, *Ann. Mat. Pura Appl.* **V** (1871), 131–162.

[31] L. Cremona, *Elements of Projective Geometry*, Dover Publications, New York, 1960.

[32] P. Deligne and D. Mumford, The irreducibility of the space of curves of given genus, *Inst. Hautes Études Sci. Publ. Math.* **36** (1969), 75–110.

[33] J. Dieudonné, *Cours de Géométrie Algébrique*, Section "Le Mathématicien" 10. 1, 11. 2, Presses Universitaires de France, Paris, 1974.

[34] I. Dolgachev, *Topics in Classical Algebraic Geometry – Part I*, available at www.lsa.umich.edu/~idolga/lecturenotes.html

[35] D. Eisenbud, *Commutative Algebra – with a View Toward Algebraic Geometry*, Grad. Texts Math. 150, Springer-Verlag, New York, 1995.

[36] F. Enriques and O. Chisini, *Lezioni sulla Teoria Geometrica delle Equazioni e delle Funzioni Algebriche*, Vol. I and III, Zanichelli, Bologna, 1915, 1924.

[37] F. Enriques, Sulle singolarità che nascono per proiezione di una superficie o varietà algebrica, *Scritti Matematici offerti a Luigi Berzolari*, Pavia, 1936, 351.

[38] G. Fano and A. Terracini, *Lezioni di Geometria Analitica e Proiettiva*, G. B. Paravia & Co., Torino, 1930.

[39] A. Franchetta, Sulla curva doppia della proiezione di una superficie generale dello S_4 da un punto generico su un S_3, *Atti Accad. Naz. Lincei. Rend. Cl. Sci. Fis. Mat. Nat.* **II** (1947), 276–279.

[40] T. Fujita, Impossibility criterion of being an ample divisor, *J. Math. Soc. Japan* **34** (1982), 355–363.

[41] T. Fujita, *Classification Theories of Polarized Varieties*, London Math. Soc. Lecture Note Ser. 155, Cambridge University Press, Cambridge, 1990.

[42] W. Fulton, *Algebraic Curves—An Introduction to Algebraic Geometry*, W.A. Benjamin, Inc., New York, 1969.

[43] D. Gallarati, Ancora sulla differenza tra la classe e l'ordine di una superficie algebrica, in *Collected Papers of Dionisio Gallarati*, ed. by A.V. Geramita, Queen's Papers in Pure and Applied Mathematics 116, Kingston, Canada, 2000, 221–236.

[44] D. Gallarati, Una proprietà caratteristica delle rigate algebriche, in *Collected Papers of Dionisio Gallarati*, ed. by A. V. Geramita, Queen's Papers in Pure and Appl. Math. 116, Kingston, Canada, 2000, 195–198.

[45] D. Gallarati, *La geometria analitico-proiettiva – dalla Rivoluzione francese alla prima guerra mondiale*, Collana di Studi e Ricerche XL, Accademia Ligure di Scienze e Lettere, Genova, 2006.

[46] P. A. Griffiths and J. Harris, *Principles of Algebraic Geometry*, Wiley-Interscience, New York, 1978.

[47] G. H. Halphen, Mémoires sur la classification des courbes gauches algébriques, *J. École Polytechnique* **52** (1882), 1–200.

[48] J. Harris, *Algebraic Geometry – A First Course*, Grad. Texts in Math. 133, Springer-Verlag, New York, 1992.

[49] R. Hartshorne, *Foundations of Projective Geometry*, Lecture Notes, W.A. Benjamin, Inc., New York, 1967.

[50] R. Hartshorne, *Algebraic Geometry*, Grad. Texts in Math. 52, Springer-Verlag, New York, 1977.

[51] H. Hironaka, Resolution of singularities of an algebraic variety over a field of characteristic zero, I; II, *Ann. of Math.* **79** (1964), 109–203; 205–326.

[52] W. V. D. Hodge and D. Pedoe, *Methods of Algebraic Geometry*, Vol. 1, 2 and 3, Cambridge University Press, Cambridge, 1994, 2004, 1994.

[53] H. P. Hudson, *Cremona Transformations – In Plane and Space*, Cambridge University Press, Cambridge, 1927; ibid., new edition with a preface by V. A. Iskovskikh and M. Reid, to appear.

[54] A. Hurwitz, Über algebraische Gebilde mit eindeutigen Transformationen in sich, *Math. Ann.* **41** (1893), 403–442.

[55] I.-J. Igusa, Arithmetic varieties of moduli of genus two, *Ann. of Math.* **72** (1960), 612–649.

[56] B. Jaffe and D. Ruberman, A sextic surface cannot have 66 nodes, *J. Algebraic Geom.* **1** (1997), 151–168.

[57] J.-P. Jouanolou, *Théorèmes de Bertini et Applications*, Progr. Math. 42, Birkhäuser, Boston, Mass., 1983.

[58] J. L. Kelley, *General Topology*, The University Series in Higher Mathematics, D. Van Nostrand, New York, 1961.

[59] G. R. Kempf, *Algebraic Varieties*, London Math. Soc. Lecture Note Ser. 172, Cambridge University Press, Cambridge, 1995.

[60] E. Kunz, *Introduction to Commutative Algebra and Algebraic Geometry*, with a preface by D. Mumford, Birkhäuser, Boston, Mass., 1985.

[61] S. Lang, *Introduction to Algebraic Geometry*, Interscience Tracts in Pure Appl. Math. 5, Interscience Publisher, Inc., New York, 1958.

[62] S. Lang, *Algebra*, , 3rd ed., Addison-Wesley, Reading, Mass., 1993.

[63] A. Lanteri and R. Mallavibarrena, Osculatory behavior and second dual varieties of Del Pezzo surfaces, *Adv. Geom.* **1** (2002), 345–363.

[64] B. Levi, Risoluzione delle singolarità puntuali delle superficie algebriche, *Atti Reale Accad. Sci. Torino* **33** (1897/98), 56–76.

[65] Q. Liu, *Algebraic Geometry and Arithmetic Curves*, 2nd enlarged edition, Oxford University Press, Oxford, 2006.

[66] B. K. Mlodziejowsky, A new proof of Noether's inequality, *Moscou M. Soc. Recueil* **29** (1914), 269–275.

[67] D. Mumford and J. Fogarty, *Geometric Invariant Theory*, 2nd enlarged edition, Ergeb. Math. Grenzgeb. 34, Springer-Verlag, Berlin, 1982.

[68] D. Mumford, *The Red Book of Varieties and Schemes*, Lecture Notes in Math. 1358, Springer-Verlag, Berlin, 1988.

[69] C. Musili, *Algebraic Geometry for Beginners*, Texts Read. Math. 20, Hindustan Book Agency, New Delhi, 2001.

[70] M. Noether, *Zur Grundlegung der Theorie der algebraischen Raumcurven*, Universitext, Verlag der Königlichen Akademie der Wissenschaften, Berlin, 1883.

[71] D. G. Northcott, *Ideal Theory*, Cambridge University Press, Cambridge, 1953.

[72] C. Peskine, *An Algebraic Introduction to Complex Projective Geometry I*, Cambridge Stud. Adv. Math. 47, Cambridge University Press, Cambridge, 1996.

[73] P. Del Pezzo, Sulle superficie dell' n^{mo} ordine immerse nello spazio di n dimensioni, *Rend. Circ. Mat. Palermo* **1** (1887), 241–271.

[74] M. Reid, *Undergraduate Algebraic Geometry*, London Math. Soc. Student Texts 12, Cambridge University Press, Cambridge, 1994.

[75] M. Reid, *Undergraduate Commutative Algebra*, London Math. Soc. Student Texts 29, Cambridge University Press, Cambridge, 1995.

[76] L. Robbiano and M. Kreuzer, *Computational Commutative Algebra 1*, Springer-Verlag, Berlin, 2000.

[77] J. Rosanes, Ueber diejenigen rationalen Substitutionen welche eine rationale Umkehrung zulassen, *J. Reine Angew. Math.* **73** (1871), 97–110.

[78] H. Schubert, Anzahl-Bestimmung für lineare Räume beliebiger Dimension, *Acta Math.* **8** (1866), 97–118.

[79] B. Segre, Sulle curve algebriche le cui tangenti appartengono al massimo numero di complessi lineari indipendenti, *Memorie Accad. Naz. Lincei* (6) **2** (1928), 578–592.

[80] B. Segre, *Lezioni di Geometria Moderna, Vol. I: Fondamenti di Geometria sopra un Corpo qualsiasi*, Zanichelli, Bologna, 1948.

[81] B. Segre, *Prodromi di Geometria Algebrica*, Edizioni Cremonese, Roma, 1972.

[82] C. Segre, Sulle rigate razionali in uno spazio lineare qualunque, *Atti Reale Accad. Sci. Torino* **XIX** (1883–1884), 265–282.

[83] C. Segre, Considerazioni intorno alla geometria delle coniche di un piano e alla sua rappresentazione sulla geometria dei complessi lineari di rette, *Atti Reale Accad. Sci. Torino* **XX** (1884–1885), 487–504.

[84] C. Segre, Sulle varietà normali a tre dimensioni composte di serie semplici razionali di piani, *Atti Reale Accad. Sci. Torino* **XXI** (1885–1886), 95–115.

[85] C. Segre, Sulle varietà che rappresentano le coppie di punti di due piani o spazi, *Rend. Circ. Mat. Palermo* **V** (1891), 192–204.

[86] C. Segre, Introduzione alla geometria sopra un ente algebrico semplicemente infinito, *Ann. Mat.* (2) **XXII** (1894), 41–142.

[87] C. Segre, Le molteplicità nelle intersezioni delle curve piane algebriche con alcune applicazioni ai principi della teoria di tali curve, *Giorn. Mat. Battaglini* (II) **XXXVI** (1898), 1–50.

[88] C. Segre, Un'osservazione relativa alla riducibilità delle trasformazioni cremoniane e dei sistemi lineari di curve piane per mezzo di trasformazioni quadratiche, *Atti Reale Accad. Sci. Torino* **XXXVI** (1900–1901), 377–383.

[89] C. Segre, Gli ordini delle varietà che annullano i determinanti dei diversi gradi estratti da una data matrice, *Atti Reale Accad. Lincei* (5) **IX** (1900), 253–260.

[90] C. Segre, Su una classe di ipersuperficie degli iperspazi legate colle equazioni lineari alle derivate parziali di $2°$ ordine, *Atti Reale Accad. Sci. Torino* **XLII** (1906–1907), 559–591.

[91] C. Segre, *Opere*, a cura dell'Unione Matematica Italiana e col contributo del Consiglio Nazionale delle Ricerche, Vol. I, Edizioni Cremonese, Roma, 1957.

[92] J. G. Semple and L. Roth, *Introduction to Algebraic Geometry*, Clarendon Press, Oxford, 1985.

[93] E. Sernesi, Una breve introduzione alle curve algebriche, *Atti del Convegno di Geometria Algebrica* (Genova-Nervi, April 12–17, 1984), ed. by D. Gallarati, Tecnoprint, Bologna, 1984, 7–38; available at http://www.mat.uniroma3.it/users/sernesi/nervi.pdf.

[94] E. Sernesi, *Deformations of Algebraic Schemes*, Grundlehren Math. Wiss. 334, Springer-Verlag, Berlin, 2006.

[95] J. P. Serre, *Groupes Algébriques and Corps de Classes*, Hermann, Paris, 1959.

[96] F. Severi, Sulle intersezioni delle varietà algebriche e sopra i loro caratteri e singolarità proiettive, Memorie Reale Accad. delle Scienze di Torino, (2) **52** (1902), 61–118.

[97] F. Severi, *Lezioni di Geometria Algebrica*, Draghi, Padova, 1908.

[98] F. Severi, Sulla varietà che rappresenta gli spazi subordinati di date dimensioni immersi in uno spazio lineare, *Ann. Mat. Pura Appl.* (3) **24** (1915), 1–32.

[99] F. Severi, Una rapida ricostruzione della Geometria sopra una curva algebrica, *Atti Reale Istituto Veneto di Scienze e Lettere* **79** (1920), 929–938.

[100] F. Severi, *Trattato di Geometria Algebrica*, Zanichelli, Bologna, 1926.

[101] F. Severi, *Vorlesungen über Algebraische Geometrie*, 1st ed., Teubner, Leipzig, 1921; Reprint, Johnson Reprint Corp., New York, 1968.

[102] F. Severi and G. Scorza Dragoni, *Lezioni di Analisi*, Vol. II, Zanichelli, Bologna, 1944.

[103] I. R. Shafarevich, *Basic Algebraic Geometry*, Vol. 1 and 2, 2nd ed., Springer-Verlag, Berlin, 1994.

[104] T. A. Springer, *Linear Algebraic Groups*, 2nd ed., Progr. Math. 9, Birkhäuser, Boston, Mass., 1998.

[105] L. Szpiro, *Lectures on equations defining space curves*, Notes by N. Mohan Kumar, Tata Institute of Fundamental Research, Bombay; published by Springer-Verlag, Berlin, New York, 1979.

[106] E. G. Togliatti, Alcuni esempi di superficie algebriche degli iperspazi che rappresentano un'equazione di Laplace, *Comment. Math. Helv.* **1** (1929), 255–272.

[107] E. G. Togliatti, Una notevole superficie di 5° ordine con soli punti doppi isolati, Festschrift R. Fueter (Zürich), Fretz AG, Zürich 1940, 127–132.

[108] C. Turrini, On the classes of a projective manifold, in *Seminars in Complex Analysis and Geometry* (Rende, 1987), ed. by J. Guenot and D. C. Struppa, Editoria Elettronica, Rende, 1988, 131–154.

[109] A. van den Essen, *Polynomial Automorphisms and the Jacobian Conjecture*, Progr. Math. 190, Birkhäuser, Basel, 2000.

[110] G. Veronese, La superficie omaloide normale a due dimensioni e del quarto ordine dello spazio a cinque dimensioni e le sue proiezioni nel piano e nello spazio ordinario, *Atti Reale Accad. Naz. Lincei* (3) **19** (1884), 344–371.

[111] E. Vesentini, Sul comportamento effettivo delle curve polari nei punti multipli, *Ann. Mat. Pura Appl.* (IV) **34** (1953), 219–245.

[112] R. J. Walker, Reduction of singularities of an algebraic surface, *Ann. of Math.* **36** (1935), 336–365.

[113] R. J. Walker, *Algebraic Curves*, Princeton University Press, Princeton, N.J., 1970.

[114] C. T. C. Wall, *Singular Points of Plane Curves*, London Math. Soc. Student Texts 63, Cambridge University Press, Cambridge, 2004.

[115] F. L. Zak, *Tangents and Secants of Algebraic Varieties*, Transl. Math. Monogr. 127, Amer. Math. Soc., Providence, R.I., 1993.

[116] O. Zariski, The reduction of the singularities of an algebraic surface, *Ann. of Math.* **40** (1939), 639–689.

[117] O. Zariski, A simplified proof for the resolution of singularities of an algebraic surface, *Ann. of Math.* **43** (1942), 583–593.

[118] O. Zariski, Reduction of the singularities of algebraic three-dimensional varieties, *Ann. of Math.* **45** (1944), 472–542.

[119] O. Zariski, The concept of a simple point of an abstract algebraic variety, *Trans. Amer. Math. Soc.* **62** (1947), 1–52.

[120] O. Zariski and P. Samuel, *Commutative Algebra*, Vol. 1 and 2, reprint of the 1958–60 edition, Springer-Verlag, New York, 1979.

Index

absolute invariant of a quadruple
  of points, 2
adjoint curve, 261
  canonical adjoint curve, 264, 286
  completeness of the series g_d^r
    cut out on a curve, 262–264
  degree of a series g_d^r
    cut out on a curve, 262
  existence of, 261
  mutually residual divisors with
    respect to the adjoints
    of given order, 264
  regularity of the linear system of
    the adjoints, 262, 263
  speciality of the series g_d^r cut out
    on a curve, 262, 263
adjunction formula, 291
algebraic complex, 431
algebraic complex, 403
  degree, 403
  of S_k in $\mathbb{P}^n$, 431
algebraic congruence, 407
  of lines in $\mathbb{P}^3$, 429
  of the chords of a curve C^3 in $\mathbb{P}^3$,
    429
algebraic correspondence, 4, 274, 395
  between curves, 274
  bihomogeneous form, 4
  double points, 275
  fixed point, 4
  membership point-hyperplane, 395
  of given valence, 275
  of indices (m, n), 4, 274
  ramification points, 275
algebraic envelope, 115
  adhering hypersurface of the
    envelope, 117
  characteristic point, 116
  class, 115
  of the tangent hyperplanes
    to a hypersurface, 117
algebraic independence, 55
algebraic ruled surface, 408
  developability criterion, 409
  of order 3, 412
  order, 409
algebraic series, 253
algebraic set
  affine, 15
  affine cone, 33
  codimension, 54
  coordinate ring, 18, 32
  cylinder, 20
  dimension, 54
  hyperplane, 16, 32
  hypersurface, 16, 32
  irreducible, 15, 32
  irreducible components, 43, 44
    reduced decomposition, 43
  non-singular, 54
  product, 20
  projection, 20
  projective, 31
  quadric (or hyperquadric), 16, 32
  singular locus, 54
  unirational, 46
algebraic system, 168, 437
  dimension, 168
  index, 172
  irreducible, 168
  linearity condition, 172
  non-linear, 437
  pure, 168
  reduced, 168
algebraic variety, 51
  affine, 18

476   Index

projective, 32
  rational transform, 40
  section by tangent hyperplanes, 113
apparent boundary (of a surface from a point), 156
  carried over a plane, 156
  theorem of the apparent boundary, 156
arithmetic Cremona group, 308
automorphism
  group of automorphisms of $\mathbb{A}^1$, 45
  of a curve of genus ≥ 2, 284
  of an affine algebraic set, 45

base point, 170
  s-fold base point, 170
  isolated base point, 171
  ordinary, 171
  simple base point, 170
  tangent cones at a base point, 174
base variety, 170
  isolated, 171
  multiple, 171
  ordinary, 171
Bertini's first theorem, 70, 175
Bertini's second theorem, 184
Bézout's theorem, 89, 102
  for plane curves, 90
  strong form, 103
  weak form, 103
bihyperplanar double line, 155
bihyperplanar double point, 155
birational automorphisms of a non-hyperelliptic curve, 277, 283
birational correspondence between curves, 255, 285
birational equivalence
  of algebraic affine sets, 28
  of algebraic projective sets, 38

of an algebraic set with a hypersurface, 29, 42, 79
of hyperelliptic curves, 281
birational isomorphism (or birational transformation), 187
  between curves, 255
  of algebraic affine sets, 28
  of algebraic projective sets, 38
birational model in $\mathbb{P}^5$ of an algebraic surface, 151
bitangent line, 118, 158
bitangent plane, 158
  to a space curve in $\mathbb{P}^3$, 200
blow-up, 194, 395
  as projection of the graph of the Segre morphism, 395
  center, 194
  exceptional divisor, 194, 395
  exceptional variety
    as linear space, 196
    as locus of Veronese varieties, 196
  of $\mathbb{P}^n$ along a line, 195
  of $\mathbb{P}^n$ along a linear space, 196
  of $\mathbb{P}^n$ at a point, 194
  of the plane at a point, 304
    proper transform, 304
    total transform, 305
branch curve, 437
Brianchon's theorem, 332

canonical curve of given genus, 271
  of genera $p = 3, 4, 5$, 272
canonical divisor, 264, 269
  degree, 269
canonical model of a curve, 270, 271
canonical series
  canonical divisor, 264, 269
  canonical model, 270
  dimension, 269
  is a birational invariant, 269
  of a curve on a quadric, 286
  of a non-singular curve in $\mathbb{P}^n$, 269

Index   477

of a plane model, 264
of a singular curve, 273
order, 264, 269
Weierstrass point, 276
Cartier divisor on a curve, 244
effective, 244
group of Cartier divisors, 244
Castelnuovo's theorem, 187
Cayley–Brill correspondence principle, 275
Chasles' correspondence principle, 4, 205
Chasles' theorem, 162
circular helix, 424
circular points, 210
class, 138, 277
of a (non-developable) ruled surface, 163
of a curve in $\mathbb{P}^2$, 205, 208
of a curve in $\mathbb{P}^3$, 199
of a hypersurface with only isolated singularities, 140
of a non-singular hypersurface, 139
of an algebraic envelope, 115
Clebsch's diagonal surface, 380
Clifford quadric, 225
Clifford's theorem, 270
Clifford's theorem (polarity associated to a curve C^n in $\mathbb{P}^n$), 225
closed mapping, 14, 49
codimension, 69
compact space, 14
complete linear series, 248, 251
complex projective space, 4
condition imposed on a linear system, 168
algebraic condition, 170
algebraic condition of dimension d, 168
linear condition, 168
linear condition of dimension d, 168

conical hypersurface (or cone) with given vertex, 111
connected space, 14
connected component, 14
continuous mapping, 14
at a point, 14
contracted ideal, 20
coordinate ring, 18, 32
coordinates, 1
projective coordinates on S_n, 4
projective coordinates on the line, 1, 3
transformation formula, 1
coresidual divisors, 264
correspondences V, I
affine case, 15–17
projective case, 31–33
Cremona equations, 308
Cremona transformation, 292
between planes, 309, 341
analytic determination of the exceptional lines, 312
correspondent of an infinitely near point in a given direction, 312
exceptional curve corresponding to a point, 312
exceptional curve of the second net, 312
inversion formulas, 310
order, 309
the associated homaloidal nets have the same number of base points, 315
between projective spaces of dimension 3, 318
by way of a planar representation of a rational surface, 382
of type (2, 2) (or quadratic transformations), 319
of type (2, 3), 320
of type (2, 4), 321

478   Index

of type (n, n') between
    r-dimensional projective
    spaces, 317
  proper transform of a line, 318
cross ratio, 2
cubic surface in $\mathbb{P}^3$
  configuration of the lines, 449
  contains twenty-seven lines, 158,
    370, 448
  has forty-five tritangent planes, 370
  planar representation, 449
  tangent planes, 158
  with four nodes, 161, 441, 442
    planar representation, 445
curve, 68
  as (set-theoretic) intersection of
    three surfaces, 236
  birational automorphisms of a
    non-hyperelliptic curve, 277
  birational classification, 278, 281
  canonical curve of given genus, 271
  canonical model, 271
  complete intersection, 198, 199, 388
  contact, 364
    of higher order, 365
  containing two linear series $g_d^1, g_{d'}^1$,
    286
  dual curve, 117, 437
  elliptic, 261, 270, 286
    birational plane model, 278
    modulus, 278
    projection of an elliptic normal
      curve, 287
  elliptic normal, 287
  embedded, 197
  genus, 205, 256, 260
  group of Cartier divisors, 244
  group of Weil divisors, 243
  hyperelliptic, 270, 271
    birational plane model, 279
  linearly normal, 254, 460, 462
  locally complete intersection, 200
  maximum number of double points,
    210
  non-complete intersection, 153
    ideal not of principal class, 153
  non-degenerate, 197
  not complete intersection, 82
  on a quadric cone, 218
  order, 198, 199
  ordinary model, 242
  osculating hyperplane, 198
  osculating plane, 198
  osculating space, 198
  Picard group, 251
  rational, 206, 256
  rational is algebraic, 220
  rational normal, 221
    associated ideal, 235
    associated polarity, 225
    through $n + 3$ points in $\mathbb{P}^n$, 224
  set theoretic complete intersection,
    82
  surface of the trisecants, 200, 259
  trigonal, 273, 459
  with general moduli, 283
curve (of type) (α, β) on a quadric of $\mathbb{P}^3$,
  214, 387
curve on a quadric cone, 218
cuspidal double curve, 151, 153
cuspidal point, 153

deficiency of an algebraic curve, 206,
  300
De Jonquières transformation
  (or monoidal transformation),
  310
Del Pezzo cone, 450
Del Pezzo surface, 192, 368, 370
derived point, 66
developable ruled surface
  regression edge, 160
diagonal, 44, 45
dilatation (or blow-up), 194

Index   479

dimension
    invariant for birationally equivalent varieties, 58
    of a hypersurface, 56
    of a projective variety, 68, 69
    of an algebraic affine set, 56
    of the intersection of a projective variety with a hypersurface, 70
    of the intersection of two varieties
        affine case, 71
        projective case, 71
    of the product of varieties, 60
    pure, 70
direct similitude (or rotohomothety), 327
discrete valuation ring, 96
divisor, 69
    coresidual, 264
    isolated, 256
    linear system of divisors
        on a variety, 174
    of a linear series on a curve, 246
    on a curve, 243
        Cartier divisor, 244
        degree, 244
        effective, 243
        support, 244
        Weil divisor, 243
        zero divisor, 251
double biplanar point, 150
double nodal curve, 150
dual character of cusp and flex, 118, 437
duality principle, 8
    dual property, 8

Eckardt point, 345, 380, 452
elimination of more variables, 20, 100
    geometric meaning, 100
elimination of one variable, 84
    Kronecker's procedure, 101

    the homogeneous case, 86
elliptic cubic cone, 335
equianharmonic quadruple, 2
essential parameters, 63
Euler's formula, 60
Euler–Sylvester resultant of two polynomials, 84, 91, 309
    isobaric of given weight, 86
exceptional curve
    associated to a base point of a linear system in $\mathbb{P}^2$, 346
    is rational normal, 346
    of a Cremona transformation between planes, 312
    of a quadratic transformation between planes, 295
exceptional line
    of a quadratic transformation between planes, 295
    of the blowing up of the plane at a point, 304
extended ideal, 20

factorial ring, 83
    irreducible element, 83
family of curves (α, β), 214
fiber, 40
    of a morphism (or regular map), 40, 41
    of a rational map (or rational transformation), 40
field of fractions, 24, 36
first order (or linear) part of the Taylor series development of a polynomial, 53
first order neighborhood, 218, 301
    of a point on a quadric, 214
flex of a curve in $\mathbb{P}^3$, 199
flex of an elliptic curve, 276
form of the first type, 4
Fréchet space, 19
Frobenius morphism, 44

fundamental $(n+1)$-hedron
    (or pyramid), 5
fundamental curve (of a linear system
    in $\mathbb{P}^2$), 345
fundamental point (of a rational surface),
    345
Fundamental Theorem for Projectivities,
    9

Geiser's involution, 183
generators of the ideal of curves in $\mathbb{P}^3$,
    388
    of the cubic, 388
    of the quartic, 389
generic (or general) object, 34
    generic conic, 35
    generic linear space, 35, 80
    generic point, 35
genus
    of a cone, 285
    of a curve on a quadric cone, 218
    of a linear system of plane curves,
        341
    of a variety locus of ∞^1 linear
        spaces, 285
    of an involution on an algebraic
        curve, 275
    of the envelope of tangents, 209
genus of an algebraic curve, 205, 256
    according to Riemann, 212, 260
    according to Weierstrass, 256, 260
    Halphen–Castelnuovo's bound,
        290
    is a birational invariant, 205, 256
    plane curve with only ordinary
        singularities, 209, 261
    the complete intersection case, 291
geometric Cremona group, 308
graded ring, 30
    associated to a ring with respect to
        an ideal, 74
    homogeneous elements of given
        degree, 30

graph, 49, 394
    of a morphism, 49
    of a rational transformation, 394
Grassmann variety (or Grassmannian),
    419
    defined by quadratic forms, 418
    dimension, 331, 419
    dual Grassmann coordinates, 417,
        421
    embedding space, 419
    Grassmann coordinates, 415
    hyperplane section, 422
    ideal of the Plücker relations, 418
    indices, 419
    is a factorial variety, 421
    is a rational variety, 421
    is not complete intersection, 421
    of the lines in $\mathbb{P}^n$, 424, 425
    order, 422
    systems of linear spaces on, 422
Grassmann's formula, 7
group of Cartier divisors, 244
group of divisor classes of a curve, 251
group of projectivities, 3, 420
    product, 3

Halphen–Castelnuovo's theorem
    (maximum genus of a curve),
        290
harmonic group, 2
harmonic polar, 149
harmonic quadruple, 2
harmonic set, 2
Hausdorff space, 19
Hessian hypersurface, 137
    as the locus of double points for
        some first polar, 136
    equation, 136
    multiplicty at a singular point, 137
    of a plane curve, 143
    order, 137

Index   481

Hilbert Nullstellensatz
    affine case, 17
    projective case, 32, 307
homaloidal net, 307
    Jacobian curve, 332
    of conics, 293, 318
        base points, 293
    of cubics, 318
    of plane curves of order n, 307
    of quartics, 319
homaloidal system, 317
homeomorphism, 14
homogeneous ideal, 30
homography (or collineation) (*see also* projectivity), 8
hyperosculating plane (to a space curve), 233
hyperplane, 16, 32
    affine, 16
    double tangent hyperplane, 368
    projective, 32
    section of varieties by tangent hyperplanes, 113
    tangent to a hypersurface, 113
    tangent to a variety, 113
    tangent to the hypersurface of a pencil at a base point, 175, 451
hypersurface, 16, 32, 51, 68, 106
    s-fold point, 109
    bihyperplanar double line, 155
    bihyperplanar double point, 155
    class, 138
    cone with given vertex, 111
    dual hypersurface, 117
    in $\mathbb{P}^4$ with a double line, 155
    intersection with its Hessian, 138
    irreducible, 107
    irreducible components, 107
    node, 139
    non-singular (or regular, or simple) point, 52
    open subset of non-singular points, 53
    order, 106
    parabolic point, 138
    principal tangent, 111, 156
    projective tangent cone, 109
    reduced, 107
    reducible (or split), 107
    section by tangent hyperplanes, 113
    sections by linear spaces, 107
    singular point, 52, 61, 62
    tangent cone in a point of a double line, 451
    tangent cone in a point of a double curve, 451
    tangent cone in a point of an s-fold variety, 153, 154
    tangent to a hypersurface, 113
    tangent to a variety, 113
    unihyperplanar double point, 155
        locus of the unihyperplanar double points, 155

ideal of principal class, 153
ideal of the denominators (of a rational function), 26
ideal of the Plücker relations, 418
incidence condition for dual dimensional subspaces, 416
independent polynomials, 62
index of speciality of a linear series, 257, 457
    of a divisor, 257
infinitesimal line (of a quadric), 214
inflectional line, 118
intersection multiplicity, 103
    of a curve with a linear branch, 244
    of a surface with a linear branch, 245
    of two curves at a point: the simple case, 90, 303

482 Index

of two plane curves at a point, 90, 96, 103, 303
   independent of the system of projective coordinates, 98
inversion (or transformation by reciprocal radius vectors), 322
involution of given genus, 275, 285
involution of given order on a line, 4, 178
   double points, 4, 178
irreducible polynomial, 16
irrelevant ideal, 32
isomorphism
   of affine algebraic sets, 22
isomorphism (or birational transformation)
   between curves, 285

Jacobian criterion, 55, 114
Jacobian group of a g_d^1, 178, 276, 431
   on $\mathbb{P}^1$, 178
   on a curve C, 203
   order, 205, 276
Jacobian matrix, 45, 52, 63, 177, 186, 341, 449
Jacobian variety, 177, 436
   dimension and order, 178
Jordan canonical form, 9, 10

Klein group, 283
Klein quadric, 401
Klein quartic curve, 283
Kummer surface, 162

Laplace equation (represented by a surface), 449
lemniscate of Bernoulli, 210
linear branch (of a plane curve), 241
   intersection multiplicity with a curve, 244
   origin, 241
   tangent, 242
linear branch (of a space curve), 245
   intersection multiplicity with a surface, 245
linear complex, 404
   curve of a linear complex, 423
   general, 404
   of S_k in $\mathbb{P}^n$, 431
   of lines in $\mathbb{P}^4$, 427, 428
      center-plane, 428
      general, 428
      special, 428
   of lines in $\mathbb{P}^n$, 428
      center, 428, 429
   pencil of, 406
   point of $\mathbb{P}^5$ that represents it, 404
   reduced equation, 406
   special, 404
      axis, 404
linear congruence, 407
   degenerate, 407, 408
   directrix (or axis), 407
   general, 407
   special, 407, 408
      of the tangent lines to a quadric at the points of a given line, 408
linear ruled surface, 410
   general, 411
   as array of lines of a quadric in $\mathbb{P}^3$, 411
linear series, 167, 203, 246
   s-fold point of a g_d^1, 178, 203, 206
   birational invariance, 202
   canonical series, 264, 273
   complete, 254
      defined by a divisor, 248, 251
      of given order, 248
   completeness of the residual series, 250
   composed with an involution, 253, 347
   non-rational, 285
   corresponding points, 204

Index 483

cut out by a linear system of
    hypersurfaces, 246
    dimension, 247
fixed divisor, 174, 202
fixed point, 246
index of speciality, 257
neutral divisor, 253
non-special, 257
number of $(r+1)$-fold points of a series g_d^r, 276, 281
on a curve, 202, 246
    dimension, 202, 246
    order, 202, 246
    partially contained in another, 202
on a planar quintic with a node, 458
projective image, 252
residual series of a divisor, 250
residual series of a generic point, 247
simple, 253
special, 257
sum and difference of complete linear series, 250, 251
zero series, 251
linear system, 167
    base point, 170
    base variety, 170
    birational invariants, 341
    complete with respect to a base group, 347
    composed with a congruence, 182, 187, 189
    composed with a pencil, 183
    composed with an involution, 182, 187, 348, 437
    curves having multiple points that are not base points, 344
    degree, 171, 188, 211, 341, 348
    dimension, 167
    exceptional curve associated to a base point, 346
    fixed component, 183, 345
    fundamental curve, 345
    genus, 341
    homaloidal, 317
    irreducible, 183
    Jacobian variety, 184
    net, 167
    of divisors on a variety, 174
    of the plane cubics tangent to a conic in 3 fixed points, 349
    order, 167
    partially contained in a given linear system, 345
    pencil, 167
    projective image, 186, 341
        of a composite linear system, 189
        order, 188
    reducible, 183
    regular, 169
    section by a subspace, 173
    section by a subvariety
        dimension, 174
    simple, 182, 187, 341
    superabundant, 169
    with fixed component, 183
linear system of divisors on a variety
    dimension, 174
local parameters, 52, 58
local ring (of an algebraic set at a point), 36
Lüroth's theorem, 47, 187, 218, 275

minimal order of a projective variety, 104
    of a curve in $\mathbb{P}^r$, 104
Möbius transformation
    of the first kind, 327
    of the second kind, 327
model of a plane curve with only ordinary singularities, 301
moduli space of curves, 277
    dimension, 278
modulus of an elliptic curve, 278

of a plane cubic, 147, 149
monoid, 380, 444
    hyperplane representation, 444
    linearly normal, 380
morphism (or regular map)
    fiber, 41
    from an open set of an affine variety, 28
    graph, 49
    graph morphism, 49
    of affine algebraic sets, 21, 22
        associated homomorphism of coordinate rings, 23
        bijective but not isomorphism, 22
        dominant, 27
        finite, 41
        isomorphism, 22
    of projective algebraic sets, 38
        finite, 42
multiplicity
    of a hypersurface in a fundamental point, 111
    of a hypersurface in a point, 73, 109
    of a hypersurface in a singular point of a hyperplane section, 115
    of a linear section of a hypersurface in a singular point, 110
    of a projective variety in a point, 80

net, 167, 310
    irreducible, 307
nodal double curve, 153
node, 73, 259
Noether's reduction theorem, 266, 460
Noether's $Af + B\varphi$ theorem, 305
Noether–Castelnuovo theorem, 316
Noether–Rosanes inequality, 315
non-degenerate projective correspondence, 5
non-singular point, 54, 61
non-singular spatial model and plane model with only nodes for an algebraic curve, 259, 261, 443
normal lines (number of), 277
normal singularities (for a surface in $\mathbb{P}^3$), 150
null system (or null polarity), 10

open mapping, 14
order
    of a curve on a rational surface, 347
    of a hypersurface, 72
    of a projective variety, 79
    of a proper intersection of projective varieties, 104
ordinary model (of a curve), 242
osculating hyperplane (to a curve), 198, 224
osculating space (of given order) to a surface, 449
    2-osculating space, 450
osculating space (to a curve), 198
oxnode, 302

parabola of order ≥ 2, 240
parabolic curve, 138
parabolic point, 138
    locus of the parabolic points, 138
    of a plane algebraic curve, 138
    of a surface of $\mathbb{P}^3$, 138
parametric representation, 190
    of a rational curve, 220
    of a rational normal curve, 235
    of a rational surface, 341, 382
    of a twisted cubic, 228
    of a unirational variety, 186
    of the Veronese variety, 191
Pascal's theorem, 147
pencil, 167
    of linear complexes, 406
    of projectively referred hyperplanes, 222

of quadric cones, 439
pencils of projectively
 referred hyperplanes, 222
permutability theorem for polars, 122
Picard group of a curve, 251
pinch-point (or cuspidal point), 150, 153,
 156, 164, 364, 366, 367, 375
planar projection of a curve in $\mathbb{P}^3$, 200
plane cubic, 140
 equianharmonic, 148
 harmonic, 148
 modulus, 147, 149
 tritangent to a conic, 349, 435
plane curve
 adjoint, 261
 approximating parabola, 240, 241
 as set of linear branches, 242
 contact of given order with
  a curve, 240
 linear branch, 241
  tangent, 242
 neighborhood of an s-fold point, 242
 node, 259
plane curve (*see also* structure
 of a singular point of a
 plane curve)
 s-fold ordinary point, 140
 condition for being a projection, 288
 cubic, 140, 146, 147, 149
  abelian group structure, 149
  tangential of a point, 146, 276
 flex, 140, 143, 276
  of first kind, 140
  of given type, 140
  point of undulation, 140
 Hessian, 143
 ordinary cusp, 73, 141, 142
 ordinary flex, 144
 ordinary point
  node, 73, 140
 quartic, 145

sextatic points of an elliptic cubic,
 276
plane projection of a curve in $\mathbb{P}^3$
 bitangent, 200
 cusp, 201
 double point, 200
 flex, 200
 tacnode, 201
 triple point, 201
Plückerian homogeneous (line)
 coordinates, 401
Plücker's formulas, 145, 158
Plückerian singularities
 (for plane curves), 144
point of contact, 240
 between two hypersurfaces, 113
 of a hypersurface with a variety, 113
 order, 240
polar (*see also* polar hypersurface), 119
polar hypersurface
 0th polar, 121
 double method for reading the
  equation of a polar, 125
 first polar, 119
  projective meaning, 120
 local behavior of the first polar of a
  surface in $\mathbb{P}^3$, 151
 multiplicity of two successive
  polars at generic points, 136
 polar groups, 121
 polar hyperplane, 121
 polar quadric, 121
 second mixed polar, 121
 second pure polar, 121
 self-conjugate point, 122
 singularity and indeterminacy of the
  polars, 126
 tangent cone, 126
 successive polar, 121, 135
  equation, 124
  multiplicity and tangent cone in
   an s-fold point, 135

486   Index

    of an s-fold point, 135
    passes through a singular point, 135
    tangent hyperplane as polar of maximal order, 126
polarity (*see also* reciprocity of $\mathbb{P}^n = \mathbb{P}^n(K)$), 10
    associated to a rational normal curve, 225
polynomial function, 21
primary decomposition (of an ideal), 44, 76
principal closed, 18
principal (or basic) open, 18, 33
    isomorphic to an affine algebraic set, 28
    of $X \subset \mathbb{A}^n$, 20
principal ideal domain, 93
projection
    of a variety from a point, 42
    of a surface, 350
    of a variety from a linear space, 77, 190
projection from a linear space, 8, 42, 49, 77
    center, 8
    multiple, 78
    projecting cone, 77
        generator, 77
        vertex, 77
    projecting space, 8
    simple, 78
projective coordinate, 1
projective generation of a curve C^n in $\mathbb{P}^n$, 222, 224
projective image of a linear series, 252
projective image of a linear system, 186, 341
    of a composite linear system, 189
    order, 188
Projective Restsatz, 265
projective scheme, 72

projective space
    complementary subspaces, 7
    condition of incidence point-hyperplane, 8
    dual reference system, 8
    duality principle, 8
    intersection of subspaces, 6
    join of subspaces, 7
    linear subspace, 6
        dual, 7
        hyperplane, 6, 7
        line, 6
        plane, 6
        point, 6
    linearly independent points, 6
    projecting space, 8
    projective homogeneous coordinates, 4
    skew subspaces, 7
    sum of points, 5
projectively referred lines, 397
projectivity, 5
    between lines, 1
    biaxial homography, 11
        axis, 11
        characteristic, 11
        induced projectivity on a fixed line, 11
    fundamental theorem for projectivities, 9
    harmonic biaxial homography, 11
    harmonic homology, 11
    homography (or collineation), 8
        fixed point, 8
        subspace of fixed points, 9
    homology, 11
        axis, 11
        center, 11
    involutory reciprocity (or involutory correlation), 10
    non-degenerate, 5

Index   487

of a line into itself, 3
　　characteristic, 3
　　fixed points, 3
　of general type, 11
　reciprocity (or correlation), 8
projectivity of $\mathbb{P}^n = \mathbb{P}^n(K)$
　singular (or degenerate) of a given
　　　type, 428
　singular space, 428
proper intersection, 102

quadrangle (or quadrangular set), 5
　complete, 5
　diagonal points, 5
　diagonal triangle, 5
　sides, 5
　　opposite, 5
　vertices, 5
quadratic transformation between
　　planes, 294, 295
　exceptional lines, 295
　fundamental points, 295
　of type I, 296
　of type II, 297
　of type III, 297
　pairs of corresponding points
　　that determine it, 444
　proper transform of a curve,
　　298, 299
　　order, 300
　　singularity, 300
　total transform of a curve, 299
quadric (or hyperquadric) in $\mathbb{A}^n$, 16
quadric (or hyperquadric) in $\mathbb{P}^n$,
　　32, 126
　$S_{\lambda-1}$-quadric cone, 127
　λ times specialized, 127–129
　associated matrix, 127
　associated polarity, 128
　canonical form, 131
　diagonal form, 131

　maximal dimension of linear
　　spaces contained in, 132, 133
　　the case n even, 134
　　the case n odd, 133
　non-degenerate
　　(or non-specialized), 127
　polar hyperplane, 128
　　as locus of the conjugate harmonic
　　　points, 129
　polar space of a given subspace, 129
　pole (of the polar hyperplane), 129
　reciprocal (or conjugate)
　　hyperplanes, 129
　reciprocal (or conjugate) points, 129
　self-polar $(n+1)$-hedron, 130
　singular point, 127
　singular space, 127
　tangent hyperplane, 129
　tangential equation, 119
quadric in $\mathbb{P}^3$
　pencil of quadric cones, 439
　projection from one of its points, 38
　stereographic projection, 212, 437
　tangent to a cubic surface, 442
quadric in $\mathbb{P}^5$, 402, 442
　lines and planes, 402
　system of planes, 403
quadrilateral (or quadrilateral set), 5
quasi-projective variety, 48

radical (of an ideal), 16
radical ideal, 16
rank of a curve in $\mathbb{P}^3$, 199
rank of a space cubic, 227
rational function, 24, 35
　domain of definition, 25, 26, 36
　local representation, 25
　regular in a point, 24, 36
　　as element of the local ring of the
　　　point, 36, 46
　regular on a neighborhood, 24

regular on an algebraic set is a
    polynomial function, 26
rational map (or rational transformation),
    26, 37, 185, 341
  composition of, 26
  domain of definition, 26, 37, 38
  equivalence relation with respect
    to a variety, 40
  exceptional set, 40
  fiber, 40, 186
  graph, 394
  local representation, 40
  of affine algebraic sets, 26
    domain, 26
    dominant, 26–28
  of projective algebraic sets, 38, 40
    dominant, 38
    regular, 37
  regular at the points of the domain,
    26
  representative of, 40
rational surface, 341
  curves on a rational surface, 347
  embedded, 342
  fundamental point, 345
  linearly normal, 349, 378
    projection, 350
  multiple points under the planar
    representation, 348
  normal, 350
  normal in $\mathbb{P}^r$, 350
  of general type, 358
rational variety, 39, 187
  parametric representation, 43
reciprocity (or correlation)
    of $\mathbb{P}^n = \mathbb{P}^n(K)$, 8
  as linear section of the Segre variety
    $\varphi(\mathbb{P}^n \times \mathbb{P}^n)$, 330, 337, 397
  associated to a curve C^n in $\mathbb{P}^n$, 225
  involutory reciprocity, 10
  null polarity (or null system), 10
    associated to a twisted cubic, 423

polarity, 10
  polar hyperplane, 128
  pole (of the polar hyperplane), 129
reciprocal points, 10
  represented by a bilinear equation,
    10
reciprocity theorem for polars, 123
reference system, 3, 5
  associated reference system (in the
    dual space), 8
  fundamental points, 3, 5
  projective coordinates on the line, 3
  unit point, 3, 5
regular pair (of a correspondence), 64
  induces projectivity, 65
regular parameter, 96
residual divisor with respect to the
    adjoints of given order, 264
  coresidual divisor, 264
Restsatz, 251
resultant ideal, 85
  of a given ideal, 100, 236
  of two polynomials, 85
resultant system, 102
Riemann–Roch theorem, 268, 269
ruled space, 402
ruled surface, 159, 162, 355, 408
  Cayley's cubic, 164, 376, 414
  criterion for developability, 159
  developable, 159, 409
  directrix, 159, 357
  general cubic (or cubic of general
    type), 164, 413
  generator, 159
    condition for singularity, 159
    multiple, 159, 412
    simple, 412
    singular, 159, 161
  non-developable, 163
    double directrix, 163
  of minimal order, 356

Index   489

of the lines supported by three curves in $\mathbb{P}^3$, 448
of the tangents to a curve in $\mathbb{P}^3$, 199, 232
singular point of the singular generator, 162

Salmon's theorem (on elliptic curves), 147
Salmon's theorem (on ruled surfaces), 448
scheme (projective)
    reduced (or reduced variety), 72
    support, 72
section theorem for polars, 123
sectional genus, 285
Segre morphism, 390
    graph (as image of the Veronese variety), 396
Segre product, 392
    of projective spaces, 393
Segre variety, 390
    as determinantal variety, 392
    diagonal subvariety of, 392
    dimension, 393
    order, 391, 393
    projections on the product factors, 391
Segre's surface, 369, 378
Segre's three dimensional cubic, 397
sextatic point (of an elliptic cubic), 276
singular point, 52, 54, 61
    s-fold point (for a hypersurface), 109
skew-symmetric matrix, 10
space cubic, 226, 435, 438
    locus of the tangents, 436
spatial model of an algebraic curve having only ordinary singularities with coplanar tangents, 242

star, 7
    center (or axis), 7
    dimension, 7
stationary point of a curve in $\mathbb{P}^3$, 199, 233, 276
Steiner surface, 364, 366, 369, 435, 456
    has a reducible nodal double curve, 151
stereographic projection
    exceptional line, 213
    fundamental point, 213
structure of a singular point of a plane curve, 302
    actual point, 302
    cusp of the second kind, 302
    cusp of the third kind, 302
    dual character of cusps and flexes, 118
    flex, 140
    neighborhood of successive order, 283, 301, 302
    node, 73, 302
    ordinary cusp, 73, 302
    oxnode, 302
    plane curve
        first order neighborhood, 301
    tacnode, 201, 302, 446
superabundance, 169, 457
superposed projective spaces, 8
support
    of a divisor (on a curve), 244
    of a projective scheme, 72
surface, 68
    containing ∞^2 conics, 354
    cuspidal double curve, 151, 155
    cuspidal double line, 155
    double biplanar point, 150
    double nodal curve, 150
    maximum number of isolated double points, 161, 162, 451
    normal (or ordinary) singularities, 150

490   Index

of minimal order, 354, 355
pinch-point (or cuspidal point), 150, 161
tangent cone in a point of a double curve, 152
tangent cone in a point of a double line, 156
that represents a Laplace equation, 449, 450
triplanar triple point, 150
uniplanar double point, 150
symbolic power of an ideal, 153

tacnode, 201, 302, 446
tangent cone
   intrinsic nature, 74
   projective tangent cone, 76, 80, 109
      order, 80
   to a hypersurface, 73, 126
   to a hypersurface in a point of a double line, 155, 451
   to a hypersurface in a point of an s-fold variety, 154
   to a hypersurface in an s-fold point of a variety, 153
   to a surface in a point of a double curve, 152
   to an affine algebraic set, 74
   to hypersurface in a point of a double curve, 451
tangent line
   corresponding to a tangent vector, 65
   principal tangent, 111, 156
   to an affine hypersurface, 51, 52
      tangent vector, 52
tangent space
   as span of the derived points, 63
   dimension, 54
   intrinsic nature, 58
   tangent hyperplane, 113

to a projective algebraic set, 62
to a projective hypersurface, 62
to an affine algebraic set, 53, 54, 57
to an affine hypersurface, 51
tangent vector, 66
Togliatti surface (quintic surface in $\mathbb{P}^3$ with 31 nodes), 162, 451
Togliatti surface (that represents a Laplace equation), 450
topological product, 14
topology, 12
   axioms for closed subsets, 13
   axioms for open subsets, 12
   base of open subsets, 13
   closed subsets, 13
   closure, 13
      adherent points, 13
   dense subset, 13
   fineness relation, 13
   interior, 13
      interior points, 13
   less fine (or coarser) topology, 13
   neighborhood, 13
   open subsets, 12
   product topology, 14
   quotient topology, 15
   relative topology, 13
   subspace, 13
   topological space, 12
transcendence basis, 56
transcendence degree, 55, 70
transitive group, 420
transversal intersection, 102
   along a component of the intersection, 102
   at a point, 102
   of divisors in a point, 59
   of hypersurfaces along a curve, 198
trigonal curve, 459
triplanar triple point, 150
tritangent conic to a cubic, 349, 435
tritangent line, 118

Index   491

twisted cubic, 388
    locus of the tangents, 436

uniformising parameter, 96
unihyperplanar double point, 155
    locus of the unihyperplanar double points, 155
uniplanar double point (or cuspidal point, or pinch-point), 150
unirational variety, 46, 186
    parametric representation, 43, 186
upper semi-continuous function, 54

valence of an algebraic correspondence, 275
Valentiner group, 284
Valentiner sextic curve, 284
variety of minimal degree, 105
variety of the chords of an algebraic curve, 259, 431, 442, 444
    of C^4 in $\mathbb{P}^4$, 232
    of a curve C^3 in $\mathbb{P}^3$, 226
    of a curve C^n in $\mathbb{P}^n$, 431
Veronese morphism (or embedding), 191
Veronese image, 191
Veronese surface, 67, 192, 351, 354, 355, 364, 365, 382, 441
    as locus of points corresponding to ∞^2 doubly degenerate conics, 360
    as variety of the chords of a curve C^3 in $\mathbb{P}^3$, 430
    contains ∞^2 conics, 362
    does not contain lines, 351, 361
    homaloidal system of the quadrics that contain it, 338
    parametric representation, 351
    variety of the chords, 361, 366, 367
    contains two ∞^2 families of planes, 361
Veronese variety, 190
    as diagonal of the Segre morphism, 394
    as locus of the zeros of quadratic forms, 191
    embedding space, 191
    order, 191
    parametric representation, 191
    projection of, 190

Weierstrass point, 276
Weil divisor on a curve, 243
    group of Weil divisors, 243

Zariski topology, 18, 33
    affine neighborhood, 48
    base for the open subsets, 20, 33
    closure, 21
    finer than the product topology, 21
    on $\mathbb{A}^n$, 18
    on $\mathbb{P}^n$, 33
    on $X \subset \mathbb{A}^n$, 19
    on $X \subset \mathbb{P}^n$, 34
        standard affine charts, 34
    on $\mathbb{P}^n$
        standard affine charts, 33
    on a product, 20, 21
    on the product $\mathbb{P}^n \times \mathbb{P}^m$, 390
    principal closed, 18
    principal open, 18, 33
    projective closure, 41
    standard affine charts, 5
Zeuthen's formula, 275